Physics of
Mammographic Imaging

IMAGING IN MEDICAL DIAGNOSIS AND THERAPY
William R. Hendee, Series Editor

Quality and Safety in Radiotherapy
Todd Pawlicki, Peter B. Dunscombe, Arno J. Mundt,
and Pierre Scalliet, Editors
ISBN: 978-1-4398-0436-0

Adaptive Radiation Therapy
X. Allen Li, Editor
ISBN: 978-1-4398-1634-9

Quantitative MRI in Cancer
Thomas E. Yankeelov, David R. Pickens, and
Ronald R. Price, Editors
ISBN: 978-1-4398-2057-5

Informatics in Medical Imaging
George C. Kagadis and Steve G. Langer, Editors
ISBN: 978-1-4398-3124-3

**Adaptive Motion Compensation in
Radiotherapy**
Martin J. Murphy, Editor
ISBN: 978-1-4398-2193-0

Image-Guided Radiation Therapy
Daniel J. Bourland, Editor
ISBN: 978-1-4398-0273-1

Targeted Molecular Imaging
Michael J. Welch and William C. Eckelman, Editors
ISBN: 978-1-4398-4195-0

Proton and Carbon Ion Therapy
C.-M. Charlie Ma and Tony Lomax, Editors
ISBN: 978-1-4398-1607-3

**Comprehensive Brachytherapy:
Physical and Clinical Aspects**
Jack Venselaar, Dimos Baltas, Peter J. Hoskin,
and Ali Soleimani-Meigooni, Editors
ISBN: 978-1-4398-4498-4

Physics of Mammographic Imaging
Mia K. Markey, Editor
ISBN: 978-1-4398-7544-5

Forthcoming titles in the series

Emerging Imaging Technologies in Medicine
Mark A. Anastasio and Patrick La Riviere, Editors
ISBN: 978-1-4398-8041-8

**Physics of Thermal Therapy:
Fundamentals and Clinical Applications**
Eduardo Moros, Editor
ISBN: 978-1-4398-4890-6

Informatics in Radiation Oncology
Bruce H. Curran and George Starkschall, Editors
ISBN: 978-1-4398-2582-2

**Cancer Nanotechnology: Principles and
Applications in Radiation Oncology**
Sang Hyun Cho and Sunil Krishnan, Editors
ISBN: 978-1-4398-7875-0

Monte Carlo Techniques in Radiation Therapy
Joao Seco and Frank Verhaegen, Editors
ISBN: 978-1-4398-1875-6

Image Processing in Radiation Therapy
Kristy Kay Brock, Editor
ISBN: 978-1-4398-3017-8

Stereotactic Radiosurgery and Radiotherapy
Stanley H. Benedict, Brian D. Kavanagh, and
David J. Schlesinger, Editors
ISBN: 978-1-4398-4197-6

Cone Beam Computed Tomography
Chris C. Shaw, Editor
ISBN: 978-1-4398-4626-1

IMAGING IN MEDICAL DIAGNOSIS AND THERAPY

William R. Hendee, Series Editor

Physics of Mammographic Imaging

Edited by
Mia K. Markey

CRC Press
Taylor & Francis Group
Boca Raton London New York

CRC Press is an imprint of the
Taylor & Francis Group, an **informa** business

A TAYLOR & FRANCIS BOOK

First published 2013 by CRC Press

Published 2019 by CRC Press
Taylor & Francis Group
6000 Broken Sound Parkway NW, Suite 300
Boca Raton, FL 33487-2742

First issued in paperback 2020

ISBN 13 : 978-0-367-57664-6 (pbk)
ISBN 13 : 978-1-4398-7544-5 (hbk)

Library of Congress Cataloging-in-Publication Data

Physics of mammographic imaging / editor, Mia K. Markey.
    p. ; cm. -- (Imaging in medical diagnosis and therapy)
  Includes bibliographical references and index.
  ISBN 978-1-4398-7544-5 (hardcover : alk. paper)
  I. Markey, Mia Kathleen. II. Series: Imaging in medical diagnosis and therapy.
  [DNLM: 1. Mammography. 2. Breast Neoplasms--diagnosis. WP 815]

618.1'907572--dc23   2012030325

Visit the Taylor & Francis Web site at
http://www.taylorandfrancis.com

and the CRC Press Web site at
http://www.crcpress.com

For PurpleGrape, the best grad student I never had.

Contents

SECTION I   Image Formation

SECTION II   Image Interpretation

SECTION III Image Analysis and Modeling

SECTION IV Complementary Approaches

SECTION V Future Directions

Series Preface

Advances in the science and technology of medical imaging and radiation therapy are more profound and rapid than ever before, since their inception over a century ago. Further, the disciplines are increasingly cross-linked as imaging methods become more widely used to plan, guide, monitor, and assess treatments in radiation therapy. Today, the technologies of medical imaging and radiation therapy are so complex and so computer-driven that it is difficult for the persons (physicians and technologists) responsible for their clinical use to know exactly what is happening at the point of care, when a patient is being examined or treated. The persons best equipped to understand the technologies and their applications are medical physicists, and these individuals are assuming greater responsibilities in the clinical arena to ensure that what is intended for the patient is actually delivered in a safe and effective manner.

The growing responsibilities of medical physicists in the clinical arenas of medical imaging and radiation therapy are not without their challenges, however. Most medical physicists are knowledgeable in either radiation therapy or medical imaging and are experts in one or a small number of areas within their discipline. They sustain their expertise in these areas by reading scientific articles and attending scientific talks at meetings. In contrast, their responsibilities increasingly extend beyond their specific areas of expertise. To meet these responsibilities, medical physicists periodically must refresh their knowledge of advances in medical imaging or radiation therapy, and they must be prepared to function at the intersection of these two fields. How to accomplish these objectives is a challenge.

At the 2007 annual meeting of the American Association of Physicists in Medicine in Minneapolis, this challenge was the topic of conversation during a lunch hosted by Taylor & Francis Publishers and involving a group of senior medical physicists (Arthur L. Boyer, Joseph O. Deasy, C.-M. Charlie Ma, Todd A. Pawlicki, Ervin B. Podgorsak, Elke Reitzel, Anthony B. Wolbarst, and Ellen D. Yorke). The conclusion of this discussion was that a book series should be launched under the Taylor & Francis banner, with each volume in the series addressing a rapidly advancing area of medical imaging, or radiation therapy of importance to medical physicists. The aim would be for each volume to provide medical physicists with the information needed to understand technologies driving a rapid advance and their applications to safe and effective delivery of patient care.

Each volume in the series is edited by one or more individuals with recognized expertise in the technological area encompassed by the book. The editors are responsible for selecting the authors of individual chapters and ensuring that the chapters are comprehensive and intelligible to someone without such expertise. The enthusiasm of volume editors and chapter authors has been gratifying and reinforces the conclusion of the Minneapolis luncheon that this series of books addresses a major need of medical physicists.

Imaging in Medical Diagnosis and Therapy would not have been possible without the encouragement and support of the series manager, Luna Han of Taylor & Francis Publishers. The editors and authors, and most of all I, are indebted to her steady guidance of the entire project.

William Hendee
Series Editor
Rochester, Minnesota

Preface

Imaging of the breast is essential for detection, diagnosis, treatment planning, and monitoring of breast cancer. Mammography, two-dimensional x-ray projection imaging of the breast, has been the dominant modality for this purpose for many years. However, mammographic imaging is undergoing dramatic changes, with an increasing number of mammographic exams being performed digitally. The migration to digital imaging is enabling the development of technologies such as contrast enhancement and decision support systems that can increase the effectiveness of mammography. Moreover, the advent of digital imaging has launched new x-ray-based approaches that retain information of the three-dimensional nature of the breast, which was lost by traditional mammography. These changes in mammographic imaging raise new questions and engender novel research directions in a variety of areas, not only in imaging physics but also in fields such as engineering and psychology.

The primary audience for this book is medical physicists and biomedical engineers; radiologists involved in breast imaging research will also benefit from it. The book is intended to be valuable either for use in a graduate level course or for self-study by practitioners. This book characterizes the current role, and future potential, of x-ray-based digital breast imaging. The goal of the text is to provide technical details, but in the context of clinical needs, complementary approaches, and ongoing research, such that the reader will be empowered to identify key areas for future research.

The text is divided into five sections. Section I covers the fundamentals of image formation in digital x-ray-based mammographic imaging modalities: digital mammography, contrast-enhanced digital mammography, stereo mammography, breast tomosynthesis, and breast CT. Section II presents key topics in image interpretation, including subjects such as display and perception. Section III pertains to issues in analysis and modeling, such as phantoms, observer models, and computer-aided diagnosis. Section IV summarizes complementary imaging modalities in widespread use (ultrasound, MRI, and nuclear medicine techniques). The text concludes with Section V on future directions, including both emerging modalities and analysis methods.

This is an exciting time in the development of medical imaging, with many new technologies poised to make a substantial impact on breast cancer care; however, as impressive as these accomplishments are, they are not enough. There are many who depend upon us to discover, create, and engineer even better imaging systems. To quote Lance Armstrong, "We have two options, medically and emotionally: give up, or fight like hell." I hope this book will inspire young scientists and engineers to fight like hell to advance mammographic imaging and that it will provide them with the fundamental knowledge needed to do so.

Acknowledgments

I am honored that William R. Hendee invited me to serve as editor for a volume of this outstanding series from Taylor & Francis. It has proven to be a wonderful opportunity to work with the impressive group of scientists, engineers, and physicians who have carefully prepared the chapters contributed to this book. I also thank the reviewers who provided their feedback on the initial book proposal and in doing so assisted me in preparing a stronger text. I am extremely grateful to the patient staff at Taylor & Francis, who provided the utmost support throughout this project. Special thanks are due in particular to Luna Han and Joselyn Banks-Kyle.

I freely admit I would never have been able to complete this project without the support of my administrative associate, Michael Don. Michael's organizational superpowers and attention to detail were indispensible.

In bringing this book to fruition, I benefited from helpful discussions with many of my colleagues and friends, but I would especially like to acknowledge the valuable counsel of Dr. Gary J. Whitman, Dr. Margie E. Snyder, and Dr. Michelle Butler. Naturally, I am also indebted to my former mentors who guided my development in medical imaging, particularly Joseph Y. Lo, Georgia D. Tourassi, and the late Carey E. Floyd Jr. Thanks are also due to Gautam S. Muralidhar for helping select a cover image and to Kailyn Morarend for assistance in the proof review.

I cannot begin to express how much I have relied upon (and occasionally tried) the patience and encouragement of my family, especially my husband, Eric, and our children: Henry, Fiona, and Leia. Only 35 more years left to plan our fabulous 50th wedding anniversary trip!

Finally, I must acknowledge the strength and grace of the women whose fight with breast cancer inspires and motivates all of us who work in mammographic imaging. My highest regards for Melanie B., Lynda G., Rita S., Patty C., Eric's aunt, Harry's mother, Joey W., and ....

About the Editor

Dr. Mia K. Markey is an associate professor of biomedical engineering at The University of Texas at Austin and adjunct associate professor of imaging physics at The University of Texas MD Anderson Cancer Center. A 1994 graduate of the Illinois Mathematics and Science Academy, Dr. Markey earned her B.S. in computational biology (1998) from Carnegie Mellon University and her Ph.D. in biomedical engineering (2002), along with a certificate in bioinformatics, from Duke University. Dr. Markey's laboratory designs decision support systems for clinical decision making and scientific discovery using artificial intelligence and signal processing technologies. Her research portfolio also includes projects in biometrics. Dr. Markey has been recognized for excellence in research and teaching with awards from organizations such as the American Medical Informatics Association, the American Society for Engineering Education, and the American Cancer Society.

Contributors

Margaret Adejolu
King's College Hospital
London, United Kingdom

Salavat Aglyamov
The University of Texas at Austin
Austin, Texas

Aldo Badano
U.S. Food and Drug Administration
Silver Spring, Maryland

Predrag R. Bakic
The University of Pennsylvania
Philadelphia, Pennsylvania

François Bochud
Lausanne University Hospital
Lausanne, Switzerland

Richard Bouchard
The University of Texas MD Anderson
    Cancer Center
Houston, Texas

Alan C. Bovik
The University of Texas at Austin
Austin, Texas

Rachel F. Brem
The George Washington University
Washington, DC

Ying (Ada) Chen
Southern Illinois University
Carbondale, Illinois

Caitrin Coffey
The George Washington University
Washington, DC

Stanislav Emelianov
The University of Texas at Austin
Austin, Texas

David J. Getty
Brigham and Women's Hospital
Boston, Massachusetts

Iulia Graf
The University of Texas at Austin
Austin, Texas

Melissa L. Hill
University of Toronto
Toronto, Ontario, Canada

Nehmat Houssami
University of Sydney
Sydney, Australia

Raunak Khisty
The University of Texas MD Anderson
    Cancer Center
Houston, Texas

Elizabeth Krupinski
University of Arizona
Tucson, Arizona

Deanna L. Lane
The University of Texas MD Anderson
    Cancer Center
Houston, Texas

Ashley M. Laughney
Darthmouth College
Hanover, New Hampshire

Huong (Carisa) Le-Petross
The University of Texas MD Anderson
    Cancer Center
Houston, Texas

Lihua Li
College of Life Information Science and
    Instrument Engineering
Hangzhou Dianzi University, China

Ernest L. Madsen
University of Wisconsin-Madison
Madison, Wisconsin

Andrew D. A. Maidment
The University of Pennsylvania
Philadelphia, Pennsylvania

Mia K. Markey
The University of Texas at Austin
Austin, Texas

Michael A. Mastanduno
Dartmouth College
Hanover, New Hampshire

Kelly E. Michaelsen
Dartmouth College
Hanover, New Hampshire

Gautam S. Muralidhar
The University of Texas at Austin
Austin, Texas

Keith D. Paulsen
Dartmouth College
Hanover, New Hampshire

Brian W. Pogue
Dartmouth College
Hanover, New Hampshire

Xin Qian
Columbia University Medical Center
New York, New York

Jocelyn Rapelyea
The George Washington University
Washington, DC

Jennifer Rusby
Royal Marsden NHS Foundation Trust
Sutton, United Kingdom

Chris C. Shaw
The University of Texas MD Anderson
    Cancer Center
Houston, Texas

R. Jason Stafford
The University of Texas MD Anderson
    Cancer Center
Houston, Texas

Tanya W. Stephens
The University of Texas MD Anderson
    Cancer Center
Houston, Texas

Jessica Torrente
The George Washington University
Washington, DC

Gary J. Whitman
The University of Texas MD Anderson
    Cancer Center
Houston, Texas

Robin Wilson
Royal Marsden NHS Foundation Trust
Sutton, United Kingdom

Martin J. Yaffe
Sunnybrook Research Institute
University of Toronto
Toronto, Ontario, Canada

Bin Zheng
University of Pittsburgh
Pittsburgh, Pennsylvania

I

Image Formation

Fundamentals of Digital Mammography

Xin Qian
*Columbia University
Medical Center*

1.1 Introduction

Mammography is a radiographic examination that is specially designed for detecting breast cancer. Digital mammography is currently the most effective screening and diagnostic tool for early detection of breast cancer. The details are in Chapter 7. It has been attributed as a major factor in reducing breast cancer mortality rate in recent years (Pisano et al. 2005). Until recently, screen-film mammography was the standard breast imaging tool used in conventional mammography. New developments in detector technology and computers are altering the landscape of mammography imaging. Digital mammography, also called full-field digital mammography (FFDM), offers the promise of revolutionizing the practice of mammography through its superior dose and contrast performance. The overall diagnostic accuracy of digital and film mammography as a means of screening for breast cancer are similar, but digital mammography is more accurate in women younger than 50 years, women with radiographically dense breasts, and premenopausal or perimenopausal women (Pisano et al. 2004). In addition, advanced applications made possible through digital imaging, such as computer-aided diagnosis (Chapters 14 and 21), dual energy, contrast enhancement (Chapter 2), and 3-D tomosynthesis (Chapter 4) are expected to further improve diagnostic sensitivity and specificity.

Digital imaging systems entered the radiology departments about 15 years ago using photostimulable phosphors (PSP), charge coupled device (CCD), and photoconduction (Thoravision) detectors. Recently introduced, flat panel x-ray detectors can offer extremely high quantum efficiency and high resolution. These digital detectors will translate into patient dose reduction and image quality improvement, and make new imaging techniques possible, such as tomosynthesis and contrast mammography. Digital detectors for mammography can be categorized as indirect and direct conversion detectors. Indirect conversion detectors utilize a method of imaging x-rays, similar to screen-film, wherein a scintillator absorbs the x-rays and generates a light. Light is then detected by an array of photon detectors. Indirect conversion detectors suffer from resolution degradation caused by light scatter in the scintillator, and from poor quantum efficiency caused by the use of thin scintillators. Direct conversion detectors utilize a direct-conversion method of imaging, wherein the x-rays are absorbed and electrical signals are created in one step.

The active area of the x-ray detector for FFDM systems must be similar to the size of the largest screen-film cassette commonly used in screen film mammographic imaging, 24 × 29 cm. For the U.S. demographics, up to 30% of all women require the larger field of view. Important detector properties for digital mammography also include: geometrical characteristics, quantum efficiency, sensitivity, spatial resolution, noise characteristics, dynamic range, uniformity, and acquisition speed. For dynamic studies, such as tomosynthesis, the rate at which sequential images can be acquired is also important. Contrast enhancement of the digital image and the wide dynamic range of digital detectors will improve visibility of mammographic features. The digital image will provide image archiving and retrieval advantages over film, and will facilitate the use of computer-aided diagnosis. Systems with high quantum efficiency, especially at increased x-ray energies, offer the possibility of decreased breast compression. This can potential reduce patient motion caused by painful breast compression.

1.2 Conventional Mammography

The film in mammography is used to record the image, display the image, and provide archival storage. All radiographic films consist of two main parts: the base and the emulsion. Other components of the film include the adhesive layer, which is between the base and the emulsion to ensure proper contact and integrity especially during processing. The emulsion of the film contains silver halide crystals. The latent image forms by the action of x-ray photons and light photons on the silver halide crystal. In mammography the emulsion is coated on one side of the base only (Lange 2002). Conventional film systems use intensifying screens to capture x-rays. These screens are often constructed of rare earth phosphors such as gadolinium oxy-sulfide (Gd_2O_2S) that output light upon absorption of x-rays. The resultant light scintillation creates a number of light photons that spread and illuminate the film in a distribution cloud, and the image is obtained by exposing the film. The resolution of screen-film mammography depends on the screen phosphor size, the phosphor layer thickness, and the phosphor concentration. As the phosphor size and layer thickness decrease, resolution increases, but the radiation dose to patients increases, too. Increasing phosphor concentration will result in better recorded detail while reducing patient dose. Therefore, it is impossible to offer a screen-film system simultaneously offering the highest possible resolution and lowest possible radiation dose. This trade-off between radiation dose and image quality must be optimized for the specific clinical application. Because x-rays are absorbed with a decaying exponential spatial distribution, in a screen-film mammography system, the film is placed at the entrance surface of the scintillating screen.

While a screen-film mammography system offers several advantages (Beutel and Kitts 1996), for instance, relatively low-cost, high-limiting spatial resolution (20 lp/mm), and inherent logarithmic compression of dynamic range onto the available optical densities of the film, there are also significant disadvantages and limitations of this type of system. Films can easily cost departments thousands of dollars per year, plus additional costs

for storage. Moreover, films are easily lost, misfiled, misplaced, or stolen. Film also must be physically transported to the physician for viewing. Another major disadvantage of film is that rigid exposure factors are required to produce optimum images, resulting in repeated imaging since it is not possible to adjust for over or under exposure. The fixed gradient (contrast) of screen-film imaging can provide a challenge for mammographers because of the need to optimize contrast to aid interpretation (Gater 2002). Film does not have a linear sensitivity to photon flux, and there is a narrow range over which it can detect small differences in contrast. In particular, tissue areas of high and low density are often suboptimally imaged. Film granularity can affect detective quantum efficiency at high optical densities and visibility of microcalcifications.

1.3 Digital Mammography

Digital mammography depends on computer technology to produce digital images of the breast and can offer the potential for several advances in mammography. Because images are captured as digital signals, electronic transfer and storage of images is possible, eliminating physical storage and distribution required by film. Digital systems offer a large dynamic range of operation, improving visualization of all areas of the breast and increasing exposure latitude. Also, the digital format allows gray scale adjustment to optimize contrast for every imaging task. Softcopy reading, computer-aided diagnosis, and three-dimensional imaging offer additional and potentially important opportunities for improvement in mammographic interpretation.

1.4 Detectors for Digital Mammography

There are two types of image capture methods used in digital mammography, which represent different generations of technology: indirect conversion and direct conversion.

1.4.1 Indirect-Conversion Digital Detectors

The earliest digital mammography systems available in the U.S.A. used indirect conversion detectors. Indirect digital systems use a two-step process similar to screen-film systems for x-ray detection (Lange 2002). A scintillator, such as cesium iodide (CsI) doped with thallium (Tl), absorbs the x-rays and generates a light scintillation that is then detected by an array of thin-film diodes (TFDs) sometimes called photodiodes. The TFDs convert the light photons to electronic signals that are captured using thin film transistors (TFTs). Some systems use CCDs as an alternative to the TFTs light collecting array and readout method. Signals from the TFD or TFT are then transferred to computer readout. As the scintillator is made thicker, light spread increases, resulting in decreased resolution. Because of its columnar structure, CsI(Tl) does not create as much light scatter as other screens. However, compromise between resolution and sensitivity still exists. The placement of the scintillator is more problematic in indirect conversion digital detectors than with screen-film systems. As with film screens, more x-rays are absorbed near the entrance of the scintillation layer than the exit. While film is placed near the entrance side of the scintillator, a photodiode/transistor array is not transparent to x-rays and the array must be placed on the exit surface of the scintillator. This causes degradation in spatial resolution compared to screen-film. Typical thicknesses of CsI(Tl) used in mammography detectors ranges from 150 to 250 μm, and these indirect conversion digital detectors exhibit light spreading similar to screen-film systems (Carlton and Adler 2001, Feig and Yaffe 1995). Examples of indirect conversion detectors are the Fischer Imaging SenoScan CsI/CCD-based detector and the GE Senograph CsI/TFT detector (Borasi et al. 2003, Samei and Flynn 2002).

1.4.2 Direct-Conversion Digital Detectors

Direct-conversion digital detectors represent a technological advance, eliminating problems associated with light scatter inherent in indirect conversion systems (Yorker et al. 2002). In direct systems, x-rays are absorbed by the detector and the electrical signal is created in a single step. Under the influence of an external electric field, holes (or electrons, depending upon the polarity of the applied field) drift towards a pixel electrode and are collected on a pixel capacitor. Because the electrons and holes travel along the direction of the electric field lines, they move without lateral charge spreading. This results in an exceptionally narrow point spread response of about 1 micron (Bissonnette et al. 2005). Selenium (Se) flat-panel detectors are used in direct systems. Se is ideal for mammographic systems because of its high x-ray absorption efficiency, high resolution, low signal to noise ratio, and dose efficiency. Spatial resolution is only limited by pixel size. The optimal pixel size is determined by the limitations of the display system, the information system, and the manufacturing process. In direct-conversion detectors, the response function maintains its sharpness even as the thickness of the Se layer is increased, so there is no trade-off between radiation stopping power and spatial resolution. In practice, the photoconductor is made sufficiently thick in order to stop the majority of the incident x-rays, and this can be done without adversely affecting the spatial resolution, an important consideration in mammography with its dual needs of high resolution and low radiation exposure. An x-ray not striking perpendicular to the detector can result in negligible blur. TFT arrays are used to transfer the electronic signals from the Se photoconductor to a computer. Using amorphous Se as the photoconductor, a thickness of 250 μm is adequate to stop more than 95% of the x-rays in the mammographic energy range. Standard screens for use in film mammography only have about 50%–70% quantum efficiency, and the scintillator CsI(Tl) used in indirect-conversion digital detectors exhibits about 50%–80% quantum efficiency. In comparison, systems using amorphous-Se can achieve almost complete quantum efficiency. Examples of direct conversion detectors are the Hologic/Lorad Selenia detectors (Ren et al. 2005, Yorker et al. 2002).

1.5 Specifications of Digital Detectors

A typical configuration for a flat panel detector system is comprised of a large number of individual detector pixels, each one capable of storing charge in response to x-ray exposure. Each detector pixel has a light-sensitive region, and a small corner of it contains the electronics. The brightness or intensity of the pixel is determined by the computer generated number stored in that pixel. The values stored in the pixels are transferred to a computer during a readout sequence. This is known as direct readout, a function of all digital systems, and should not be confused with direct conversion digital detection. TFT arrays are the active electronic readout mechanism commonly used in both direct and indirect digital mammography systems. The arrays are typically deposited onto a glass substrate in multiple layers, beginning with readout electronics at the lowest level, followed by charge collector arrays at higher levels. The composition of the top layer depends upon the type of detector. If the system uses indirect conversion detection, both x-ray pixels and light-sensitive elements are deposited on the top layer. Direct conversion detectors do not require conversion of x-rays to light, so light sensitive elements are not necessary for these systems. Semiconductor arrays for direct conversion detectors are much easier to fabricate than arrays for indirect conversion detectors because Se-based detectors do not require a photodiode structure on top of the TFT. CCDs are an alternative to TFT arrays in indirect conversion systems. Basic CCD-based systems consist of a series of metal oxide semiconductor capacitors that are fabricated very close together on the semiconductor surface. These systems use fiber optics to capture images from light emitted from scintillators or intensifying screens, but suffer from light loss and added complexity due to the fiber optics (Borasi et al. 2003).

1.5.1 Digital Detector Pixels

Digital detector matrix refers to the layout of pixels in rows and columns. Each pixel corresponds to a specific location in the image. A 100 × 100 matrix of pixels has less information stored per pixel than a 1000 × 1000 matrix of pixels. Each pixel has a bit depth, which represents the dynamic range assigned to the pixel. The higher the bit depth, the higher the contrast. Pixel size is important because as the pixel size decreases the amount of data contained in the images rapidly increase, but the electronic noise will also increase. Spatial resolution also depends on pixel size. Mammographic imaging requires the detection and classification of extremely small objects. In particular, microcalcifications can be as small as 100 to 200 μm. Any useful digital mammography system must be able to image small microcalcifications. Digital detectors are comprised of arrays of pixels that are 100 μm or smaller (Samei and Flynn 2002). However, the electronics of each detector pixel takes up a certain amount of the area; for flat panels with smaller pixels, a large fraction of each pixel's area is not sensitive to light. Therefore, the light collection efficiency decreases as the detector pixel get smaller. The

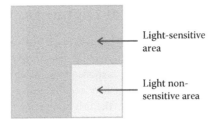

FIGURE 1.1 Within a single pixel, there are light sensitive area and light nonsensitive area.

ratio of the light sensitive area to the entire area of each detector element is called the fill factor, as shown in Figure 1.1. Since light photons that are not detected do not contribute to the image, a high fill factor is needed. The choice of the detector pixel dimensions requires a tradeoff between spatial resolution and contrast resolution. Pixel size will also affect the storage, transmission time, and archival of the image (Bushong 1997, Vyborny et al. 2000). The size of the image matrix is determined by the characteristics of the imaging equipment and by the capacity of the computer. The matrix size can often be selected by the operator and is sometimes called the field of view (FOV). For the same FOV, the spatial resolution is better with a smaller pixel size.

1.5.2 Digital Detector Field of View

FOV can be an operator-defined parameter that controls the apparent size of the area to be imaged, whereas pixel size is the operator-defined parameter that controls the resolution in the image. FOV is important in mammography because to image most of the average female population, the FOV should be the size of the largest screen-film cassette, 24 × 30 cm (Bushong 1997). A digital detector of size 18 × 24 cm (the smaller film cassette size) is inadequate to image approximately 20% to 30% of U.S. women. If the breast is too large to image on the detector in one exposure, multiple exposures that "tile" the breast are required. This solution has disadvantages such as requiring multiple additional compressions, additional setup and imaging times, breast regions that suffer from repeated radiation exposure, and the difficulty of the radiologist's review of multiple images. Use of a large detector for digital mammography is an important design issue. The one technical challenge in the use of a large detector for small breasts is positioning; however, the use of smaller compression paddles in conjunction with the large detector solves this problem. The larger detector can be used to advantage when imaging smaller breasts. The imaging can be performed in a magnification mode, creating the effect of smaller pixels and reducing scattered radiation at the same time.

1.6 Digital Mammography Data Acquisition

For image acquisition, there are a variety of system-level considerations with digital mammography systems.

1.6.1 X-Ray Source

Spatial resolution is the ability of an imaging system to allow two adjacent structures to be visualized as being separate. Light diffusion in the receptor phosphor screen and relative motion of the x-ray source, the breast, and the image receptor during the exposure can contribute to image blurring. In addition, spatial resolution losses are also caused by geometric factors, such as the x-ray source focal spot size and the magnification of a given structure of interest. Digital mammographic detectors offer improved performance characteristics, particularly better dynamic ranges. It is possible that higher x-ray energies may permit lower dose or higher image quality with digital mammography, particularly for patients with dense breasts. Detectors that have high intrinsic quantum efficiency allow the use of higher energies. The typical energies used in mammography are in range of 22–49 keV (Batignani et al. 1991, Diekmann et al. 2007). Heel effect is an intrinsic problem of x-ray source intensity nonuniformity. Since the anode of an x-ray tube is angled, the intensity of x-ray beam along the longitudinal axis of the tube varies. Figure 1.2 illustrates the heel effect of an x-ray source. Consequently, the intensity of the x-ray beam is greater on the cathode side than on the anode side.

1.6.2 Scatter Rejection Methods

X-ray scatter escaping the breast without useful anatomical information is recorded by the image detector. The scatter to primary ratio characterizes the amount of image contrast loss and apparent sharpness reduction. The preferred method to reduce scatter is to employ the use of radiation antiscatter grids interposed between the patient and the detector. Antiscatter grids are regularly used to absorb scattered x-rays in mammography. Problems with such a grid include partial absorption of direct radiation, partial transparency to scattered x-rays, grid line artifacts, poor efficiency with increased patient dose, the need to design mechanical movements, and the use of toxic material. Moreover, grids required almost fixed direction of radiation and, thus, prevent 3-D or stereo mammographic imaging. There are two common methods of grid design. Standard linear grids are constructed of long thin slats of radio-opaque materials

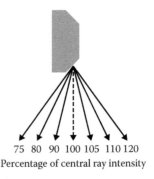

75 80 90 100 105 110 120
Percentage of central ray intensity

FIGURE 1.2 Heel effect of x-ray source. Consequently, the intensity of the x-ray beam is greater on the cathode side than on the anode side.

separated by relatively radiolucent spacer materials. These grids effectively reduce scattered photons, but also block some of the primary beam, mainly from absorption in the spacers. An alternative method of grid construction utilizes a crossed air grid, known as a high-transmission cellular grid. This grid absorbs scattered radiation in two directions, as opposed to the linear grid, which removes scattered radiation only in one direction. This grid absorbs less of the primary beam, because the intersepta material is air (Almeida and Barnes 1999). Slot scan geometries (Nishikawa et al. 1987) and a representative of a commercial digital mammography system (Fischer Imaging Systems, Denver) were compared against the grid technologies (Boone et al. 2002). The benefit of a slot scan system is that primary photons that pass through the breast are not attenuated.

1.6.3 Automatic Exposure Controls

Optimized exposure plays an important role and is a requirement in digital mammography. Automatic Exposure Control (AEC) performance is linked to this fundamental requirement, and the AEC should provide consistent signal values as breast thickness is varied over the range of x-ray tube potentials. Both the quality of the mammography image and the dose efficiency are critically dependent on the x-ray spectrum. The shape of the spectrum is determined by the anode material, the applied potential to the tube, and the amount and type of added metallic filtration in the beam. For digital mammography, it is no longer necessary to have a separate AEC sensor as part of the system because the detector can serve as a multielement sensor. To use the detector in this way, it is necessary that the detector can be read out quickly to determine what the optimum exposure factors should be. Optimization of exposure can consider various statistics from a short test pulse made prior to the actual image acquisition. These might include the minimum signal from the most attenuating region of the breast. The algorithm can require that in the actual exposure, this signal has to be greater than some preset value. Determination of the optimum x-ray spectrum for mammography involves a careful compromise between image contrast, radiation dose, and image statistical noise.

1.6.4 Breast Compression and Position

Maximizing the amount of exposed breast tissue is important for effective breast cancer detection and diagnosis, as illustrated in Figure 1.3. If the breast is not properly positioned, all of the breast tissue may not be imaged. Imaging the anterior breast tissues is simple, but imaging the breast tissue adjacent to the chest wall is difficult. Given that the volume of the breast increases toward the chest wall, failure to image these tissues could exclude large areas where cancer might develop. A high-quality mammography system has an advanced compression device and does not have any breast positioning limitations. Breast compression contributes to digital image quality by immobilizing the breast (reducing motion unsharpness), producing a more uniform, thinner tissue which can lower scattered radiation,

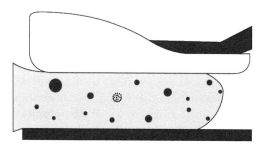

FIGURE 1.3 A breast phantom is used to illustrate how breast compression paddle works.

geometric blurring, and anatomical super-position, and, thus, reduce radiation dose.

1.7 Performance of Digital Mammography Systems

Image quality in mammography is affected by the shape, size, and x-ray absorption properties of the anatomic region or the lesion to be radiographed in addition to x-ray beam quality, geometric unsharpness, and the resolution, characteristic curve, and noise properties of the imaging system (Haus et al. 1977). The parameters that most fully characterize the overall image quality obtained on mammography systems are: modulation transfer function, noise power spectrum, and detective quantum efficiency.

1.7.1 Modulation Transfer Function

The modulation transfer function (MTF) is the most widely accepted measure of the spatial resolution response of imaging system performance. MTF is a measure of signal transfer over a range of spatial frequencies and quantifies spatial resolution, demonstrated in Figure 1.4. There have been a few methods used to measure the MTF over the past decades (Bradford et

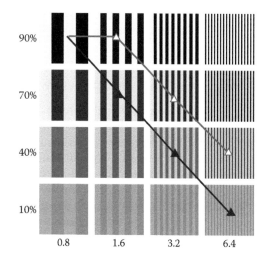

FIGURE 1.4 MTF is a measure of signal transfer over a range of spatial frequencies and quantifies spatial resolution.

al. 1999, Fetterly et al. 2002, Fujita 1992, Samei et al. 1998). The slit (Bradford et al. 1999) and edge (Samei et al. 1998) response methods are the two most commonly used methods for MTF calculation. The edge response method is an International Electrotechnical Commission (IEC) standard for MTF calculation (Commission 2003). For quick evaluation of an imaging system's spatial resolution, a line pair phantom is often employed, such as CIRS 016A and 016B. The ultimate resolution limit of a digital imaging system is determined by its pixel size. A scattered photon in an indirect-conversion detector can spread over several pixels, which further limits the spatial resolution of the system. Direct conversion systems do not suffer from this limitation.

1.7.2 Noise Power Spectrum

The noise in images is recognized as an important factor in determining image quality. Noise power spectra (NPS) provide the means of characterizing image noise and play a central role in the ultimate measure of image quality. The noise power spectrum (NPS) provides a quantitative description of the amount and frequency of the noise fundamentally produced within a particular imaging system (Dobbins et al. 1995, Flynn and Samei 1999, Giger et al. 1984, Stierstorfer and Spahn 1999). The goal of the NPS measurement is to use the limited amount of image data acquired to get a smooth and accurate NPS estimation with the finest frequency resolution. The NPS is defined as the Fourier transform of the auto-correlation function of the noise-only image, which is an image without an object in the x-ray trajectory path. The NPS provides an estimation of the spatial frequency dependence of the pixel-to-pixel fluctuation presented in the image. The following equation shows the discrete NPS of a discrete random variable

$$W_d(u,v) = \lim_{\substack{N_x \to \infty \\ N_y \to \infty}} \left\langle \frac{x_0 y_0}{N_x N_y} \left| \sum_{m,n} \Delta\tilde{a}[m,n] e^{-2\pi i(umx_0 + vny_0)} \right|^2 \right\rangle$$

where N_x and N_y are the number of pixels along the axes in the region of interest.

1.7.3 Detective Quantum Efficiency

Like MTF of imaging system, Detective Quantum Efficiency (DQE) also provides a quantitative measurement of imaging system performance. DQE measures signal-to-noise ratio, contrast resolution, and dose efficiency. Both MTF and DQE measurements are used to determine how well a system captures information over a range of spatial frequencies. Even with a high MTF at high spatial frequencies, small objects can still get lost in the noise of the system. Increasing signal and decreasing noise in the system can increase visibility of small objects. The experimental determination of the DQE is usually done by measurements of the MTF and the output NPS, the results of which are, after proper normalization, combined to obtain the DQE.

In the past few years, DQE evaluations have been reported for a number of digital detector systems (Ren et al. 2005, Borasi et al. 2003, Samei and Flynn 2002). A comparison of the results is sometimes difficult because slightly different assessment methods have been used and it is unclear if observed DQE differences are caused by these experimental variations or are due to real differences of the detectors investigated. To standardize the DQE measurement and calculation for digital x-ray imaging systems, IEC published the international standard IEC 62220-1 in 2003. The following equation shows the calculation of the detector DQE

$$\mathrm{DQE}(f) = \frac{\Phi[k\,\mathrm{MTF}(f)]^2}{\mathrm{NPS}(f)}$$

where k is the gain factor of the system and Φ is the photon fluence.

1.7.4 Imaging Quality Analysis

Analysis of image quality has meaning only in the context of a particular imaging task required by mammography (ICRU 1996). Specifically, the imaging tasks unique to digital mammography that determine the essential characteristics of a high quality mammogram are its ability to visualize the following features of breast cancer: the characteristic morphology of a mass, the shape and spatial configuration of calcifications, distortion of the normal architecture of the breast tissue, asymmetry between images of the left and right breast, and the development of anatomically definable new densities when compared to prior studies. Figure 1.5 shows the standard acrylic phantom used for American College of Radiology (ACR) accreditation and Mammography Quality Standards Act (MQSA) inspections. The phantom is approximately equivalent in x-ray absorption to a 4.2 cm thick compressed breast consisting of 50% glandular and 50% adipose tissue. The phantom has fibers with diameters of 1.56, 1.12, 0.89, 0.75, 0.54, and 0.40 mm; specks with diameters of 0.54, 0.40, 0.32, 0.24, and 0.16 mm; and masses with decreasing diameters of 2.00, 1.00, 0.75, 0.50, and 0.25 mm.

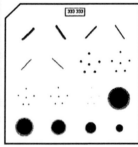

FIGURE 1.5 A standard acrylic phantom used for ACR accreditation and MQSA inspections, the features embedded in the phantom are shown on the right side.

1.8 Digital Mammographic Image Display and Storage

Digital mammograms can be printed on film or displayed on a monitor. Typically, laser-printed film can display $4,000 \times 5,000$ pixels at 12-bit gray scale (Pisano et al. 2000). The disadvantages of hard-copy image display at digital mammography are obvious. Once an image is printed, it can no longer be manipulated, and any information available in the digital data but not captured on the printed image will therefore be lost. With the currently available high-luminance high-spatial-resolution monitors ($2,000 \times 2,500$ pixels), any digital mammographic images can be displayed. A monitor must render mammographic details with sufficient luminance and grayscale range to prevent the loss of contrast details or the presence of contour artifacts. A two-monitor portrait setup is recommended for minimizing head and shoulder rotation, keeping the body and arms in ergonomic positions, and avoiding near vision deficits. Eyeglasses, when required, should be specifically selected for the viewing distance. The display monitor should present the images in portrait geometry, maintaining the original aspect ratio of the acquisition device.

Digital images are captured and stored on a computer. A digital image is a 2-D matrix of detector pixels, which is defined by its size and bit depth. The size of an image is its length in pixels multiplied by its width in pixels, and the bit depth is the number of shades of gray that can be displayed. Signals are digitized into one of 2^n intensity levels within each pixel, where n is the number of bits of digitization and is typically 12 or 14, depending on the detector's design. The demand for high spatial resolution in mammography requires smaller pixel size and higher digitization. Therefore, image size depends on the digital detector pixel size, the number of pixels per image, and the type of digitization. If the detected digital signals are digitized to 12 bits, it implies 2^{12} or 4,096 signal values stored per pixel. Whether the system provides 12- or 14-bit digitization (8 bits = 1 byte), a digital detector of N pixels requires $2N$ bytes of storage (2 bytes per pixel).

1.9 Advanced Applications

Digital mammography offers some advanced applications that are not practical or possible with screen-film mammography.

1.9.1 Digital Breast Tomosynthesis

Digital tomosynthesis (Chapter 4) is a new kind of breast screening and diagnostic system in which multiple x-ray pictures of each breast are taken from different viewing angles. The breast is positioned the same way as it is in a conventional mammogram, but only a little pressure is applied—just enough to keep the breast in a stable position during the procedure. The x-ray tube moves in an arc around the breast. Then the information is sent to a computer, where it is assembled to produce clear, highly focused 3-D images throughout the breast. Preliminary studies (Andersson 2008) show that these acquisitions can be performed with doses similar to screening mammography, and that

the images offer diagnostic information not available with 2-D imaging. High quantum efficiency will be important for these applications, to reduce the total dose given that multiple images are required. As the first digital breast tomosynthesis scanner approved by FDA early this year, Hologic Selenia Dimensions scanners are commercially available.

1.9.2 Telemammography

Telemammography can become very effective with digital mammography, since it allows underserved and geographically remote populations to access the latest in breast care. It is anticipated that telemammography, with concurrent digital acquisition and remote review of images, will allow complete evaluation of patients in a single visit at a remote location.

1.9.3 Digital Image Processing

The digital format of mammographic images allows digital image processing (Chapter 2). One such processing technique is contrast enhancement, a process whereby the contrast of different structures in the breast is altered to improve detectability. Also, image processing involves edge enhancement or smoothing the image and zooming in on a suspicious region in an image for better viewing. All these are possible without additional exposure to the patient. Dual-energy imaging is linked to this advance. Two images are acquired at different x-ray energies. Because of the differing x-ray attenuation characteristics of glandular tissue, adipose tissue and microcalcifications, processing of the dual images can enhance the visibility of certain structures (Lemacks et al. 2002).

1.9.4 Computer-Aided Diagnosis

Early studies on quantitative analysis of medical images by computer were reported in the 1960s (Lodwick et al. 1963). Since then, computer aided diagnosis (CAD) has been an active research topic (Feig and Yaffe 1995, Vyborny et al. 2000). Please see Chapters 14 and 21 for details. The digital image is examined by software, and suspicious areas are highlighted for further scrutiny by the radiologist. The challenge for these systems is to find the proper balance between sensitivity and specificity. Increasing sensitivity can result in too many false positives marked on an image. Conversely, if not enough true positives are marked, the system will be offering little help to the radiologist. However, it is expected that eventually these types of systems will become routine, since it is very easy to perform the CAD procedure on the digitally acquired image.

1.10 Summary

As digital mammography has shown to significantly lower the radiation dose, reduce the recall and biopsy rates, it is likely that access to digital mammography will continue to increase. More than 60% of U.S. breast imaging facilities offer digital mammography and more are acquiring digital services each year. Digital mammography systems offer improvements over screen film imaging systems, especially in imaging dynamic range, digital acquisition, storage, display and postprocessing. Technical developments in digital mammography are still occurring rapidly and will translate into more effective and efficient patient care.

References

Almeida AD, Barnes G. 1999. Mammography grid performance. Radiology 210: 227–232.

Andersson I. 2008. Breast tomosynthesis and digital mammography: a comparison of breast cancer visibility and BIRADS classification in a population of cancers with subtle mammographic findings. Eur Radiol 18: 9.

Batignani G et al. 1991. A detailed Monte Carlo study of the performance of a silicon crystal for X-ray detection in the diagnostic energy range. Nucl Instrum Method A 305: 574.

Beutel J, Kitts E. 1996. The image quality characteristics of a novel film/screen system for mammography. Proc SPIE 2708: 8.

Bissonnette M et al. 2005. Digital breast tomosynthesis using an amorphous selenium flat panel detector. In Flynn MJ, ed. Medical Imaging 2005: Physics of Medical Imaging, Proc SPIE, Vol. 5745. San Diego, CA: SPIE.

Boone JM, Seibert JA, Lane SM. 2002. Grid and slot scan scatter reduction in mammography: comparison by using Monte Carlo techniques. Radiology 222: 519–527.

Borasi G, Nitrosi A, Ferrari P, Tassoni D. 2003. On site evaluation of three flat panel detectors for digital radiography. Med Phys 30: 1719–1731.

Bradford CD, Peppler WW, Waidelich JM. 1999. Use of a slit camera for MTF measurements. Med Phys 26: 2286–2294.

Bushong SC. 1997. Radiologic Science for Technologist. Mosby.

Carlton RR, Adler AM. 2001. Principles of Radiographic Imaging: An Art and a Science. Albany, NY: Delmar.

Commission IE. 2003. Medical Electrical Equipment: Characteristics of Digital X-ray Imaging Devices. Part 1: Determination of the Detective Quantum Efficiency. Geneva, Switzerland.

Diekmann F, Sommer A, Lawaczeck R, Diekmann S, Pietsch H, Speck U, Bick U. 2007. Contrast-to-noise ratios of different elements in digital mammography: evaluation of their potential as new contrast agents. Invest Radiol 42: 319–325.

Dobbins JT, Ergun DL, Rutz L, Hinshaw DA, Blume H, Clark DC. 1995. DQE(f) of four generations of comouted radiography acquisition devices. Med Phys 22: 1581–1593.

Feig SA, Yaffe MJ. 1995. Digital mammography, computer-aided diagnosis and telemammography. Radiol Clin North Am 36: 1205–1230.

Fetterly KA, Hangiandreou NJ, Schueler BA, Ritenour ER. 2002. Measurement of the presampled two-dimensional modulation transfer function of digital imaging systems. Med Phys 29: 913–921.

Flynn MJ, Samei E. 1999. Experimental comparison of noise and resolution for 2K and 4K storage phosphor radiography systems. Med Phys 26: 1612–1623.

Fujita EA. 1992. Simple method for determining the modulation transfer function in digital radiography. IEEE Trans Med Imaging 11: 34–39.

Gater L. 2002. Digital mammography: state of the art. Radiol Technol 173: 10.

Giger ML, Doi K, Metz CE. 1984. Investigation of basic imaging properties in digital radiography. 2. Noise Wiener spectrum. Med Phys 11: 797–805.

Haus AG, Doi K, Metz CE, Bernstein J. 1977. Image quality in mammography. Radiology 125: 77–85.

ICRU. 1996. Medical Imaging—The Assessment of Image Quality. Report 54. Bethesda, MD: International Commission on Radiation Units and Measurements.

Lange POA. 2002. Review of Mammography. New York: McGraw-Hill.

Lemacks MR, Kappadath SC, Shaw CC, Liu X, Whitman GJ. 2002. A dual-energy subtraction technique for microcalcification imaging in digital mammography: a signal-to-noise analysis. Med Phys 29: 1739–1751.

Lodwick GS, Haun CL, Smith WE. 1963. Computer diagnosis of primary bone tumor. Radiology 80: 273–275.

Nishikawa RM, Mawdsley GE, Yaffe MJ. 1987. Scanned-protection digital mammography. Med Phys 14: 717–727.

Pisano ED, Cole DB, Yaffe MJ, Aylward SR. 2000. Radiologists' preferences for digital mammographic display. Radiology 216: 820–830.

Pisano ED, Yaffe MJ, Kuzmiak CM. 2004. Digital Mammography. Philadelphia: Lippincott Williams and Wilkins.

Pisano ED, Gatsonis C, Hendrick E, Yaffe M, Baum JK et al. 2005. Diagnostic performance of digital versus film mammography for breast-cancer screening. N Engl J Med 353: 1773–1783.

Ren B, Ruth C, Stein J, Smith A, Shaw I, Jing Z. 2005. Design and performance of the prototype full field breast tomosynthesis system with selenium based flat panel detector. Proc SPIE Phys Med Imaging 5745: 550–561.

Samei E, Flynn MJ. 2002. An experimental comparison of detector performance for computed radiography systems. Med Phys 29: 447–459.

Samei E, Flynn MJ, Reimann DA. 1998. A method for measuring the presampled MTF of digital radiographic systems using an edge test device. Med Phys 25: 102–113.

Stierstorfer K, Spahn M. 1999. Self-normalizing method to measure the detective quantum efficiency of a wide range of X-ray detector. Med Phys 26: 1312–1319.

Vyborny CJ, Giger ML, Nishikawa RM. 2000. Computer-aided detection and diagnosis of breast cancer breast imaging. Radiol Clin North Am 38: 725–740.

Yorker JG, Jeromin LS, Lee DLY, Palecki EF, Golden KP, Jing ZX. 2002. Characterization of a full field digital mammography detector based on direct X-ray conversion in selenium. In Antonuk LE, Yaffe MJ, eds. Medical Imaging 2002: Physics of Medical Imaging, Proceeding of SPIE. San Diego, CA: SPIE.

Contrast-Enhanced Digital Mammography

Melissa L. Hill
University of Toronto

Martin J. Yaffe
University of Toronto

2.1 Motivating Ideas

Although mammography screening (Chapter 1) has been shown to contribute to reducing mortality from breast cancer by up to 30% (Tabár et al. 2011), its accuracy can be limited by both the obscuring effects of healthy tissue overlying and underlying a lesion or, in some cases, the lack of intrinsic radiographic contrast between malignant and normal fibroglandular tissue. Superposition of tissue results from the projection nature of mammograms where images are formed by transmitted x-rays that must traverse all tissues along a path through the thickness of the breast before detection. Overlap of the more attenuating fibroglandular tissues in particular, can produce complex patterns and regions of considerably increased opacity in a mammogram (Chapter 7). This means that image interpretation can be particularly difficult in breasts with a high proportion of fibroglandular tissue, often referred to as radiographically "dense" breasts (Boyd et al. 2007). The sensitivity of cancer detection (fraction of cancers that are detected) has been shown in one study to be only 63% for women with dense breasts, as compared to 87% for fatty breasts (Carney et al. 2003). An estimated 40% of women undergoing mammographic screening (i.e., routine examination of asymptomatic women) have dense breast tissue, so this limitation affects a large portion of the screening population (Jackson et al. 1993). Two promising techniques that may overcome the superposition of breast tissue observed in digital mammography are breast tomosynthesis (Chapter 4) and dedicated breast computed tomography (CT) (Chapter 5).

However, even with tomographic slice images from tomosynthesis or breast CT, there remains a limited contrast between healthy fibroglandular tissue and cancer that is fundamental to the x-ray attenuation properties of breast tissues. Figure 2.1 illustrates the small difference between the linear attenuation coefficients measured from samples of infiltrating ductal carcinoma as compared to that measured from normal fibroglandular tissue (Johns et al. 1987). The use of digital mammography rather than screen-film allows for manipulation of the image contrast after acquisition, which partially helps overcome this problem in women with dense breasts (Pisano et al. 2005), but it does not change the inherent tissue x-ray attenuation properties. One means of increasing the x-ray attenuation of a lesion relative to the surrounding tissue is by the use of an exogenous x-ray contrast agent.

For more than 100 years, it had been recognized by surgeons that tumors are generally more vascular than the normal surrounding tissue (Warren 1979). However, for many years, it was believed that the increased blood volume was due to dilation of existing vessels (Coman et al. 1946). In 1971, Folkman proposed that tumor growth is dependent on the formation of new blood vessels, the process of neo-angiogenesis. Decades later, it is well established that tumors that grow larger than about 1 mm in diameter require additional supplies of oxygen and nutrients for survival, and transmit molecular signals to recruit the sprouting of new blood vessels (Folkman 2000). Studies in breast cancer patients have shown that angiogenesis is an independent prognostic marker as it positively correlates with the degree of

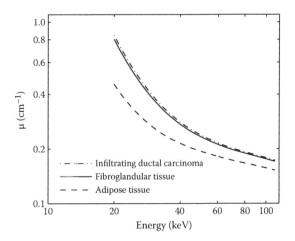

FIGURE 2.1 Measured linear x-ray attenuation coefficients from adipose tissue, fibroglandular tissue and an infiltrating ductal carcinoma. (From Johns PC, Yaffe MJ. 1987. *Physics in Medicine and Biology* 32(6): 675–695. With permission.)

metastasis, tumor recurrence and survival rates (Weidner et al. 1991, 1992).

These neoangiogenic vessels tend to be poorly formed and organized, being tortuous, thin-walled, and leaky, allowing intravenous contrast agents to pool in the extravascular space of tumors (Fukumura et al. 2008). Early studies by Chang, Watt, and Ackerman in the 1980s each demonstrated that angiogenesis could potentially be an imaging biomarker for breast cancer, as they showed that malignant lesions enhance more strongly than benign lesions following injection of an iodinated contrast medium (Chang et al. 1982, Watt et al. 1986). Chang's technique used a whole-body CT scanner, requiring a large radiation dose for the examination (Chang et al. 1982). Watt and Ackerman used a lower radiation dose with their digital subtraction angiography approach, but it involved an invasive catheterization procedure to the superior vena cava (Watt et al. 1986). Therefore, neither of these approaches was appropriate for consistent use in the clinical setting.

In the mid-1980s, a technique for contrast-enhanced breast magnetic resonance imaging (MRI) (Chapter 18) was developed, where lesion signal enhancement was achieved by the use of Gd-based contrast agents (Heywang et al. 1986, Kaiser 1985). By following the pharmacokinetics of the contrast agent over time in dynamic contrast-enhanced MRI (DCE MRI), investigators found that signal enhancement and wash-out could be used to differentiate malignant and benign lesions (Buadu et al. 1996). An early, rapid uptake and then fast washout of contrast agent was found to be generally indicative of a malignant lesion, while a slower steadily increasing uptake of contrast agent was associated with benign lesions (Heywang et al. 1989, Mussurakis et al. 1995). The sensitivity achieved in breast DCE-MRI is excellent with results reported between 88% and 100%, however, this technique can suffer from low specificity of 30% to 70% (Zakhireh et al. 2008). Currently, contrast-enhanced breast MRI

often utilizes information about both the kinetics and the degree of uptake and morphology of the area of enhancement to arrive at a diagnosis.

The advent of digital detector technology has enabled several advanced mammography techniques owing to the easy manipulation of digital data, including contrast-enhanced digital mammography (CEDM). In CEDM, intravenous injection of an iodinated contrast agent takes advantage of the properties of tumor angiogenesis to provide increased image contrast between tumor tissue and the surrounding normal tissue. The problem of tissue superposition is mitigated by an image subtraction process, which largely suppresses the contrast due to the normal parenchyma. The motivations for the use of CEDM are (1) it provides a low-cost and high-resolution alternative to breast MR such that imaging could be performed in breast cancer centers without access to MR, (2) it provides concordant information to initial mammographic findings, (3) it provides an alternative for patients who might be unable or unwilling to undergo an MR examination (e.g., claustrophobia, obesity or body habitus, metallic implants, allergy to Gd), (4) use of CEDM can potentially reduce unnecessary biopsy rates for equivocal mammograms, and (5) CEDM may better delineate extent of disease and indicate multifocality.

Two main approaches to CEDM have been used, namely temporal subtraction and dual-energy imaging. In the following sections we will describe these techniques, the clinical experience with CEDM and the future directions of this imaging approach.

2.2 Temporal Subtraction CEDM

In temporal subtraction CEDM the patient is positioned using light breast compression, then an initial precontrast, or "mask" image is acquired, followed by administration of a monophasic intravenous injection, and then one or several postcontrast images are acquired at different time points. The contrast-to-noise ratio of iodine in the images is usually too low for direct interpretation, so the precontrast image is logarithmically subtracted from the postcontrast images to produce images with bright regions that depict contrast agent uptake. The protocols used for clinical studies have typically been modeled on MR protocols, with between three and six postcontrast images at 1- to 3-min intervals to capture the contrast agent uptake kinetics (Jong et al. 2003, Diekmann et al. 2005, Dromain et al. 2006). Figure 2.2 presents an example of image acquisition timing and the resulting unsubtracted and subtracted images from a five time-point temporal subtraction CEDM examination. To ensure that the image acquisitions throughout the examination remain registered to the precontrast image, the patient must be comfortable during the examination and the breast must be immobilized. As a result, most investigators have used the craniocaudal (CC) view, which allows the patient to be seated during the examination. Typically the breast compression force used in digital mammography is about 10 daN (Hendrick et al. 2010).

Exam event	Time (min)	Unsubtracted image	Subtracted image
Precontrast image	N/A	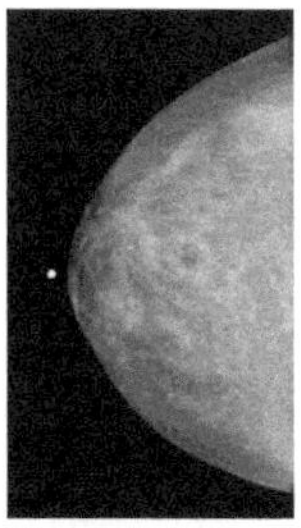	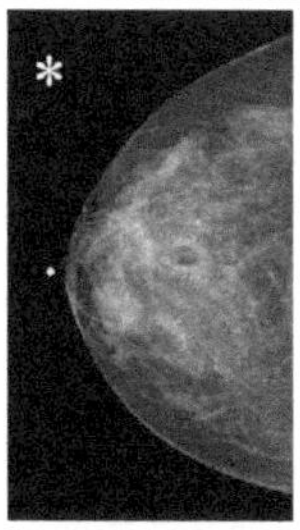
Contrast injected	0	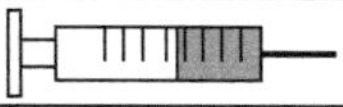	
Postinjection image #1	1	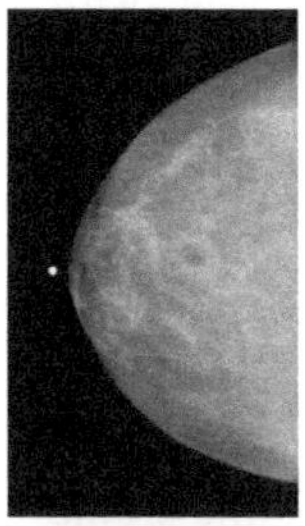	
Postinjection image #2	3	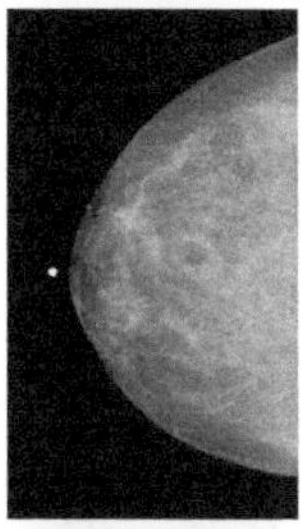	
Postinjection image #3	5	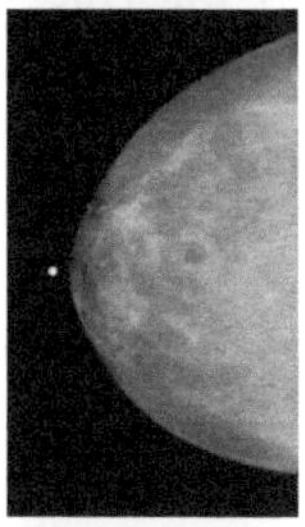	
Postinjection image #4	7	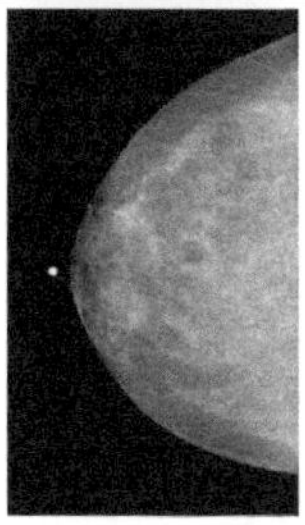	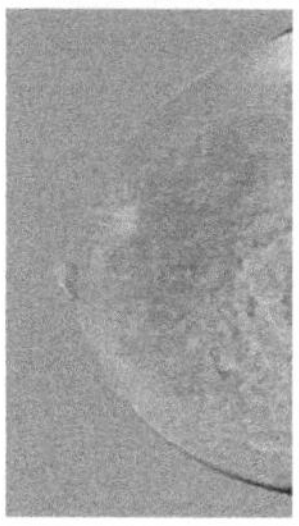

FIGURE 2.2 Example timing and images from a five time-point CEDM examination with temporal subtraction (craniocaudal view). The lesion (histology proven 2 cm infiltrating ductal carcinoma) is more easily identified on the subtracted postcontrast images. *Mammogram acquired with conventional technique for comparison with the appearance of the high-energy unsubtracted images. (Images courtesy of Dr. R.A. Jong.)

To avoid reduced blood flow, a light compression force of less than 6 daN was used in the CEDM trials led by Drs. Jong and Dromain (Jong et al. 2003, Dromain et al. 2006).

Standard iodinated, nonionic, small-molecule contrast agents have been administered by intravenous injection, typically via catheter into the antecubital vein contralateral to the breast of concern, at a dosage of about 1–1.5 ml/kg body weight, over a period of between 30 and 60 s, usually with a power-injector. Iodinated contrast agents were selected because they are clinically available, have acceptable patient tolerance, and they have x-ray attenuation characteristics such that a relatively large attenuation difference between iodine and normal breast tissues can be achieved in an x-ray energy range achievable with minimal modifications to standard mammographic equipment. A comparison of the x-ray linear attenuation coefficients of breast tissue and the tissue with a typical concentration of iodine (1 mg/ml) in Figure 2.3, illustrates that an iodine-based contrast agent increases the x-ray attenuation beyond its k-edge at 33.2 keV.

To take advantage of the relatively large x-ray attenuation difference between iodine and breast tissues beyond the iodine k-edge, the x-ray spectra used for temporal CEDM imaging are shaped such that the mean spectrum energy lies above the k-edge, which is demonstrated in the example spectrum in Figure 2.3. A spectral optimization performed by Skarpathiotakis et al. (2002) using the ratio of the square of the signal-difference-to-noise ratio (SDNR) to the mean glandular dose (MGD) as an index of performance, determined that the optimum tube potentials for molybdenum and rhodium targets and a variety of breast thickness and tissue compositions ranged from 45 to 49 kV when the x-ray spectrum is filtered by 0.3 mm copper. An example optimization for imaging a 4 cm thick breast with a molybdenum anode with copper filtration is shown in Figure 2.4.

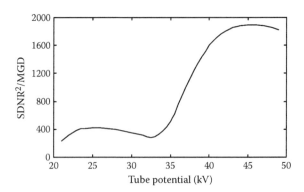

FIGURE 2.4 Plot of the ratio of the square of the signal-difference-to-noise ratio (SDNR) to the mean glandular dose (MGD) per mm² as a function of x-ray tube potential in kilovolts (kV) for a molybdenum target with 0.3 mm copper filtration. The plot peaks at 47 kV for this example of imaging a 4 cm breast composed of 50% fibroglandular tissue, and 50% adipose tissue. (From Skarpathiotakis M, Yaffe MJ, Bloomquist AK et al. 2002. *Medical Physics* 29(10): 2419–2426. With permission.)

The radiation dose absorbed in the breast is related to the product of the fluence (number) of incident x rays and the dose deposited per incident x-ray photon. As the x-ray energy is increased, each photon can deposit a larger amount of energy to the breast. On the other hand the breast becomes much more radiolucent with increasing energy (see Figure 2.1) so the fluence of incident photons required to form an image drops markedly. The overall effect is a reduction of dose compared to that required in producing a conventional mammogram at lower energies. Thus, investigators have been able to design image acquisition protocols to ensure that the total radiation dose from all images collected as part of the CEDM examination acquisitions is less than the dose of a conventional two-view mammography, while maintaining acceptable image quality. For example, in the pilot study by Dromain et al. (2006) the total mean glandular radiation doses ranged from 1 to 4 mGy for 7 images, and in the work by Diekmann et al. (2005) doses varied between 0.5 to 2.5 mGy for 4 images.

2.2.1 Temporal Subtraction CEDM Theory

A simplified model can be used to illustrate the physics behind logarithmic subtraction in temporal subtraction CEDM. Figure 2.5 provides a schematic representation of x-ray transmission through a model of a breast containing a lesion with and without the presence of a contrast agent.

For the simplified case of a monoenergetic beam, no scattered radiation, and an ideal detector, the absorbed x-ray fluences before and after uptake of iodine, Φ_B and Φ_I, at detector element positions (i, j), are:

$$\Phi_B(i,j) = \Phi_0(i,j)e^{-\mu_b T} \qquad (2.1)$$

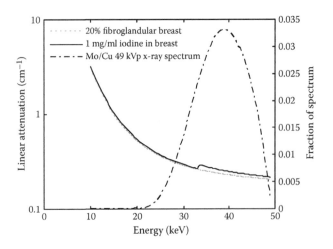

FIGURE 2.3 X-ray linear attenuation coefficients of breast tissue composed of 20% fibroglandular and 80% adipose tissue, and 1 mg/ml iodine in the breast tissue (left y-axis), along with a typical high-energy x-ray spectrum used in temporal subtraction CEDM (right y-axis).

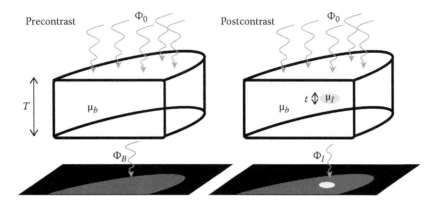

FIGURE 2.5 Schematic diagram illustrating x-ray transmission of an incident x-ray fluence Φ_0, through a compressed breast of thickness T, and x-ray attenuation μ_b, with a lesion of size t, and transmitted x-ray fluences before and after uptake of iodine contrast agent, Φ_B and Φ_I, with attenuation μ_I.

and

$$\Phi_I(i,j) = \Phi_0(i,j)e^{-\mu_b(T-t)-\mu_I t}, \qquad (2.2)$$

where Φ_0 is the incident x-ray fluence, T is the compressed breast thickness, μ_b is the linear attenuation coefficient of the breast tissue, μ_I is the linear attenuation coefficient of iodine, and t is the thickness of the lesion containing iodine. Linear subtraction would result in an image dependent on the breast tissue attenuation and thickness:

$$\Phi_I(i,j) - \Phi_B(i,j) = \Phi_0(i,j)e^{-\mu_b T}\left(e^{-(\mu_I-\mu_b)t} - 1\right). \qquad (2.3)$$

This result depends on the incident x-ray fluence and the overall attenuation of the breast. If the thickness of the breast varied from location to location this would be reflected in the image. Using logarithmic subtraction the image dependence on breast properties can be canceled as follows:

$$S = \ln \Phi_I(i,j) - \ln \Phi_B(i,j) = -(\mu_I - \mu)t. \qquad (2.4)$$

The result of logarithmic subtraction is a signal, S, that varies linearly with iodine concentration and the thickness of the lesion, more closely related to the variables of interest. The relationship is not as simple for polyenergetic x-ray beams, and real detectors, where calculation of the absorbed fluence requires integrating over the x-ray spectrum to account for the spectral shape and the detector absorption efficiencies at each energy. However, Skarpathiotakis et al. (2006) have shown that if a relatively narrow x-ray spectrum is implemented through appropriate choice of target material, filter and tube potential, signal linearity with iodine content is achieved. Figure 2.6 presents results from modeling to demonstrate that logarithmic subtraction of images acquired with a typical polyenergetic spectrum used in clinical application, results in CEDM subtraction images with a signal as a function of iodine areal density that differs from a linear fit by less than 2% from its regression.

This signal linearity with iodine content could be an important advantage of CEDM over DCE-MRI, in that CEDM images provide quantitative information about the contrast agent

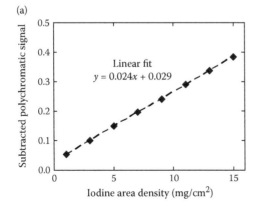

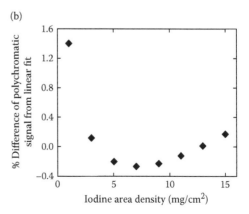

FIGURE 2.6 (a) Logarithmic subtraction of modeled polychromatic signals versus iodine area density and linear regression. (b) the percent difference between the signal in (a) and its linear regression. (From Skarpathiotakis M, Yaffe MJ, Bloomquist AK et al. 2002. *Medical Physics* 29(10): 2419–2426. With permission.)

concentration. Paramagnetic contrast agents such as Gd are not by themselves detectable with MRI, but are visible because they shorten T1 and T2 relaxation times of the nearby hydrogen nuclei (Chapter 18). Therefore, Gd concentration cannot be measured directly and there is no linear relationship between the degree of signal enhancement and the Gd concentration in DCE-MRI (Landis et al. 2000).

2.2.2 Clinical Experience with Temporal Subtraction CEDM

In the past decade technical and clinical experience has been gained in temporal subtraction CEDM through a series of three published clinical pilot studies (Jong et al. 2003, Diekmann et al. 2005, Dromain et al. 2006). These first studies demonstrated that CEDM is useful as an adjunct to mammography to depict contrast enhancement due to the presence of tumor angiogenesis. Table 2.1 summarizes the subject characteristics and the clinical findings for each study. The image acquisition protocols for these clinical studies are comparable. All three studies used a commercial digital mammography system (Senographe 2000D, GE Healthcare) with a cesium iodide–amorphous silicon flat-panel detector, modified for the acquisition of high energy (45–49 kVp) exposures by the addition of a copper filter. Light compression in the CC view was used to immobilize the breast throughout the examination. The contrast agent was administered intravenously with about 100 ml of clinically available, nonionic, small-molecule iodinated medium, at a rate of between 1.7 and 4 ml/s

via a catheter into the arm contralateral to the breast of concern. Between three and six postcontrast images were acquired at time intervals from 1 to 3 min, covering a total time period of between 3 and 10 min postcontrast administration. From these images, contrast agent signal-time curves were plotted to evaluate the contrast uptake kinetics according to the classifications shown in Figure 2.7. The total radiation dose of all image acquisitions was selected to be below the total dose of a conventional mammography screening examination in two views.

Jong et al. (2003) examined 22 women with a mean age of 59 years who were scheduled to undergo biopsy for a suspicious lesion seen on imaging. Among these women, there were 10 cancers, eight of which enhanced on the subtracted CEDM images. Of the two nonenhancing cancers, one was an invasive ductal carcinoma and one was a ductal carcinoma in situ. In 12 subjects with benign lesions, seven had no enhancement and five showed nodular enhancement. The five with enhancement included three fibroadenomas and two cases of fibrocystic change with focal intraductal hyperplasia. The contrast uptake kinetics did not allow consistent differentiation of benign and malignant lesions because most uptake curves were of the constant increase or plateau type, indicative of benign or indeterminate lesions according to the breast MRI literature (Kuhl et al. 1999).

Diekmann et al. (2005) imaged 21 subjects originally scheduled for biopsy to confirm a suspicious lesion seen on imaging. Among 15 malignant lesions, enhancement was observed in 14. The one false positive was in a low-grade ductal carcinoma in situ. Of an additional 11 noncancerous lesions, six enhanced. We

TABLE 2.1 Select Study and Subject Characteristics and the Associated Findings from Published Temporal Subtraction CEDM Clinical Pilot Studies Compared to Breast MRI Findings

Characteristic	Jong et al. (2003)	Diekmann et al. (2005)	Dromain et al. (2006)	Pooled CEDM Results	MRI (Kuhl et al. 1999)
Number of subjects (without technical difficulties)	22	21	18	61	230
Mean patient age (range)	59 (40–74)	N/A	63 (42–80)	—	45.5
Number of postcontrast images	5	3	6	—	9
Time of last postcontrast image	10 min	3 min	7 min	—	6.3 min
Number of malignant lesions	10	15	20	45	101
Mean lesion size in mm (range)	19 (7–45)	39 (8–98)	16 (5–35)	24	—
% enhancing (true positives)	80 (8/10)	93 (14/15)	80 (16/20)	84	—
% Type 1 (constant increase)	38 (3/8)	36 (5/14)	44 (7/16)	41	9
% Type 2 (plateau)	38 (3/8)	57 (8/14)	25 (4/16)	41	34
% Type 3 (washout)	25 (2/8)	7 (1/14)	25 (4/16)	18	57
% Type 4 (decrease)	N/A	N/A	6 (1/16)	—	N/A
Number of benign lesions	12	10	0	22	165
% enhancing	42 (5/12)	50 (5/10)	N/A	45	—
% Type 1 (constant increase)	60 (3/5)	80 (4/5)	N/A	70	83
% Type 2 (plateau)	20 (1/5)	20 (1/5)	N/A	20	11
% Type 3 (washout)	20 (1/5)	0	N/A	10	6

Sources: Jong RA, Yaffe MJ, Skarpathiotakis M et al. 2003. *Radiology* 228: 842–850. With permission. Diekmann F, Diekmann S, Jeunehomme F et al. 2005. *Invest Radiol* 40(7): 397–404. With permission. Dromain C, Balleyguier C, Muller S et al. 2006. *Am J Roentgenol* 187(5): 528–537. With permission. Kuhl CK, Mielcareck P, Klaschik S et al. 1999. *Radiology* 211(1): 101–110.

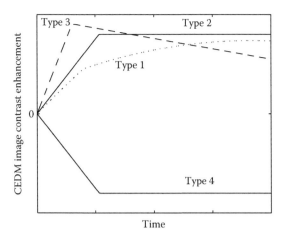

FIGURE 2.7 Classifications of time-signal intensity, or kinetic, curves as applied in the temporal subtraction CEDM clinical pilot studies: constant increase (type 1), plateau (type 2), washout (type 3), and decrease (type 4).

note that in Table 2.1, to simplify analysis we have omitted one noncancerous lesion from the benign category that was determined to be of a borderline type at histology. Of the enhancing noncancerous lesions one was an adenomyoepithelioma (borderline), one had complete remission after chemotherapy, and four were fibroadenomas. The majority of lesions in this study displayed kinetic curves with either a constant increase or plateau. Thus, the authors suggest that differentiation of lesions based on the presence or absence of enhancement is more appropriate than separation based on dynamic curves.

Dromain et al. (2006) performed temporal CEDM in 20 patients who had suspicious lesions and were scheduled for surgery. In two cases where the lesions were located close to the pectoral muscle, motion between the precontrast and postcontrast image acquisitions caused the lesion to move out of the field of view and therefore, not be visible in the subtracted images. Table 2.1 summarizes the results for the remaining 18 successfully imaged cases. Of 20 malignant lesions, 16 enhanced. The four lesions without enhancement were all infiltrating ductal carcinomas. Tumor sections obtained from all surgical specimens were evaluated for intratumoral microvessel density (MVD) based on CD34-immunostaining. No statistically significant correlation was found between MVD and CEDM image enhancement. However, comparing the mean MVD of four nonenhancing lesions at 56.4 microvessels/mm^2 to the mean MVD of the enhancing lesions at 79.2 microvessels/mm^2, there is a suggestion (difference is not statistically significant) that enhancement is associated with a greater MVD. The authors acknowledged the limitations of MVD as a metric for correlation with contrast enhancement since the degree of enhancement would also be related to functional parameters of the vasculature such as permeability. No correlation was identified between the rate of contrast enhancement and the histological type of malignant tumors, suggesting that a large

degree of heterogeneity exists between tumors of the same type and their ability to induce angiogenesis. This heterogeneity of enhancement within a particular tumor type is consistent with the breast MRI literature (Buadu et al. 1996, Frouge et al. 1994). Good correlation (R = 0.86) was observed between the size of the region of enhancement measured on the subtracted CEDM images and the size of the tumor measured at pathology, indicating that CEDM could be useful in determining the extent of disease. Similar to the results from the malignant lesions in the Jong and Diekmann studies, a constant increase in enhancement was the most common kinetic curve observed in this study. Although one case with a decrease in enhancement (type 4) was documented, the authors hypothesized that the apparent signal decrease could be due to the presence of motion artifact in this examination.

In an analysis of the pooled results from these three clinical studies outlined in Table 2.1, one observes that most (84%) of malignant lesions enhanced in the temporal CEDM exams, while slightly fewer than half (45%) of the benign lesions enhanced. Unlike breast MRI findings, most malignant lesions had either an increasing or plateau uptake curve. However, in a similar manner to breast MRI, a washout pattern had a relatively high positive predictive value for cancer. Only 10% (1/10) of benign enhancing lesions had a washout pattern, whereas 18% (7/38) of enhancing malignant lesions had a washout curve. Since very few benign lesions have been imaged with CEDM, it would be instructive to perform further clinical studies with women scheduled for biopsy to confirm the diagnostic value of a washout pattern.

One hypothesis to explain the lack of washout in most cancers depicted by CEDM is the projection nature of this technique. Unlike breast MRI where the images are tomographic slices of a plane of tissue, in CEDM the signal from each pixel represents a summation of signals from a column of tissue from the entire breast thickness. Although the uptake of contrast in normal tissue is low, it tends to increase with time over the time scale of a typical CEDM examination. Thus, the washout from a small lesion may be masked by the summation of iodine signal from the several centimeters of normal tissue within which the lesion resides.

Another hypothesis is that differences in the sensitivity of x-ray imaging and MRI to their respective contrast agents may result in perceived differences in the uptake kinetics. In CEDM nearly 300 times the amount of iodine is injected compared to the amount of gadolinium administered in breast MRI. For example, a typical CEDM contrast injection is about 100 ml of iodinated agent administered at a rate of 3 ml/s giving a total iodine dose of about 30 g, while in MRI, approximately 15 ml of Gd-based contrast is injected at a rate of 2 ml/s, resulting in a gadolinium dose of about 1.1 g (Kuhl et al. 1999). The effect of a more protracted contrast bolus in CEDM compared to breast MRI (~30 vs. ~8 s) combined with a much larger amount of iodine that must accumulate for visualization means that there could be a longer time to peak image enhancement in CEDM compared to breast MRI. The distribution of dynamic curves of malignant lesions observed in the study by Diekmann et al. (2005) where the last postcontrast image was acquired at 3 min, compared to the distribution in the studies of Jong et al. (2003) and Dromain et al. (2006) where longer imaging periods were used, supports this hypothesis. While for each of the Jong and Dromain studies, 25% of the enhancing cancers had washout patterns, only 7% of the cancers in the Diekmann study were of this kinetic curve type. Also, in the Diekmann study, relatively more malignancies were observed with a plateau uptake pattern compared to the Jong and Dromain studies, suggesting that if postcontrast images had been acquired at a later time point in the Diekmann study, some of the lesions classified with a plateau curve may have actually been observed to wash out.

A third hypothesis is that the differences between the contrast agent uptake kinetics observed in CEDM compared to breast MRI are due to the use of breast compression in CEDM which may alter blood flow, despite the reduced compression compared with screening mammography. Carp et al. (2008) have demonstrated with optical measurements of compressed breast tissue that blood flow is reduced even with a compression force of only one-third of that of the typical compression forced use for screening exams. This compression force resulted in a drop in total hemoglobin of about 15% compared to a baseline measurement without compression.

2.3 Dual-Energy CEDM

Dual-energy imaging takes advantage of the differences in x-ray attenuation properties of materials of different elemental compositions to produce material-selective images. The principles of dual-energy imaging have been described extensively in the literature (Alvarez et al. 1976, Brody et al. 1981, Lehmann et al. 1981, Johns et al. 1985, Boone et al. 1990, Lemacks et al. 2002), with many applications in digital radiology where it is important to remove normal anatomical structure to permit visualization of a pathology (Odagiri 1992, Fischbach 2003, Ducote 2006). In dual-energy CEDM, the large difference in x-ray attenuation properties of the iodinated contrast agent at energies above and below the k-edge of iodine, compared to the small change in breast tissue attenuation between these energies, is exploited. A pair of images is acquired after contrast medium administration at a low-energy and a high-energy. Typically the high-energy image is acquired with a tube potential between 45 and 49 kVp, and the low-energy image is acquired at tube potentials within a conventional mammographic imaging range (26 to 30 kVp). Figure 2.8 illustrates with an example pair of high-energy and low-energy x-ray spectra, how the selection of the x-ray beam energies can provide differential iodine contrast.

The same types of iodinated contrast agents and injection protocols used in temporal subtraction CEDM have been applied in dual-energy CEDM (Lewin et al. 2003, Dromain et al. 2011). Prior to breast compression, standard iodinated, nonionic, contrast agents are administered by intravenous injection via catheter into the antecubital vein contralateral to

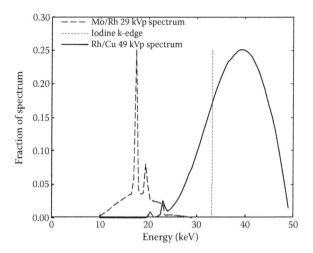

FIGURE 2.8   Low-energy (Mo/Rh, 29 kVp), and high-energy (Rh/Cu, 49 kVp) x-ray spectra used to produce good differential iodine contrast in dual-energy CEDM images.

the breast of concern, at a dosage of about 1–1.5 ml/kg body weight, over a period of between 30 and 60 s. Investigators have allowed the contrast medium to circulate within the vasculature for about 2 min before image acquisition during which time the breast is compressed with a force equal to that applied during a screening mammogram (Lewin et al. 2003, Dromain et al. 2011). Owing to the ability to rapidly image the low- and high-energy image pairs, there is usually only minimal motion artifact present in the subtracted dual-energy CEDM images (Dromain et al. 2011).

2.3.1 Dual-Energy CEDM Theory

The recombination of dual-energy images to cancel breast tissue can be understood by again referring to the simplified model (assuming monoenergetic x rays at two different energies) presented for CEDM temporal subtraction in Figure 2.5, with the modification that two postcontrast images are acquired, one low-energy image with incident fluence, Φ_0^L, and a high-energy image with incident fluence, Φ_0^H, and no precontrast image is needed. The transmitted low-, Φ^L, and high-energy, Φ^H, x-ray fluences that exit the breast after attenuation by normal breast tissue and a lesion are respectively:

$$\Phi^L(i,j) = \Phi_0^L(i,j)e^{-\mu_b^L(T-t)-\mu_l^L t} \tag{2.5}$$

and

$$\Phi^H(i,j) = \Phi_0^H(i,j)e^{-\mu_b^H(T-t)-\mu_l^H t}, \tag{2.6}$$

where subscripts denote materials, and superscripts denote the imaging energy. Weighted logarithmic subtraction of the high- and low-energy images results in the following form:

$$\ln\left(\Phi^H(i,j)\right) - w_I \ln\left(\Phi^L(i,j)\right) = \ln\left(\Phi_0^H(i,j)e^{-\mu_b^H(T-t)-\mu_I^H t}\right)$$

$$- w_I \ln\left(\Phi_0^L(i,j)e^{-\mu_b^L(T-t)-\mu_I^L t}\right)$$

$$= k - \mu_b^H(T-t) - \mu_I^H t$$

$$- w_I\left(-\mu_b^L(T-t) - \mu_I^L t\right), \quad (2.7)$$

where k is a constant resulting from the high- and low-energy x-ray fluencies. A suitable choice of weighting parameter, w_I, as the ratio of breast tissue linear attenuation coefficients allows for the cancelation of breast tissue from the resulting recombined image. To create an iodine image, we therefore choose the weighting parameter to be of the form:

$$w_I = \frac{\mu_b^H}{\mu_b^L}. \quad (2.8)$$

Substitution of Equation 2.8 into Equation 2.7 gives:

$$\ln\left(\Phi^H(i,j)\right) - w_I \ln\left(\Phi^L(i,j)\right) = k - t\left(\mu_I^H - w_I \mu_I^L\right), \quad (2.9)$$

where the choice of the high- and low-energy imaging x-ray spectra are such that $\mu_I^H > \mu_I^L$, and $w_I < 1$, resulting in a recombined image dominated by the linear attenuation of iodine at the high-energy. In practice, when a polyenergetic x-ray spectrum is used, there is a nonlinear relationship between the logarithmically subtracted signals and the amount of iodine present. Additionally, the presence of scattered radiation can be important in both quantitative and qualitative evaluations if not removed. For example, Arvantis et al. (2009) observed a 33% underestimation of projected iodine thickness without scatter correction applied.

2.3.2 Clinical Experience with Dual-Energy CEDM

In an early study by Lewin et al. (2003) on dual-energy CEDM, 26 women scheduled for biopsy or with greater than 50% chance of malignancy were recruited for imaging. Images were acquired precontrast and postcontrast in two separate breast compressions in the mediolateral oblique (MLO) view. There was about 150 s delay after contrast administration before postcontrast image acquisition and about 30 s between low- and high-energy image exposures. High-energy images were acquired at 44 kVp with 8 mm aluminum filtration. The low-energy images were acquired with either Mo/Mo at 30 kVp or 33 kVp Rh/Rh at a total examination dose about 1.2 times that of conventional two-view digital mammography. In the weighted log-subtracted images there was strong enhancement, moderate enhancement, and weak enhancement reported in eight, three, and three of 14 malignant lesions, respectively. In 12 patients with benign lesions, 10

had no enhancement while two were rated as having weak enhancement, corresponding to one atypical ductal hyperplasia and one fibrocystic change. Thus, the presence of enhancement was strongly predictive of malignancy, with a positive predictive value of 87.5% and a sensitivity of 100%.

In a more recent study, Dromain et al. (2011) enrolled 120 women for dual-energy CEDM imaging from recalls from screening and unresolved cases from mammography and ultrasound. In this study two different views of the affected breast were acquired. Two minutes after the initiation of the contrast agent administration, the breast was compressed in an MLO view and a pair of low- and high-energy images was acquired. The breast was then compressed in the CC position and a new pair of low- and high-energy exposures was performed 4 min after the initiation of contrast agent administration. The total radiation dose delivered to the patient was about 1.2 times the dose delivered for a standard two-view screening mammogram. Sensitivity improved significantly from 78% to 93% when mammograms were read in combination with dual-energy CEDM.

2.4 Comparison of Temporal Subtraction and Dual-Energy CEDM

In summary, temporal subtraction CEDM imaging has the advantage that contrast agent uptake kinetics can be measured via repeated imaging at multiple time points within a reasonable dose, which is possibly helpful to improve specificity. However, the clinical pilot studies reviewed here have shown that the assumption that the lesion enhancement is the same between modalities may be false. For example, the majority of malignant lesions demonstrated continuous enhancement in each of the Jong and Dromain studies, compared to breast MRI, where a washout curve was much more likely to be observed. This difference may be due to the use of compression in this image acquisition approach, or the projection nature of the imaging technique. A drawback of the temporal subtraction approach is the restriction to one unilateral view. To image both breasts, or to acquire two views of the same breast, would require a delay between successive exams to allow for washout of contrast agent, and to prevent renal toxicity.

Although the compression used in temporal subtraction CEDM is light, the force may cut off, or reduce blood flow in the breast, altering the contrast uptake kinetics. However, this compression is required because motion artifact degrades the CEDM image quality. Even with compression, for a long examination (~15 min), this motion can be quite large, meaning that an image registration algorithm must be used for image interpretation. It is not clear how compression would alter the contrast agent kinetics in dual-energy CEDM. Although the breast is not under compression during contrast administration, in the published clinical studies a full compression force was applied and this has been shown to reduce blood flow.

Interpretation of temporal subtraction CEDM images requires simultaneous evaluation with mammography. In dual-energy

CEDM it may be possible to use the low-energy image as a surrogate for a conventional digital mammogram.

2.5 Potential Clinical Application of CEDM

Presently, only two studies have been published that report Receiver Operating Characteristic curve (ROC) analyses of the diagnostic potential of CEDM. One study evaluated temporal subtraction CEDM in 70 subjects (Diekmann et al. 2011), and the other assessed dual-energy CEDM in 120 women (Dromain et al. 2011). In both cases a statistically significant increase in sensitivity was achieved when CEDM images were read in conjunction with mammograms (Diekmann et al. 2011, Dromain et al. 2011). Dromain et al. (2011) showed a sensitivity of 93% for CEDM when read with mammograms, in comparison to 78% with mammography alone, without a decrease in specificity. This suggests that CEDM is useful as an adjunct to mammography. Due to the use of an intravenous contrast agent and the additional risk this imposes on a patient, CEDM would not be applied for population screening, but it could be helpful in diagnostics. In particular, it may be useful in cases of equivocal mammography, or when lesions are occult on mammography. For example, Diekmann et al. (2011) demonstrated an increased number of malignant lesions depicted with the addition of temporal subtraction CEDM images compared to mammography alone. Diekmann et al. (2011) also showed that CEDM is effective in the evaluation of the dense breast where specificity was increased from only 0.35 to 0.59. This improvement in imaging performance could help to reduce unnecessary biopsies and improve cancer localization.

CEDM also has the potential to provide accurate information on the extent of disease to improve staging. Dromain et al. (2011) found that the combination of CEDM read with mammography provided the most accurate size estimate of lesions as validated by histopathology, compared with mammography alone, or ultrasound. In fact, there was no significant difference between the size measured on CEDM with mammography compared to histology. This suggests that CEDM may also be useful in the evaluation of residual disease after a lumpectomy with positive margins and the assessment of breast cancer recurrence. It is also possible that, like breast MRI, CEDM could be useful to test for contralateral disease and for the presence of a cancer when axillary metastasis is found (Saslow et al. 2007, Zakhireh et al. 2008).

Although nonionic iodinated contrast agent is generally well tolerated, CEDM cannot be used in patients with a contraindication to iodinated contrast agents. This will include pregnant patients, those with impaired kidney function, and those with a history of reaction to a contrast agent (ACR Committee on Drugs and Contrast Media 2010). CEDM would not be indicated for the assessment of response to neoadjuvant chemotherapy because a nonirradiating method would be preferred when several successive examinations are required (Dromain et al. 2009).

Overall, the most promising application of CEDM is as an alternative method to breast MRI, since the potential indications for CEDM are similar to those for breast MRI (Saslow et al. 2007, Zakhireh et al. 2008). In centers where there is limited access to breast MRI, this might be especially advantageous. CEDM is a fast, high-resolution imaging technique, which could be particularly useful in emerging rapid diagnostic units since there is immediate availability of CEDM in the mammography suite.

2.6 Future Directions of CEDM

2.6.1 Areas for Development in CEDM

Although CEDM is now approved for clinical use in many countries, a number of research opportunities exist for both physics optimization and clinical development of this mammographic modality. In the following sections we briefly review a variety of current topics of research in CEDM and suggest some areas for future improvement of CEDM imaging.

2.6.1.1 Optimization of CEDM Clinical Protocol

2.6.1.1.1 *Image Acquisition Technique Factors*

Previously, optimization of CEDM has been done based on signal-difference-to-noise ratio (SDNR), where the only noise sources considered were quantum and detector noise (Skarpathiotakis et al. 2002, Ullman et al. 2005, Arvanitis et al. 2009, Palma et al. 2008). This may be appropriate for imaging a particularly fatty breast, but it is well known that another source of image variation, referred to as "clutter" or anatomical noise, is an important factor in lesion detectability (Burgess et al. 2001, Rolland et al. 1992). Recently, it has been recognized that due to incomplete tissue cancelation in the image subtraction or recombination process, and due to contrast agent uptake in normal tissue, there can be an influence of anatomical noise in contrast-enhanced breast imaging (Hill et al. 2010b, Fredenberg et al. 2010). Figure 2.9 illustrates qualitatively the appearance of breasts in temporal subtraction and dual-energy CEDM with a range of normal tissue contrast agent uptake. Investigators have quantified the anatomical noise in mammographic images by the power law exponent, β; the slope of a fit to the spectral content of the image signal as a function of spatial frequency on a log-log plot (Burgess 1999, Heine et al. 2000). This methodology has also been applied to CEDM images, demonstrating that an "uptake clutter" is produced by normal tissue contrast agent extravasation (Hill et al. 2011). The power law exponent associated with each of the CEDM images shown in Figure 2.9 is listed. The magnitude of β appears to correlate with the texture of the breast parenchyma in these images.

Fredenberg et al. (2010) recently developed a theoretical framework for CEDM imaging which included both quantum noise and anatomical noise, as quantified by a power-law description. Their work demonstrated that the optimal choice of kV for CEDM acquisition with a photon-counting detector changed according to the amount of anatomical noise considered.

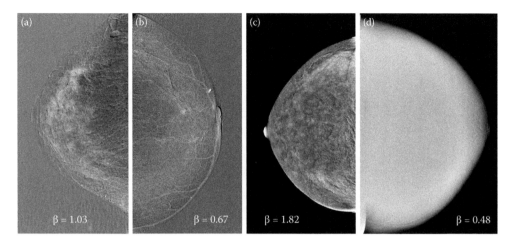

FIGURE 2.9 Example CEDM images with varying amounts of noise from normal tissue contrast agent uptake. Images (a) and (b) are examples of temporal subtraction CEDM (Courtesy Dr. R.A. Jong), and (c) and (d) are examples of dual-energy CEDM. The power-law exponent, β, which indicates the amount of anatomical noise, is given for each image. (Courtesy Dr. C. Dromain.)

Preliminary work on lesion detectability in temporal subtraction CEDM has also shown that lesion detection performance is affected by the clutter produced by contrast agent uptake in the normal tissue (Hill et al. 2010b). Therefore, imaging performance of temporal subtraction CEDM and dual-energy CEDM acquired with energy integrating detectors would likely benefit from system optimization performed with the effects of anatomical noise included in the analysis.

2.6.1.1.2 Image Acquisition Timing

Data in the literature on iodinated contrast agent kinetics in the breast include the very early studies done on contrast-enhanced breast imaging (Chang et al. 1980), the temporal subtraction CEDM clinical pilot studies reviewed in Section 2.2.2, and some recent multirow detector computed tomography (MDCT) studies on the breast (Inoue et al. 2003, Miyake et al. 2005, Tozaki et al. 2006, Perrone et al. 2008). Each of these studies is limited in some way for determination of the true underlying iodinated contrast agent uptake kinetics in a CEDM examination. The early studies have poor image quality, the CEDM trials are confounded by the projection nature of the imaging, and the MDCT exams have relatively coarse temporal resolution. However, the MDCT studies probably provide the most useful kinetic information since the data are volumetric and no breast compression is used. In particular, data from Perrone et al. (2008) confirm the hypothesis that the time of peak enhancement of malignant lesions is later with iodinated contrast agents and x-ray imaging than with gadolinium contrast agents in breast MRI. In breast MRI the time to peak enhancement is within 2 min of contrast administration (Kuhl et al. 1999). However, Perrone et al. (2008) show that that time to peak enhancement of malignant lesions in MDCT is roughly 3 min. One problem with applying these data to CEDM is that even in the dual-energy technique where the contrast agent is administered without breast compression, findings from Carp et al. (2008) using optical imaging methods

indicate that blood flow is affected by compression, and that each serial compression reduces blood flow further. It is not clear how the compression(s) in dual-energy CEDM would affect the contrast agent time-course in the breast, although this information could be helpful to a radiologist in the interpretation of these images. It would be useful to have high-quality data on the kinetics of iodinated contrast agents in malignant lesions, benign lesions, and normal breast tissue with and without breast compression so that the timing of CEDM image acquisition could be selected to maximize lesion sensitivity. However, to collect this sort of data would require a large study that would likely be difficult to implement in practice. Perhaps further characterization of blood flow in the breast with and without compression would allow for predictive modeling of iodinated contrast material kinetics.

2.6.1.2 Artifact Reduction

2.6.1.2.1 Image Registration Algorithms for CEDM

Image registration is the process of determining a transformation that maps the points in one image to the corresponding location in another image. This process is important in temporal subtraction CEDM where there can be relatively large motion between successive image acquisitions during an examination of up to 10 min in duration. For most of the temporal subtraction CEDM clinical pilot studies, simple image registration algorithms were applied with the assumption of rigid motion (Jong et al. 2003). Rigid motion image registration algorithms account for rotation and translation, but we know that a breast may deform in a nonrigid manner where planes of tissue may shift with respect to one another. The application of nonrigid registration algorithms to DCE-MRI data where the breast is under only very light compression compared to mammography, has been quite successful (Li et al. 2009). Because one of the greatest drawbacks of the temporal subtraction CEDM method is the potential for motion artifacts, it may be worth revisiting

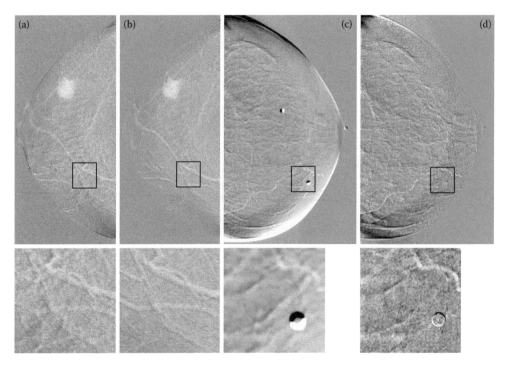

FIGURE 2.10 Comparison of log-subtracted images from two temporal subtraction CEDM cases with rigid registration algorithm applied (a) and (c) (These images are the same as those presented to the radiologists in the Jong et al. (2003) study); and with a nonrigid registration algorithm applied (b) and (d). The zoomed regions in the lower row illustrate the differences between the image appearance with the two registration algorithms applied. Some of the distracting motion artifacts in images (a) and (c) have been eliminated with the nonrigid registration algorithm. The images are displayed with the same contrast settings for each image registration algorithm implementation. Images courtesy Dr. R.A. Jong.

the images acquired in clinical studies to reevaluate the diagnostic accuracy when more advanced image registration algorithms are applied.

Figure 2.10 presents a previously unpublished comparison of the appearance of temporal subtraction CEDM images of two cases from the clinical study by Jong et al. (2003) when a rigid transform (left) was applied and a nonrigid transform was applied (right). A subject with comparatively little motion and a subject with greater motion were selected for analysis. Although truth is unknown, the nonrigid registration algorithm appears to perform better than the rigid algorithm for both CEDM subjects.

2.6.1.2.2 *Effect of X-Ray Scatter in Dual-Energy CEDM*

It is well known that x-ray scatter contamination in low-energy and high-energy images can lead to errors in quantification of a dual-energy signal (Shaw et al. 1987, Wagner et al. 1988, Kappadath et al. 2004). In a recent dual-energy CEDM clinical study by Dromain et al. (2011), signal quantification was not of interest, but a white region near the breast edge was identified as an image artifact, which negatively impacted the analysis of the peripheral area of the breast. Figure 2.11 shows a clinical case with this artifact present at the breast periphery. This artifact results from the different characteristics of x-ray scatter in the low-energy and high-energy images. This leads to incomplete cancellation of the scattered radiation signal in the recombined image. The scattered radiation artifact is nonuniform across the image due to regional variations in the amount of scattered

x-rays, especially at the breast periphery where the x-ray scatter-to-primary ratio sharply increases due to a large component of scatter from the breast compression paddle and support plate (Sechopoulos et al. 2007). In the more central region of the breast

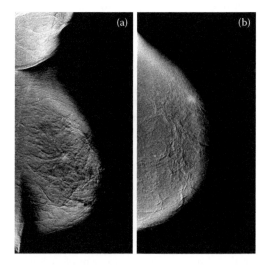

FIGURE 2.11 Recombined dual-energy CEDM images of a 67-year-old woman with an enhancing malignant lesion visible as a mass in the (a) mediolateral view and (b) craniocaudal view. The whiter regions near the breast edge correspond to a scattered radiation artifact. Images courtesy Dr. C. Dromain.

this component of scattered radiation is largely attenuated by the breast itself before it reaches the detector.

An enhancing malignant lesion is visible in the breast periphery in the CC view in Figure 2.11. This subject illustrates the importance of a reduction in the scattered radiation artifact because a more weakly enhancing lesion could have been missed within the scatter artifact in the CC view. A number of scatter reduction techniques exist in the literature, which have been successfully applied to mammography (Seibert et al. 1988, Baydush et al. 2000, Trotter et al. 2002) and to dual-energy mammography (Kappadath et al. 2005), and one of these could potentially be applied in dual-energy CEDM to reduce the artifact in the breast periphery.

2.6.1.3 Quality Control Considerations

Most image quality concerns for CEDM are in common with those for conventional digital mammography. However, what are of particular interest in CEDM are those factors that affect the iodine detectability and/or quantitative measurement. A review of Equations 2.7 and 2.9 indicates that the most important factors that determine dual-energy CEDM image quality are likely to be the x-ray beam quality and the performance of the image subtraction, or dual-energy recombination algorithm. The x-ray beam quality directly influences the iodine signal-difference-to-noise ratio, so this should be monitored regularly. There is not yet enough experience in CEDM to indicate how frequently a beam quality test should be done, although in digital mammography the half-value layer (HVL) has been shown to be quite stable over time with a mean coefficient-of-variation of 0.05 for at least three repeated measurements of an individual system over 45 digital mammography units (Bloomquist et al. 2006).

At our institution we currently use a solid iodine contrast-detail phantom, developed in-house for monitoring the performance of a prototype dual-energy CEDM system based on a modified commercial mammography system (Senographe DS, GE Healthcare). A dual-energy recombined image of this epoxy-based phantom with iodinated disks is shown in Figure 2.12

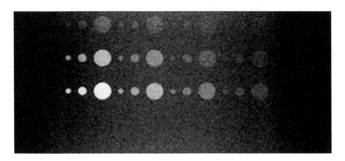

FIGURE 2.12 Example dual-energy CEDM recombination image of an iodine contrast-detail phantom. An epoxy-based phantom with 3, 5, and 10 mm iodinated disks of varying thickness embedded in the phantom. From top to bottom row, the disk thicknesses are 1, 3, and 5 mm. Iodine aerial densities are 0.3, 0.5, 1.0, and 2.0 mg/cm² iodine in the bottom row (5 mm depth).

(Hill et al. 2009). This phantom is useful for evaluation of signal-difference-to-noise ratio and also provides a qualitative test of disk detectability to ensure that the mammography system and dual-energy recombination algorithm performance are adequate.

2.6.2 Advanced Contrast-Enhanced Breast Imaging Techniques

2.6.2.1 Contrast-Enhanced Breast Tomosynthesis

The use of contrast-enhancement with a tomographic imaging technique has been proposed to address the limitation of tissue superposition in CEDM, which can cause the iodine in normal tissue to mask the appearance and kinetics of the contrast uptake in a lesion (Chen et al. 2007). Figure 2.13 demonstrates the potential of dual-energy contrast-enhanced digital breast tomosynthesis (DBT) for the suppression of out-of-plane structures, lesion localization, and sensitivity to iodine. Please refer to Chapter 4 for a description of the physics of DBT. Both CEDM approaches described previously, of temporal subtraction and dual-energy imaging have been shown in clinical pilot studies to be feasible with DBT (Chen et al. 2007, Carton et al. 2010). Additionally, the use of a slot-scanned imaging technique, with a photon-counting detector has been studied for application in dual-energy contrast-enhanced breast tomosynthesis (Schmitzberger et al. 2011).

Chen et al. (2007) imaged 13 women with Breast Imaging Reporting and Data System (BI-RADS) category 4 or 5 lesions using temporal subtraction contrast-enhanced tomosynthesis acquired using nine projection images over a 50 degree angular range. It was noted that tomosynthesis allowed for lesion localization, a detailed assessment of breast cancer morphological features, and vascular assessment through contrast-enhancement that was consistent with results from breast MRI performed on the same subjects (Chen et al. 2007). In many cases the lesion margins were visualized better than by digital mammography (Chen et al. 2007).

A limitation of patient motion during the long scan time of the temporal subtraction DBT technique was identified by Chen et al. (2007), which is addressed by the use of dual-energy (Carton et al. 2010, Schmitzberger et al. 2011). In a feasibility study of the image quality achievable in dual-energy contrast-enhanced DBT compared to the temporal-subtraction approach, Carton et al. (2010) imaged one patient with a malignant lesion with a precontrast high-energy DBT and then separate high- and low-energy postcontrast DBT image acquisitions were each made at three time points. A total radiation dose of 6.35 mGy was delivered, approximately twice that used in two-view screening mammography (Carton et al. 2010). Suspicious enhancement of the lesion was observed in each of the temporal subtraction and dual-energy contrast-enhanced DBT and this correlated well with observations on breast MRI (Carton et al. 2010). However, it was noted that less motion artifact was apparent in the dual-energy images compared with the temporal subtraction images,

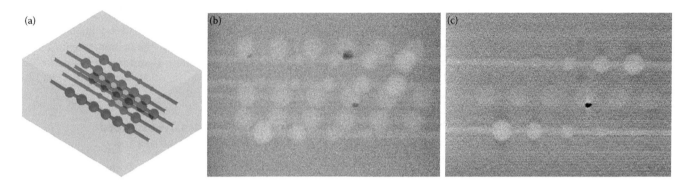

FIGURE 2.13 (a) Schematic of experimental breast-equivalent plastic phantom. (b) High-energy (Rh/Cu, 49 kV) projection image (0°) of phantom in (a) with 1.0 mg/ml iodine in lesion spheres illustrates tissue superposition. (c) Central slice at 2 cm depth of dual-energy CE-DBT reconstruction (1.0 mg/ml iodine in lesion) demonstrates suppression of out-of-plane structure and sensitivity to iodine contrast. Black artifacts are air bubbles.

resulting in superior visualization of the tumor in the dual-energy images (Carton et al. 2010).

The first two studies were done with a typical signal-integrating detector. Recently, Schmitzberger et al. (2011) used a photon-counting detector to perform dual-energy contrast-enhanced DBT in 10 women, nine with malignant lesions, and one with a "high-risk" lesion. Their prototype system used a commercial slot-scanned mammography system (MicroDose L30, Royal Philips Electronics, Eindhoven, The Netherlands) with a detector that uses two energy thresholds to split detected photons into low-energy and high-energy images. They made 21 projection acquisitions at 40 kVp over an 11 degree angular range, reconstructing with a 3 mm slice thickness and 100 μm in-plane resolution. Tomosynthesis datasets were acquired prior to contrast administration and at 120 s and 480 s postcontrast with light compression force applied throughout the examination. About 1.26 mGy average glandular dose was delivered for the entire examination to a standard breast. Of nine visible lesions (one was out of FOV), seven (all cancers) were rated by radiologists as BI-RADS four or higher when dual-energy contrast-enhanced DBT was read with mammography, while two cases (one cancer, one borderline) were rated as borderline diagnoses (Schmitzberger et al. 2011). An important finding of this study was that radiologists categorized all lesions with visible enhancement as having washout (Schmitzberger et al. 2011). These contrast agent kinetics observed using DBT provide evidence to support the hypothesis that when tissue superposition is removed, the uptake kinetics might behave more closely to those measured in breast MRI.

2.6.2.2 Contrast-Enhanced Dedicated Breast CT

Dedicated breast computed tomography (CT) is an emerging mammographic technique that will be described in detail in a later chapter (see Chapter 5). However, we wanted to briefly note here the important developments in the combination of breast CT with the use of a contrast agent. Breast CT offers the advantage of tomographic slice imaging with isotropic voxels and the

potential for quantitative measurement of contrast uptake in a lesion, which may aid in diagnosis.

Recently Prionas et al. (2010) reported on a study of 46 women at a mean age of 53 years who had 54 mammographically or ultrasonogaphically identified BI-RADS category 4 or 5 lesions. A single breast CT scan delivers a radiation dose equivalent to a two-view screening mammogram, and reconstructed voxel dimensions are about 0.25 mm. Each breast was scanned before and after contrast administration, with the unaffected breast scanned first, then the affected breast, followed by intravenous injection of 100 ml of iodinated contrast agent at 4 ml/s, and then postcontrast scans of the affected breast (~96 s postcontrast) and finally unaffected breast (~247 s postcontrast) (Prionas et al. 2010). Images were interpreted without subtraction. Quantitative analysis was performed on 52 lesions, where enhancement was observed in all 29 malignant lesions at a mean of 55.9 HU, significantly greater than the 17 of 23 enhancing benign lesions at a mean enhancement of 17.6 HU (Prionas et al. 2010). Radiologists rated the contrast-enhanced lesions as significantly more conspicuous than in the unenhanced images.

2.6.3 Novel Contrast Agents for CEDM

Although iodine works well as a contrast agent in most diagnostic radiology applications, its x-ray attenuation properties are not ideal at standard mammographic imaging energies. It would be ideal to acquire contrast-enhanced images at conventional mammographic energies which provide excellent visualization of microcalifications and morphology, while at the same time achieving excellent enhancement from the contrast media. In this implementation it is possible that due to the presence of the contrast agent the contrast-enhanced image would provide images with equal diagnostic information, if not superior image quality to conventional mammograms. Perhaps these images would not have to be read in combination with conventional mammograms, and the total radiation dose delivered to the patient from all diagnostic exams could be reduced. A

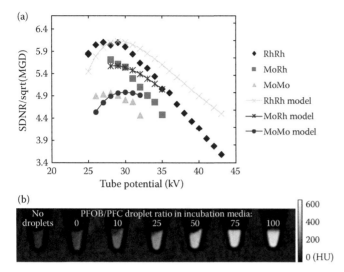

FIGURE 2.14 (a) Experimental measurements and modeled results (right y axis) of optimal imaging technique in mammography (Senographe 2000D) of 6 mm depth well plate with 3% PFOB solution under 5 cm thick 50/50 breast-tissue equivalent plastic. (b) Murine alveolar macrophage cells were incubated with 400 μl of varying combinations of perflurocarbon and perflurooctyl bromide (%) emulsion for 2 h and imaged with cone-beam CT at 30 keV mean x-ray beam energy.

number of elements have been identified that could achieve this excellent low-energy contrast-enhancement, including bismuth (Diekmann et al. 2007; Lawaczeck et al. 2003), and zirconium (Lawaczeck et al. 2003). Using these elements would mean that a contrast agent has to be developed exclusively for mammography, which is expensive and difficult due to the pharmacological studies that would be required (Diekmann et al. 2007). Lawaczeck et al. (2003) also investigated bromine for low-energy CEDM, which showed moderate contrast potential compared to zirconium, however, clinical implementation of this element as a contrast agent may be feasible due to existing clinical usage of a brominated perfulorocarbon, perflurooctyl-bromide (PFOB) for artificial breathing and as a blood substitute. PFOB has good biocompatibility and has been shown to provide good contrast in a mammographic energy range. Figure 2.14b shows the results of an in vitro study to demonstrate contrast-enhancement of PFOB in CT imaging done with a beam with about 30 keV mean x-ray energy. In Figure 2.14a, the results of an experimental optimization are shown for the use of PFOB in mammography according to the maximization of SDNR per square-root average glandular dose. This plot shows that to image a 6 mm lesion in a 5 cm thick compressed breast with 50% fibroglandular composition, the optimum mammographic technique is imaging with Rh/Rh at 29 kVp.

A series of emerging contrast agents can also serve as molecular probes, that either take advantage of endogenous targeting or that have the potential to be targeted to a particular cell-surface receptor (Hill et al. 2010a, Cheung et al. 2010, Zheng et al. 2006, Karathanasis et al. 2008, Samei et al. 2009, Ghaghada et al. 2011).

With appropriate ligand selection it may be possible in CEDM to visualize the up-regulation of cell-surface markers associated with breast cancer. This will be an important area of future work to improve the specificity of breast imaging.

References

ACR Committee on Drugs and Contrast Media. 2010. ACR Manual on Contrast Media. American College of Radiology.

Alvarez RE, Macovski A. 1976. Energy-selective reconstructions in X-ray computerised tomography. Phys Med Biol 21(5): 733–744.

Arvanitis CD, Speller R. 2009. Quantitative contrast-enhanced mammography for contrast medium kinetics studies. Phys Med Biol 54(20): 6041–6064.

Baydush AH, Floyd CE. 2000. Improved image quality in digital mammography with image processing. Med Phys 27(7): 1503–1508.

Bloomquist, Aili K, Yaffe MJ, Pisano ED et al. 2006. Quality control for digital mammography in the ACRIN DMIST trial: Part I. Med Phys 33(3): 719–736.

Boone JM, Shaber GS, Tecotzky M. 1990. Dual-energy mammography: A detector analysis. Med Phys 17(4): 665–675.

Boyd NF, Guo H, Martin LJ et al. 2007. Mammographic density and the risk and detection of breast cancer. N Engl J Med 356(3): 227–236.

Brody WR, Butt G, Hall A et al. 1981. A method for selective tissue and bone visualization using dual energy scanned projection radiography. Med Phys 8(3): 353–357.

Buadu LD, Murakami J, Murayama S et al. 1996. Breast lesions: correlation of contrast medium enhancement patterns on MR images with histopathologic findings and tumor angiogenesis. Radiology 200(3): 639–649.

Burgess, AE. 1999. On the detection of lesions in mammographic structure. Proc SPIE 3663: 419–437.

Burgess AE, Jacobson FL, Judy PF. 2001. Human observer detection experiments with mammograms and power-law noise. Med Phys 28: 419–437.

Carney PA, Miglioretti DL, Bonnie Y et al. 2003. Individual and combined effects of age, breast density, and hormone replacement therapy use on the accuracy of screening mammography. Ann Int Med 138(3): 168–175.

Carp SA, Selb J, Fang Q et al. 2008. Dynamic functional and mechanical response of breast tissue to compression. Opt Exp 16(20): 16064–16078.

Carton, A-K, Gavenonis SC, Currivan JAC et al. 2010. Dual-energy contrast-enhanced digital breast tomosynthesis—A feasibility study. Br J Radiol 83(988): 344–350.

Chang, CH, Nesbit DE, Fisher DR et al. 1982. Computed tomographic mammography using a conventional body scanner. Am J Roentgenol 138(3): 553–558.

Chang, CH, Sibala JL, Fritz SL et al. 1980. Computed tomography in detection and diagnosis of breast cancer. Cancer 46 (4 Suppl): 939–946.

Chen SC, Carton, A-K, Albert M et al. 2007. Initial clinical experience with contrast-enhanced digital breast tomosynthesis. Acad Radiol 14(2): 229–238.

Cheung ENM, Alvares RDA, Oakden W et al. 2010. Polymer-stabilized lanthanide fluoride nanoparticle aggregates as contrast agents for magnetic resonance imaging and computed tomography. Chem Mater 22(16): 4728–4739.

Coman DR, Sheldon WF. 1946. The significance of hyperemia around tumor implants. Am J Pathol 22: 821–831.

Diekmann F, Diekmann S, Taupitz M et al. 2003. Use of iodine-based contrast media in digital full-field mammography—Initial experience. RoFo Fortschr Geb Rontgenstr Nuklearmed 175(3): 342–345.

Diekmann F, Diekmann S, Jeunehomme F et al. 2005. Digital mammography using iodine-based contrast media: Initial clinical experience with dynamic contrast medium enhancement. Invest Radiol 40(7): 397–404.

Diekmann F, Freyer M, Diekmann S et al. 2011. Evaluation of contrast-enhanced digital mammography. Eur J Radiol 78(1): 112–121.

Diekmann F, Sommer A, Diekmann S et al. 2007. Contrast-to-noise ratios of different elements in digital. Invest Radiol 42(5): 319–325.

Dromain C, Balleyguier C, Muller S et al. 2006. Evaluation of tumor angiogenesis of breast carcinoma using contrast-enhanced digital mammography. Am J Roentgenol 187(5): 528–537.

Dromain C, Thibault F, Muller S et al. 2011. Dual-energy contrast-enhanced digital mammography: Initial clinical results. Eur Radiol 21(3): 565–574.

Dromain C, Balleyguier C. 2009. Contrast-enhanced digital mammography. In Bick U, Diekmann F, eds. Digital Mammography, 216. New York: Springer.

Folkman J. 1971. Tumor angiogenesis: therapeutic implications. N Engl J Med 285(21): 1182–1186.

Folkman J. 2000. Incipient angiogenesis. J Natl Cancer Inst 92(2): 94–95.

Fredenberg E, Hemmendorff M, Cederström B et al. 2010. Contrast-enhanced spectral mammography with a photon-counting detector. Med Phys 37(5): 2017–2029.

Frouge C, Guinebretière JM, Contesso G et al. 1994. Correlation between contrast enhancement in dynamic magnetic resonance imaging of the breast and tumor angiogenesis. Invest Radiol 29(12): 1043–1049.

Fukumura D, Jain RK. 2008. Imaging angiogenesis and the microenvironment. Acta Pathol Microbiol Immunol Scand 116(7–8): 695–715.

Ghaghada KB, Badea CT, Karumbaiah L et al. 2011. Evaluation of tumor microenvironment in an animal model using a nanoparticle contrast agent in computed tomography imaging. Acad Radiol 18(1): 20–30.

Heine JJ, Velthuizen RP. 2000. A statistical methodology for mammographic density detection. Med Phys 27(12): 2644–2651.

Hendrick RE, Pisano ED, Averbukh A et al. 2010. Comparison of acquisition parameters and breast dose in digital mammography and screen-film mammography in the American College of Radiology Imaging Network digital mammographic imaging screening trial. Am J Roentgenol 194(2): 362–369.

Heywang SH, Hahn D, Schmidt H et al. 1986. MR imaging of the breast using gadolinium-DTPA. J Comput Assist Tomogr 10(2): 199–204.

Heywang SH, Wolf A, Pruss E et al. 1989. MR imaging of the breast with Gd-DTPA: use and limitations. Radiology 171(1): 95–103.

Hill ML, Corbin IR, Levitin RB et al. 2010a. In vitro assessment of poly-iodinated triglyceride reconstituted low-density lipoprotein: initial steps toward CT molecular imaging. Acad Radiol 17(11): 1359–1365.

Hill ML, Mainprize, JG. Jong RA et al. 2011. Design and validation of a mathematical breast phantom for contrast-enhanced digital mammography. Proc SPIE 7961: 79615E.

Hill ML, Mainprize G, Mawdsley GE et al. 2009. A solid iodinated phantom material for use in tomographic x-ray imaging. Med Phys 36(10): 4409–4420.

Hill ML, Mainprize JG, Yaffe MJ. 2010b. An observer model for lesion detectability in contrast-enhanced digital mammography. Lect Notes Comput Sci 6136: 720–727.

Inoue M, Sano T, Watai R et al. 2003. Dynamic multidetector CT of breast tumors: diagnostic features and comparison with conventional techniques. Am J Roentgenol 181(3): 679–686.

Jackson VP, Hendrick RE, Feig SA et al. 1993. Imaging of the radiographically dense breast. Radiology 188(2): 297–301.

Johns PC, Yaffe MJ. 1985. Theoretical optimization of dual-energy x-ray imaging with application to mammography. Med Phys 12(3): 289–296.

Johns PC, Yaffe MJ. 1987. X-ray characterisation of normal and neoplastic breast tissues. Phys Med Biol 32(6): 675–695.

Jong RA, Yaffe MJ, Skarpathiotakis M et al. 2003. Contrast-enhanced digital mammography: Initial clinical experience. Radiology 228: 842–850.

Kaiser W. 1985. MRI of the female breast. First clinical results. Arch Int Physiol Biochim 93(5): 67–76.

Kappadath SC, CC. Shaw. 2004. Quantitative evaluation of dual-energy digital mammography for calcification imaging. Phys Med Biol 49(12): 2563–2576.

Kappadath SC, CC. Shaw. 2005. Dual-energy digital mammography for calcification imaging: Scatter and nonuniformity corrections. Med Phys 32(11): 3395–3408.

Karathanasis E, Chan L, Balusu SR et al. 2008. Multifunctional nanocarriers for mammographic quantification of tumor dosing and prognosis of breast cancer therapy. Biomaterials 29(36): 4815–4822.

Kuhl, CK, Mielcareck P, Klaschik S et al. 1999. Dynamic breast MR imaging: Are signal intensity time course data useful for differential diagnosis of enhancing lesions? Radiology 211(1): 101–110.

Landis, CS, Li X, Telang FW et al. 2000. Determination of the MRI contrast agent concentration time course in vivo following bolus injection: Effect of equilibrium transcytolemmal water exchange. Magn Reson Med 44(4): 563–574.

Lawaczeck R, Diekmann F, Diekmann S et al. 2003. New contrast media designed for x-ray energy subtraction imaging in digital mammography. Invest Radiol 38(9): 602–608.

Lehmann LA, Alvarez RE, Macovski A et al. 1981. Generalized image combinations in dual KVP digital radiography. Med Phys 8(5): 659–667.

Lemacks MR, Kappadath SC, Shaw CC et al. 2002. A dual-energy subtraction technique for microcalcification imaging in digital mammography—A signal-to-noise analysis. Med Phys 29(8): 1739–1751.

Lewin JM, Isaacs PK, Vance V et al. 2003. Dual-energy contrast enhanced digital subtraction mammography: Feasibility. Radiology 229(1): 261–268.

Li X, Dawant BM, Welch EB et al. 2009. A nonrigid registration algorithm for longitudinal breast MR images and the analysis of breast tumor response. Magn Reson Imaging 27(9): 1258–1270.

Miyake K, Hayakawa K, Nishino M et al. 2005. Benign or malignant? Differentiating breast lesions with computed tomography attenuation values on dynamic computed tomography mammography. J Comput Assist Tomogr 29(6): 772–779.

Mussurakis S, Buckley DL, Bowsley SJ et al. 1995. Dynamic contrast-enhanced magnetic resonance imaging of the breast combined with pharmacokinetic analysis of gadolinium-DTPA uptake in the diagnosis of local recurrence of early stage breast carcinoma. Invest Radiol 30(11): 650–662.

Palma BA, Brandan ME. 2008. Analytical optimization of digital subtraction mammography with contrast medium using a commercial unit. Med Phys 35(12): 5544–5557.

Perrone A, Lo Mele L, Sassi S et al. 2008. MDCT of the breast. Am J Roentgenol 190(6): 1644–1651.

Pisano ED, Gatsonis C, Hendrick E et al. 2005. Diagnostic performance of digital versus film mammography for breast-cancer screening. N Engl J Med 353(17): 1773–1783.

Prionas ND, Lindfors KK, Ray S et al. 2010. Contrast-enhanced dedicated breast CT: Initial clinical experience. Radiology 256(3): 714–723.

Rolland, JP, Barrett HH. 1992. Effect of random background inhomogeneity on observer detection performance. J Opt Soc Am A 9(5): 649–658.

Samei E, Saunders RS, Badea CT et al. 2009. Micro-CT imaging of breast tumors in rodents using a liposomal, nanoparticle contrast agent. Int J Nanomed 4: 277–282.

Saslow D, Boetes C, Burke W et al. 2007. American Cancer Society guidelines for breast screening with MRI as an adjunct to mammography. Cancer J Clin 57(2): 75–89.

Schmitzberger FF, Fallenberg M, Lawaczeck EM, Hemmendorff R, Moa E et al. 2011. Development of low-dose photon-counting contrast-enhanced tomosynthesis with spectral imaging. Radiology 259(2): 558–564.

Sechopoulos I, Suryanarayanan S, Vedantham S et al. 2007. Computation of the glandular radiation dose in digital tomosynthesis of the breast. Med Phys 34(1): 221–232.

Seibert JA, Boone JM. 1988. X-ray scatter removal by deconvolution. Med Phys 15(4): 567–575.

Shaw, C-G, Plewes DB. 1987. Effects of scattered radiation and veiling glare in dual-energy tissue–bone imaging: A theoretical analysis. Med Phys 14(6): 956–967.

Skarpathiotakis M, Yaffe M, Bloomquist AK et al. 2002. Development of contrast digital mammography. Med Phys 29(10): 2419–2426.

Tabár L, Vitak B, Chen TH-H et al. 2011. Swedish Two-County Trial: impact of mammographic screening on breast cancer mortality during 3 decades. Radiology 260(3): 658–663.

Tozaki M, Kobayashi T, Uno S et al. 2006. Breast-conserving surgery after chemotherapy: Value of MDCT for determining tumor distribution and shrinkage pattern. Am J Roentgenol 186(2): 431–439.

Trotter DEG, Tkaczyk JE, Kaufhold J et al. 2002. Thickness-dependent scatter correction algorithm for digital mammography. Proc SPIE 4682: 469–478.

Ullman G, Sandborg M, Dance DR et al. 2005. A search for optimal x-ray spectra in iodine contrast media mammography. Phys Med Biol 50(13): 3143–3152.

Wagner FC, Macovski A, Nishimura DC. 1988. Dual-energy x-ray projection imaging: Two sampling schemes for the correction of scattered radiation. Med Phys 15(5): 732.

Warren, BA. 1979. The vascular morphology of tumors. In Tumor Blood Circulation: Angiogenesis, Vascular Morphology and Blood Flow Of Experimental Human Tumors, ed. H-I Peterson, 1–47. Boca Raton, FL: CRC Press.

Watt CA, Ackerman LV, Windham JP et al. 1986. Breast Lesions: Differential Diagnosis Using Digital Subtraction Angiography. Radiology 159(1): 39–42.

Weidner N, Folkman J, Pozza F et al. 1992. Tumor angiogenesis: a new significant and independent prognostic indicator in early-stage breast carcinoma. J Natl Cancer Inst 84(24): 1875–1887.

Weidner N, Semple JP, Welch WR et al. 1991. Tumor angiogenesis and metastasis—Correlation in invasive breast carcinoma. N Engl J Med 324(1): 1–8.

Zakhireh J, Gomez R, Esserman L. 2008. Converting evidence to practice: A guide for the clinical application of MRI for the screening and management of breast cancer. Eur J Cancer 44(18): 2742–2752.

Zheng J, Perkins G, Kirilova A et al. 2006. Multimodal contrast agent for combined computed tomography and magnetic resonance imaging applications. Invest Radiol 41(3): 339–348.

3

Stereo Mammography

David J. Getty
Brigham and Women's Hospital

In the current standard mammography approach to breast cancer detection, the radiologist reads two orthogonal 2-D x-ray images of each breast, a top down, craniocaudal (CC) view and a sideways mediolateral oblique (MLO) view. In each view, all of the three-dimensional tissue and structure contained within the volume of the breast is superimposed and collapsed onto a 2-D projection image. Detecting cancer in this way is a daunting and often extremely difficult task because of the complexity in the many ways cancer can present and the many ways in which it can be hidden (see Chapter 7). Most daunting of all is that detection accuracy depends heavily on the radiologist's ability to discern how the various patches of dense tissue or particles of calcification visible in the separate 2-D images are related across the separate images and, ultimately, how they are positioned in the breast volume and locally shaped or organized in depth. What is daunting is that with the present 2-D technique all of that abstraction of volumetric information requires cognitive merging of information collected separately from the two images. The critical contribution of stereo imaging is that it enables the reader to acquire that volumetric information directly, and gather it much more precisely—immediately perceiving the structure within the breast volume in depth.

3.1 Standard 2-D Mammography versus 3-D Stereo Mammography

3.1.1 Limitations of Standard 2-D Mammography

There are several limitations inherent to working with 2-D mammography (see Chapter 1). A true focal finding may often go undetected when masked in the 2-D projections by overlying or underlying normal tissue (Meeson et al. 2003, Burgess et al. 2001, Burgess et al. 1997, Ma et al. 1992). Second, the chance alignment of normal fibro-glandular tissue, or isolated elements of calcium, at different depths within the breast may mimic a true focal lesion (Blanchard et al. 2006, Lehman et al. 1999, Sickles 1998), resulting in false positive callbacks of patients.

An additional limitation is the inability to easily derive information about the volumetric morphology of a detected lesion, information particularly important in suggesting the presence of architectural distortion and spiculations as well as the spatial distribution of micro-calcifications. With standard 2-D mammography, volumetric information can be obtained only in a relatively limited way by cognitive merging of information derived separately from the two standard craniocaudal (CC) and mediolateral oblique (MLO) projections.

3.1.2 Addressing the Limitations of 2-D Mammography Using Stereo Mammography

With the advent of full-field digital mammography (FFDM) (Pisano et al. 2005, Lewin et al. 2002), stereoscopic digital mammography (SDM) is now a practical possibility, providing direct in-depth views of the internal structure of the breast and the potential for increased screening accuracy (Getty et al. 2008, Getty 2007, Getty and Green 2007, Chan et al. 2005, Getty 2004, Getty 2003, Getty et al. 2001, Hsu et al. 1995). Masking of true lesions may be reduced with SDM since the lesion may be seen as separate from normal tissue aligned with it but at different depths in the breast volume. Focal areas of micro-calcifications may be more easily detected when segregated in depth from other unrelated calcification and dense tissue at other locations within the breast volume. SDM may also contribute to more accurate diagnosis of a suspect lesion, particularly for micro-calcifications and architectural distortion where the volumetric arrangement of the elements of micro-calcifications or radially oriented spiculations comprising the suspect lesion are critical to the interpretation.

3.2 Stereoscopic Vision

As spoken by Roland, Stephen King's hero in *The Dark Tower*: "a one-eyed man sees flat. It takes two eyes, set a little apart from each other, to see things as they really are." Interestingly, evolution has led to the development of two quite distinct types of visual systems in animals. Prey animals typically have laterally placed eyes with little or no overlap of the visual fields of the two eyes, resulting in a lack of stereo vision but a very wide field of view (Figure 3.1a). In contrast, predatory animals (including humans) have forward facing eyes with almost complete overlap of the visual fields, permitting stereoscopic vision at the expense of a reduced total field of view (Figure 3.1b). While not everyone has excellent depth perception, most people do—especially radiologists. Anyone with significant visual anomalies or limitations is not likely to choose a profession, such as radiology, in which viewing images is central to their work.

3.2.1 Stereopsis

Because our two eyes are separated horizontally, on average by about 65 mm, each has a slightly different view of the world.

You can easily appreciate this difference by holding up a finger in front of you and alternately closing one and then the other eye without moving your finger. You will see that the location of your finger relative to the background changes as you alternate eyes. The basis of stereo vision, stereopsis ("solid seeing"), is the angular horizontal disparity between corresponding points of an object in the two retinal images. When you fixate a point in a scene, your eyes rotate, or "converge," to bring the point of fixation onto the fovea of each retina. There is zero retinal disparity in the depth "plane" defined by the point of fixation. A point on an object that lies farther away from you than the fixation point creates images on the two retinas that have positive or uncrossed retinal disparity, determined by the angular difference of the corresponding points from the fovea on the two retinas. Similarly, a point on an object that lies closer to you than the fixation point creates retinal images that have negative or crossed retinal disparity. The sign and magnitude of the retinal disparity are sufficient for the visual system to determine an object's depth relative to the point of fixation. The consequences of the three different types of disparity for stereo display are discussed below in Section 3.5.1.

3.2.2 Depth Acuity

Our ability to discriminate the relative depth of objects in normal vision—to say which is nearer, for example—is remarkably good. Studies have shown that when the objects are side-by-side vertical line segments, we can detect a difference in depth corresponding to as little as 0.2 mm when viewed at a distance of 22 inches—the typical viewing distance from a stereo display screen (Goodsitt et al. 2002, Wong et al. 2002, Goodsitt et al. 2000).

3.2.3 Creating Depth Perception from Pairs of 2-D Images

In a famous paper delivered in 1838, Wheatstone observed that one can create the perception of depth in an artificial visual scene by presenting separately to each eye a 2-D image corresponding to what would be seen by that eye as it observed that scene. He demonstrated his insight by creating a number of drawings of objects distributed in depth, as would be seen by the left and right eyes, and displaying them in a mirror-based stereoscope

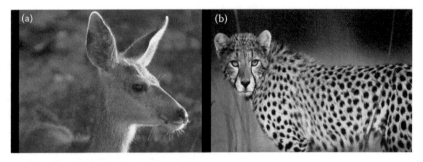

FIGURE 3.1   (a) Prey animals with laterally placed eyes and no stereo vision. (b) Predatory animals with forward facing eyes and stereo vision.

that he invented. The same principle applies to the acquisition and display of stereo pairs of medical images, including stereo mammograms.

3.3 Acquisition of Stereo Pairs of Mammograms

To acquire a stereo pair of mammograms, two sequential x-ray exposures are made of the breast with the x-ray source rotated by a small angular amount (typically 4 to 8 degrees) between the two exposures, as shown in Figure 3.2. The greater the amount of rotation, the greater will be the amount of depth perceived in the displayed stereo breast image. However, angular rotations greater than about 6 degrees will lead to increasing difficulty in visual fusion of the two images and increasing visual strain and discomfort.

Several examples of stereo mammograms are shown below. A benign cyst is seen in Figure 3.3, malignant architectural distortion is present in Figure 3.4, and several areas of microcalcification indicating the presence of malignancy are visible in Figure 3.5. If you are able to cross your eyes, it is possible to view them in stereo, perceiving the depth in the fused stereo image. By crossing your eyes your right eye will be viewing the left-hand image and your left eye will be viewing the right-hand image. You may be aided in this process by holding a pencil such that the point lies at the page surface between the two images. Slowly raise the pencil away from the page surface, towards yourself, while continuing to look at the pencil point. You should reach a

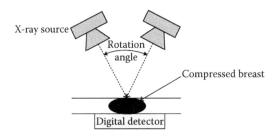

FIGURE 3.2   Acquisition of a stereo pair of mammograms.

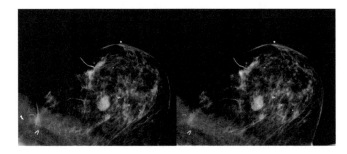

FIGURE 3.3   Stereo pair of mammographic images. The white sphere in the center of the breast is a benign mass. The wire at the upper left was placed on the skin by the technologist to indicate the location of a prior biopsy. The small white "BB" at the top marks the location of the nipple. The white linear segments at the lower left are surgical clips from a prior biopsy or excision.

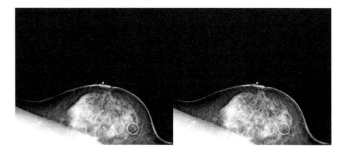

FIGURE 3.4   Stereo mammogram containing malignant architectural distortion, the white tissue located to the left of the nipple marker. Note how the malignant strands of tissue ("spiculations") radiate out from the center of the lesion in depth. The circular marker at the lower right was placed on the skin by the technologist to indicate the location of a mole or other benign skin abnormality.

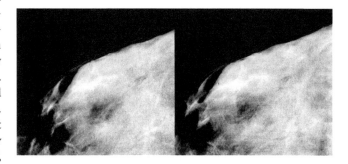

FIGURE 3.5   Stereo mammogram containing several clusters of microcalcifications associated with malignancy. Note how the central spherical cluster can be seen to lie in front of a linear series of calcifications lining a milk duct located farther back in depth.

point where the resulting crossing of your eyes will enable fusion of the two images.

3.3.1 Relationship of the X-Ray Source, Compressed Breast, and Digital Detector

There are several important considerations in acquiring a stereo pair of mammograms. First, it is critical that the breast being imaged not move or deform in any way between the two x-ray exposures. Any such movement or deformation will result in two images that will be difficult or impossible for the viewer to fuse into a single in-depth image. This requirement also implies that the compression table holding the breast to be imaged must remain fixed in position, independent of the movement of the x-ray source.

Second, ideally, the x-ray detector located beneath the breast should also remain fixed, in an unchanging relationship to the breast, as the x-ray source is moved between the two exposures. In most current digital mammography systems, the x-ray source, the compression table and the detector are all mounted on a gantry in a fixed relationship such that independent rotation of the x-ray source is not possible, ruling out the possibility of acquiring stereo mammograms.

Interestingly, breast tomosynthesis systems currently being developed do allow for the compressed breast to remain fixed in position as the yoked x-ray source and detector rotate around the breast, acquiring multiple low-dose images that are then used to reconstruct a volumetric stack of slices through the breast (see Chapter 4; Helvie 2010, Park 2007). Such systems could also be used to create stereo mammograms by two different methods, one using the raw acquisition images and another using the reconstructed slice images.

In the first method, a spaced pair of the raw, low-dose acquisition projection images (separated by 4 to 8 degrees of rotation) are displayed as a stereo pair. In a recent study involving a case set of normal and malignant mass cases (Muralidhar et al., presentation at RSNA 2011), two readers viewed a single stereo pair of raw projection images for each case. They reported moderate to excellent depth perception for most cases. The two readers achieved good sensitivity (86.9%, 91.3%) and specificity (79.1%, 83.3%) measures for mass detection, in spite of the low dose of the raw images. Of interest, it is also possible for a viewer to experience rotation of the breast, seen in depth, by displaying successive stereo pairs of images (for example, images 1 and 4, followed by 2 and 5, and so on). In a recent study, detection of masses using multiview stereo images (22 successive stereo pairs of raw images) was compared to single image standard mammography of the same cases (Webb et al. 2011). The mean AUC for five readers was 0.614 for standard mammography and 0.778 for multiview stereo mammography, a significant difference.

One potential problem with this method arises from the fact that since both the x-ray source and the detector rotate together, the resulting pair of images will suffer from some amount of keystone distortion, as shown in Figure 3.6. There will be noncorresponding vertical magnification in the two images, greatest near the left and right edges. This could potentially make it difficult or impossible for a viewer to fuse the pair into a single stereo image because the visual system is not very tolerant of vertical disparity, something that does not occur in normal vision. The situation could be remedied, however, by applying a mathematical transformation to each digital image to undo the keystone

distortion—in effect to project each image back to a rectangular plane. This transformation would result in some loss of spatial resolution in the corrected stereo pair of images. However, in the Muralidhar study reported above, there was no correction for keystone distortion, yet, the readers still reported good stereo depth perception.

The second method uses the volume consisting of the set of reconstructed image slices. A stereo pair of projection images can be created through the volume, separated by 4 to 8 degrees of horizontal separation, and displayed on a stereo display. Furthermore, the viewer can be given control over the stereo point of view, permitting the viewer to effectively rotate the displayed breast volume around both horizontal and vertical axes.

3.3.2 Double the X-Ray Dose?

If each mammogram in a stereo pair is acquired with a full standard x-ray dose, then each stereo view would require double the x-ray dose compared to a standard 2-D mammogram. However, there is reason to believe that the x-ray dose for each image of a stereo pair can be reduced significantly without loss of lesion detection accuracy. Each of the two images in the stereo pair contains an independent sample of quantum noise. It is known that in fusing the two images into a single perceived image, the visual system very effectively averages the noise in the two images, resulting in a reduced noise level in the fused image (Maidment et al. 2003). In a reader study using a breast phantom, Maidment found that the dose of each image in a stereo pair could be reduced almost by half before detection accuracy was affected significantly. In addition, new x-ray detectors currently under development will permit lower x-ray doses without increased quantum noise levels. In particular, Fujifilm Medical Systems has developed a stereoscopic acquisition and display system, based on a modified AspireHD Plus FFDM system, that acquires stereo mammograms with an average glandular dose that is equivalent to that of conventional approved 2-D FFDM systems in use (Fujifilm press release, November 27, 2011, http://fujifilmusa.com/press/news/display_news?newsID= 880211). This stereo mammography system is currently in clinical use in Japan and in Europe, and clinical trials to seek FDA approval in the U.S. will begin shortly.

3.4 Display of Stereo Mammograms

The goal in displaying a stereo image pair is to channel the image intended for the left eye solely to the left eye and to channel the image intended for the right eye solely to the right eye, while maintaining precise alignment and minimizing cross-talk between the two images. A number of methods have been developed to accomplish this, and they can be categorized in several different ways (Held et al. 2011). A major distinction is whether or not the observer has to wear special glasses or other headgear. Most stereoscopic display systems do require the observer to wear glasses or other gear. Other systems, referred to as autostereoscopic display systems, permit the observer to view the stereo

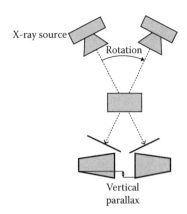

FIGURE 3.6 Keystone distortion resulting from yoked rotation of the x-ray tube and the detector.

image without headgear, but with constraints on the viewing position of the observer.

3.4.1 Passive Glasses Stereo Display Systems

The best current approach to stereo display of mammograms is the passive glasses system in which both left and right eye images are presented simultaneously through independent channels to the intended eye. The approach uses two separate display monitors, each displaying just one of the two images. This approach is exemplified by the prototype stereo display (SD2250) developed by Planar Systems, Inc. (Fergason et al. 2005, Planar Systems, Inc, "Planar Stereoscopic Displays," http://www.planar3d.com/), shown in Figure 3.7. This stereo display consists of two 5 megapixel, grayscale monitors mounted one above the other with an angular separation of 110 degrees between the two faces. The two images comprising a stereo pair, each displayed on one of the two monitors, are cross-polarized. A glass plate with a half-silvered coating is placed between the two monitor faces, bisecting the 110 degree angle between them. The image presented on the lower (vertical) monitor is transmitted through the glass plate, while the image presented on the upper (angled) monitor is reflected from the top surface of the glass plate. The radiologist wears lightweight passive cross-polarized glasses with the result that each eye sees only the intended one of the images. The radiologist's visual system then fuses the two images into a single, directly visible in-depth image of the breast. Stereo mammograms are necessarily presented in a horizontal orientation rather than the vertical orientation typically used in standard mammography. This is necessary because of the direction of the movement of the x-ray tube in acquiring the stereo pair of images which dictates the direction of horizontal disparity.

A second possible approach to simultaneous presentation is found in head-mounted displays in which two very small, high-resolution monitors are mounted in a lightweight piece of headgear together with optics that deliver each monitor's image to

FIGURE 3.7 Viewing of a stereo mammogram displayed on a planar stereo display system (SD2250).

only one eye. Image brightness is not a limitation with this type of system, but because of the small size of the monitors, image resolution is currently somewhat limited and not adequate for display of high-resolution stereo mammograms which result in images of 5 Megapixels or more.

3.4.2 Active Glasses Stereo Display Systems

Another method for displaying stereo digital image pairs makes use of temporal multiplexing. In this case, both left and right eye images are displayed on a single LCD monitor, with the two images presented alternately on successive frames. The trick, of course, is to find a way to deliver each image to the appropriate eye and only to that eye. These systems use special glasses equipped with LCD lenses that act as optical shutters (e.g., NVIDIA Corporation, "3D Vision," http://www.nvidia.com/object/3d-vision-main.html). The glasses receive an infrared synchronization signal from the display controller card which renders each lens alternately clear and then opaque, in opposition, on successive frames. Perceived image brightness is reduced considerably by the glasses both because the lens is not entirely clear even when the lens is open, but also because each eye sees only half of the frames. In order to avoid perceived flicker of the image, the monitor must be driven at a very high refresh rate—typically on the order of 120 Hz. While this technology has widespread use in commercial contexts, and now in home 3-D TVs, there are difficulties in driving very high-resolution LCD monitors (5 MP or more) at a sufficiently high refresh rate without "ghosting." Consequently, there are no current stereo mammography display systems using this technology.

3.4.3 Autostereoscopic Display Systems

Autostereoscopic display systems are different in that they do not require the observer to wear special headgear or glasses. On the other hand, the observer is constrained to view the display from a particular lateral position and distance relative to the display. Although a number of different technologies have been developed to accomplish this (3D Forums, "Autostereoscopic Displays," http://www.3d-forums.com/autostereoscopic-displays-t1.html), the two most commonly used are parallax barrier and lenticular lens techniques. In both cases, the left and right eye images are interleaved on the display, typically an LCD display, such that successive columns of pixels on the display alternate between left and right eye images.

In parallax barrier systems, a grid plate with a series of vertical slits, at half the spacing of the pixels, is placed in front of the LCD elements. When the observer is seated directly in front of the display, at a specified distance, the solid vertical strips of the grid block the right eye's view of the left-eye pixel columns and, similarly, the left eye's view of the right-eye pixel columns.

Other systems use a lenticular lens sheet placed over the LCD matrix. The sheet consists of a horizontal series of vertically oriented cylindrical lenses, each the width of two pixel columns. The lenses bend light from the left-eye columns of the LCD slightly to the left and light from the right-eye columns slightly to the right.

A significant limitation of all autostereoscopic systems is the restriction of the viewer to a particular location in front of the display. Several groups have worked on adding eye tracking to the display system, using two video cameras mounted on top of the display, for example, to dynamically determine the observer's current eye location and to dynamically adjust the parallax grid to channel the alternating pixel columns to the appropriate eye. A second limitation of this type of display is the limited horizontal resolution. Since left and right eye columns of pixels alternate on the display, the horizontal pixel count is only half that of the display, although with no similar reduction vertically. None of the currently available autostereoscopic systems is capable of displaying 5MP stereo mammographic images.

3.5 Manipulation of the Displayed Stereo Mammogram

In addition to control over grayscale windowing (including grayscale inversion) that is available in controlling standard 2-D mammographic displays, there are other aspects of a displayed stereo mammogram that can be manipulated and controlled by the radiologist. We discuss these aspects in the sections that follow.

3.5.1 Horizontal Parallax

In a stereo display, the retinal disparity that leads one to perceive depth is created by horizontal parallax in the displayed image—the horizontal separation of corresponding points in the left- and right-eye images on the display screen (Getty 2003). There are three types of parallax, illustrated below in Figure 3.8. If a point belonging to an object is displayed at exactly the same position in the left- and right-eye images, then it is said to have zero parallax. The perceptual effect is that the object is seen to lie at the surface of the display screen.

In the other two cases, a point belonging to an object is displayed at different locations in the left- and right-eye images. If

the right-eye point is displaced to the right of the left-eye point, called uncrossed or positive parallax, then the object will be perceived to lie behind the screen surface. The larger the separation, the farther the object will be behind the screen surface.

In the third case, if the right-eye point is displaced to the left of the left-eye point, called crossed or negative parallax, then the object will be perceived to lie in front of the display surface. Again, the larger the separation, the farther the object will be from the screen surface, towards the observer.

3.5.2 Location of the Displayed Breast Volume

A second aspect of the viewed volume that can be manipulated is the location of the displayed volume in depth relative to the display screen surface. If one shifts the right-eye image to the right while holding the left-eye image fixed, then the horizontal parallax of all points will be changed in the direction of uncrossed parallax (Figure 3.8). Points originally with uncrossed parallax will have larger uncrossed parallax, and points with crossed parallax will have decreased crossed parallax. The perceived effect is to shift the entire viewed volume backward in depth, away from the observer, with the amount of shift in depth proportional to the amount of right lateral shift of the right-eye image. Similarly, shifting the right-eye image in the other direction, to the left, will shift the viewed volume towards the viewer relative to the screen surface. It is only the amount of relative shift of the two images that matters, so one could just as well make shifts to the left-eye image or to both. In fact, splitting a desired amount of shift equally between the two images will minimize the amount of stereo image lost at the left and right edges of the display.

People are typically more comfortable with a displayed volume that starts at the screen surface and extends back behind the monitor. However, if the total range of parallax is large, as would occur with a large rotational angle at stereo image acquisition, then one would like to minimize the maximal amount of both crossed and uncrossed parallax. This is accomplished by locating the displayed volume so that it is half behind and half in front of the display surface.

3.5.3 Inversion of Depth

While the stereo point of view of the imaged breast is predetermined by the point of view at the time of image acquisition, there is another aspect of the viewed volume that the observer can manipulate (Getty 2003). In particular, one can invert depth by swapping the two images, presenting the left-eye image to the right eye and the right-eye image to the left eye. Tissue originally seen at the front of the breast volume will now be seen at the back of the volume, and vice versa. If, in addition to swapping the two images, one also spins each image 180 degrees about a vertical axis, then the inverted depth image is seen as if one had walked around the breast to view it from the backside.

Inverting depth can be important in stereo viewing. It is typically easier to attend to objects seen in the foreground compared

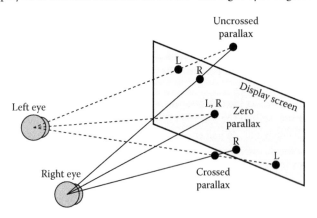

FIGURE 3.8 Perceptual result of three different cases of horizontal parallax in displaying corresponding points.

to those seen in the background through the possibly distracting clutter of objects in the foreground. By allowing a radiologist to invert depth, tissue originally at the back of the displayed breast can be moved to the front of the breast, possibly making it easier to detect lesions and perceive structure.

3.5.4 Stereo Cursor

It is possible to create a stereo cursor that the radiologist can control with a mouse, to move it horizontally, vertically and in depth throughout the displayed stereo volume. The desired cursor shape is drawn in both left and right eye images with horizontal and vertical location in both images controlled by corresponding mouse movements. The location of the cursor in depth is determined by the amount of horizontal parallax—the number of pixels of horizontal separation between the drawing of the cursor in the two images. Control of this separation can be given to the radiologist using the scroll wheel. It is worth pointing out that whatever the shape of the cursor, it should have strong vertical components. The reason is that every point along the length of a vertical straight or curved line segment creates horizontal disparity that leads to depth perception, whereas only the two endpoints of a horizontal line segment create identifiable points with perceivable horizontal disparity. Interior points have ambiguous correspondence between the two images.

3.5.5 Depth Quantization

Because we are working with digital images, the amount of horizontal parallax between pairs of corresponding points in the two images is necessarily an integer multiple of the spacing between pixels. Consequently, the perceived location of points in depth will also occur in quantized depth planes. The actual functional relationship between pixel spacing and depth plane spacing depends also on the distance of the observer from the display screen, as we will discuss next.

3.5.6 Distance of the Radiologist from the Display Screen

The amount of perceived depth in a stereo image depends on the distance of the observer's eyes from the display screen. If one wants the amount of perceived depth to be the same as the actual depth present in the imaged breast, then the observer needs to be at a distance from the screen such that the angle, α, formed between the observer's two eyes and a central point on the screen, is equal to the separation angle used in acquiring the stereo image pair. For an acquisition separation angle of 6 degrees, the appropriate viewing distance is about 62 cm (about 24 inches). If the observer is farther away than this, perceived depth will be greater than the actual depth of the breast; if the observer is closer, then perceived depth will be less than actual depth. The relationship is given by

Correct viewing distance = 3.25 cm/tan (acquisition angle/2),

where 3.25 cm represents half the average interocular spacing of 6.5 cm. The smaller the separation angle at acquisition, the larger is the correct viewing distance. In practical terms, one's perception of this change in depth with viewing distance is small for the range of distances that a radiologist finds comfortable and useful, and is not really an issue of much consequence.

3.6 Clinical Study Comparing Stereo Mammography to Standard 2-D Mammography

A prospective screening study was conducted at Emory University comparing standard full-field digital mammography (FFDM) with stereoscopic digital mammography (SDM) for the detection of cancer (D'Orsi et al., in press. Here I discuss only a comparison with regard to detection of actionable findings.). An actionable breast finding is one determined at a callback diagnostic workup examination to require either accelerated follow-up (BI-RADS 3 rating) or biopsy (BI-RADS 4 or 5 rating). The goal of the study was to determine if SDM can improve on FFDM accuracy for the detection of actionable findings at screening.

3.6.1 Study Design

This prospective screening study utilized a paired design with each enrolled patient receiving both a standard FFDM examination (2-D CC and MLO views) and an independent stereo examination (stereo CC and MLO views) at the same visit. Only patients at elevated risk for development of breast cancer were enrolled, with the intention of obtaining a higher yield of suspicious findings. The FFDM examination was performed on a clinical GE Senographe 2000D mammography unit; the SDM examination was performed on a research GE Senographe 2000D mammography unit, modified to permit off-axis image acquisition and with modified x-ray collimation, which restricted the x-ray field to the image receptor. For the SDM examination, the x-ray tube was rotated by 10 degrees (+5 and −5 degrees from 0) between the two acquisitions of each stereo view. Each of the two acquisitions for a stereo view was acquired with approximately the same dose as that of the FFDM examination.

FFDM images were viewed on a standard, FDA-approved, dual-monitor GE Review Workstation. SDM images were viewed on the prototype medical stereo display, the Planar StereoMirror SD2250 described above in Section 3.4.1.

Five mammographers participated in the clinical study. All readers had functional stereo vision and excellent measured depth discrimination acuity, determined using a Randot Stereo Acuity Test (Stereo Optical Co., Inc., "Stereo Tests," http://www.stereooptical.com/products/stereotests#randot). For each patient, the FFDM was read by one mammographer and the SDM examination was read by a different mammographer, on the same day without knowledge of each other's reading results. Clinical histories and prior mammograms were available for comparison. For each reading, the mammographer filled out a study form indicating the presence, nature, location and assessment of any findings using the BI-RADS lexicon, and also a rating of breast density using the BI-RADS scale. If either reader reported any findings for a case, the two readers met to determine finding concordance or discordance and to determine the nature of workup. All reported findings were subjected to standard diagnostic film-screen workup examinations to determine truth. Cases for which both readers reported the examination as normal (BI-RADS 1) or containing only benign findings (BI-RADS 2), truth was determined by the results of the subsequent "yearly" screening examination.

3.6.2 Results

The final analyzed sample included 1326 examinations. Mean patient age was 58.3 years with a standard deviation of 11.5 years while the median age was 58 years with a range from 32–91. There were 330 findings reported by one or both modalities (173 by FFDM alone, 119 by SDM alone, and 38 by both). Standard (nonstereo) diagnostic workup examinations determined that 244 of these findings were nonactionable (184 BI-RADS 1, 60 BI-RADS 2) while 86 were actionable (41 BI-RADS 3, 41 BI-RADS 4, 4 BI-RADS 5).

Radiologists using FFDM alone detected and reported 53 of the 86 actionable findings (61.6%), while radiologists using SDM alone detected and reported 59 (68.6%) (Figure 3.9). While radiologists using SDM did, in fact, detect more of the actionable findings than FFDM, the difference is not statistically significant ($p > 0.25$).

Of the 244 findings that were determined at workup to be nonactionable, radiologists using FFDM reported 158 (64.8%) while radiologists using SDM reported only 98 (40.2%) (Figure 3.10). This 38% reduction in false positive reports is highly significant ($p < 0.0001$).

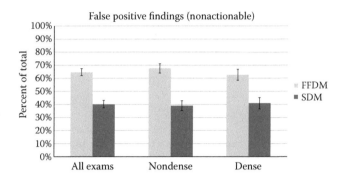

FIGURE 3.10 Percentage of total findings reported by FFDM and by SDM that were proven to be nonactionable at workup examinations, for all breasts and separately for nondense and dense breasts.

The same pattern of results was found when we examined the subsets of 183 dense and 147 nondense breasts. Dense breasts were defined as those receiving a BI-RADS density rating of 3 or 4, while nondense breasts were defined as those receiving a BI-RADS density rating of 1 or 2.

Regarding actionable findings, for nondense breasts, radiologists using FFDM detected 28 of 42 actionable findings (66.7%) while radiologists using SDM detected 26 (61.9%), a nonsignificant difference ($p > 0.4$) (Figure 3.9). For dense breasts, radiologists using FFDM detected 25 of 44 actionable findings (56.8%) while radiologists using SDM detected 33 (75.0%), a nonsignificant improvement ($p > 0.1$).

Regarding nonactionable findings, for nondense breasts, radiologists using FFDM reported 71 of 105 nonactionable findings (67.6%) while radiologists using SDM reported only 41 (39.0%) (Figure 3.10). This 42% reduction in false positive reports by SDM is highly significant ($p < 0.002$). For dense breasts, FFDM reported 87 of 139 nonactionable findings (56.8%) while SDM reported 57 (41.0%). This 34% reduction in false positive reports by SDM is also highly significant ($p < 0.006$).

3.7 Discussion

This clinical study demonstrated that stereo mammography, compared to standard 2-D mammography, reduced false positive patient recalls by 38% while maintaining comparable sensitivity for detection of actionable findings. If stereo mammography were to replace standard 2-D mammography for screening, this would result in a large reduction in medical costs resulting from false positive patient recalls and in the number of women facing the anxiety of being called back for diagnostic workup examinations that are, in fact, unnecessary.

It is worth mentioning anecdotally another important type of potential saving with stereo mammography. All of the mammographers in the Emory study agreed that with a small amount of experience, they were able to read the stereo mammogram more quickly and more confidently than a standard mammogram.

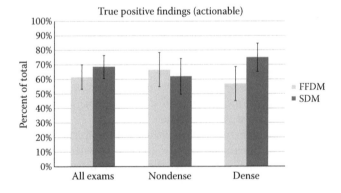

FIGURE 3.9 Percentage of total findings reported by FFDM and by SDM that were proven to be actionable at workup examinations, for all breasts and separately for nondense and dense breasts (error bars represent 95% confidence intervals).

References

Blanchard K, Colbert JA, Kopans, DB et al. 2006. Long-term risk of false-positive screening results and subsequent biopsy as a function of mammography use. Radiology 240(2): 335–342.

Burgess AE, Li X, Abbey CK. 1997. Visual signal detectability with two noise components: anomalous masking effects. J Opt Soc Amer: A Opt, Image Sci Vis 14(9): 2420–2442.

Burgess AE, Jacobson L, Judy PF. 2001. Human observer detection experiments with mammograms and power-law noise. Med Phys 28(4): 419–437.

Chan HP, Goodsitt MM, Helvie MA et al. 2005. ROC study of the effect of stereoscopic imaging on assessment of breast lesions. Med Phys 32(4): 1001–1009.

Chelberg DM, Babbs CF, Pizlo Z, Delp EJ. 1994. Digital stereomammography. In Proceedings of the 2nd International Workshop on Digital Mammography. York, England, 181–190.

Fergason JL, McLaughlin CW et al. 2005. An innovative beamsplitter-based stereoscopic/3D display design. In Proceedings of the SPIE. 2005, pp. 488–492.

Getty DJ. 2003. Stereoscopic and biplane imaging. In: Samei E, Flynn MJ, eds. Advances in Digital Mammography: Categorical Course in Diagnostic Radiology Physics. pp. 199–209. Oak Brook, IL: RSNA.

Getty DJ. 2004. Stereoscopic digital mammography. In Proceedings of 1st Americas Display Engineering and Applications Conference (ADEAC 2004). Ft. Worth, TX, Ch. 2.3.

Getty DJ. 2007. Clinical medical applications for stereoscopic 3D displays. J Soc Inf Display 15: 377–384.

Getty DJ, D'Orsi CJ, Pickett RM. 2008. Stereoscopic digital mammography: improved accuracy of lesion detection in breast cancer screening. In Krupinski EA, ed. International Workshop on Digital Mammography 2008. pp. 74–79. Berlin, Heidelberg: Springer Verlag.

Getty DJ, Green PJ. 2007. Clinical medical applications for stereoscopic 3D displays. J Soc Inf Display 15: 377–384.

Getty DJ, Pickett RM, D'Orsi CJ. 2001. Stereoscopic digital mammography: improving detection and diagnosis of breast cancer. In Lemke H, ed. Computer Assisted Radiology and Surgery. pp. 506–511. Amsterdam: Elsevier.

Goodsitt MM, Chan HP, Darner KL, Hadjiiski LM. 2002. The effects of stereo shift angle, geometric magnification and display zoom on depth measurements in digital stereomammography. Med Phys 29(11): 2725–2734.

Goodsitt MM, Chan HP, Hadjiiski L. 2000. Stereomammography: Evaluation of depth perception using a virtual 3D cursor. Med Phys 27(6): 1305–1310.

Held RT, Hui TT. 2011. A guide to stereoscopic 3D displays in medicine. Acad Radiol 18: 1035–1048.

Helvie MA. 2010. Digital mammography imaging: Breast tomosynthesis and advanced applications. Radiol Clin North Amer 48(5): 917–929.

Hsu J, Chelberg DM, Babbs CF, Pizlo Z, Delp EJ. 1995. Preclinical ROC studies of digital stereomammography. IEEE Trans Med Imag 14(2): 318–327.

Lehman CD, White E, Peacock S, Drucker MJ, Urban N. 1999. Effect of age and breast density on screening mammograms with false-positive findings. Amer J Roentgenol 173(6): 1651–1655.

Lewin JM, D'Orsi CJ, Hendrick RE et al. 2002. Clinical comparison of full-field digital mammography and screen-film mammography for detection of breast cancer. Amer J Roentgenol 179(3): 671–677.

Ma L, Fishell E, Wright B, Hanna W, Allan S, Boyd NF. 1992. Case-control study of factors associated with failure to detect breast cancer by mammography. J Nat Cancer Inst 84(10): 781–785.

Maidment AD, Bakic PR, Albert M. 2003. Effects of quantum noise and binocular summation on dose requirements in stereoradiography. Med Phys 30(12): 3061–3071.

Meeson S, Young KC, Wallis MG, Cooke J, Cummin A, Ramsdale ML. 2003. Image features of true positive and false negative cancers in screening mammograms. Br J Radiol 76(901): 13–21.

Park JM, Franken EA Jr, Garg M, Fajardo LL, Niklason LT. 2007. Breast tomosynthesis: Present considerations and future applications. Radiographics 27(Suppl 1): S231–S240.

Pisano ED, Gatsonis C, Hendrick E et al. 2005. Diagnostic performance of digital versus film mammography for breast-cancer screening. N Engl J Med 353(17): 1773–1783.

Sickles EA. 1998. Findings at mammographic screening on only one standard projection: Outcomes analysis. Radiology 208(2): 471–475.

Webb LJ, Samei E, Lo JY et al. 2011. Comparative performance of multiview stereoscopic and mammographic display modalities for breast lesion detection. Med Phys 38(4): 1972–1980.

Wong BPH, Woods RL, Peli E. 2002. Stereoacuity at distance and near. Optometry Vis Sci 79(12): 771–778.

4

Breast Tomosynthesis

Ying (Ada) Chen
Southern Illinois University

4.1 Introduction

4.1.1 Clinical Motivation

Over the last two decades, breast imaging has dramatically changed. Early detection is viewed as the best hope to decrease breast cancer mortality by allowing intervention at an earlier stage of cancer progression (Nass et al. 2001, Cherney and Reminton 1999, Kopans 1997, Bushberg et al. 1994). It is universally accepted that mammography is the most important and efficacious tool for the early detection of breast cancer (Bassett et al. 2005).

However, limitations of mammography have been well publicized, such as many callbacks from screening, and low positive predictive value of about 15–34% from biopsy (Kuntzen and Gisvold 1993, Kopans 1992). About 30% of breast cancers are still missed in traditional mammography (Wu et al. 2004).

Improving breast imaging technologies may permit breast cancer to be detected at a smaller size and earlier stage, thereby reducing the number of women who die each year from breast cancer. Thus, tremendous efforts have been made in the incremental improvements in imaging technologies in the field of breast cancer detection.

4.1.2 Breast Tomosynthesis Background

Attempts to develop three-dimensional imaging methods to separate objects from overlying anatomical structure go back to the early 20th century (Dobbins and Godfrey 2003). In 1917, Radon introduced the famous Radon transformation of tomography, describing the mathematics of generating internal object planes from two-dimensional projection data (Radon, 1917). In 1932, Ziedses des Plantes led the pioneering effort in conventional linear tomography and Ernest Twining contributed to its clinical prominence (Dobbins and Godfrey 2003, Ziedses des Plantes 1932).

Early tomography systems utilized a linear, opposing motion of the x-ray tube and the film receptor to generate a focal plane.

The procedure had to be repeated if more than one focal plane was needed. This led to high dose to the patients. Second, the modality was not capable of sufficiently suppressing out-of-plane blur (Dobbins and Godfrey 2003). In the late 1990s, the advent of digital x-ray acquisition technology made it possible to acquire a series of low-dose projection images from different locations of an x-ray source to provide the depth information for tomosynthesis reconstruction (Chen et al. 2005, Dobbins and Godfrey 2003).

Tomosynthesis imaging has already been investigated in many applications including chest imaging, joint imaging, angiography, dental imaging, and breast imaging (Mertelemeier et al. 2006, Rakowski and Dennis 2006, Chen et al. 2006, Godfrey et al. 2006, Maidment et al. 2006, Zhang et al. 2006, Duryea et al. 2003, Godfrey et al. 2003, Godfrey and Dobbins 2002, Badea et al. 2001, Godfrey et al. 2001, Warp et al. 2000, Suryanarnyannan et al. 1999, Sone et al. 1995, Dobbins 1990, Sklebitz and Haendle 1983). In the field of breast imaging, a few investigations and efforts have been made by several research groups and manufacturers, including Niklason and colleagues' publication in 1997 of a tomosynthesis method with the x-ray tube moved in an arc above the stationary breast and detector (Niklason et al. 1997), Wu et al.'s report in 2003 of the maximum likelihood iterative algorithm (MLEM) to reconstruct the three-dimensional distribution of x-ray attenuation in the breast (Wu et al. 2003), matrix inversion tomosynthesis (MITS) technique in breast tomosynthesis (Chen et al. 2004), filtered back-projection (FBP) algorithms (Mertelemeier et al. 2006, Chen et al. 2005, Wu et al. 2004, Stevens et al. 2001, Lauritsch and Haerer 1998, Matsuo et al. 1993, Grant 1972), tuned-aperture computed tomography (TACT) reconstruction methods developed by Webber and investigated by Suryanarayanan et al. (1999), algebraic reconstruction techniques (ART) (Zhang et al. 2006), etc.

Major manufacturers are actively involved in readying commercial tomosynthesis devices for public sale. Several prototype

Breast Tomosynthesis scanners have been manufactured by commercial vendors including GE, Hologic and Siemens. They have been investigated by several research groups (Chen et al. 2007, Maidment et al. 2006, Zhang et al. 2006, Bissonnette et al. 2005, Wu et al. 2003).

Considerable effort by different research groups and manufacturers has been made to improve breast tomosynthesis systems. Experiments with phantoms, mastectomy specimens, cadaver specimens, and human subjects with different techniques were designed and performed manually and automatically in the last few years by researchers to implement and optimize imaging configurations and image reconstruction algorithms, as well as to help determine the best tomosynthesis acquisition strategy.

The designs of most current breast tomosynthesis scanners are similar, based on a full-field digital mammography (FFDM) system. To generate the limited angle series of projection images, a conventional x-ray tube mounted on a rotating arm moves along an arc above the compressed breast with a partial isocentric motion where the path of the tube does not lie in a plane that is parallel to the detector plane. Mechanical design, related control software, and an accurate angle measurement device are necessary to ensure that the x-ray tube rotates automatically with respect to the rotation center and to record the x-ray tube's angular location precisely for each projection view. A typical tomosynthesis scan takes about 20 seconds to more than 1 minute, depending on number of views acquired. In order to speed up the scan time, binning mode can be implemented by some manufacturers to average neighboring pixels at the expense of image quality and resolution (Bissonnette et al. 2005, Ren et al. 2005).

Recently, the U.S. Food and Drug Administration (FDA) approved the commercialization of a breast tomosynthesis device Selenia Dimensions made by Hologic, Inc., one of the leading manufacturers of premium diagnostic products, medical imaging systems and surgical products dedicated to serving the healthcare needs of women. This system upgrades the full-field digital mammography (FFDM) system to enable the x-ray tube's rotation along an arc above the digital detector. Based on the fast computation speed and good structure reconstruction, filtered back-projection algorithm is used for this system to reconstruct three-dimensional information of the breast.

Some scientists are also conducting research on stationary breast tomosynthesis imaging systems. A novel multibeam x-ray source was developed by Zhou et al. recently (Zhou et al. 2010, Yang et al. 2008, Lalush et al. 2006, Zhang et al. 2005). This breast tomosynthesis system was built up with fixed multibeam field-emission x-ray (MBFEX) sources based on unique properties of carbon nanotube electron emitters. A parallel imaging configuration was applied to a multibeam x-ray system design where the path of the x-ray tube lies in a plane that is parallel to the detector plane. The total scan time for a typical acquisition of 25 views is about 11.2 seconds (Yang et al. 2008). Compared with other typical prototype systems, it may have the potential to increase the tomosynthesis imaging speed with a simplified system design and reduce patients' discomfort and the motion blur associated with the x-ray tube's movement as is typical in other systems.

In this chapter, we will introduce and discuss breast tomosynthesis imaging background, reconstruction, and optimization. We hope to contribute to the advancement of breast tomosynthesis imaging reconstruction and optimization for better breast cancer detection.

4.2 Breast Tomosynthesis Reconstruction and Optimization

4.2.1 Breast Tomosynthesis Reconstruction Algorithms

Various efforts have been made to investigate different breast tomosynthesis reconstruction algorithms by the academic community and medical imaging industry. In this section, we will discuss a few representative breast tomosynthesis imaging reconstruction algorithms, including the traditional shift and add (SAA) algorithm, Niklason and colleagues' image stretching shift-and-add (ISSAA) algorithm, point-by-point back-projection (BP), maximum likelihood expectation maximization (MLEM) iterative method, matrix inversion tomosynthesis (MITS), filtered back projection (FBP), and simultaneous algebraic reconstruction technique (SART).

4.2.1.1 Traditional Shift and Add Reconstruction Algorithm

Traditional shift and add (SAA) reconstruction algorithm is a common mathematical method to line up the in-focus structures along the direction of x-ray tube's motion. It typically also serves as a foundation of several other algorithms with deblurring functions (Chen et al. 2007, Chen et al. 2004, Suryanarnyannan et al. 1999, Niklason et al. 1997).

For breast tomosynthesis imaging, when the x-ray tube is located at different positions, objects at different heights above the detector are projected onto the detector at positions depending on the relative height of the objects.

With parallel imaging configurations, the x-ray tube's positions are located in a plane that is parallel to the detector plane. The magnification of objects depends only on the height of the object. Projection images acquired during the tube's movement may be shifted and added together to line up and bring into focal structures in a certain plane for reconstruction.

For the commonly used breast tomosynthesis partial isocentric motion, the x-ray tube typically moves along an arc above the stationary digital detector. The path of tube movement is not parallel to the detector plane. Under this condition, the projected positions on the detector from central points of each plane can be used to calculate the relative shift required to bring an object of interest into focus. One can then add together those shifted planes to emphasize the object of interest and blur away structures elsewhere.

In Figure 4.1, plane i is the reconstruction plane containing the object of interest. P_i is the projected position of the middle point of this plane, and z is the plane height above the detector. Point O is the axis of rotation of the x-ray tube arm. In order to

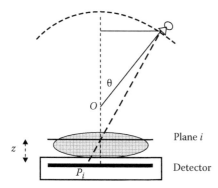

FIGURE 4.1 Shift-and-add (SAA) breast tomosynthesis algorithm.

make an object at the center of plane i in focus, the necessary shift amount for the projection image from angle θ is:

$$\text{shift}_k(z) = P_i = L \cdot \sin\theta \cdot \frac{z}{L \cdot \cos\theta + (R - z)}$$

where L is length of the rotation arm, and R is the height of the rotation axis from the detector. We can obtain the reconstruction plane as the average of all N shifted projection images:

$$T_z(x, y) = \frac{1}{N} \sum_{k=1}^{N} I_k(x', y) \otimes \delta[x' - \text{shift}_k(z)]$$

As mentioned earlier, SAA is a mathematical method to reconstruct tomosynthesis planes without further deblurring algorithms. Based on this simple method, one can generate an arbitrary set of reconstruct slices passing through the object to provide three-dimensional information. Some out-of-plane artifacts will exist on the reconstructed planes with SAA reconstruction.

4.2.1.2 Niklason's Imaging Stretching Shift-and-Add (ISSAA)

Niklason and colleagues modified the shift-and-add technique for breast tomosynthesis imaging to address the isocentric motion issue (Niklason et al. 1997). With this reconstruction method, compared with the traditional shift-and-add (SAA) algorithm, the projected positions on the detector are adjusted according to the partial isocentric motion geometry first. The shift amount for each projection image is calculated based on the adjusted "shrunk" image. Figure 4.2 shows the principle of the ISSAA algorithm.

In Figure 4.2, point A is in the plane i in the breast object. P_i is the projected position of point A at angle θ.

If we assume the tube moves in a line that is parallel to the detector plane, the projected position on the detector will be changed to be P'_i.

$$p = \frac{(L \cdot \cos\theta + R - z) \cdot P_i - z \cdot L \cdot \sin\theta}{L \cdot \cos\theta + R}, P'_i = p + z \cdot \frac{L \cdot \tan\theta + p}{L + R - z}$$

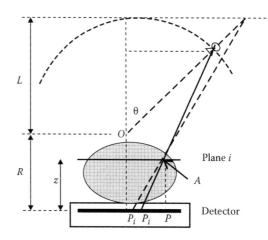

FIGURE 4.2 Niklason (ISSAA) breast tomosynthesis algorithm.

where L is length of the rotation arm, and R is the height of the rotation axis from the detector. According to the equation, we can use the original P_i to calculate the point A's position p, then "stretch" it to P'_i to satisfy the parallel motion assumption.

The new shift amount for each plane will be $\text{shift}_k(z) = P'_i$. Then the shifted planes can be added together for reconstruction.

4.2.1.3 Point-by-Point Back-Projection (BP)

Most current commercial breast tomosynthesis prototype systems utilize the partial isocentric motion of the tube, where the x-ray sources move along an arc above the stationary detector. The path of the tube does not lie in a plane that is parallel to the detector plane and the amount of magnification varies with tube angle.

While the effects due to isocentric motion are small for most objects, the use of SAA methods introduces morphological distortions with small objects such as microcalcifications (Chen et al. 2007). Therefore, neither a simple SAA reconstruction algorithm nor the ISSAA algorithm is entirely suitable for breast tomosynthesis.

With the point-by-point back-projection (BP) algorithm, shift amounts for every pixel location on each reconstructed plane are computed, taking into account the 2-D arc projection location of reconstructed objects in each plane.

For BP reconstruction algorithm, the pixel value of a point in the tomosynthesized reconstruction is calculated as $\frac{1}{N} \sum_{i=1}^{N} I(P_i)$, where $I(P_i)$ is the pixel value at a given location on the ith projection image, and N is the total number of projection images.

Computation times for the point-by-point BP algorithm are roughly comparable to the SAA method. With point-by-point BP, the shift amount is calculated according to the exact location of the pixel in the reconstruction slices, thereby, addressing this issue of variable magnification.

4.2.1.4 Maximum Likelihood Expectation Maximization

The maximum likelihood expectation maximization (MLEM) iterative method was widely used in other imaging fields such as emission or transmission computed tomography, single photon emission computed tomography (SPECT), and electron microscopy (Chidlow and Möller 2003, Bouwens et al. 2001, Lange and Carson 1984, Crowther et al. 1970).

Wu et al. (2004) applied the iterative method of maximum likelihood expectation maximization algorithm (MLEM) to reconstruct the 3-D distribution model of x-ray attenuation in the breast. This iterative method of MLEM has been demonstrated to be effective for reconstructing three-dimensional information of human breasts.

The likelihood function L of the probability of getting the projection value in the experiment is the objective to be maximized during iterations. The 3-D object space is expressed as a 3-D attenuation coefficient model μ (Wu et al. 2004, Wu et al. 2003, Lange and Fessler 1995).

The probability of getting the projections Y in the experiment can be expressed as $L = P(Y \mid \mu)$.

Under Poisson distribution assumption, $P(Y) = \dfrac{e^{-\lambda} \cdot \lambda^{Y}}{Y!}$, where Y_i is the measured number of x-ray photons at projection pixel i on the image. λ is the mean value of the distribution, here representing the computed number of photons based on the attenuation model μ.

The joint likelihood for all pixels involved can be expressed as:

$$L = \prod_{i}\left\{ \frac{e^{-\lambda} \cdot \lambda^{Y_i}}{\lambda!} \right\}, \quad \lambda = D_i \cdot e^{-\langle l, \mu \rangle_i}, \quad \langle l, \mu \rangle_i = \sum_{j} l_{i,j} \cdot \mu_j$$

where j is the individual voxel in the 3-D attenuation μ model. $\langle l, \mu \rangle_i$ represents the total attenuation along the projection ray to pixel i. $l_{i,j}$ is the intersection of projection line i passing through the individual voxel j. D_i is the number of incident x-ray photons before attenuation at pixel i, or the number of incident photons on the "flat surface" images.

Figure 4.3 represents the ray tracing method along the three-dimensional attenuation model. When an individual x-ray projection line passes through the 3-D attenuation model, it generates two intersection points A and B on two neighboring layers of p and q. In order to calculate the total attenuation along an individual projection line, the intersection $l_{i,j}$ between the projection beam i and each voxel j should be computed.

Maximizing the joint likelihood function is the same as maximizing the possibility of the occurrence of getting the measured projection image. Lange and Fessler's expectation maximization (EM) convex algorithm (Lange and Fessler 1995) was used to maximize the joint likelihood function. The attenuation coefficients in the model are updated as (Wu et al. 2003):

$$\mu_j^{(n+1)} = \mu_j^{(n)} + \Delta\mu_j^{(n)} = \mu_j^{(n)} + \frac{\mu_j^{(n)} \cdot \sum_i l_{i,j} \cdot (D_i \cdot \exp(-\langle l, \mu^{(n)} \rangle_i - Y_i)}{\sum_i (l_{i,j} \langle l, \mu^{(n)} \rangle_i D_i \exp(-\langle l, \mu^{(n)} \rangle_i)}$$

Nonnegativity and a maximum attenuation coefficient ($0 \leq u \leq 1$ cm^{-1}) were used for constraints. Ray tracing can be used to calculate the total attenuation along each projection line.

4.2.1.5 Matrix Inversion Tomosynthesis (MITS)

MITS (Chen et al. 2004, Dobbins and Godfrey 2003) used linear algebra to enable high-speed reconstruction of a finite set of planes. The conventional shift-and-add tomosynthesis of n reconstructed planes can be represented as the convolution of the actual structure s_i in plane i and the blurring function f_{ij} due to the structures on other planes j.

$$t_1 = s_1 \otimes f_{11} + s_2 \otimes f_{12} + \ldots + s_n \otimes f_{1n}$$
$$t_2 = s_1 \otimes f_{21} + s_2 \otimes f_{22} + \ldots + s_n \otimes f_{2n}$$
$$\ldots$$
$$t_n = s_1 \otimes f_{n1} + s_2 \otimes f_{n2} + \ldots + s_n \otimes f_{nn}$$

$$\begin{pmatrix} T_1 \\ T_2 \\ \ldots \\ T_n \end{pmatrix} = \begin{pmatrix} F_{11} & F_{12} & \ldots & F_{1n} \\ F_{21} & F_{22} & \ldots & F_{2n} \\ \ldots & & & \\ F_{n1} & F_{n2} & \ldots & F_{nn} \end{pmatrix} \cdot \begin{pmatrix} S_1 \\ S_2 \\ \ldots \\ S_n \end{pmatrix}$$

In Fourier frequency space, the convolution can be expressed as above matrix calculation, which can also be written as $T = M \cdot S$, where M is the matrix of Fourier Transform of blurring functions, and S is the vector of the Fourier Transform of actual structures. Thus, we can solve for the actual structures by multiplying by the inverse matrix M^{-1} and taking the inverse Fourier Transform $s = FT^{-1}(M^{-1} \cdot T)$.

Based on the SAA algorithm and the matrix of blurring functions, we can acquire the real structure by MITS reconstruction. The blurring function can be calculated according to the geometries of each reconstruction plane and object plane to generate the matrix of blurring functions.

Figure 4.4 shows how to calculate the blurring function f_{ij}. When the structure on plane j is reconstructed, blur from the

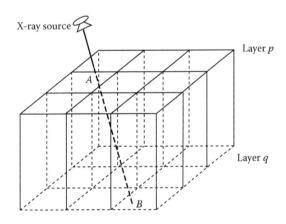

FIGURE 4.3   Ray tracing along 3-D attenuation model.

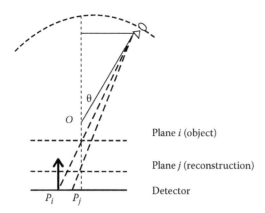

FIGURE 4.4 MITS breast tomosynthesis algorithm.

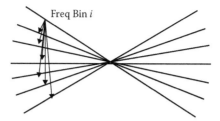

FIGURE 4.5 Sampling density calculation in spatial frequency domain.

structure on plane *i* will exist. After the conventional shift-and-add computation to line up the in-plane structures, the δ function at P_i will be shifted to the position of $P_i - P_j$. Therefore, the blurring function can be calculated according to the geometries of each reconstruction plane and object plane to generate the matrix of blurring functions.

A technique named "slabbing" can be applied to the MITS reconstructed images to reduce high frequency noise and residual artifacts from out-of-plane objects. MITS was originally developed by Dobbins. Godfrey and Dobbins applied it to the reconstruction of pulmonary nodules (Godfrey 2005). In MITS reconstruction for breast imaging, several adjacent reconstruction planes can be averaged to suppress low-contrast low-frequency artifacts that arise from the partial-pixel sharing of digital image data.

It has been demonstrated that MITS reconstruction shows good high-frequency reconstruction performance by providing sharp edges of small target lesions such as microcalcifications.

4.2.1.6 Filtered Back-Projection (FBP)

FBP algorithms are used widely in computed tomography (CT). For tomosynthesis reconstruction, based on the central slice theorem, a parallel projection samples the object on a plane perpendicular to the projection plane. In frequency space, the main limitation of the tomosynthesis reconstruction is the incomplete sampling.

Filtered backprojection as a concept is relatively easy to understand. A problem of back-projection reconstruction is blurring. One would expect that a high-pass filter could be used to eliminate blurring (Amini et al. 1997). The optimal way to eliminate these patterns in the noiseless case is through a ramp filter. The combination of back-projection and ramp filtering is known as filtered back-projection.

For most FBP algorithms, ramp filters are designed based on the frequency sampling density, which can be calculated as the inverse of the shortest distance from a sampled point in Fourier space to sampled points from another view (Stevens et al. 2001, Lauritsch and Haerer 1998, Feldkamp et al. 1984).

Figure 4.5 shows a ramp filter concept designed based on the sampling density. Other filters such as Hamming and Gaussian

filters can also be applied to reduce the amplification of high frequency noise (Mertelemeier et al. 2006, Chen et al. 2005, Wu et al. 2004).

Generally, the FBP algorithm involves the following steps:

1. Fourier Transform the original projection images into frequency space.
2. Multiply the inverse sampling density ramp filters with the Fourier Transform of the projection images.
3. Apply other filters to reduce high frequency noise.
4. Inverse Fourier Transform the filtered projection images.
5. Back-project the filtered projection in spatial domain for reconstruction.
6. Postimage processing.

4.2.1.7 Other Reconstruction Algorithms

There are also quite a few other reconstruction algorithms, such as the simultaneous algebraic reconstruction techniques (SART) (Zhang et al. 2006), ordered-subset convex iterative reconstruction method (Lalush et al. 2006), and other blended algorithms to blend the filtering and iterative algorithms together to improve the reconstructed images. Some of those algorithms are developed based on general foundations. Some are not used widely. Therefore, we will not discuss each of them one by one in this section.

Before we end this section on breast tomosynthesis imaging reconstruction algorithms, we will illustrate tomosynthesis reconstruction results with a prototype breast tomosynthesis system (Bissonnette et al. 2005).

Figure 4.6 shows the phantom reconstruction investigation with different tomosynthesis reconstruction algorithms.

As shown in Figure 4.6b, 250 ml water was added to the phantom and absorbed by the sponge phantom. A fragment of the American College of Radiology (ACR) phantom and three beans were embedded into the phantom to simulate a cluster of five calcifications and three masses, respectively.

Figure 4.6a is the middle (0°) projection image of the sponge phantom for a tomosynthesis sequence of 21 projection images with an angular range of ±10° at the rotation center. A 26 KVp Mo/Mo spectrum was used. Figure 4.6a is similar to a low dose standard mammogram. Two squares on Figure 4.6(a) represent the locations of one simulated mass (bean) and an ACR cluster of calcifications. One can barely see the simulated mass and calcifications on Figure 4.6a.

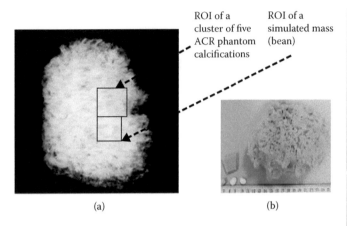

ROI of a cluster of five ACR phantom calcifications

ROI of a simulated mass (bean)

(a)   (b)

FIGURE 4.6 (a) Middle projection image, similar to standard mammography; (b) photograph of the water-sponge phantom.

Figure 4.7 shows the reconstruction images from five different algorithms including SAA, ISSAA, MITS, FBP, and MLEM, respectively. The image reconstructions were performed on 2048 × 2048 pixels projection images with 85 μm pixel size. Because of the intensive computation of the iterative MLEM method, the projection images were down-sampled to 256 × 256 first for MLEM reconstruction. The MLEM reconstructed ROI (region of interest) is 16 times smaller than those of the other four methods.

All five algorithms (Figure 4.7) depicted the objects better than with the low dose standard mammogram (Figure 4.6a). SAA and ISSAA reconstructed the simulated lesions clearly, but had blurry artifacts, especially for the edges of lesions. MITS and FBP had sharper edges and few artifacts. This is due to the deblurring algorithms included in the MITS and FBP techniques. The MLEM reconstruction was from down-sampled projection images. However, it still generated a good reconstruction of the simulated mass (bean).

Figure 4.8 shows a reconstructed plane of a human subject containing a lesion at a height of 7.5 mm above the surface plate from different algorithms: SAA, MITS, FBP, and MLEM, respectively.

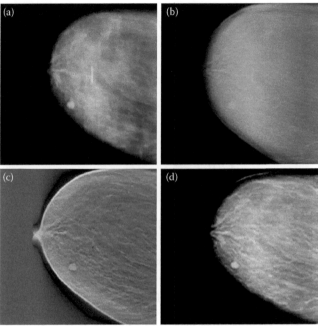

FIGURE 4.8 DBT reconstruction plane of a lesion at $Z = 7.5$ mm above the surface plate: (a) SAA, (b) MITS, (c) FBP, (d) MLEM with down-sampled projection images.

It has been demonstrated that breast tomosynthesis methods are capable of providing better reconstructions as compared with standard mammography and improved conspicuity of anatomical structures to enable early breast cancer detection.

4.2.2 Breast Tomosynthesis Imaging Configurations Optimization

It is widely recognized that research on optimization is important to evaluate and compare breast tomosynthesis reconstruction algorithms for different imaging configurations.

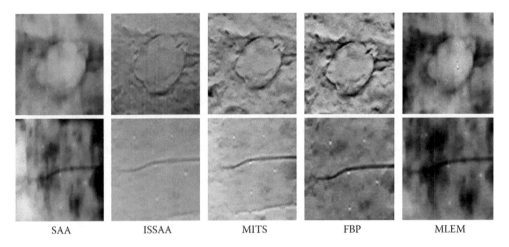

SAA   ISSAA   MITS   FBP   MLEM

FIGURE 4.7 Water-sponge phantom reconstructions (ROI) for five tomosynthesis algorithms. The bean simulating a mass is shown at the upper row, and a cluster of micro-calcifications is shown at the bottom.

The selection of optimal acquisition parameters and reconstruction algorithm plays an important role in producing better performance. However, there are several factors involved in this task and some of them are not individually independent. An effective methodology will enable one to optimize selection of acquisition parameters and compare algorithms for various digital breast tomosynthesis methods.

In this section, we will discuss optimization methods based on image quality analysis in the frequency domain, focusing on the methodology of noise equivalent quanta NEQ(f) analysis for different acquisition parameters and reconstruction algorithms (Chen et al. 2007, Chen et al. 2006, Chen et al. 2005).

The NEQ(f) combines the modulation transfer function (MTF) of signal performance and the noise power spectrum (NPS) of noise characteristics. It enables one to evaluate the performance of different acquisition parameters and algorithms for comparison and optimization purposes.

The relative NEQ(f) is defined as:

$$NEQ_{relative}(f) = \frac{RECONMTF_{relative}^2(f) \cdot MTF_{system}(f)^2}{SubNPS_{tomo}(f)}.$$

The MTF and the NPS are included to evaluate the performance of image reconstruction algorithms and the tomosynthesis imaging configurations.

The MTF values at other frequencies are typically normalized by the MTF value at zero frequency. The RECONMTF$_{relative}$(f) is the relative MTF associated with the specific image reconstruction algorithm and imaging configuration parameters. Computer simulations and reconstructions can be used to calculate the RECONMTF$_{relative}$(f) on a defined reconstruction plane.

The MTF$_{system}$(f) is the measured MTF of the imaging system. The SubNPS$_{tomo}$(f) is the mean-subtracted NPS on the same reconstruction plane.

For each combination of imaging configuration and reconstruction algorithm, data sets of tomosynthesis projection sequences with several different simulated impulse locations can be simulated and then reconstructed. One should note that different location of simulated impulse will results in different relative reconstruction MTF results.

In order to measure the SubNPS$_{tomo}$(f), data sets of tomosynthesis sequences should be acquired for each imaging configuration. Mean-subtracted images should be applied to the NPS analysis to remove fixed pattern noise, including structured noise and system artifacts.

The averaged RECONMTF$_{relative}$(f), measured MTF$_{system}$(f), and measured SubNPS$_{tomo}$(f) should be combined together to generate the relative NEQ(f) for each combination of imaging configuration and image reconstruction algorithm.

In order to match the different samplings along the frequency axes, linear interpolations should be performed to combine the RECONMTF$_{relative}$(f) and MTF$_{system}$(f) together as a relative total MTF.

Details of relative MTF(f), SubNPS$_{tomo}$(f), and relative NEQ(f) measurements are shown in the following sections.

4.2.2.1 Modulation Transfer Function (MTF) Analysis

The MTF measurement should include two parts: (1) the system MTF of the detector and (2) the relative reconstruction MTF associated with specific reconstruction algorithm and imaging configurations.

The system MTF describes the measured MTF of the detector. In this section, a relative reconstruction MTF is calculated as the magnitude of the Fourier Transform of the impulse response associated with a specific reconstruction algorithm and acquisition parameters.

4.2.2.1.1 System MTF Measurement

In order to measure the system MTF of the detector, an edge method can be applied at a range of tube angles to see if there is any difference with angle. One can use a MW2 technique proposed by Saunders et al. (2005) for the system MTF measurement with tungsten/rhodium target/filtration spectrum (Table 4.1).

Figure 4.9 shows the setup of the system MTF measurement experiment. A 0.1-mm Pt-Ir edge is placed in contact with the detector and oriented at a 1°–3° angle with respect to the pixel array. Edge images should be acquired with the MW2 technique and 2 mm aluminum filtration. A previously published routine (Saunders et al. 2005, Saunders and Samei 2003) can be used to analyze the edge images in a region around the edge to compute the presampled system MTF.

As an illustration, Figure 4.10 shows the edge-method measured system MTF when the edge center was close to the chest wall and the x-ray tube was at angles of 0°, ±15°, ±25° with a prototype breast tomosynthesis system (Bissonnette et al. 2005). Three data sets were measured and averaged for each angular location of the x-ray tube.

In Figure 4.10, the MTF varied little with different angles. The MTF curve of –25° is a little lower than that of other angles. This may be caused by the x-ray tube's motion and velocity difference at the –25° location, which is the beginning position of the x-ray tube during the tomosynthesis sequence.

4.2.2.1.2 Relative Reconstruction MTF

The relative reconstruction MTF describes the calculated relative MTF associated with specific algorithm and imaging

TABLE 4.1 System MTF Measurement Technique

Technique	Acquisition Mode	kV	Spectrum	Number of Views (N)	Total mAs	mAs for a Single Projection
MW2	B0XD11 (slow speed)	28	W/Rh	11	303	≈27.5

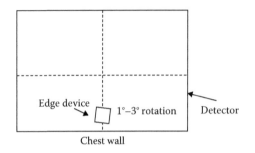

FIGURE 4.9 Setup of system MTF measurements.

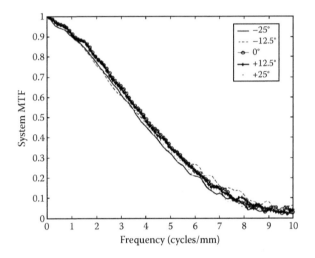

FIGURE 4.10 System MTF with the edge located near the chest wall.

configurations. A simulated delta function can be used as a standard signal input, and is placed in the proper locations on a series of simulated projection images, based on a given set of tomosynthesis acquisition parameters.

A ray-tracing simulation method can be used to project the single delta function onto the detector to simulate the tomosynthesis sequence of projection images. Figure 4.11 shows the ray-tracing method to simulate the projection images of a delta function.

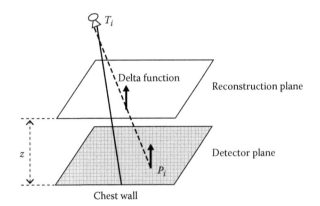

FIGURE 4.11 Ray-tracing simulation.

In Figure 4.11, a single delta function at defined distance above the detector is simulated. The delta function is projected onto the detector at P_i when the x-ray tube moves to location T_i. Partial pixel interpolation should be performed if P_i falls into a noninteger pixel location. With this ray-tracing method, for a specific group of acquisition parameters, a data set of projection images from different x-ray tube locations can be simulated.

Data sets of projection images without background can be simulated according to the projected locations of the delta function with different x-ray tube locations.

One should consider two different impulse locations: (1) the impulse is located away from the chest wall at the defined distance above the detector, and (2) the impulse was located near the chest wall at the defined distance above the detector. The simulated projection data sets should then be reconstructed by the reconstruction algorithms for comparison and optimization.

A one-dimensional Fourier Transform through the column of pixels with the greatest impulse response on the defined reconstruction plane can be performed to compute the relative reconstruction MTF along the tube's motion direction. During impulse simulations, if the impulse is projected onto a noninteger location on the projection image, a linear interpolation among neighboring pixels should be performed. Therefore, the reconstruction MTF is affected by the real location of the simulated impulse. In order to include this factor in the calculation, the relative reconstruction MTF from multiple locations inside a pixel should be considered.

The relative reconstruction MTF along the x-ray tube's motion direction can be calculated as the magnitude of the Fourier Transform of the impulse response on the defined reconstruction plane where the simulated delta function is located.

In order to illustrate the impulse response of different reconstruction algorithms and acquisition modes, results from a representative example of impulse location inside a pixel are presented in this section.

Figures 4.12 and 4.13 show the in-plane impulse response of the point-by-point BP reconstruction algorithm with a simulated breast tomosynthesis configuration of 25 projection images and a 50° view angle. The x and y axes give the pixel location on the reconstruction plane. Only a 40 × 40 pixel region close to the impulse is shown for clarity. Figure 4.12 shows the in-plane response when the impulse was located about 4 cm away from the chest wall and 40.5 mm above the detector. Figure 4.13 shows the in-plane response when the impulse was close to chest wall (20 pixels from chest wall) and 40.5 mm above the detector. One can see that the BP algorithm can reconstruct the simulated impulses to give sharp in-plane responses.

Figure 4.14 illustrates the calculated relative reconstruction MTF for nine sample locations of simulated impulses that are evenly spaced inside a pixel. Results from point-by-point BP reconstruction algorithm with simulated breast tomosynthesis configuration of 25 projection images and 50° view angle is presented here as an example. The X axis represents the frequency bins. The Y axis represents the magnitude of the Fourier Transform of the impulse response at the defined reconstruction

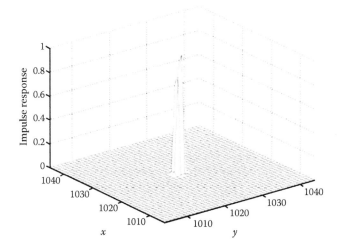

FIGURE 4.12 In-plane impulse response of point-by-point BP: impulse was located about 4 cm away from the chest wall.

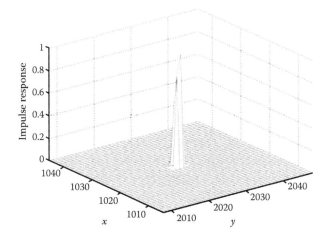

FIGURE 4.13 In-plane impulse response of point-by-point BP: impulse was located near chest wall.

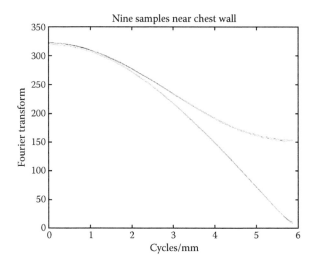

FIGURE 4.14 Relative reconstruction MTF of nine samples of point-by-point BP: the impulse was located near chest wall.

plane where the simulated impulse was located (40.5 mm above the detector). The nine calculated relative MTFs from nine different impulses' locations inside a pixel can be averaged to generate the relative MTF for each algorithm and acquisition mode.

After the characterization of system MTF and relative reconstruction MTF, one can combine these two parts to generate the MTF signal performance of the image reconstruction algorithm and imaging configurations being evaluated. The next step is to characterize the noise property with the same settings.

4.2.2.2 Noise Power Spectrum (NPS) Analysis

The NPS is typically used to characterize the noise properties of digital imaging systems. In this section, we introduce the NPS measurement to evaluate the noise performance of breast tomosynthesis image reconstruction algorithms and imaging configurations.

Figure 4.15 shows BR12 phantom slabs of different thicknesses. The BR 12 phantom can be used to mimic breast tissue equivalent attenuation and scattered radiation. For example, one can use two identical phantom slabs of BR12 (47% water/53% adipose equivalent) for a total of 4 cm thickness in the NPS experiments and put them directly on the surface plate (detector cover) to mimic the breast tissue.

The mean-subtracted NPS should be measured to quantitatively evaluate the noise performance. The purpose of studying the mean-subtracted NPS is to remove fixed pattern noise, including structured noise and system artifacts (Dobbins 2000).

In this section, we will use the back-projection (BP) algorithm with a breast tomosynthesis configuration of 49 projection images and a 50° view angle as an illustration. For mean-subtracted NPS measurement, ten identical tomosynthesis sequences of flat images with the phantom slabs on the detector were acquired. The tomosynthesis sequences were then reconstructed. Mean-subtracted image reconstruction data sets were analyzed and compared on a defined reconstruction plane with the same height (40.5 mm above the detector plane) as described in the reconstruction MTF measurement (Section 4.2.2.1). If one needs to compare different imaging configurations, the same cumulative tube output (mAs) should be kept consistent for all tomosynthesis experiments.

For a typical image size of 2048 × 2048, a published method (Godfrey et al. 2009, Dobbins 2000, Dobbins et al. 1995) can be

FIGURE 4.15 NPS measurement.

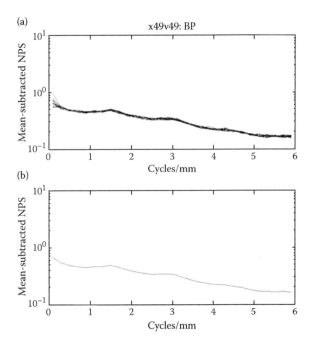

FIGURE 4.16 Mean-subtracted NPS of point-by-point BP: (a) 10 individual mean-subtracted NPS; (b) averaged mean-subtracted NPS.

applied to each mean-subtracted image using 64 ROIs (region of interests) of size 128 × 128 to examine the noise response. The $SubNPS_{tomo}(f)$ should be calculated as the average of these ten mean-subtracted NPS for each imaging configuration.

As an illustration, Figure 4.16 shows an example of the mean-subtracted NPS results of point-by-point BP with the breast tomosynthesis imaging configuration of 49 projection images and ±25° view angle of a prototype tomosynthesis system.

Now we have both MTF and NPS evaluations of tomosynthesis imaging configurations and image reconstruction algorithms. One can combine them together to compute the relative NEQ(f) for comparison and optimization.

4.2.2.3 Relative Noise-Equivalent Quanta Analysis

The relative noise-equivalent quanta (NEQ(f)) is an image quality metric that combines both signal and noise properties to compare different acquisition techniques and reconstruction algorithms in tomosynthesis.

The relative NEQ(f) can be calculated by equation:

$$NEQ_{relative}(f) = \frac{RECONMTF_{relative}^2(f) \cdot MTF_{system}(f)^2}{SubNPS_{tomo}(f)}$$

As shown in Section 4.2.2.1, $MTF_{system}(f)$ is the measured system MTF of the tomosynthesis imaging system. $RECONMTF_{relative}(f)$ should be calculated as the magnitude of

Fourier Transform of the impulse response on the defined reconstruction plane. As shown in Section 4.2.2.2, $SubNPS_{tomo}(f)$ represents the averaged mean-subtracted NPS on the same defined reconstruction plane. One can combine them together to calculate the relative NEQ(f).

For each breast tomosynthesis imaging configuration and image reconstruction algorithm, the same inputs should be considered in all cases (the same simulated impulse magnitude for relative reconstruction MTF measurement and same accumulative tomosynthesis sequence exposure level for NPS measurement). Therefore, one can make relative comparisons to evaluate performance of different algorithms and acquisition modes with the same inputs.

In this section, we use point-by-point BP as an illustration. The relative NEQ(f) along the x-ray tube's motion direction of a commercial prototype breast tomosynthesis system was examined (Bissonnette et al. 2005). As shown in Table 4.2, seven different available imaging configurations were investigated.

Figures 4.17, 4.18, and 4.19 show the $RECONMTF_{relative}(f)$, $SubNPS_{tomo}(f)$ and $NEQ_{relative}(f)$ results of point-by-point BP, respectively.

Compared with other imaging configurations, the B0-bin-A50-P25 (25 projections, ±25° angular range, binning mode) has lower high frequency noise characteristics for BP (Figure 4.18). This is due to the 2 × 1 binning mode to sum up the neighboring 2 pixels along the tube's motion direction to speed up the acquisition time. High frequency components and noise were smoothed under this binning mode.

Among the other six nonbinning acquisition modes, point-by-point BP performed slightly worse for B0-A44-P13 (13 projections, ±22.5° angular range) and A25-P13 (13 projections, ±12.5° angular range) (Figure 4.19).

In Figure 4.17, compared with nonbinning acquisition modes, point-by-point BP algorithm also showed lower relative RECON MTF for B0-bin-A50-P25 (25 projections, ±25° angular range, binning mode). The binning mode introduced blur

TABLE 4.2 Seven Different Available Imaging Acquisition Modes

Imaging Configuration*	Projection Number	Angular Range	Scrub ("Bo")**	Binning***
B0-A44-P13	13	≈±22°	Yes	No
A25-P13	13	±12.5°	No	No
A50-P25	25	±25°	No	No
B0-A50-P25	25	±25°	Yes	No
B0-bin-A50-P25	25	±25°	Yes	Yes
A25-P25	25	±12.5°	No	No
A50-P49	49	±25°	No	No

*Imaging configurations were named based on angular range and number of projection images for clarity. For example, B0-bin-A50-P25 means angular range = 50°, projection images = 25, with "Bo" scrub and binning mode; A44-P13 means angular range = 44°, projection images = 13.

**Scrub ("Bo") means a simple dark current image between each frame.

***Binning mode means to sum up the neighboring 2 pixel values along the x-ray tube's motion to speed up the tomosynthesis scan.

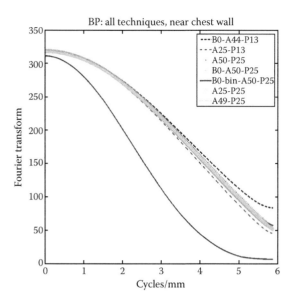

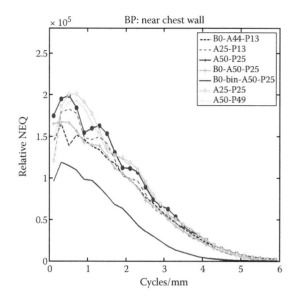

FIGURE 4.17 RECONMTF$_{relative}$(f) of point-by-point BP: impulse was located near chest wall.

FIGURE 4.19 NEQ$_{relative}$(f) of point-by-point: impulse was located near chest wall.

and smoothing effects on the projection images. Therefore, the relative RECON MTF was damaged by this smoothing effect. One cannot compare one imaging configuration against another purely based on only MTF or SubNPS results. However, relative NEQ(f) analysis combining MTF and NPS together can provide a fair comparison with the same signal and noise inputs.

Figure 4.19 is relative NEQ(f) results for the point-by-point BP with all investigated imaging configurations, respectively. As shown in Figure 4.19, point-by-point BP performed slightly

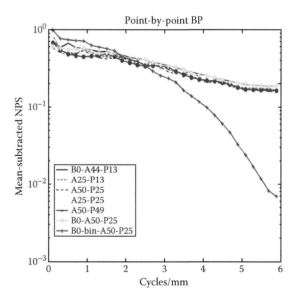

FIGURE 4.18 SubNPS$_{tomo}$(f) of point-by-point BP for all acquisition modes.

better for nonbinning modes with more projection numbers such as A50-P49 (49 projections, ±25° angular range), A25-P25 (25 projections, ±12.5° angular range), and A50-P25 (25 projections, ±25° angular range) by providing slightly higher relative NEQ curves.

Point-by-point BP has the lowest relative NEQ values with B0-bin-A50-P25 binning modes (Figure 4.19) at all frequencies. One possible reason is that the relative RECON MTF was calculated by computer simulation, while the SubNPS was measured experimentally. During experimental measurements, noise was smoothed by the binning mode. The binning mode provides a trade-off between reducing the resolution by averaging neighboring pixel values and improving the speed of read-out.

In order to evaluate and optimize the breast tomosynthesis imaging configurations and image reconstruction algorithms, one should consider fair comparisons and investigations in spatial and frequency domains based on physical measurements and computation as well.

4.3 Conclusions and Future Work

Breast tomosynthesis is an innovative digital breast imaging technology that enables physicians to detect smaller tumors at the early stages of breast cancer. It has been considered as a challenge to the current mammography technology.

Tremendous efforts have been made to develop tomosynthesis reconstruction algorithms and to optimize the imaging configurations. Traditional shift-and-add (SAA) and Niklason's image stretching shift-and-add (ISSAA) are mathematical methods to line up in-plane structures. Maximum likelihood expectation maximization (MLEM) reconstructs the attenuation coefficient of the voxel model inside the breast based on the Poisson joint distribution assumptions. However, due to the iterative nature and computation of intersections between voxels and x-ray projection lines during ray-tracing calculation, the computation of MLEM is time-consuming. MITS uses linear algebra to calculate the matrix of blurring functions of tomosynthesized planes and involves a matrix inversion in the frequency domain to approach the real structure on the reconstruction slice. FBP applies the inverse filter to the projection images to include the sampling density information in the frequency domain and back-projects the filtered projection images in spatial domain to generate the reconstruction slice. Compared with SAA and ISSAA, MITS and FBP use deblurring algorithms to remove and suppress out-of-plane blur, thus having better performance in image reconstruction. Compared with MLEM, MITS and FBP reconstruct the breast at a fast speed that is convenient and realistic for clinical usage, especially for hospitals with limited computation budgets or environments and emergency cases. There are also quite a few other reconstruction algorithms that are applicable in the breast tomosynthesis imaging fields.

Optimized image acquisition techniques are essential to provide better tomosynthesis reconstruction images with different image reconstruction algorithms. None of the imaging acquisition parameters are independent of each other. In order to theoretically and experimentally compare and optimize the imaging acquisition parameters, various measurements and analysis have been performed in the last few years, including studies based on computer simulation, such as the impulse response analysis of different algorithms and imaging configurations, NPS measurements, reconstruction slice spacing, and different algorithms, and methodology of NEQ analysis. Experimental measurements, simulations and calculations have been made by different groups to measure and combine the above physical characteristics. In this chapter, a relative NEQ(f) analysis was described to combine the signal and noise characteristics of the imaging system and the image reconstruction algorithms.

Future work is needed to improve and validate novel image reconstruction algorithms with better image quality, to optimize the breast tomosynthesis design, and to develop novel imaging system with fast speed. It is also necessary to conduct expanded human observer studies and clinical trials to carefully and thoroughly evaluate breast tomosynthesis against standard conventional mammography.

The breast tomosynthesis technology has been FDA approved for one manufacturer Hologic, Inc. Clinical trials using Hologic tomosynthesis system showed measureable benefits including improved lesion and margin visibility and the ability to accurately localize structures three-dimensionally in the breast (Hardy 2011). It has been reported that breast tomosynthesis improves the characterization and margin assessment of masses. For clinical detection and characterization of microcalcification, there is limited literature. The engineering design, reconstruction and optimization of different devices may lead to different results and conclusions. Much more work is necessary before conclusions can be drawn regarding the detection of microcalcifications with breast tomosynthesis imaging (Helvie 2010).

References

Amini B, Björklund M, Dror R, Nygren A. 1997. Tomographic Reconstruction of SPECT Data [Online]. Rice University. Available: http://www.owlnet.rice.edu/~elec539/Projects97/cult/report.html.

Badea C, Kolitsi Z, Pallikarakis N. 2001. A 3D imaging system for dental imaging based on digital tomosynthesis and cone beam CT. Proc Int Feder Med Biol Eng 2: 739–741.

Bissonnette M, Hansroul M, Masson E, Savard S, Cadieux S, Warmoes P, Gravel D, Agopyan J, Polischuk BT, Haerer WH, Mertelmeier T, Lo JY, Chen Y, Dobbins JT III, Jesneck JL, Singh S. 2005. Digital breast tomosynthesis using an amorphous selenium flat panel detector. Proc SPIE 5745: 529–540.

Boone JM, Nelson TR, Lindfors KK, Seibert JA. 2007. Dedicated breast CT: Radiation dose and image quality evaluation. Radiology 221: 657–667.

Bouwens L, Van de Walle R, Gifford H, King MA, Lemahieu I, Dierckx RA. 2001. Resolution recovery for list-mode reconstruction in SPECT. Phys Med Biol 46(8): 2239–2253.

Bushberg JT, Seibert JA, Leidholdt EM Jr, Boone JM. 1994. The Essential Physics of Medical Imaging. Baltimore: Lippincott Williams and Wilkins.

Chen B, Ning R. 2002. Cone-beam volume CT breast imaging: feasibility study. Med Phys 29: 755–770.

Chen Y. 2007. "Digital breast tomosynthesis (DBT) – A novel imaging technology to improve early breast cancer detection: Implementation, comparison and optimization" (PhD diss., Duke University).

Chen Y, Lo JY, Dobbins JT III. 2004. Matrix inversion tomosynthesis (MITS) of the Breast: Preliminary Results. In RSNA 90th Scientific Assembly.

Chen Y, Lo JY, Dobbins JT III. 2005. Impulse response analysis for several digital tomosynthesis mammography reconstruction algorithms. Proc SPIE 5745: 541–549.

Chen Y, Lo JY, Baker JA, Dobbins JT III. 2006. Gaussian frequency blending algorithm with matrix inversion tomosynthesis (MITS) and filtered back projection (FBP) for better digital breast tomosynthesis reconstruction. Proc SPIE 6142: 122–130.

Chen Y, Lo JY, Baker JA, Dobbins JT III. 2006. Noise power spectrum analysis for several digital breast tomosynthesis reconstruction algorithms. Proc SPIE 6142: 1677–1684.

Chen Y, Lo JY, Dobbins JT III. 2007. Importance of point-by-point back projection (BP) correction for isocentric motion in digital breast tomosynthesis: Relevance to morphology of microcalcifications. Med Phys 34(10): 3885–3892.

Chen Y, Lo JY, Ranger NT, Samei E, Dobbins JT III. 2007. Methodology of NEQ(f) analysis for optimization and comparison of digital breast tomosynthesis acquisition techniques and reconstruction algorithms. Proc SPIE 6510: 65101–6510I.

Cherney T, Reminton P. 1999. Breast cancer modality continues decline in Wisconsin. Wisconsin Med J 98(4): 47–49.

Chidlow K, Möller T. 2003. Rapid emission tomography reconstruction. TokioIn Workshop on Volume Graphics (VG03). Tokio.

Crowther RA, Derosier DJ, Klug A. 1970. The reconstruction of a three-dimensional structure from projections and its application to electron microscopy. Proc Roy Soc Lond A 317: 319–340.

Dobbins JT III. 1990. Matrix inversion tomosynthesis improvements in longitudinal x-ray slice imaging. U.S. Patent #4,903,204. Assignee: Duke University.

Dobbins JT III. 2000. Image quality metrics for digital systems. In Beutel J, Kundel HL, Van Metter RL, eds. Handbook of Medical Imaging. Physics and Psychophysics SPIE, Vol. 1, pp. 161–222.

Dobbins JT. III, Ergun DL, Rutz L, Hinshaw DA, Blume H, Clark DC. 1995. DQE(f) of four generations of comouted radiography acquisition devices. Med Phys 22(10): 1581–1593.

Dobbins JT III, Godfrey DJ. 2003. Digital x-ray tomosynthesis: current state of the art and clinical potential. Phys Med Biol 48: 65–106.

Dobbins JT III, Powell AO, Weaver YK. 1987. Matrix inversion tomosynthesis: Initial image reconstructions. Abstract Summaries of the RSNA 165(P): 333.

Duryea J, Dobbins JT III, Lynch JA. 2003. Digital tomosynthesis of hand joints for arthritis assessment. Med Phys 30: 325–333.

Feldkamp LA, Davis LC, Kress JW. 1984. Practical come-beam algorithm. J Opt Soc Am A 1: 612–619.

Glick SJ. 2007. Breast CT. Annu Rev Biomed Eng 9: 501–526.

Godfrey DJ. 2005. "Optimization and clinical implementation of matrix inversion tomosynthesis (MITS) for the detection of subtle pulmonary nodules" (PhD diss., Duke University).

Godfrey DJ, Dobbins JT III. 2002. Optimization of matrix inversion tomosynthesis via impulse response simulations. In RSNA 88th Scientific Assembly.

Godfrey DJ, McAdams HP, Dobbins JT III. 2006. Optimization of the matrix inversion tomosynthesis (MITS) impulse response and modulation transfer function characteristics for chest imaging. Med Phys 33(3): 655–667.

Godfrey DJ, McAdams HP, Dobbins JT III. 2009. Stochastic noise characteristic in matrix inversion tomosynthesis (MITS). Med Phys 36(5): 1521–1532.

Godfrey DJ, Rader A, Dobbins JT III. 2003. Practical strategies for the clinical implementation of matrix inversion tomosynthesis. Proc SPIE 5030: 379–390.

Godfrey DJ, Warp RL, Dobbins JT III. 2001. Optimization of matrix inverse tomosynthesis. Proc SPIE 4320: 696–704.

Grant DG. 1972. Tomosynthesis: a three-dimensional radiographic imaging technique. IEEE Trans Biomed Eng BME-19: 20–28.

Hardy K. 2011. Finding the role for digital breast tomosynthesis. Radiol Today 12(11): 32.

Helvie MA. 2010. Digital mammography imaging: Breast tomosynthesis and advanced applications. Radiol Clin North Am 48(5): 917–929.

Highnam R, Brady M. 1999. Mammographic Image Analysis. Dordrecht, The Netherlands: Kluwer Academic Publishers.

Holland R, Mravunac M, Hendriks J, Bekker BV. 1982. So-called interval cancers of the breast. Cancer 49: 2527–2533.

Knutzen AM, Gisvold JJ. 1993. Likelihood of malignant disease for various categories of mammographically detected, nonpalpable breast lesions. Mayo Clin Proc 68: 454–460.

Lalush DS, Quan E, Rajaram R, Zhang J, Lu J, Zhou O. 2006. Tomosynthesis reconstruction from multi-beam x-ray sources. Proc 2006 IEEE Int Symp Biomed Imaging, 1180–1183.

Lange K, Carson R. 1984. EM reconstruction algorithms for emission and transmission tomography. J Comput Assist Tomogr 8(2): 306–316.

Lange K, Fessler JA. 1995. Globally convergent algorithms for maximum a posteriori transmission tomography. IEEE Trans Image Process 4(10): 1430–1438.

Lauritsch G, Haerer W. 1998. A theoretical framework for filtered back-projection in tomosynthesis. Proc SPIE 3338: 1127–1137.

Maidment AD, Ullberg C, Lindman K, Adelöw L, Egerström J, Eklund M, Francke T, Jordung U, Kristoffersson T, Lindqvist L, Marchal D, Olla H, Penton E, Rantanen J, Solokov S, Weber N, Westerberg H. 2006. Evaluation of a photon-counting breast tomosynthesis imaging system. Proc SPIE 6142: 89–99.

Matsuo H, Iwata A, Horiba I, Suzumura N. 1993. Three-dimensional image reconstruction by digital tomosynthesis using inverse filtering. IEEE Trans Med Imag 12: 307–313.

Mertelemeier T, Orman J, Haerer W, Dudam MK. 2006. Optimizing filtered backprojection reconstruction for a breast tomosynthesis prototype device. Proc SPIE 6142: 131–142.

Nass SJ, Henderson IC, Lashof JC. 2001. Mammography and Beyond: Developing Technologies for the Early Detection of Breast Cancer. Washington, DC: National Academy Press.

Niklason LT, Christian BT, Niklason LE, Kopans DB, Castleberry DE, Opsahl-Ong BH, Landberg CE, Slanetz PJ, Giardino AA, Moore R, Albagli D, DeJule MC, Fitzgerald PF, Fobare DF, Giambattista BW, Kwasnick RF, Liu J, Lubowski SJ, Possin GE, Richotte JF, Wei CY, Wirth RF. 1997. Digital tomosynthesis in breast imaging. Radiology 205: 399–406.

Pisano ED, Yaffe MJ, Kuzmiak CM. 2004. Digital Mammography. Philadelphia: Lippincott Williams and Wilkins.

Radon J. 1917. Über die bestimmung von funktionen durch ihrd intergralwerte l ängs gewisser mannigfaltigkeiten. Ber Verch Saechs Akad Wiss Leipzig Math Phys Kl 69: 262–267.

Rakowski JT, Dennis MJ. 2006. A comparison of reconstruction algorithms for C-arm mammography tomosynthesis. Med Phys 33(8): 3018–3032.

Ren B, Ruth C, Stein J, Smith A, Shaw I, Jing Z. 2005. Design and performance of the prototype full field breast tomosynthesis system with selenium based flat panel detector. Proc SPIE 5745: 550–561.

Saunders RS, Samei E. 2003. A method for modifying the image quality parameters of digital radiographic images. Med Phys 30: 3006–3017.

Saunders RS, Samei E, Jesneck J, Lo JY. 2005. Physical characterization of a prototupe selenium-based full field digital mammography detector. Med Phys 32(2): 588–599.

Sklebitz H, Haendle J. 1983. Tomoscopy: Dynamic layer imaging without mechanical movements. AJR 140: 1247–1252.

Sone S, Kasuga T, Sakai F, Kawai T, Oguchi K, Hirano H, Li F, Kubo K, Honda T, Haniunda M, Takemura K, Hosoba M. 1995. Image processing in the digital tomosynthesis for pulmonary imaging. Eur Radiol 5: 96–101.

Stevens GM, Fahrig R, Pelc NJ. 2001. Filtered backprojection for modifying the impulse response of circular tomosynthesis. Med Phys 28: 372–380.

Stiel GM, Stiel LSG, Klotz E, Nienaber CA. 1993. Digital flashing tomosynthesis: A promising technique for angiocardiographic screening. IEEE Trans Med Imag 12(2): 314–321.

Suryanarnyannan S, Karellas A, Vedantham S, Glick SJ, Orsi CJ, Webber RL. 1999. Comparison of contrast-detail characteristics of tomosynthetic reconstruction techniques for digital mammography. Radiology 213: 368–369.

Warp RJ, Godfrey EJ, Dobbins JT III. 2000. Applications of matrix inverse tomosynthesis. Proc SPIE 3977: 376–383.

Wu T, Moore RH, Rafferty EA, Kopans DB. 2004. A comparison of reconstruction algorithms for breast tomosynthesis. Med Phys 9: 2636–2647.

Wu T, Stewart A, Stanton M, McCauley T, Phillips W, Kopans DB, Moore RH, Eberhard JW, Opsahl-Ong B, Niklason L, Williams MB. 2003. Tomographic mammography using a limited number of low-dose cone-beam projection images. Med Phys 30: 365–380.

Yang G, Rajaram R, Cao G, Sultana S, Liu Z, Lalush DS, Lu J, Zhou O. 2008. Stationary digital breast tomosynthesis system with a multi-beam field emission x-ray source array. Proc SPIE 6913: 69131A.

Zhang J, Yang G, Lu J, Zhou O. 2006. Multiplexing radiography using a carbon nanotube based X-ray source. Appl Phys Lett 89: 064106-1–064106-3.

Zhang Y, Chan H, Sahiner B, Wei J, Goodsitt MM, Hadjiiski LM, Ge J, Zhou C. 2006. A comparative study of limited-angle cone-beam reconstruction methods for breast tomosynthesis. Med Phys 33(10): 3781–3795.

Zhou W, Qian X, Lu J, Zhou O, Chen Y. 2010. Multi-beam x-ray source breast tomosynthesis reconstruction with different algorithms. Proc SPIE 7622H: 1–8.

Ziedses des Plantes BG. 1932. Eine neue methode zur differenzierung in der roentgenographie (planigraphie). Acta Radiol 13: 182–192.

5

Breast CT

Chris C. Shaw
*The University of Texas MD
Anderson Cancer Center*

Gary J. Whitman
*The University of Texas MD
Anderson Cancer Center*

5.1 History of Development

The desire to use computed tomography (CT) for breast imaging is not new. The efforts to develop breast CT (BCT) began not long after computed tomography (CT) was developed and commercialized and involved the use of a specially designed CT system (CT/M) (Chang et al. 1980, Gisvold et al. 1979, Chang et al. 1978, Chang et al. 1977, Gisvold et al. 1977) as well as a conventional body scanner (Muller et al. 1983, Chang et al. 1982). Because of poor image quality and long scanning time, the CT/M system did not prove to be practical. The use of a body scanner for breast imaging, on the other hand, required the entire chest to be scanned, resulting in increased patient dose and poorer resolution of the reconstructed breast images. Due to technological limitations, the concept of a dedicated BCT scanner was put on the backburner until the early 2000s.

In the 1990s, various flat panel detectors were developed and used to construct digital x-ray systems for general radiography as well as mammography applications. The availability of these detectors facilitated research and development of cone beam computed tomography (CBCT) techniques in academic institutions. With these techniques, a cone shaped beam, like those used in radiography, fluoroscopy or mammography, is used to rotate around and scan the patient. One of these efforts was aimed to implement dedicated breast CT using the pendant geometry, with which the patient lies prone on the table with one of her breasts protruding downward through an opening and scanned by a CBCT system underneath the table. The use of CBCT techniques for dedicated BCT was first proposed independently by Boone et al. (2001) at the University of California at Davis and Ning et al. (2002) at the University of Rochester (O'Connell et al. 2010, Lindfors et al. 2008, Ning et al. 2007, Boone et al. 2006, Chen and Ning 2002, Boone et al. 2001). The former constructed the first patient imager for clinical evaluation. The latter led to the first effort to commercialize the BCT technology. Patient studies conducted to date have shown that BCT images are superior to mammography in detecting and visualizing breast anatomy and soft tissue masses but are more limited in imaging small microcalcifications. A third group at the Duke University has recently constructed a patient imager with a specially designed quasimonoenergetic x-ray source to improve the image quality (McKinley et al. 2005, Tornai et al. 2005). Two other groups, one at the University of Massachusetts and the other at The University of Texas MD Anderson Cancer Center, have constructed bench top experimental systems to image mastectomy breast specimens in an effort to investigate the imaging properties of BCT (O'Conner et al. 2008, Lai et al. 2007, O'Conner et at. 2007, Yang et al. 2007). The latter has also proposed and investigated the use of a collimated x-ray beam to scan a preselected volume-of-interest to allow smaller microcalcifications or other details to be seen without increasing the integral breast dose (Chen et al. 2009, Lai et al. 2009, Chen et al. 2008). To enhance the visibility of cancer tissue, BCT with contrast injection has been explored and investigated (Prionas et al. 2010). Research groups in USA as well as Europe have also begun to develop and investigate improved BCT techniques in an effort to bring BCT into practical use (Russo et al. 2010, Russo et al. 2010, Kyriakou et al. 2008, Ning et al. 2006, von Smekal et al. 2004).

5.2 General Design

Like a regular CT system, a dedicated BCT system consists of an x-ray tube and a detector assembly mounted on a gantry which rotates around the breast during the scan. However, a dedicated BCT system scans and exposes the breast only. This helps spare the rest of the chest from radiation exposure, thus reducing the patient dose. In addition, since the breast occupies only a smaller volume, the reconstruction volume is substantially reduced, thus allowing for greater pixel density and resulting in higher resolution in the reconstructed images.

In mammography, the x-ray field is collimated and oriented away from the chest so that one side of the x-ray field is pointing straight down along the chest wall while the breast is drawn outward, compressed and held between the table/detector and a compression plate. The orientation can be altered by rotating the gantry holding the x-ray tube and the table/detector around the breast and recompressing the breast to allow a second mammogram to be taken at a different view angle. Dedicated BCT follows a similar approach except that the patient lies prone on a table with one of her breasts protruding downward through an opening and scanned by the x-ray source-detector assembly rotating around the breast over 360° underneath the table. This is often referred to as "pendant geometry" (Figure 5.1). For BCT, it is unnecessary and even disadvantageous to compress the breast into a slab. Actually, the quality of the reconstructed images would be better optimized if the breast could be molded into the shape of a long, skinny half-ellipsoid. It may also be possible to use a bowl shaped holder to lightly compress the breast into the shape of a hemi-ellipsoid. Either way, the discomfort or pain from compression in mammography is eliminated. While other configurations may be possible, this is how all current dedicated BCT systems are designed.

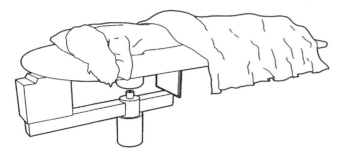

FIGURE 5.1 Pendant geometry for dedicated breast CT with which the patient lies prone on the table with one of her breasts protruding downward and scanned by the x-ray source and detector assembly rotating around the breast underneath the table.

5.2.1 Gantry and Table

The gantry and table configuration is very similar to that of a stereotactic biopsy system except that the gantry holding the x-ray tube and the detector needs to be rotated around the breast at a reasonably fast speed over 360°. Due to the combined large weight of the detector and x-ray tube, a robust heavy duty gantry and motor drive system must be used. One design issue is how to route the power, interface and data cables from the x-ray tube to the generator and from the detector to the controller/acquisition computer while allowing the x-ray tube and detector to rotate around the breast for over 360°. Commercial flexible cable holders are available to coil the cables during the scan and to uncoil the cables afterward (Boone et al. 2006). A more elegant but more complex and expensive approach is to use a slip ring which allows the power, interface signals, and data to be connected between the rotating gantry and the stationary sources without using cables (Ning et al. 2006). This eliminates the need to coil and uncoil cables and, furthermore, allows the gantry to be rotated continuously for multiple revolutions.

The patient table needs to be designed for patient comfort and the ability to allow the entire breast and even part of the axillary tissue to protrude downward for the BCT scan. The area around the opening may be specially contoured for these purposes. Since the opening needs to accommodate either the left or the right breast, the table needs to be designed to be either wider or longer than usual to allow the patient to shift sideways or lie down in two opposite orientations. Due to the presence of the rotating gantry in the middle of the table, the patient needs to climb up from either end of the table.

5.2.2 X-Ray Source

While low energy x-rays have been successively used in mammography, it is impractical to use them for BCT because they cannot penetrate an uncompressed breast and they incur an excessively high breast dose. 49–80 kVp x-rays generated with a tungsten target are typically used in BCT (Weigel et al. 2010, Crotty et al. 2007, Glick et al. 2007, Chen and Ning 2002, Boone et al. 2001). These x-rays penetrate the breast well but produce lower contrast in the projection images. However, the low contrast of tissue structures and small calcifications are easily restored after image reconstruction. The x-ray source used in BCT has two functional requirements: first, the focal spot needs be as close as possible to one end of the tube housing. With some tubes used for BCT, the focal spot is located 4–5 cm below one end of the housing, allowing the rotating x-ray beam to be placed and aligned slightly lower than 4 cm underneath the table to maximize the coverage of the breast on the chest wall side. Second, the focal spot size must be small enough to allow a reasonably small pixel size to be used for image acquisition. The typical focal spot size used for BCT is 0.3–0.4 mm. With these requirements, the power of the x-ray tube is generally on the lower side. Fortunately, the source-to-detector distance can be made short to increase the x-ray flux while the breast, generally small compared to other parts of the body and containing no bony structures, is less attenuating. Like mammographic x-rays,

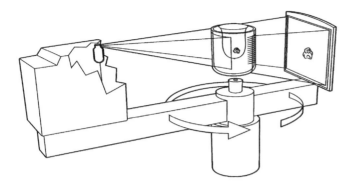

FIGURE 5.2 As in mammography, a half cone beam is used to scan the breast. The position of the focal spot inside the tube determines how close the top of the half cone beam is from the table and how much the breast can be covered. The breast is usually positioned at the center of rotation. The source to isocenter and detector to isocenter distances should be selected to match focal spot blurring with detector blurring and the pixel size to optimize the spatial resolution.

the x-rays for BCT need to be collimated into a half cone beam to maximize the breast coverage on the chest side while allowing the x-ray tube to be rotated underneath the table (Figure 5.2). A bow tie filter, which varies in thickness from the edges toward the central axis of the breast (or the rotating axis), may be used to compensate for the variation of x-ray attenuation from the breast center to the periphery and equalize the transmitted x-ray intensity at the detector input. This helps reduce the breast dose and avoid signal saturation near the border of the breast.

5.2.3 Detector

For BCT, an image detector with efficient x-ray absorption, smaller pixel size, low noise level and a fast framing rate is required. For this reason, amorphous silicon/cesium iodide (aSi/CsI) flat panel detectors have been widely used for BCT and other CBCT applications. With this type of detector, the transmitted x-rays are converted into visible light in the CsI scintillator layer. The light propagates through the needle structures of CsI crystals, created to minimize lateral light spread, and reaches a 2-D array of image elements where it is converted into charges by the built-in photodiodes and read out later as image signals. The flat panel detectors have been developed for various applications, ranging from digital radiography, digital mammography, small animal imaging, to cone beam CT for use in radiation treatment procedures (Stock et al. 2009, Conover et al. 2005, Lee et al. 2003, Floyd et al. 2001, Vedantham et al. 2000). For BCT, a large field of view is required to accommodate a magnified (by a factor of as large as 2) large breast. A fast acquisition rate (7.5–30 fps) is required to minimize the scanning time and the motion artifacts. A detector with minimal inactive space on one edge, like those designed for digital mammography, is very desirable as it would allow for more coverage of the breast on the chest wall side.

The quality of a flat panel detector is largely determined by its pixel size, data depth, and efficiency of x-ray absorption. The pixel size reflects the spatial resolution capability of the detector. The data

depth and efficiency of x-ray absorption are directly linked to the contrast sensitivity or the ability to image low contrast objects. For BCT, a flat panel detector designed for fluoroscopy is more suitable than the high resolution detectors designed for mammography applications. A detector widely used for BCT as well as other CBCT applications is the Paxscan 4030CB by Varian Medical Systems, Salt Lake City, UT. It has a native pixel size of 194 μm which, with a magnification factor of 1.33, would lead to a voxel size of 146 μm for 3-D breast imaging. This is considerably larger than the minimum pixel size of 100 μm for digital mammography. However, the use of higher resolution detectors is simply impractical due to consideration of the breast dose, system alignment, and breast motion caused by pulsating blood flow. In fact, with many BCT systems, the detectors have to be operated in the binning mode to minimize the noise level at a reasonable exposure level and to reduce the data size for speedier image acquisition, reconstruction, and manipulation. However, binning results in a larger effective pixel size (398 μm or larger) which degrades the resolution.

With the dose level commonly kept under the mean glandular dose limit for two view mammograms, the spatial resolution become less important. However, the use of smaller pixel sizes is essential to imaging smaller calcifications, although higher exposure levels would be required. Similarly, larger data depth may be important in imaging bony structures or large breasts. The data depth of a flat panel detector could be 12, 14, or 16 bits, corresponding to a maximum gray scale value of 4095, 16383, or 65535. It is usually determined by the ratio of the maximum unsaturated signal to the root mean square dark current noise, which occurs as the result of current leakage in the photodiodes and electronic noise in the preamplifiers. Since the breast is relatively small compared to the chest, abdomen or head and does not contain bony structures, the 14 bit data depth of the aforementioned detector should be sufficient. The readout gains are usually variable to boost the signal size for low exposures. However, the visibility of low contrast objects is still limited by the dark current noise level which would be amplified as well.

5.2.4 Image Reconstruction

Although iterative algorithms have recently been shown to have the potential advantage of dose reduction, they generally take much longer time to complete the reconstruction then the filtered back projection (FBP) algorithms. Thus, the well known Feldkamp (FDK) algorithm is often used for image reconstruction. This algorithm is readily available and can be easily implemented on various computing platforms. The voxel size for the reconstruction is generally selected by projecting the pixel size back to the isocenter. Thus, a 194 μm pixel size would result in a voxel size of 146 μm with a magnification factor of 1.33. This allows a 14 cm diameter breast to be covered with 1000 × 1000 slice images. The projection images would have to be 1000 pixels wide as well, resulting in an enormous amount of projection data to be reconstructed. The projection images may be acquired in the binning mode to reduce the image size, lessening the computation task of image reconstruction. A multiple CPU/core PC,

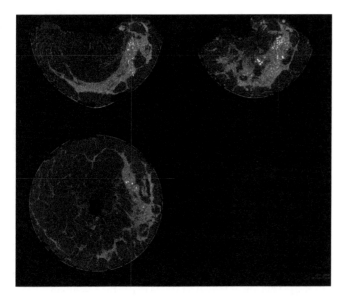

FIGURE 5.3 Examples of BCT images in cranio-caudal (upper left), sagittal view (upper right) and coronal view (lower left). Notice the clear separation of dense tissues (white) from the adipose tissues (gray) and the calcifications (bright white) in the dense tissues. The gray lines in each image indicate the positions of the slices in the other two images.

PC cluster or even a graphic processing unit (GPU) based workstation may be used to keep the reconstruction time down to several minutes or even shorter. A reconstruction filter is used to smooth out the noise and improve the visualization of the tissue structures. Figure 5.3 shows examples of BCT images in three different views: cranio-caudal (axial), sagittal, and coronal. Figure 5.4 shows the corresponding mammograms in the

craniocaudal (CC) and mediolateral oblique (MLO) views. Notice that the dense tissue structures are well separated from the adipose tissue in the BCT images with a slice thickness of 145 μm. The microcalcifications are easily visible in the BCT images with their high contrast although they may appear larger than they really are, and smaller microcalcifications may not be visible. Microcalcifications appear smaller and sharper in the mammograms, reflecting the superior spatial resolution of mammography. However, the overlapping of microcalcifications with dense tissue structures makes it difficult to see all microcalcifications in the CC view mammogram.

5.3 Image Quality

5.3.1 Image Contrast

Image contrast in BCT images, like that in regular CT images, is linked to the difference of CT numbers between the object and the background materials. Thus, microcalcifications have much higher contrast than soft tissues. However, small microcalcifications may be subjected to partial pixel effect during image acquisition and produce lower contrast in the projection images. This translates into lower CT numbers and contrast for the microcalcifications, which could be reduced to such a low level that the microcalcifications are no longer visible. Since the CT numbers are proportional to the ratio of the difference of the linear attenuation coefficients between the object and water to that of water, both the CT numbers and image contrast are less affected by the x-ray kVp. The variations become even less if a filtered x-ray beam is used. For the same reason, image contrast is less affected by radiation scattering and beam hardening although the CT numbers themselves are subjected to significant negative biases, resulting in so called "cupping artifacts."

5.3.2 Spatial Resolution

Spatial resolution is essential to x-ray breast imaging, in particular to imaging small microcalcifications. Spatial resolution characterizes the ability of an imaging system to resolve small details of an object. In x-ray breast imaging, such details could be small microcalcifications, the shape and dimensions of a microcalcification cluster, or the detailed shape of a soft tissue mass. The spatial resolution of a BCT system correlates highly with the spatial resolution quality of the projection images, which is determined by the modulation transfer function of the detector, focal spot size and imaging geometry (where the x-ray source, object, and detector are placed in relation to each other).

The spatial resolution capability of a flat panel detector may be characterized by its native pixel size. Theoretically, the spatial resolution of the flat panel detector is also determined by image blurring in the CsI layer. A thicker CsI layer, commonly used in digital radiography systems, tends to result in greater blurring, leading to lower spatial resolution, while a thinner CsI layer, commonly used in digital mammography systems, tends to result in less blurring, leading to higher spatial resolution.

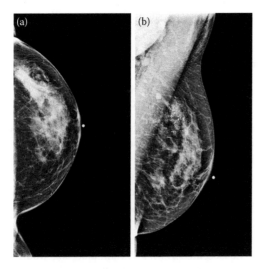

FIGURE 5.4 (a) Craniocaudal (CC) view and (b) mediolateral oblique (MLO) view mammograms of the same breast imaged in Figure 5.3 for comparison. These two views correspond approximately to the superior–inferior and sagittal views in Figure 5.3. Notice that the detailed appearance of the microcalcifications reflect the spatial resolution quality of the mammogram but their overlapping with the dense tissue structures make then more difficult to visualize.

However, to contain the cost of the detector and subsequent image processing, storage, transmission and display, the pixel size of the detector is usually chosen to match the optical resolution of the light image that falls onto the image elements. Thus, the pixel size serves as a good indicator for the spatial resolution capability of the detector. For imaging tissue structures and large microcalcifications, a detector with a pixel size of ~200 μm provides more than adequate spatial resolution. For imaging small microcalcifications, a detector with a smaller pixel size is necessary to allow the microcalcifications to be seen individually without merging with each other and to allow larger microcalcifications or microcalcification clusters to be imaged with their dimensions and shapes accurately visualized.

As in projection imaging, the positions of the x-ray source (focal spot in the x-ray tube) and the detector relative to the breast affect the spatial resolution capability of the BCT system. When an object is projected onto the detector, it is magnified with a factor equal to the ratio of the source-to-detector distance to the source-to-object distance. This magnification is referred to as geometric magnification, which helps increase the apparent size of the object, thus improving the spatial resolution capability of a projection imaging system. However, there is a limit on how much the magnification can be increased. The x-ray focal spot is projected through each point of the object onto the detector and creates a shadow in the image. The image of the object is then blurred with this shadow acting as the kernel or point spread function. This is referred to as the focal spot blurring effect which increases with the magnification factor, though not proportionally. Thus, the optimal geometric factor is the one with which the size of the focal spot shadow equals the pixel size and the spatial resolution performance peaks. In a CT scan, an off-center object is magnified with different factors in different projection views. However, the magnification for an object at the center of rotation (isocenter) is usually used to estimate the average magnification factor in assessing the average effects of magnification and the overall spatial resolution quality in the reconstructed images. In fact, the native (in nonbinning mode) or effective (in binning mode) pixel size of the detector is usually divided by the geometric magnification factor to select the optimal voxel size for image reconstruction. This voxel size is usually used to indicate the spatial resolution quality of the reconstructed images.

Different geometries have been used in designing BCT systems. With some BCT systems, the detector array and the x-ray focal spot are placed at about the same distance from the isocenter, resulting in a geometric factor of close to 2 and a focal spot shadow of similar size as the focal spot itself. For instance, a 0.4 mm wide focal spot would be projected onto the detector with a 0.4 mm wide shadow in the projection image. This sets a resolution limit which makes it unnecessary to use a high resolution detector with a pixel size smaller than 400 μm. Thus, when a high resolution detector (e.g., Paxscan 4030CB with a 194 μm pixel size) is used, it is usually operated in the binning mode to achieve a larger effective pixel size (388 μm with 2 × 2 binning) for speedier image acquisition and reconstruction.

The imaging geometry can also be configured with smaller geometric magnification to take full advantage of the resolution capability of the detector. This involves moving the detector closer to the isocenter, to which the object is centered, thus producing a focal spot shadow similar in size to the native pixel size of the detector. For instance, with a source-to-isocenter distance of 76 cm and a source-to-detector distance of 114 cm, the geometric magnification factor is reduced to 1.5 and a 0.4 mm focal spot would produce a 0.2 mm wide shadow, which is similar in size to the native pixel size of the aforementioned detector. This would help improve the spatial resolution of the reconstructed images and allow smaller calcifications to be properly visualized.

5.3.3 Noise Level

In projection imaging, the ability to image low contrast objects is directly linked to the contrast signal-to-noise ratio, defined as the ratio of the contrast signal to the noise level, which, for small low contrast objects, should be similar inside and outside the object regions. However, signals in CT images are CT numbers computed as the difference of linear attenuation coefficients between the object and water normalized by the coefficient for the water. Thus, contrast signals in BCT do not vary significantly with the x-ray techniques. The ability to image small low contrast objects is dictated more by the noise level in the reconstructed images.

Noise in BCT images originates mostly from noise in the projection images. During image reconstruction, noise in the projection images is propagated to the reconstructed images. Due to normalization and logarithmic mapping performed prior to reconstruction, the noise level in the reconstructed images reflects the signal-to-noise ratios in the projection images rather than the noise levels there. Thus, as the x-ray exposure level is increased for image acquisition, both the signal size and noise level increase in the projection images. In the reconstructed images, however, the signal size remains similar while the noise level decreases.

Noise in the projection images consists of a quantum noise component and a system noise component. The former is associated with the process of x-ray absorption while the latter is associated with the process of image signal readout. The level of the quantum noise component is affected by many factors, including the x-ray kVp, mAs, imaging geometry, breast size, breast composition, and the detector used. The x-ray kVp and mAs determine the x-ray output from the x-ray source. The x-ray kVp, the breast size, breast composition, and the source-to-detector distance together determine the fluence of x-rays that transmit through the breast and reach the detector. The quantum detection efficiency of the detector determines the percentage of the x-ray fluence absorbed and used to produce signals in the projection images. Random fluctuations occur during absorption of x-ray photons in the detector and result in quantum noise in the image with the level proportional to the square root of the number of x-ray photons absorbed in each pixel. The system noise, originating from the readout electronics, is added during the image readout process. An image quality metric, referred to as the detective quantum efficiency (DQE), combines the effect of the system noise with the x-ray absorption ability of the detector and is often used to quantify the efficiency of the detector to preserve the x-ray statistics, measured as the signal-to-noise ratio squared, during the x-ray detection and image formation process.

5.3.4 Signal Accuracy and Artifacts

BCT images differ from mammograms in that the images can be easily separated into regions of dense tissue, adipose tissue, calcifications, and skin. Theoretically, each region should have similar CT numbers whose level indicates the tissue type. In reality, these numbers are nonuniform and they tend to decrease from the periphery toward the center of the breast. This nonuniformity is the result of the beam hardening effect and acceptance of x-ray scatter as part of image signals. This effect is referred to as the cupping artifact. Accurate CT numbers are not essential to reading BCT images but may be essential to quantitative image analysis, e.g., image segmentation or computer-aided diagnosis (Chapter 21). Thus, the BCT image data need to be corrected for the biases of the CT numbers if they are to be used for quantitative analysis. Figure 5.5 shows two coronal view BCT images, one with (a) and the other without (b) cupping artifact correction. With the images presented with the same window and level, the corrected image shows overall higher CT numbers than the uncorrected image, reflecting the fact that the correction algorithm restored signals (normalized attenuation coefficients) degraded due to the presence of scatter signals and beam hardening effects in the projection data. The uncorrected image also shows lower CT numbers toward the center, indicating greater effects from scatter signals and beam hardening effects toward the center.

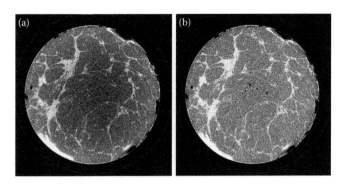

FIGURE 5.5 Coronal view BCT images (a) with and (b) without cupping artifact correction. With the images presented with the same window and level, the corrected image shows overall higher CT numbers than the uncorrected image, reflecting the fact that the correction algorithm restored signals (normalized attenuation coefficients) degraded due to the presence of scatter signals and beam hardening effects in the projection data. The uncorrected image also shows lower CT numbers toward the center, indicating greater effects from scatter signals and beam hardening effects toward the center.

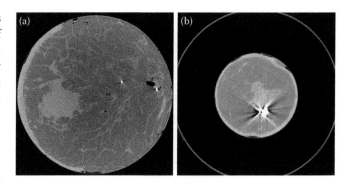

FIGURE 5.6 Metal artifacts in BCT may appear as saturated metal objects, streaks around them (a) or Morriere patterns (b).

Intrinsic to 3-D image reconstruction is the generation of artifacts as part of the reconstructed image signals. Artifacts may be generated in different ways. Some are more visible than others and may obscure the lesions or anatomy to be detected or examined. The first type of artifact is associated with excessive attenuation by dense objects like metal clips, used to indicate locations of lesions, or a large calcifications (Figure 5.6). Image reconstruction in x-ray CT relies on accurate attenuation measurement with the projection images acquired. Attenuation by a dense object could result in excessively low intensity of transmitted x-rays which may be too low to be measurable. This phenomenon not only makes it impossible to directly reconstruct CT signals for the objects themselves, but also affects reconstruction of objects or structures around them. The generated artifacts are often referred to as metal clip artifacts. They may appear as saturated images of the objects with incorrect dimensions or shapes, star shaped streaks around the objects (Figure 5.6a), or Morriere patterns all over the images (Figure 5.6b). Fortunately, the streaking artifacts and Morriere patterns can be largely reduced by replacing the erroneous projection data in the dense object regions with presumed data. These data, while being artificially inserted and incorrectly reflecting the true attenuation, help reduce the streaking artifacts and Morriere patterns. In addition, they could help restore, at least partially, the apparent dimensions and shapes of these heavily attenuating objects and the accuracy of CT signals in and around them.

Other types of artifacts could result from misalignment, defects of the detector pixels, or improper bias offsetting or gain correction of the projection image data. These types of artifacts may sometimes appear with images generated with earlier experimental prototype imagers. However, as the alignment and signal normalization problems are ironed out, they tend to be reduced to an unnoticeable level.

5.3.5 Tissue Structures

Based on the patient and specimen imaging studies to date, tissue structures are well resolved with an isocenter glandular tissue dose at the same level as the mean glandular dose limit for two view mammograms for an average size breast. Dense tissue is visually well separated from the adipose tissue. Due to the coarseness of tissue structures, the spatial resolution of current BCT systems is adequate for imaging tissue structures even though the detector may be operated in the binning mode with an effective pixel size of 388 μm. However, it is also clear that soft tissue masses or tumors cannot be distinguished from dense tissue solely based on differences in signal levels or CT numbers. Thus, if soft tissue masses or tumors are completely embedded in the dense tissue, they cannot be distinguished from the dense tissue or detected. As with mammograms, only when the masses or tumors grow out of the dense tissue and spread into the adipose tissue regions can they be visualized and identified based on the difference of their morphological appearance from that of normal tissue structures. This should be expected as the CT numbers of masses or tumors are simply too close to those of the dense tissue. Even with regular clinical CT, masses and tumors cannot be easily or consistently seen. The BCT images intrinsically have higher noise levels due to the lower exposures used and the additional noise introduced by scattered radiation which is detected as part of the image signals during image acquisition. Thus, it would be even more difficult to directly depict the difference in CT numbers between the soft tissue masses and normal dense tissue in BCT images. However, BCT is superior to mammography in depicting the morphological appearance of soft tissue masses as it provides consistently higher tissue contrast and consecutive slice images for 3-D rendition while in mammography, the morphological appearances of masses are less clear as they are obscured by overlapping tissue structures and by the integration of the tissue structures along the x-ray paths. Direct detection and visualization of soft tissue masses embedded inside dense tissue may be achieved by BCT with contrast injection (Prionas et al. 2010). This may result in images similar and comparable to contrast enhanced MRI. This is a new technique and its application being explored and investigated.

5.3.6 Microcalcifications

Although BCT provides true 3-D images of the breast and, therefore, is actually a new modality, it has often been used in the same way as or compared to mammography. Thus, it is expected to demonstrate the ability to image small microcalcifications if it is to be used to replace mammography in screening, diagnosis, or assessment of breast cancers. Early reports on patient studies suggested that BCT has ample resolution for imaging tissue structures or soft tissue masses (O'Connell et al. 2010, O'Connoer et al. 2008, Lindfors et al. 2008). However, only larger microcalcifications are consistently visible in BCT images. (see Sections 5.5.1 and 5.6.1). There are several factors to prevent current BCT systems from effectively imaging small microcalcifications. First of all, to contain the breast dose, the flat panel detectors are often operated in the binning mode, resulting in a large effective pixel size (e.g., 388 μm for 2 × 2 binning with the Paxscan 4030CB detector) and the partial pixel effect during image acquisition. Thus, although microcalcifications, with their high linear attenuation coefficient, should have

superior contrast in the background of adipose or dense tissues, the contrast may be substantially degraded through the partial pixel effect during image acquisition. The degradation may be so severe that these small microcalcifications, masked by the image noise, become invisible. This problem may be resolved by increasing the x-ray techniques and reducing the noise level in the reconstructed images, thus allowing the small microcalcifications to become visible. However, this would increase the radiation dose and hence the cancer risk to the breast. Another solution is to use higher resolution detectors to reduce the partial pixel effect. However, since the image pixels are smaller in size, they would absorb a smaller number of photons, thus lowering the signal size and the system noise may become a limiting factor. Higher exposures may be used to overcome the influence of the system noise. However, this, again, would increase the patient dose.

5.4 X-Ray Techniques and Doses

The x-ray kVp used with current BCT systems ranges from 50 to 100. However, the advantage of using lower kVps has not been experimentally demonstrated. On the contrary, higher kVps were found to be necessary to allow the x-rays to penetrate large breasts and result in projection images of reasonable quality for reconstruction. In general, 80 kVp is a more practical choice for small and medium breasts (14 cm or smaller in diameter) while higher kVps may be necessary for imaging larger breasts (14 cm or greater in diameter). The mean glandular dose for a BCT scan can be estimated by measuring the open field exposure (air kerma) at the isocenter. Monte Carlo simulation may be used to determine the factor to convert the isocenter exposure to the average breast dose or mean glandular dose for the breast (Boone et al. 2004, Thacker and Glick 2004). Although no legal limit on the breast dose in BCT has been set, it is a common practice to keep the mean glandular dose at the same as or below the mean glandular dose limit (6 mGys) for two view mammograms for a 50% adipose, 50% dense breast with a compressed thickness of 5 cm. As mentioned in the previous section, higher exposure levels may be necessary for imaging small microcalcifications. However, to date, there has not been consensus on whether there is sufficient justification for increasing the x-ray exposure level to allow small microcalcifications to be seen at the expense of higher breast dose.

5.5 Comparison to Other Modalities

5.5.1 Digital Mammography (Chapter 1)

Breast CT provides a larger data set than conventional digital mammography, requiring more data storage space and transmission time and increased interpretation times. O'Connell et al. (2010) studied 40 cone-beam breast CT data sets and compared them with the mammograms for the coverage of the breast, lesion size and lesion visibility. Digital mammography was performed on all but three subjects, who underwent film-screen mammography instead. Cone beam breast CT was performed with an average glandular dose similar to that of mammography. For mammography, the average glandular dose ranged from 2.2 to 15 mGy (mean, 6.5 mGy). For cone-beam breast CT, the average glandular dose ranged from 4 to 12.8 mGy (mean, 8.2 mGy). For heterogeneously dense and extremely dense breasts, the difference between the mean dose for mammography and that of cone-beam breast CT was not statistically significant (7.0 mGy vs. 8.1 mGy, $p = 0.06$).

The breast CT image sets and the mammograms were reviewed by three mammographers for the coverage of the breast, lesion size, and lesion visibility. The results show that mammography demonstrated statistically significantly better coverage than cone-beam breast CT in the axilla and the axillary tail ($p < 0.0001$). Breast tissue coverage was noted to be statistically significantly better with cone-beam CT than with mammography in the lateral ($p < 0.0001$), medial ($p < 0.0001$), and posterior ($p = 0.0002$) aspects of the breast. Most calcifications and all masses detected by mammography were also detected with cone-beam breast CT. Overall, 85% of calcifications less than 1 mm in size and all calcifications greater than 1 mm in size were detected on both mammography and cone-beam breast CT.

Seven calcification findings were noted with mammography but not with cone-beam breast CT. Two of these findings were in the central breast, three were in the posterior breast, one was in the inner breast, and one was in the lower breast. Five of these findings were single calcifications (<1 mm), one was a calcification cluster, and one was a two-to-four calcification group. Lesion size was measured in 55 findings, out of which there was agreement between the mammographic and cone-beam breast CT measurements in 33 findings (60%). There was a trend toward greater agreement in the measurements as the lesion size increased. In general, masses were more conspicuous in cone-beam breast CT as the breast anatomy and lesions were displayed on a slice by slice basis, eliminating overlapping with tissue structures from other slices. On the other hand, mammography was better than cone-beam breast CT in depicting the details of calcifications, especially those 1 mm and smaller in size, reflecting the lower resolution of cone-beam breast CT. A questionnaire was administered to all 23 women regarding breast compression and the comfort levels of the imaging procedures. Breast compression was tolerable for all women, and five women were unaware of breast compression. Twenty out of 23 (86.9%) women reported that cone-beam breast CT was as comfortable, or more comfortable, than mammography. For women who reported on discomfort when undergoing cone-beam breast CT, the predominant areas of discomfort were the neck, the shoulders, and the ribs.

5.5.2 Digital Tomosynthesis (Chapter 4)

Breast CT and tomosynthesis are both designed to reduce overlapping structures, leading to fewer false positive findings. In breast CT, multiple 2-D projection images are acquired and then reconstructed to form true three-dimensional (3-D) images, which can be viewed in the transverse, sagittal, and coronal planes. In tomosynthesis, a limited angle scan is used to obtain 30–60 mammograms to reconstruct thick layer tomographic images in parallel to the detector. Tomosynthesis is more similar to digital mammography than breast CT. The similarity to mammographic images will likely lead to a shorter learning curve for interpreting tomosynthesis images as compared to cone-beam breast CT images. As tomosynthesis images are obtained with limited angular sampling, concerns have been raised regarding the out-of-plane artifacts and the ability to visualize calcifications (Karellas and Vedantham 2008, Karellas et al. 2008).

Gong et al. (2006) performed a computer simulation study with simulated lesions embedded into a 3-D breast model. The authors evaluated lesion detection accuracy with digital mammography, tomosynthesis, and breast CT. The same total dose was used in the tomosynthesis and the breast CT simulations. A Receiver Operating Characteristic (ROC) study was performed with five physicist observers. The average area under the ROC curves was 0.76 for digital mammography, 0.93 for tomosynthesis, and 0.94 for breast CT. Both tomosynthesis and cone-beam breast CT provided statistically significant higher lesion detectability than digital mammography. The improved performance of breast CT and tomosynthesis compared to digital mammography may be due to decreased image clutter, which resulted in improved tumor visibility. Lesion detectabilities for tomosynthesis and cone-beam CT were not statistically significantly different from each other. It should also be noted that in the study by Gong et al. (2006), breast compression was used in digital mammography and tomosynthesis but not generally in breast CT. Without compression, higher kVp settings were needed, resulting in lower subjective breast tumor contrast.

5.5.3 Ultrasound (Chapter 15)

When cone-beam breast CT is compared to ultrasound, it appears that both modalities have their own advantages. While breast CT provides high resolution 3-D images of the breast anatomy and any lesions that may be present, ultrasound is quicker and less expensive than cone-beam breast CT. In addition, ultrasound results in no ionizing radiation to the patient, and ultrasound is well tolerated by most patients. Thus, sonography can be a useful adjunct to cone-beam breast CT, especially for identifying and distinguishing cysts from malignant masses and for guiding interventional procedures.

5.5.4 Magnetic Resonance Imaging (Chapter 16)

Breast magnetic resonance imaging (MRI) has been shown to be valuable in identifying mammographically occult malignancies. It is likely that cone-beam breast CT can also demonstrate mammographically occult malignancies (Lindfors et al. 2008). Future studies are needed to compare breast MRI to cone-beam breast CT. It will be important to determine if the identification of small, subtle mammographically occult breast cancers will require the administration of intravenous contrast material (Lindfors et al. 2008, Jong et al. 2003).

5.5.5 Breast-specific Gamma Imaging (Chapter 17)

Breast specific gamma imaging with technetium 99 m sestamibi has received considerable interest over the past several years. While breast specific gamma imaging studies will likely be easier for radiologists to interpret compared to cone-beam breast CT studies, future studies are needed to determine the false negative rates for both studies. It is likely that some subcentimeter malignancies may not be detectable with breast-specific gamma imaging but may be identified with cone-beam breast CT.

5.5.6 Positron Emission Mammography

Positron emission mammography uses planar or curved detector heads to image the breast with mild compression. Resolution is higher (2–3 mm) in planes parallel to the detector, but poorer in planes perpendicular to the detector due to the limited angle image acquisition geometry. While cone-beam breast CT can provide high resolution 3-D images of the breast, positron emission mammography has the advantage of being able to use fluorine-18-fluorodeoxyglucose (18F-FDG) or other tracers to differentiate and identify malignant lesions by their uptake (Bowen et al. 2009).

5.6 Future Work Needed

Breast CT or more specifically cone beam breast CT has undergone some clinical evaluation. The emphasis of these evaluation efforts has been on comparison with mammography and to a lesser extent with digital tomosynthesis. Current clinical prototypes have mostly been optimized for imaging tissue structures and soft tissue masses. Large microcalcifications can be imaged but there seems to be a technical hurdle for imaging small-microcalcifications. At least one company (Koning Corporation, Rochester, NY) has introduced a commercial unit and is in the process of obtaining FDA approval. Other commercial units may be in development. Further technical development may help validate breast CT for traditional imaging tasks of mammography. However, being a true 3-D imaging technique, breast CT may lead to some new applications which, in addition to technical development, also need to be explored and investigated. In this section, several research areas are envisioned for further advances and applications of BCT techniques.

5.6.1 Improved Resolution

One major drawback of current BCT techniques is the lack of resolution to image small microcalcifications that are visible on mammograms. Since microcalcifications are a major indicator for early stage breast cancers (Chapter 7), it is essential to improve the resolution capability of BCT to a level at least comparable to that of mammography in terms of imaging small microcalcifications. In a study by Lindfors et al. (2008), 10 healthy volunteers and 69 women with Breast Imaging Reporting and Data System

(BIRADS) category 4 and 5 lesions underwent screen-film mammography and cone-beam breast CT. While breast CT was significantly better than mammography for visualization of masses ($p = 0.002$), mammography outperformed BCT for visualization of calcifications ($p = 0.006$). Thus, improved spatial resolution and decreased noise will be needed for improved detection of microcalcifications.

5.6.2 Dose Reduction

If cone-beam breast CT is to be used to screen for breast cancer, then the dose should be kept within the mean glandular dose (MGD) limit of 6 mGy for two view screening mammograms. It has been demonstrated that tissue structures may be imaged with BCT with an MGD comparable to the limit set for two view screening mammograms. However, this has been achieved by using pixel binning and compromising on the spatial resolution. With current BCT technology, a higher exposure level is required to image smaller microcalcifications and a higher MGD would be incurred (Lai et al. 2006). Thus, one important area for further development are methods to achieve better image quality and resolution without increasing the exposure level or MGD to the breast. There have been many efforts in using iterative reconstruction algorithms to obtain 3-D images of reasonable quality from projection images acquired at reduced exposure levels (Pan et al. 2008). These algorithms may provide a promising approach to obtaining high quality BCT images with acceptable breast dose to image small microcalcifications. This should be a priority in BCT research in the future.

5.6.3 Contrast-Enhanced CT

As with regular CT, the attenuation difference between normal tissue and cancers is seldom sufficient to directly differentiate the two. Injection of contrast material into the bloodstream would enhance the contrast of cancerous tissues to a visible level and can potentially help detect and visualize cancers. This concept has been widely applied to CT, MRI, and to a lesser extent, digital mammography (Chapters 2 and 16) Jong et al. 2003. Thus, there is a great interest and need to develop and investigate techniques of contrast enhanced BCT. It is likely that the enhancement characteristics in using contrast enhanced cone-beam breast CT to image malignancies will be similar to those seen in contrast-enhanced breast MRI (Lindfors et al. 2008). A greater volume of the contrast agent will be needed for cone-beam breast CT as compared to breast MRI. In addition, the iodine-based x-ray contrast agent used in cone-beam breast CT is associated with a slightly higher risk of adverse reactions as compared to MRI contrast agents, but the vast majority of contrast reactions with iodine-based agents are minor, such as hives (Boone et al. 2006).

5.6.4 Volume-of-Interest Imaging

It is not unusual that mammograms are acquired multiple times throughout the entire course of screening, diagnostic work-up, and staging. Similarly, BCT scans may be repeated as needed. In situations where the cancer location and size are already known, a preselected volume-of-interest (VOI) may be imaged with collimated x-rays in conjunction with a low exposure full field scan. This would allow the volume-of-interest to be imaged with higher image quality without increasing the MGD of the breast. It is certainly not applicable to all imaging applications but a useful option in some specific situations.

Chen et al. (2008) described a VOI cone beam breast CT technique to combine high exposure projections of the VOI, acquired with collimated x-rays, with low exposure full field projections for image reconstruction. This technique may be used to obtain high quality 3-D images of the VOI with an overall breast dose similar to that with the regular cone beam breast CT technique (Lai et al. 2009). Furthermore, the projections of the VOI may be acquired in the non-binning mode or with a separate high resolution detector to enhance the resolution of the reconstructed images of the VOI (Chen et al. 2009).

5.6.5 PET-CT

The main application of BCT, like that of mammography, is to provide high resolution images of the breast anatomy to allow abnormalities to be detected and visualized by examining the morphological features. PET, on the other hand, can directly depict the location and activity of cancerous tumors albeit with rather low resolution and no anatomical information. Thus, a combination of the two would provide the benefits of both modalities. For instance, tumor activity could be depicted in high resolution 3-D breast anatomy through image registration to allow the tumor margins to be more accurately determined for surgery, radiation therapy, or monitoring of chemotherapy. Whole body PET-CT systems have been shown to have increased utility in oncologic imaging as compared to PET alone or CT alone. A dedicated PET-CT breast imaging system has been assembled and successfully used to image women with mammographic findings highly suggestive of malignancy (BI-RADS category 5) (Bowen et al. 2009).

5.6.6 Specimen Imaging

One area in breast imaging that would likely benefit from the use of cone-beam CT is specimen radiography. Specimen radiography may be performed following mastectomies and segmental mastectomies. Specimen radiography is critical following segmental mastectomies to verify that the margins of resection are free of tumor. Currently, specimen radiography is not standardized. At some centers, one view is obtained. In other practices, two orthogonal views are obtained, while at other facilities, the specimens are sliced and then radiographed to provide a crude three-dimensional (3-D) assessment. This is a tedious and labor intensive procedure. Because radiation dose is not an issue in imaging specimens, it is easier to achieve high resolution. With sufficient resolution, cone-beam CT could be used to provide true 3-D images without the need to slice the breast specimen.

5.6.7 Computer-Aided Visualization/ Detection/Diagnosis (Chapter 21)

In using cone-beam breast CT for screening or diagnosis, the radiologists will be presented with large image sets. For instance, to cover a 15 cm diameter, 7.5 cm high breast (assuming a hemiellipsoidal shape) with a voxel size of 150 μm, the volume image set will amount to 500 1000 × 1000 images. However, thin slice images may not be always optimal for detecting large structures or even calcifications. Thus, the images could be processed for more efficient and accurate review. It is also likely that the reader performance will be enhanced by the use of computer-aided detection and/or diagnosis algorithms. In mammography, computer-aided detection has been shown to be beneficial in identifying suspicious masses and calcifications, especially for inexperienced readers. Ideally, computer aided detection/diagnosis algorithms should be developed and used to analyze the entire 3-D data image set and identify suspicious lesions for further review. An alternative approach is to analyze and process the 3-D image set and select and generate images for optimized visualization and review. This may involve the generation of alternative projection views and/or thick layer images to better cover the suspicious lesions or structures.

References

Boone JM et al. 2001. Dedicated breast CT: Radiation dose and image quality evaluation. Radiology 221(3): 657–667.

Boone JM, Shah N, Nelson TR. 2004. A comprehensive analysis of DgN(CT) coefficients for pendant-geometry cone-beam breast computed tomography. Med Phys 31(2): 226–235.

Boone JM et al. 2006. Computed tomography for imaging the breast. J Mammary Gland Biol Neoplasia 11(2): 103–111.

Bowen SL et al. 2009. Initial characterization of a dedicated breast PET/CT scanner during human imaging. J Nucl Med 50(9): 1401–1408.

Chang CH et al. 1977. Computed tomography of the breast. A preliminary report. Radiology 124(3): 827–829.

Chang CH et al. 1978. Computed tomographic evaluation of the breast. AJR Am J Roentgenol 131(3): 459–464.

Chang CH et al. 1980. Computed tomography in detection and diagnosis of breast cancer. Cancer 46(4 Suppl): 939–946.

Chang CH et al. 1982. Computed tomographic mammography using a conventional body scanner. AJR Am J Roentgenol 138(3): 553–558.

Chen B, Ning R. 2002. Cone-beam volume CT breast imaging: feasibility study. Med Phys 29(5): 755–770.

Chen L et al. 2008. Feasibility of volume-of-interest (VOI) scanning technique in cone beam breast CT—A preliminary study. Med Phys 35(8): 3482–3490.

Chen L et al. 2009. Dual resolution cone beam breast CT: A feasibility study. Med Phys 36(9): 4007–4014.

Conover DL et al. 2005. Small animal imaging using a flat panel detector-based cone beam computed tomography (FPD-CBCT) imaging system. Med Imag: Phys Med Imag 5745(Pts 1 and 2): 307–318.

Crotty DJ, McKinley RL, Tornai MP. 2007. Experimental spectral measurements of heavy K-edge filtered beams for x-ray computed mammotomography. Phys Med Biol 52(3): 603–616.

Floyd CE Jr. et al. 2001. Imaging characteristics of an amorphous silicon flat-panel detector for digital chest radiography. Radiology 218(3): 683–688.

Gisvold JJ, Karsell PR, Reese EC. 1977. Clinical evaluation of computerized tomographic mammography. Mayo Clin Proc 52(3): 181–185.

Gisvold JJ, Reese DF, Karsell PR. 1979. Computed tomographic mammography (CTM). AJR Am J Roentgenol 133(6): 1143–1149.

Glick SJ et al. 2007. Evaluating the impact of X-ray spectral shape on image quality in flat-panel CT breast imaging. Med Phys 34(1): 5–24.

Gong X et al. 2006. A computer simulation study comparing lesion detection accuracy with digital mammography, breast tomosynthesis, and cone-beam CT breast imaging. Med Phys 3(4): 1041–1052.

Jong RA et al. 2003. Contrast-enhanced digital mammography: initial clinical experience. Radiology 228(3): 842–850.

Karellas A, Lo JY, Orton CG. 2008. Point/Counterpoint. Cone beam x-ray CT will be superior to digital x-ray tomosynthesis in imaging the breast and delineating cancer. Med Phys 35(2): 409–411.

Karellas A, Vedantham S. 2008. Breast cancer imaging: a perspective for the next decade. Med Phys 35(11): 4878–4897.

Kyriakou Y et al. 2008. Concepts for dose determination in flat-detector CT. Phys Med Biol 53(13): 3551–3566.

Lai C-J et al. 2006. Effects of radiation dose level on calcification visibility in cone beam breast CT: A preliminary study—Art. no. 614233. In Flynn MJHJ, ed. Medical Imaging 2006: Physics of Medical Imaging. Pts. 1–3, pp. 14233–14233.

Lai C-J et al. 2007. Visibility of microcalcification in cone beam breast CT: Effects of x-ray tube voltage and radiation dose. Med Phys 34(7): 2995–3004.

Lai CJ et al. 2009. Reduction in x-ray scatter and radiation dose for volume-of-interest (VOI) cone-beam breast CT-a phantom study. Phys Med Biol 54(21): 6691–709.

Lee SC et al. 2003. A flat-panel detector based micro-CT system: performance evaluation for small-animal imaging. Phys Med Biol 48(24): 4173–4185.

Lindfors KK et al. 2008. Dedicated breast CT: Initial clinical experience. Radiology 246(3): 725–733.

Lindfors KK et al. 2010. Dedicated breast computed tomography: The optimal cross-sectional imaging solution? Radiol Clin North Amer 48(5): 1043-+.

McKinley RL et al. 2005. Initial study of quasi-monochromatic X-ray beam performance for X-ray computed mammotomography. IEEE Trans Nucl Sci 52(5): 1243–1250.

Muller JW, van Waes PF, Koehler PR. 1983. Computed tomography of breast lesions: comparison with x-ray mammography. J Comput Assist Tomogr 7(4): 650–654.

Ning R et al. 2006. A novel cone beam breast CT scanner: Preliminary system evaluation—Art. no. 614211. In Medical Imaging 2006: Physics of Medical Imaging. Pts 1–3.

Ning R et al. 2007. A novel cone beam breast CT scanner: System evaluation. In Medical Imaging 2007: Physics of Medical Imaging. Pts 1–3.

O'Connell A et al. 2010. Cone-beam CT for breast imaging: Radiation dose, breast coverage, and image quality. Amer J Roentgenol 195(2): 496–509.

O'Connor JM et al. 2007. Characterization of a prototype, table-top X-ray CT breast imaging system. In Medical Imaging 2007: Physics of Medical Imaging. Pts 1–3.

O'Connor JM et al. 2008. Using mastectomy specimens to develop breast models for breast tomosynthesis and CT breast imaging - Art. no. 691315. In Medical Imaging 2008: Physics of Medical Imaging. Pts 1–3.

Pan X et al. 2008. Anniversary paper. Development of x-ray computed tomography: The role of medical physics and AAPM from the 1970s to present. Med Phys 35(8): 3728–3739.

Prionas ND et al. 2010. Contrast-enhanced dedicated breast CT: Initial clinical experience. Radiology 256(3): 714–723.

Russo P et al. 2010. Dose distribution in cone-beam breast computed tomography: An experimental phantom study. IEEE Trans Nucl Sci 57(1): 366–374.

Russo P et al. 2010. X-ray cone-beam breast computed tomography: Phantom studies. IEEE Trans Nucl Sci 57(1): 160–172.

Stock M et al. 2009. Image quality and stability of image-guided radiotherapy (IGRT) devices: A comparative study. Radiother Oncol 93(1): 1–7.

Thacker SC, Glick SJ. 2004. Normalized glandular dose (DgN) coefficients for flat-panel CT breast imaging. Phys Med Biol 49(24): 5433–5444.

Tornai MP et al. 2005. Design and development of a fully-3D dedicated x-ray computed mammotomography system. In Medical Imaging 2005: Physics of Medical Imaging. Pts 1 and 2.

Vedantham S et al. 2000. Full breast digital mammography with an amorphous silicon-based flat panel detector: Physical characteristics of a clinical prototype. Med Phys 27(3): 558–567.

von Smekal L et al. 2004. Geometric misalignment and calibration in cone-beam tomography. Med Phys 31(12): 3242–3266.

Weigel M, Vollmar SV, Kalender WA. 2010. Estimations of dose for dedicated breast CT at optimized settings with respect to lesion type and size. RSNA.

Yang WT et al. 2007. Dedicated cone-beam breast CT: Feasibility study with surgical mastectomy specimens. AJR Am J Roentgenol 189(6): 1312–1315.

Ying (Ada) Chen
Southern Illinois University

David J. Getty
Brigham and Women's Hospital

Melissa L. Hill
University of Toronto

Mia K. Markey
The University of Texas at Austin

Xin Qian
*Columbia University
Medical Center*

Chris C. Shaw
*The University of Texas MD
Anderson Cancer Center*

Gary J. Whitman
*The University of Texas MD
Anderson Cancer Center*

Martin J. Yaffe
University of Toronto

6

Comparison of Advanced X-Ray Modalities

6.1 Introduction

The hierarchical model of efficacy, recently reviewed and extended by Gazelle et al. (2011), provides a framework for comparing advanced x-ray modalities in terms of efficacy at several levels: technical, diagnostic accuracy, diagnostic thinking, therapeutic, patient outcome, and societal (see also Chapter 13 for a brief review of the model). However, given the rapidly developing nature of x-ray-based breast imaging, it is not possible to provide definitive comparisons of all five modalities introduced in this section. Thus, our goal in this chapter is to summarize the similarities and differences among advanced x-ray modalities for breast imaging, with an emphasis on what has been determined to date regarding their technical and diagnostic efficacies.

Contrast-enhanced digital mammography (CEDM) (Chapter 2), stereo mammography (Chapter 3), breast tomosynthesis (Chapter 4), and breast CT (Chapter 5) are being developed with the goal of addressing key technical limitations of digital mammography (Chapter 1), which are widely believed to hinder its efficacy for breast cancer detection and diagnosis (Chapter 7). One limitation of digital mammography is that it is an anatomical examination, i.e., it does not provide any functional information regarding the physiological properties of the imaged tissue. The goal of CEDM is to provide functional information regarding blood flow. The motivation is that cancerous tissues are known to exhibit increased blood flow relative to normal tissues. In most typical scenarios, the other advanced modalities discussed in this section (Chapters 3–5) are strictly anatomical examinations, although there has been some research on introducing contrast into these imaging studies as well, as discussed in Chapter 2.

Another fundamental limitation of digital mammography is that it is a 2-D projection examination of a 3-D structure. The shared goal of stereo mammography, breast tomosynthesis, and breast CT is to retain 3-D information regarding the breast structure to reduce confusion in interpretation that can result from overlapping tissues, which can falsely resemble an abnormality or mask a true abnormality. In stereo mammography, two projections are acquired at slightly different angles (e.g., angular rotation of approximately 4°–8°). Special stereo display systems, which typically include some lightweight headgear (e.g., polarized glasses), are used to assist the radiologist in fusing the stereo pairs to achieve the perception of depth. Thus, stereo mammography retains 3-D information despite collecting merely pairs rather than singleton images by capitalizing on the immense capabilities of the human visual system. In comparison, in a breast tomosynthesis examination, a series of low-dose projection images (e.g., approximately 11–49) are collected over a larger, albeit still limited, range of angles (e.g., approximately 15°–50°). Radiologists do not typically view the low-dose projection images from a breast tomosynthesis examination; rather, algorithms are used to reconstruct slices that are displayed for interpretation, oftentimes with a cine mode option. Breast CT refers to a full CT examination of the breast using a dedicated scanner, i.e., hundreds of projections over a full angular range (360°) are collected. As with any CT examination, reconstruction algorithms are used to combine the information from the projections into a stack of slice images for display. It should also be noted that stereo viewing is also being investigated for modalities that collect many projections, and thus, many possible stereo pairs, such as from breast tomosynthesis or breast CT. In addition to viewing stereo pairs of the raw acquisition projections, it is also possible to create stereo pairs of projection images from the reconstructed volume of slices from any desired point of view.

6.2 Dose

It is essential that the knowledge to be gained from an x-ray-based study be commensurate with the risk posed to the patient by exposure to ionizing radiation. This is of particular concern in breast imaging wherein healthy, asymptomatic women undergo imaging for screening purposes and a substantial number of women who are ultimately found to be disease-free are recalled for additional imaging studies. Hence, dose is an important factor to consider in comparing x-ray-based breast imaging modalities. Since dose is such a critical consideration, it is not surprising that all of the advanced modalities reviewed in this section have been designed with a goal of attaining a total dose that is comparable to that of standard mammography. For example, image acquisition protocols have been developed to ensure that the total radiation dose from all images required for CEDM is less than that of a conventional mammogram. Likewise, while early stereo mammography studies utilized stereo pairs with each image acquired at full dose, newly developed stereo mammography systems acquire stereo pairs at doses similar to standard digital mammography. Similarly, current implementations of breast tomosynthesis can provide both a standard mammogram and a tomosynthesis examination while staying within the dose permitted by the U.S. federal guidelines for screening mammography. In general, even the dose for breast CT examinations can be kept within the level of a typical two-view mammographic study. However, there are questions as to whether this dose provides all the diagnostic information desired or if higher doses should be used in order to allow detection of small microcalcifications on breast CT.

6.3 Contrast, Noise, Spatial Resolution, and Artifacts

In evaluating any imaging modality, key technical considerations include contrast, spatial resolution, noise, and artifacts. While these issues are discussed in depth for each modality in the preceding chapters, here we summarize a few points to assist the reader in appreciating the relative technical advantages and disadvantages of different x-ray-based modalities in breast imaging.

There are two broad approaches to CEDM: temporal subtraction and dual-energy imaging. Both approaches to CEDM have strengths and weaknesses; at this time, from a technical perspective, neither technique is clearly superior to the other and both continue to be investigated. For example, in the temporal subtraction approach, motion between the precontrast and the postcontrast acquisitions can result in image artifacts; on the other hand, CEDM by dual-energy imaging is subject to scattered radiation artifact. Of arguably greater interest are the differences between temporal subtraction and dual energy in terms of the imaging set up (e.g., scan time, patient positioning) as discussed in the following section. In considering the resolution of CEDM, what is probably of greatest interest is the comparison of it to other functional modalities for breast imaging; in particular, CEDM provides a key advantage of high resolution relative to breast MRI.

As implied by the name, the purpose of CEDM is to enhance the signal by use of a contrast agent. While results with conventional iodinated contrast agents are encouraging, an important area of future work in CEDM is the development of novel contrast agents that could be targeted to increase specificity.

For stereo mammography, many of the technical concerns pertain to the specialized display needed to facilitate stereo viewing, as opposed to the process of image acquisition. For example, with present technology, stereo display is limited to setups that include a lightweight headgear given the high resolution requirements of mammographic imaging. However, given the current rapid development of 3-D display technologies in the entertainment industry, one can speculate that significant advances in 3-D display will soon be coming to the medical market as well.

Historically, some people questioned whether breast tomosynthesis would prove valuable given that it is not a true 3-D modality in the sense that the spacing between reconstructed slices is coarse compared to the high in-plane resolution. However, clinical trials to date are encouraging that the 3-D information retained by breast tomosynthesis does enable more accurate detection and characterization of breast lesions than standard mammography alone. While breast tomosynthesis is now poised for widespread use with its recent approval by the U.S. FDA in 2011, there is still considerable interest in further optimizing the technology. Improving breast tomosynthesis is challenging, since one must optimize not only image acquisition parameters but also the algorithm used to reconstruct slices from the projection data. The choice of reconstruction algorithm used in breast tomosynthesis is critical since it substantially influences the quality of the image (e.g., presence of artifacts from out-of-plane objects).

A limitation of breast CT is that the contrast, noise, and spatial resolution levels that can currently be achieved at doses comparable to digital mammography are such that smaller microcalcifications are not visualized. This limits the use of breast CT to imaging soft tissue only and prevents breast CT from replacing conventional mammography for screening. Finally, we note that, as is the case for breast tomosynthesis, the design of the reconstruction algorithm used in breast CT is key for ensuring that artifacts are adequately addressed.

6.4 Imaging Setup

Characteristics of the imaging setup (e.g., compression level, scan time) are also pertinent as they can impact the practical role that can be played by a modality in breast imaging. In discussing setup factors for CEDM, it must be noted that there are some points of meaningful differences between the temporal subtraction and dual-energy approaches. For temporal subtraction CEDM, very light compression, considerably less than that of digital mammography, is needed; in comparison, published studies of dual-energy CEDM have employed standard compression levels. Dual-energy CEDM can be readily performed for both the CC and MLO views; on the other hand, temporal subtraction CEDM is typically performed for the CC view only so as to allow the patient the comfort of sitting, given the time required for the examination. Both the temporal subtraction and the dual-energy approaches require a few additional minutes for the injection and circulation of contrast, but the temporal subtraction strategy also requires time to collect approximately 3 to 6 postcontrast images at approximately 1- to 3-min intervals. Moreover, the temporal subtraction approach is limited to unilateral imaging because of the delay that would otherwise be required between successive examinations to allow for washout of contrast agent.

Stereo mammography is performed with standard positioning (CC and MLO) and compression level. In other words, typically two sets of stereo pairs are collected: one for the CC projection and one for the MLO projection. The additional time required to collect stereo pairs, rather than singleton images as in standard digital mammography, is negligible. It is worth noting that stereo imaging would benefit from minimal breast compression, resulting in increased real and perceived depth in the breast volume.

Breast tomosynthesis typically employs standard positioning and compression, i.e., a standard CC digital mammogram plus tomosynthesis CC as well as a standard MLO digital mammogram plus tomosynthesis MLO. Similar to traditional mammography, studies to date suggest that both the CC and the MLO positions are necessary for the most effective use of tomosynthesis. In tomosynthesis imaging, the design goal is to minimize not just scan time but also reconstruction time. Modern systems can reconstruct tomosynthesis slices in a matter of seconds.

Breast CT uses a pendant geometry in which the patient lies on a table with her breast hanging below through an opening in the table. This is the same basic positioning as is typically used for stereotactic biopsy or breast MRI. Compression is not used in breast CT; actually, the standard mammographic compression would result in lower-quality breast CT images. As with any CT examination, it is important to optimize reconstruction time; however, this does not limit the practicality of the modality since with modern computing hardware, efficient algorithms can reconstruct breast CT slices in a matter of minutes.

6.5 Observer Considerations and Decision Support

Modern imaging modalities, including advanced x-ray-based breast imaging modalities, hold great promise for advancing human health, but also present substantial challenges in their continued development and optimal deployment. Obviously, it is essential for both practical and ethical reasons to optimize imaging systems as much as possible in ways that limit the involvement of human participants. This is particularly of concern when developing imaging systems for screening for disease since abnormalities are, fortunately, of relatively low prevalence. Thus, both physical phantoms (Chapter 11) and digital phantoms (Chapter 12) will be of increasing importance to the future of breast imaging.

Recent advances in x-ray-based breast imaging point towards a sometimes-overlooked aspect of medical imaging: the importance of the human observer. For example, consider the fact that studies to date indicate that stereo mammography may have substantial benefits over traditional digital mammography, such as in reducing false positive interpretations. Of interest, readers in these studies reported that they could read the stereo mammogram more rapidly and with greater confidence than they could read a standard digital mammogram. Yet, stereo mammography is not dramatically different from conventional digital mammography from the point of view of image acquisition; the strength of stereo mammography is its synergy with the human visual system. This begs the question of whether stereo might be an effective viewing mode for modalities such as tomosynthesis and CT for which many stereo pairs also happen to be collected. However, it also points to the need to appreciate not only the power of the human visual system, but also its variability; humans do not all have the ability to see in stereo to the same degree. Moreover, modalities such as breast tomosynthesis and breast CT, from which a large amount of information could potentially be presented to the human observer, make apparent the need to optimize image presentation with not only reader accuracy but also reader efficiency in mind. Thus, in order for advanced x-ray-based modalities to reach their full potential in breast cancer care, there must be increased emphasis on considerations of display (Chapter 9) and human perception (Chapter 10). Furthermore, for both practical and ethical reasons, it would be desirable to have more sophisticated observer models (Chapter 13) so as to reduce the reliance on human observer studies.

A new modality that can be created by modifying hardware or software of an existing imaging system will obviously be easier to quickly adopt as compared to one that requires more de novo development. For example, consider the case of breast tomosynthesis; following recent approval by the FDA in 2011, U.S. facilities with Hologic's Selenia® Dimensions® 2-D digital mammography units could upgrade to breast tomosynthesis by a software change only. Similarly for dual-energy CEDM, after recent approval by the FDA in 2011, U.S. facilities with General Electric's Senographe DS® or Senographe Essential® digital mammography equipment could obtain CEDM as an upgrade. In the same manner, observers can more readily work effectively with images from a new modality that are similar to those of some existing, familiar modality. For example, it is widely believed that radiologists will be able to quickly learn to interpret breast tomosynthesis because of the resemblance to traditional mammography images. However, while familiarity gives an edge for early adoption, there is no guarantee that a particular new method that can quickly be adjusted to will necessarily prove to be the most effective in the long run. Moreover, while radiologists' capacity for image interpretation is impressive, it has its limits. For this reason, computer-aided detection (CAD) and related visualization tools have been in use with mammography, film screen, and later digital, for many years (Chapter 14). Advanced x-ray-based modalities bring new needs and opportunities for sophisticated visualization and decision support systems (Chapter 21).

6.6 Potential Impact

Imaging plays key roles in breast cancer screening, diagnosis, intervention (biopsy guidance), and surgical planning. Even aside from the tremendous role of imaging, such as various forms of microscopy, in pathological analysis of biopsy and surgical specimens, modalities more traditionally associated with diagnostic imaging, such as x-ray, are also regularly used for specimen imaging. X-ray mammography, particularly digital mammography as facilities increasingly modernize, is the staple of breast cancer screening. However, diagnostic breast imaging presently employs a wider range of modalities; in addition to digital mammography, ultrasound (Chapter 15) is routinely used and many women also benefit from breast MRI (Chapter 16). Similarly, no one single modality dominates for image-guided procedures and surgical planning (Chapter 8). Thus, as we consider the promise of the advanced x-ray-based methods reviewed in this section, we recognize that they should be viewed as potentially complementary rather than mutually exclusive options. Finally, we must acknowledge the potential for expanded and new roles in breast imaging of imaging modalities that are currently less established in this domain, such as nuclear medicine (Chapter 17), elastography (Chapter 18), electrical impedance (Chapter 19), and optical imaging (Chapter 20).

Of the modalities reviewed in this section, stereo mammography and breast tomosynthesis have the highest potential to revolutionize mammographic screening by enabling interpretations that are at least as accurate as digital mammography at comparable doses. It is not clear which if either will ultimately replace digital mammography as the dominant screening method, although both modalities have that potential. Some stereo mammography systems are designed such that one image of the stereo pair is equivalent to a standard mammogram, which allows the radiologist to view any case in either traditional or stereo modes. Similarly, breast tomosynthesis systems in use can collect both a standard mammogram and a tomosynthesis scan together such that the overall dose is still within the U.S. limits. Moreover, recent efforts at reconstructing the equivalent of a standard mammogram from the set of low-dose tomosynthesis projections are encouraging.

CEDM has the potential to substantially change the diagnostic workup of suspicious breast lesions by providing another route to functional information. In this manner, the emerging role of CEDM is more like the traditional role of breast MRI than that of digital mammography. Likewise, breast CT is more likely to be adopted in the future for diagnostic purposes, rather than screening, given current limitations on visualizing small microcalcifications when the exposure level is limited to keep the breast dose to screening levels. Of course, stereo mammography and breast tomosynthesis could also be useful in a diagnostic setting in addition to screening applications. Preliminary studies suggest that having access to 3-D information in the diagnostic work up, such as from an advanced x-ray-based technique, may enable more accurate assessments, e.g., measurement of tumor size. Moreover, in the future, contrast-enhanced variants of breast tomosynthesis or breast CT could be especially valuable for diagnostic imaging.

All of the advanced x-ray-based modalities reviewed in this section could enable improved or new approaches for image-guided interventions (biopsy) and surgical planning for breast cancer treatment. However, additional technological development may be needed to reach this potential. For example, different algorithms may be needed for reconstruction when a metal object such as a biopsy needle is present in order to avoid artifacts. In addition, there may also be a role for advanced x-ray modalities, such as breast CT, in specimen imaging.

6.7 Summary

This is an exciting time in the development of advanced x-ray-based modalities for breast imaging. Digital mammography, now in its second decade of widespread use (e.g., first approved by the U.S. FDA in 2000), has opened up the doors for development and deployment of advanced technologies. Several of the technologies in this section have recently become available for routine use. A dual-energy form of CEDM from General Electric was introduced in Europe in 2010 and was approved for use in the United States in late 2011. A stereo mammography system from Fujifilm is in use in Europe, and trials are scheduled to begin imminently in the United States (as of late 2011). A breast tomosynthesis system from Hologic was approved for use in the United States in early 2011. However, these achievements should be not be misconstrued as an ending; rather, these landmarks denote a new beginning for breast imaging with considerable research still ahead. There are now many questions regarding the optimal use of advanced x-ray modalities that must be investigated over the next few years. In addition, recent advancements in imaging systems raise a host of new scientific questions concerning display, perception, and modeling of both imaging systems and observers. Moreover, the interpretation challenges raised by these new modalities motivate the design of novel quantitative imaging tools and decision support systems.

Medical imaging in breast cancer care is inherently multimodality; no single technique is sufficient for the biological variability encountered in diagnosis, let alone the differences across stages of care (e.g., screening vs. treatment planning). As we consider what the future may hold, we recognize that the modalities reviewed in this section are complementary to each other with different prospects for improving healthcare. Moreover, there are intriguing opportunities for synergy. For example, CEDM and breast CT are both promising modalities independently, but perhaps a contrast-enhanced version of breast CT would have even greater impact. Similarly, perhaps the lessons learned from stereo mammography will provide the basis for a new display paradigm for breast tomosynthesis.

Reference

Gazelle GS et al. 2011. A framework for assessing the value of diagnostic imaging in the era of comparative effectiveness research. Radiology 261: 692–698.

Image Interpretation

II

7

Breast Cancer Detection and Diagnosis

Margaret Adejolu
King's College Hospital

Gary J. Whitman
The University of Texas MD Anderson Cancer Center

7.1 Epidemiology

Breast cancer is the most common cancer in females globally and is the second most common cancer worldwide after lung cancer. Recent estimates of the incidence and mortality rates of major cancers worldwide by the World Health Organization (WHO) showed that approximately 1.38 million new cases of breast cancer were diagnosed in 2008, accounting for 23% of all female cancers and 10.9% of all new cancers (Ferlay et al. 2010). The National Cancer Institute's (NCI) Surveillance, Epidemiology, and End Results (SEER) Program estimated that one in eight women born in the United States will develop breast cancer at some time in their lives (Altekruse et al. 2010). The age-adjusted incidence rate of breast cancer in the United States is approximately 124/100,000 women per year, and 24/100,000 women die from the disease yearly (Howlander et al. 2011). There has been a steady decrease in breast cancer mortality in the United States over the last two decades. Breast screening is estimated to account for 28% to 65% of the total mortality reduction with adjuvant treatment accounting for the rest of the reduction (Berry et al. 2005).

7.2 Pathology of Breast Cancer

The majority of breast cancers are adenocarcinomas originating in the epithelium of the ducts and lobules of the glandular tissue of the breast, usually in the terminal lobular ductal unit (TLDU). Noncarcinomatous breast cancers are very rare and include tumors that originate in the connective tissues of the breast. Breast cancer is described as invasive when malignant cells breach the basement membrane of ducts or lobules and invade the breast stroma. The majority of invasive breast carcinomas arises from the ductal epithelium and are categorized as invasive ductal carcinoma (IDC) of the usual ("not otherwise specified") type. Invasive lobular carcinoma (ILC) is the second most common histopathological type of breast carcinoma and constitutes approximately 10%–15% of all invasive breast cancers (Lopez et al. 2008). Carcinoma in situ of the breast is a histopathological term used to describe a process in which the malignant epithelial cells are confined within the ductal system or the lobules by an intact basement membrane, with no demonstrable invasion of the adjacent stroma. Ductal carcinoma in situ (DCIS) accounts for 20%–30% of all breast cancers detected at screening mammography. Evidence suggests that approximately 14%–75% of cases of DCIS may progress to invasive carcinoma (Raza et al. 2008). Invasive breast cancer carries a worse prognosis than carcinoma in situ due to its capacity to metastasize to distant locations.

7.3 The Role of Breast Imaging Modalities in the Detection and Diagnosis of Breast Cancer

7.3.1 Mammography

Mammography is the gold standard for screening for breast cancer (Chapter 1). The American Cancer Society (ACS) recommends that women age 40 years and older should have a screening mammogram every year and should continue to do so as long as they are in good health.

A two-view examination of each breast, taken in the mediolateral oblique (MLO) and the craniocaudal (CC) projections, is the standard for screening mammography (Figure 7.1).

Supplemental views tailored to a specific indication may be required to confirm the presence of a mass, to closely examine the morphological features, and to localize a lesion in the breast. For example, tangential views may be used to confirm the superficial location of suspected intradermal microcalcifications. Spot compression views and magnification views may help to differentiate a true architectural distortion from summation shadows.

Diagnostic mammograms are the initial imaging modality of choice for the evaluation of breast symptoms. Diagnostic mammograms usually include the MLO and CC views of each breast with supplemental views as required. Mammograms are not recommended as the initial imaging modality in women younger than 30 years because of the theoretically increased radiation risk due to mammography, the low incidence of breast cancer (less than 1%) in women in this age group, and poor mammographic visualization of benign masses, which are the predominant finding in this group of patients (Parikh 2009, Feig 1990). Breast ultrasound (Chapter 15) is the initial imaging modality of choice for evaluating clinically palpable breast lesions in women younger than 30 years of age. Studies reviewing the mammographic findings of symptomatic young women subsequently proven to have breast cancer found that the abnormality was mammographically visible in 86%–90% of cases (Jeffries 1990, Shaw de Paredes 1990, Meyer 1983). Bilateral mammography is therefore recommended if ultrasound demonstrates an abnormality suspicious for malignancy to exclude multifocal and contralateral lesions.

Digital mammography has gradually replaced conventional film-screen mammography over the last decade. A large study comparing film screen mammography with digital mammography found that the overall diagnostic accuracy of digital and film screen mammography as a means of screening for breast cancer were similar. However, the study found that digital mammography was more accurate in detecting breast cancer in women under the age of 50 years, women with mammographically dense breasts and premenopausal and perimenopausal women (Pisano et al. 2005). These findings, along with other advantages of digital mammography, such as easier access to images; improved ability to use applications such as computer-aided diagnosis (CAD) (Chapter 14); improved means of image transmission, retrieval, and storage; improved ability to enhance and manipulate images to optimize visualization of breast abnormalities; and the use of a lower radiation dose while maintaining diagnostic accuracy have made digital mammography the gold standard for both diagnostic screening and screening mammography.

CAD systems are increasingly being used for screening mammography. Studies have shown a 4.7%–19.5% increase in the detection of cancer when CAD is used with single read mammography (Ko 2006, Morton 2006, Birdwell 2005, Dean 2005). Double reading of screening mammograms instead of interpretations by a single reader has been shown to increase the rate of cancer detection by 4% to 14% in several studies (Blanks 1996). A study by a group in Europe where double reading is standard practice compared the performance of a single reader using CAD to a double reading. The results demonstrated that the cancer detection rates of a single reader using computer-aided detection and of two readers are similar (7.02 per 1000 and 7.06 per 1000, respectively (Gilbert et al. 2008).

Digital breast tomosynthesis (DBT) is a 3-D application of digital mammography in which up to 15 projection images are acquired over a total scan arc between 20° and 40° (Chapter 4).

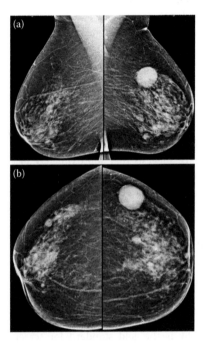

FIGURE 7.1 Standard two-view mammograms. (a) Mediolateral oblique and (b) craniocaudal mammograms are the standard views taken in screening mammography. Note the well-defined, rounded mass in the upper outer quadrant of the left breast, which was shown to represent a cyst on ultrasound.

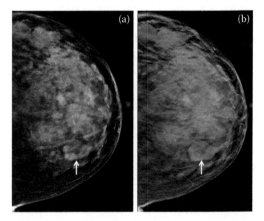

FIGURE 7.2 Digital breast tomosynthesis. (a) Craniocaudal mammogram of the left breast demonstrates a mass with indistinct margins in the inner breast (arrow). (b) Digital breast tomosynthesis shows that the mass (arrow) has sharp margins and a radiolucent halo in keeping with a benign lesion. Ultrasound showed that the mass was a cyst.

DBT allows for 3-D visualization of breast tissues and eliminates the problem of overlapping breast tissue, which may obscure breast lesions (Figure 7.2).

Studies have shown that imaging with DBT has a greater sensitivity for detection of breast cancer when compared with 2-D mammography, particularly in women with dense breasts (Andersson et al. 2008, Rafferty 2007).

7.3.2 Ultrasound

Breast ultrasound has an established role in the characterization of palpable breast abnormalities and in further evaluation of abnormalities identified on other breast imaging modalities. It provides information on the internal matrix, shape, margins, and vascular patterns of breast lesions that help to differentiate benign from malignant masses with a fairly high degree of certainty. Recent applications of breast ultrasound such as breast elastography (Chapter 18), which measures the relative stiffness of breast lesions, have been shown to improve the diagnostic value of breast ultrasound in differentiating benign from malignant masses. Ultrasound is also the most convenient modality for guiding interventional procedures such as core biopsies, fine needle aspirations, and wire localizations (Chapter 8). Ultrasound is the preferred initial imaging technique for evaluating symptomatic breast lesions in women younger than 30 years.

The use of breast ultrasound as a sole screening tool in screening of asymptomatic women is associated with unacceptably high rates of false-positives and false-negatives and is not recommended. There is evidence that addition of ultrasound to mammography in population screening of high-risk women with dense breast tissue increases the detection of breast cancer (Berg 2008, Crystal et al. 2003, Leconte et al. 2003, Kolb et al. 2002). A landmark study, The American College of Radiology Imaging Network National Breast Ultrasound Trial (ACRIN 6666), which compared the performance of screening with ultrasound and mammography versus mammography alone in women with an elevated risk of breast cancer found that a single screening ultrasound examination, added to a screening mammogram, increased detection of breast cancer when compared with mammography alone among women at increased risk for breast cancer who also had dense breast tissue. However, this was accompanied by a substantial increase in the number of false-positives, resulting in additional biopsies, and increased patient anxiety (Berg et al. 2008).

7.3.3 Dynamic Contrast-Enhanced Breast MRI

Dynamic contrast-enhanced MRI (DCE-MRI) of the breast is highly sensitive for the detection of breast cancer (Chapter 16). Breast lesions identified on mammography or ultrasound may appear as foci of enhancement (measuring <5 mm), enhancing masses, or nonmass-like areas of enhancement on DC-MRI of the breast. Malignant breast lesions demonstrate marked contrast enhancement and rapid washout of contrast as a result of

the presence of an increased number of abnormal new vessels that are large and highly permeable. The specificity of breast MRI is improved when both morphologic and kinetic features are included in the evaluation of suspicious lesions. Three types of enhancement kinetic curves have been described in breast lesions following the administration of intravenous contrast material; type 1 curves have a slow rise with persistent enhancement over time, type 2 curves have a slow or rapid initial rise followed by plateau in the delayed phase. Type 3 curves have a rapid enhancement and rapid washout of contrast. Studies have shown there is a higher incidence of malignancy in lesions that display type 2 and type 3 curves, while type 1 curves are strongly associated with benignity (Figure 7.3). Kuhl et al. (1999) showed that the incidence of malignancy in the three types of curves was 8.9% in type 1, 33.6% in type 2, and 57.4% in type 3.

The main role of DCE-MRI in breast cancer imaging is as an adjunct to other imaging techniques. DCE-MRI of the breast helps to characterize breast lesions when other modalities are inconclusive. Also, MRI can be used in to guide biopsies of breast lesions that are not clearly visible on ultrasound or mammography (Chapter 8). Other indications for performing DCE-MRI of the breast include the diagnosis of recurrent breast cancer when conventional imaging in modalities are inconclusive, identification of the primary lesion in patients presenting with metastatic disease and/or axillary lymphadenopathy where the breast lesion is occult on other modalities, determining the extent of disease in newly diagnosed breast cancer, evaluation of residual

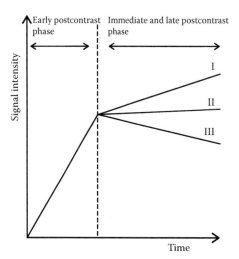

FIGURE 7.3 Time-signal intensity MRI enhancement curves. Time-signal intensity MRI enhancement curves consist of an initial upslope occurring in the first 2 minutes, which may be slow, medium, or rapid (the diagram shows a rapid upslope only). The delayed phase of contrast enhancement occurs 2 minutes or more after the injection of contrast, after the initial upslope and may show an increase, plateau, or washout of contrast enhancement. Type 1 curves show an initial slow rise and a continued rise with time. Type 2 curves show a slow or rapid initial rise followed by a plateau in the delayed phase. Type 3 curves show a rapid initial rise, followed by washout in the delayed phase.

disease after surgery, and to monitor response to neoadjuvant chemotherapy (ACR 2008).

Breast screening with yearly DCE-MRI in addition to mammography is recommended for women with more than a 20% lifetime risk of developing breast cancer, including those with a genetic predisposition to breast cancer such as BRCA1 and BRCA2 mutation carriers or individuals with a history of mantle radiation for Hodgkin disease (as MRI has been shown to be the most sensitive means of detecting breast cancer in this group of women) (Saslow et al. 2007, Warner 2004).

7.4 The Breast Imaging Reporting and Data System (BI-RADS)

In the United States, breast abnormalities detected on mammography, ultrasound, and breast MRI are reported and classified using the Breast Imaging Reporting and Data System (BI-RADS). BI-RADS is a lexicon and a quality assurance tool designed by the American College of Radiology (ACR) in order to standardize breast imaging reporting and to facilitate outcome monitoring (ACR 2003). BI-RADS provides a breast imaging lexicon with standardized terminology, a reporting system designed to provide an organized approach to image interpretation and reporting, and an assessment structure and suggested final assessment categories based on morphology and distributions of breast abnormalities. Abnormalities seen on the various breast imaging modalities are classified into one of seven categories; BI-RADS 0 are incomplete assessments in which additional imaging and/ or prior images are needed before a final assessment can be assigned. BI-RADS 1 lesions have normal findings; BI-RADS 2 lesions have benign findings, BI-RADS 3 lesions have probably benign findings. BI-RADS 4 lesions that are suspicious for cancer and require biopsy. BI-RADS 5 lesions are those with a classic imaging features of breast cancer with a very high likelihood of malignancy. BI-RADS 6 lesions are biopsy proven breast cancers.

7.5 Detection and Diagnosis of Breast Cancer

The majority of breast cancers are first identified on screening or diagnostic mammograms. Breast cancer may present mammographically as a mass, a group of microcalcifications, a focal asymmetry, or as an area of architectural distortion. Breast cancer may also be identified by ancillary findings such as skin thickening, nipple inversion, and skin dimpling. Review of previous imaging studies, when available, is essential, as lesions with static mammographic appearances over a period are unlikely to be malignant.

7.5.1 Masses

7.5.1.1 Mammography

Masses are the predominant mammographic finding in the majority of invasive breast cancers (Newstead et al. 1992) and are

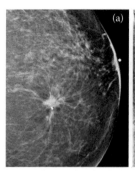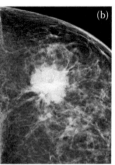

FIGURE 7.4 Malignant lesions on mammography. (a) Histologically confirmed invasive ductal carcinoma appears as a small mass with long spicules extending from it. A clip marker is noted in the masss. (b) Histologically confirmed invasive ductal carcinoma appears as a mammographically dense mass with irregular margins and an internal clip marker.

the main mammographic feature of approximately 10% of DCIS cases (Raza et al.). A mass is defined in the BI-RADS mammography lexicon as a 3-D structure demonstrating convex outward borders that is usually evident on two orthogonal views (ACR 2003). Detailed analysis of the morphology of masses detected on mammography, usually with the aid of supplemental views, such as spot compression and magnification views, is required to determine the likelihood of malignancy and the need for biopsy. The margins of a lesion, along with the density, size, and shape, help to predict the histology or the likelihood of being benign or malignant.

The margin of a breast mass is the most important factor in determining its significance. Margins can be described as circumscribed, obscured, microlobulated, ill-defined, or spiculated. Liberman (1994) found that spiculated margins and irregular shapes were associated with a high positive predictive value (PPV) for malignancy (81% and 73%, respectively) (Liberman 1994) (Figure 7.4). Histopathogical correlations show that spiculated margins are indicative of invasion of surrounding tissues or adjacent desmoplastic reaction secondary to a malignant process. An irregular shape indicates nonuniformity of growth often seen in carcinomas.

Cancers also tend to be greater in mammographic density when compared to benign lesions such as cysts or fibroadenomas.

7.5.1.2 Ultrasound

The ACR BI-RADS ultrasound lexicon describes six morphologic features of solid breast masses: shape, orientation, margin, lesion boundary, internal echo pattern, and posterior acoustic features. Analysis of these morphological features can help to classify lesions into cysts and solid lesions and help to predict the likelihood of malignancy (Figure 7.5).

Similar to mammography, the margin of the lesion is the most important factor in predicting the nature of a mass on ultrasound. The margin of a lesion may be described as circumscribed when it is well defined with an abrupt transition between the lesion and the surrounding tissue. The presence

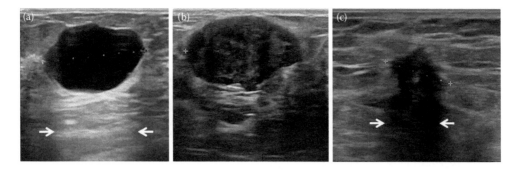

FIGURE 7.5 Characterization of breast masses with ultrasound. (a) Breast cysts typically appear as anechoic lesions (with a lack of internal echoes) with thin walls, sharp margins, and posterior acoustic enhancement (arrows). (b) The typical sonographic appearance of a benign solid mass such as a fibroadenoma is an oval circumscribed hypoechoic mass with a solid internal matrix and with an orientation parallel to the chest wall. (c) Malignant lesions typically appear as irregular, hypoechoic masses that have a nonparallel orientation relative to the chest wall and posterior acoustic shadowing (arrows).

of a circumscribed margin is associated with a high PPV (90%) for benignity. Malignant breast lesions are more likely to have noncircumscribed margins, which may be spiculated, angular, microlobulated, or indistinct. The presence of a spiculated margin suggests infiltrating growth of the lesion into the surrounding tissue and is associated with a PPV of 86% for malignancy (Hong et al. 2005). A study by Paulinelli et al. (2005) showed that a lesion with a noncircumscribed margin is 17 times more likely to be malignant than benign.

The orientation of a lesion on ultrasound is another important predictor of malignancy. A lesion that is taller than it is wide (nonparallel orientation) is suggestive of spread of the lesion through tissue plane boundaries and is more commonly seen in malignant than in benign lesions. Benign lesions grow along tissue planes and are usually wider than they are tall (parallel orientation) (Figure 7.5).

Breast lesions may demonstrate posterior acoustic shadowing, posterior acoustic enhancement, a combination of these two (complex) processes, or no posterior effects. Posterior acoustic shadowing is a suspicious finding, associated with invasive cancer, but it may also be seen with benign lesions such as with postoperative scars, complex sclerosing lesions, macrocalcifications, and in patients with dense breast tissue. Posterior acoustic enhancement is an indeterminate sonographic finding that can be associated with a variety of entities, including normal anatomic structures, simple cysts, complicated cysts, fibroadenomas, nodular sclerosing adenosis, papillomas, complex cystic masses, invasive ductal carcinomas, and lymphoma (Raza et al. 2010).

The internal echo pattern of a lesion relative to that of the subcutaneous fat within the breast should also be evaluated. Lesions that are hyperechoic are almost always, but not invariably, benign. The majority of malignant and benign solid breast masses are hypoechoic. Therefore, other features, such as margin characteristics, are required to establish the level of suspicion when hypoechoic breast lesions are identified. Masses with irregular shapes on ultrasound are more likely to be malignant while oval shaped lesions have a high PPV for benignity.

Abnormalities of the surrounding tissues such as edema, architectural distortion, or thickening of Cooper's ligaments have high PPVs for malignancy. Hong et al. 2005 showed that a lesion that was associated with thickening of the Cooper's ligaments is likely to be malignant.

Breast cancer may occasionally have benign features on ultrasound and mammography (Figure 7.6). Fourteen of 64 invasive cancers (22%) in one study appeared as fibroadenoma-like lesions that were round or oval, that had posterior acoustic enhancement or an indifferent posterior acoustic pattern. Four of these fibroadenoma-like tumors appeared almost anechoic, such that they could be mistaken for cysts (Schrading 2008).

Histological subtypes of cancers that have a tendency to have circumscribed margins include high-grade (grade III) invasive ductal carcinomas, intracystic papillary carcinomas, invasive papillary carcinomas, mucinous carcinomas, medullary carcinomas, metaplastic carcinomas, and malignant phyllodes tumors. Compliance with the BI-RADS guidelines ensures

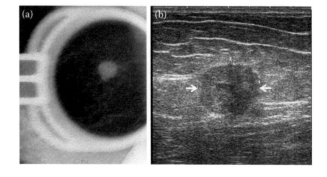

FIGURE 7.6 Malignant breast lesion with "benign" features. (a) Magnified mammography demonstrates a mass with circumscribed margins. (b) Ultrasound demonstrates a hypoechoic mass (arrows) with lobulated margins that has features similar to a fibroadenoma. Ultrasound-guided biopsy showed grade 3 invasive ductal carcinoma. Detailed inspection of the margins of the lesion on magnification mammography and ultrasound showed some irregularity.

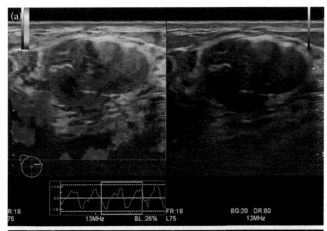

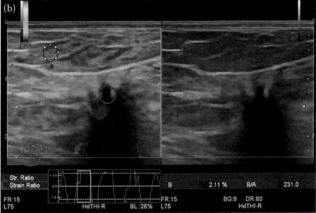

FIGURE 7.8 Ultrasound-guided core needle biopsy. Postfire images show a 14-gauge core biopsy needle within a breast mass.

FIGURE 7.7 (See color insert.) Elastography. (a) Ultrasound strain and B-mode images on split-screen mode show a histologically proven fibroadenoma. (Right) B-mode image shows a well defined hypoechoic lesion. (Left) Elastogram shows a mosaic pattern consisting predominantly of green areas (high strain) and blue areas (absence of strain), typical for a fibroadenoma. (b) Ultrasound strain and B-mode images on split-screen mode show a breast mass with a histological diagnosis of invasive ductal carcinoma. (Right) B-mode image shows a hypoechoic lesion with angular margins. (Left) Elastogram shows absence of strain information consistent with a "stiff" lesion such as a carcinoma.

that all solid lesions except those with characteristically benign ultrasound appearances, such as lipomas, fibroadenomas, and hematomas, do not become cases of missed cancers.

Elastography is a relatively new ultrasound technique that improves the diagnostic value of ultrasound by enabling evaluation of the stiffness of tissue (Chapter 18). Cancers are statistically significantly "stiffer" than benign fibroadenomas and other benign lesions (Figure 7.7).

7.5.1.3 MRI

Breast cancer presenting as a mass on mammography is likely to present as an enhancing mass on DCE-MRI. An enhancing mass on MRI is further characterized by its shape, margin, and type of enhancement (homogeneous, heterogeneous, rim enhancement, dark internal septations, enhancing internal septations, and central enhancement). As in other modalities, the nature

of the margin of a mass is the most predictive morphological feature of a lesion on MRI. Irregular or spiculated margins have a PPV of 84%–91% for malignancy (Nunes et al. 2001). Other features associated with malignancy include heterogeneous internal enhancement and enhancing internal septations. Rim-like enhancement has a PPV of 84% for breast cancer but is an uncommon finding, with a prevalence of 16% (Macura et al. 2006). Enhancement kinetics showing rapid enhancement and rapid washout of contrast (type 3 curves) are the most common type of enhancement kinetics seen in malignancy.

7.5.1.4 Imaging-Guided Biopsy

Most masses seen on mammography will be readily identified and biopsied under ultrasound guidance, which is the quickest, cheapest, and most convenient means of imaging-guided biopsy (Figure 7.8). Lesions that are better seen on mammography or MRI may require stereotactic or MRI-guided biopsy. Fine needle aspiration biopsy (FNAB) may be performed initially, however core biopsy is superior to FNAB in terms of sensitivity, specificity, and correct histological grading of palpable masses and is the recommended method for tissue sampling (ACR 2003).

7.5.2 Microcalcifications

Microcalcifications are common mammographic findings that may result from a variety of benign and malignant processes. Malignant microcalcifications represent dystrophic calcification of necrotic cancer cells filling the ducts and are the main mammographic features in approximately 95% of all DCIS cases (Holland et al. 1994). Studies have shown that 26%–38% of biopsy-proven malignant microcalcification lesions without a mass also contain an associated focus or foci of invasive carcinoma (Stomper et al. 2003). Microcalcifications associated with invasive cancer are almost always within a DCIS component (Kopans 2007). Microcalcifications may also be seen in epithelial proliferations such as atypical ductal hyperplasia (ADH), atypical lobular hyperplasia (ALH), and flat epithelial atypia (FEA), lesions that have been shown to be either nonobligate precursors of breast carcinomas or markers for increased risk of invasive cancer.

7.5.2.1 Mammography

Mammographic assessment of microcalcification involves detailed analysis of the distribution, morphology, density, number, size, and the presence or absence of interval change over a period of time. Magnification mammography is essential in the evaluation of microcalcifications as it provides greater detail of the morphology of individual calcifications and their distribution, compared to standard mammographic views.

7.5.2.2 Distribution

The distribution of microcalcification may help to predict the anatomical location of breast calcifications and also the likelihood of malignancy. Linear or branching calcifications are likely to lie within the ducts and those in a segmental distribution are usually within a duct and its branches. Linear or branching calcification and those in a segmental distribution are highly suspicious for DCIS (Figure 7.9a and b). Single or multiple rounded clusters that represent calcifications within the terminal lobular ductal unit (TLDU) may be benign or malignant. Multiple clusters in both breasts are likely to be benign, while a solitary cluster is more suspicious for malignancy (Figure 7.10). Regionally distributed calcifications that are scattered in a large volume of the breast in a nonsegmental pattern and diffuse calcifications scattered randomly throughout the breast are almost always benign (Kopans 2007).

7.5.2.3 Morphology

Casting-type microcalcifications are typically seen in high-grade DCIS. They are fine fragmented linear or branching casts of the ductal lumen (Figure 7.11) and have a PPV of 96% for malignancy (Tabar 2005).

Granular or pleomorphic microcalcifications are microcalcifications that vary in form, size, and density and are described as having the appearance of crushed stone (Figure 7.12). Granular microcalcifications are malignant in approximately 60% of cases (Tabar 2004). Benign causes of granular microcalcifications include fibrocystic changes, fibroadenomas, papillomas, and fat necrosis.

Powdery or amorphous calcifications are very fine calcifications in which the calcifications are too small to be perceived individually but have a powderish appearance when in a cluster. These microcalcifications may be associated with low grade DCIS, and some are associated with invasive cancer. This pattern of microcalcification is commonly seen in association with sclerosing adenosis.

Punctate microcalcification are round or oval microcalcifications that are uniform in size, shape, and density. The individual foci have well-defined margins and measure 0.5 mm or less. Punctate calcifications are most commonly seen in benign conditions such as fibrocystic changes but may be seen in association

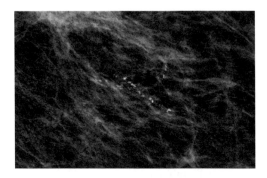

FIGURE 7.9 Linear or branching microcalcifications. Magnified mediolateral mammogram shows pleomorphic calcification in a linear distribution. Stereotactic biopsy showed high-grade DCIS.

FIGURE 7.10 Cluster of microcalcifications. Magnified mediolateral mammogram shows a cluster of pleomorphic calcifications. Stereotactic biopsy showed high-grade DCIS.

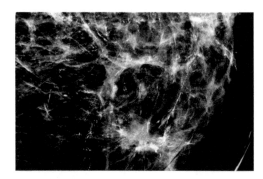

FIGURE 7.11 Casting microcalcifications. Magnified mammogram shows diffuse casting microcalcifications. Stereotactic biopsy showed high-grade DCIS.

FIGURE 7.12 Pleomorphic microcalcifications. Magnified mammogram shows microcalcifications varying in shape and size. Stereotactic core needle biopsy showed high-grade invasive ductal carcinoma.

with other types of microcalcifications in approximately 50% of cases of DCIS and may be the predominant type of calcification in up to 15% of DCIS cases (Kopans 2007).

7.5.2.4 Size

Malignant calcifications are almost always less than 0.5 mm in diameter and are thus referred to as microcalcifications. Coarse calcifications are usually but not invariably benign.

7.5.2.5 Ultrasound

Mammography is the gold standard for detecting and assessing microcalcifications. However, microcalcification can be identified on ultrasound using high frequency transducers as punctate echogenic dots that are <1 mm and do not cast acoustic shadows (Yang et al. 1997) (Figure 7.13). The advantage of detecting microcalcifications on ultrasound is that it offers the option to perform interventional procedures, such as needle biopsy and needle localization under ultrasound guidance, which is a faster and less expensive means of imaging-guided biopsy, compared to stereotactic biopsy or mammographically guided needle localization. Ultrasound may also demonstrate an associated soft tissue mass that is mammographically occult.

7.5.2.6 MRI

Breast MRI is not indicated in the detection and initial assessment of microcalcifications detected on mammography but has a role in the assessment, diagnostic workup, and follow-up of patients with proven DCIS. Malignant calcifications (especially as noted in DCIS) most commonly appear as nonmass-like areas of enhancement on breast MRI. A nonmass-like area of enhancement on breast MRI may be further characterized by the distribution (focal area, linear, ductal, segmental, regional, multiple regions, or diffuse) and pattern of internal enhancement (homogeneous, heterogeneous, stippled, punctate, clumped, reticular, or dendritic).

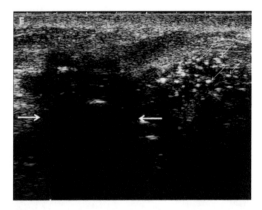

FIGURE 7.13 Ultrasound shows a hypoechoic mass with irregular, indistinct margins (horizontal arrows) and multiple punctuate bright echoes (slanted arrows), representing microcalcifications. Histology showed invasive ductal carcinoma associated with high-grade DCIS.

The most commonly reported MRI manifestation of DCIS is a clumped nonmass-like enhancement in a ductal, linear, segmental, or regional distribution Other manifestations such as a linear-ductal enhancement, segmental enhancement, focal or regional enhancement, diffuse enhancement, or an enhancing mass have also been described (Mossa-Basha et al. 2010). The reported enhancement kinetic curves in DCIS are variable. The plateau enhancement kinetic curve (Kuhl type 2) is more commonly seen in DCIS than washout or progressive enhancement kinetics (Jansen et al. 2007).

7.5.2.7 Image-Guided Biopsy

Vacuum assisted core biopsy using an 11-gauge needle is the method of choice for sampling microcalcifications with stereotactic guidance. Vacuum assisted core biopsy with an 11-gauge device with stereotactic guidance is reliable as it results in a higher frequency of calcification retrieval and hence a lower frequency of histological underestimation in samples with atypical ductal hyperplasia and DCIS, when compared with 14-gauge spring-loaded or 14-gauge vacuum-assisted biopsy devices (O'Flynn 2010). Ultrasound localization of mammographically detected microcalcifications may be attempted in cases with suspicious microcalcifications associated with a focal mass or a focal asymmetry (Comstock et al. 2009).

7.5.3 Architectural Distortion

Breast cancer may manifest as an area of architectural distortion resulting from desmoplastic response of host tissues to the malignancy (Sickles 1984). Architectural distortion may present clinically, as an area of skin dimpling or nipple inversion. The majority of lesions presenting as architectural distortions are benign; however, it is impossible to reliably distinguish between benign and malignant causes on imaging alone. Therefore, these lesions are normally subjected to image-guided biopsy.

7.5.3.1 Mammography

Architectural distortion is the main mammographic feature in 9% of invasive cancers (Sickles 1986) and 7%–13% of cases of DCIS (Raza et al. 2008). On mammography, architectural distortion appears as convergence of tissue spicules toward a focal point without an associated mass (Figure 7.14a). Skin dimpling and nipple inversion may be evident on mammography.

7.5.3.2 Ultrasound

Ultrasound may be performed at the expected location of an area of architectural distortion identified on mammography in an attempt to identify an ultrasound correlate. Ultrasound may show an area of distortion in which convergence of tissues toward a focal point may be perceived when scanning in real time. Mammographic distortions may also appear as irregular hyperechoic mass-like lesions on ultrasound (Figure 7.14b) (Newell et al. 2010).

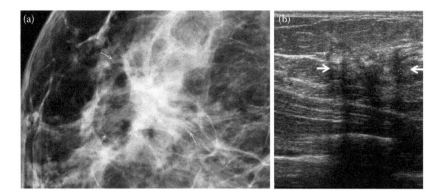

FIGURE 7.14 Architectural distortion. (a) Magnified mammogram shows long linear spicules (arrows), converging toward a focal point. No definite central mass is seen. (b) On ultrasound, the area of architectural distortion corresponds to an ill-defined heterogeneous area of altered echotexture (arrows). Ultrasound-guided biopsy showed invasive ductal carcinoma.

7.5.3.3 MRI

In the majority of cases, a final BI-RADS assessment regarding a mammographic distortion can be made with further mammographic views or ultrasound. MRI has a role in the evaluation of a small proportion of architectural distortions where the mammographic findings are subtle or equivocal and ultrasound is negative or inconsistent with the mammographic finding. In these situations, breast MRI, due to its sensitivity and high negative predictive value (NPV) for the detection of invasive tumors, may confirm the presence of malignancy or provide reassurance that a perceived architectural distortion represents summation artifacts. A distortion caused by malignancy may be seen on MRI as an enhancing mass (Moy et al. 2009) or a nonmass-like area of enhancement (Thomassin-Naggara et al. 2011).

7.5.3.4 Image-Guided Biopsy

Biopsy of an area of architectural distortion should be performed on ultrasound if there is a visible correlate. Stereotactic core biopsy should be performed in the absence of an ultrasound

correlate. MRI-guided biopsy is reserved for the minority of cases where visualization on mammography and ultrasound is suboptimal (Newell et al. 2010).

7.5.4 Focal Asymmetry

The BI-RADS mammography lexicon defines asymmetries as mammographic opacities that are planar, lack convex borders, usually contain interspersed fat and lack the conspicuity of a 3-D mass. Asymmetry involving greater than a quadrant of the breast is termed "global asymmetry" while the term focal asymmetry refers to a smaller area. The majority of focal and global asymmetries represents an islands of fibroglandular breast tissue and are usually secondary to hormonal influences. A small proportion of breast cancers present as focal asymmetries. Therefore, new focal asymmetries must be viewed with suspicion and further evaluation is required to exclude malignancy (Figure 7.15). The presence of a clinically palpable mass, associated architectural distortion, or clusters of calcifications increases the likelihood of malignancy.

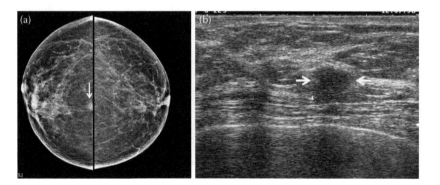

FIGURE 7.15 Focal asymmetry. (a) Craniocaudal mammograms show a focal asymmetry in the right breast (vertical arrow). (b) Ultrasound shows a hypoechoic mass (arrows) with indistinct margins, correlating with the mammographic abnormality. Ultrasound-guided core needle biopsy revealed invasive ductal carcinoma.

A review of the literature shows that malignancy can be found in 0%–14% of asymmetric breast tissue biopsies (Sperber 2007). A study that consisted of developing densities only (focal asymmetries that are new, larger, or of increased density when comparison is made with a previous mammogram) found that a developing density suspicious enough to lead to a recommendation for biopsy had a PPV of 42.9% for malignancy (Leung et al. 2007).

7.5.4.1 Ultrasound

Focal asymmetries do not have a characteristic ultrasound appearance. They may be occult on ultrasound or appear as a solid mass particularly when due to malignancy. Sperber (2007) studied the mammographic and ultrasound appearances of 97 focal asymmetries. There was no ultrasound correlate in 55 (88.5%) of the 61 patients who had ultrasound examinations. Only one of these sonographically occult lesions was malignant on final histology. Seven solid masses were detected sonographically. A mass lesion was identified in seven (11.5%) of 61 ultrasound examinations and two of four masses that appeared to be malignant on ultrasound were malignant on pathology as well. The results of this study suggest that a focal asymmetry with no demonstrable correlate on sonography is likely to be benign. Identification of a solid mass on ultrasound carries a higher probability of malignancy. The absence of a sonographic correlate does not exclude malignancy especially in the setting of a "developing asymmetry." Five (23.8%) of 21 malignant lesions in the study of developing densities by Leung and Sickles (2007) had no ultrasound correlate.

7.5.4.2 MRI

Malignant lesions presenting as asymmetric densities may present as irregular enhancing masses or nonmass-like areas of enhancement.

7.6 Ancillary Mammographic Features Suggestive of Breast Cancer

7.6.1 Axillary Lymph Node Enlargement

Axillary lymph nodes are the main lymphatic drainage of the breast, and enlarged axillary lymph nodes may be the only feature of occult breast cancer on mammography (Figure 7.16a–d). MLO views of the breast provide a limited view of the axilla and the lower axillary lymph nodes. Normal axillary nodes are oval in shape and have lucent centers that represent the fatty hilar regions. Normal or reactive lymph nodes usually measure less than 2 cm in their long axis dimension. Axillary lymph nodes that are of increased mammographic density, enlarged, lacking a lucent center, and round in shape are suspicious for malignant infiltration (Figure 7.16a).

Ultrasound has an established role in the characterization of abnormal lymph nodes. Ultrasound features suggestive of malignant infiltration of lymph nodes are an increase in size (≥2 cm), round shape, focal, or generalized cortical thickening of more than 2 mm, and replacement of the fatty hilum (Figure 7.16b). Malignant lymph nodes may also show increased peripheral blood flow rather than the hilar vascularity seen in normal lymph nodes (Whitman et al. 2011). All patients with suspicious breast lesions should have an ultrasound examination of the axilla with ultrasound-guided FNAB or core biopsy of any suspicious lymph nodes.

7.6.2 Nipple Inversion/Skin Retraction

Nipple and skin retraction may be a secondary sign of malignancy on clinical examination or on mammography. Nipple and/or skin retraction is usually associated with large cancers that are readily seen on mammography.

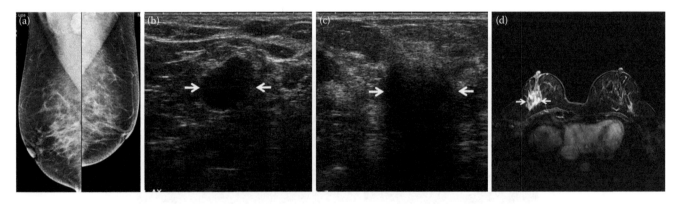

FIGURE 7.16 Breast cancer presenting as right axillary lymph node enlargement on screening mammography. (a) Bilateral mediolateral oblique mammograms show an enlarged lymph node in the right axilla. (b) Ultrasound of the right axilla demonstrates a morphologically abnormal lymph node (arrows) with no demonstrable fatty hilum. (c) Ultrasound reveals a hypoechoic solid mass (arrows) in the breast that proved to be invasive ductal cancer on ultrasound-guided core needle biopsy. (d) DCE-MRI of the breast demonstrates the invasive ductal carcinoma as an irregular enhancing mass (arrows).

7.6.3 Skin Thickening

Generalized skin thickening may be a manifestation of inflammatory carcinoma of the breast, a rare type of cancer that accounts for 1% of breast cancers. Thickening is due to tumor emboli permeating the dermal lymphatics. Skin thickening may also be a late manifestation of noninflammatory breast cancer.

References

Altekruse SF, Kosary CL, Krapcho M et al. (eds). 2010. SEER Cancer Statistics Review, 1975-2007, National Cancer Institute. Bethesda, MD, http://seer.cancer.gov/csr/1975_2007/, based on November 2009 SEER data submission, posted to the SEER web site.

American College of Radiology. 2003. Breast Imaging Reporting and Data System Atlas (BI-RADS Atlas). Reston, VA: American College of Radiology.

Andersson I, Ikeda DM, Zackrisson S, Ruschin M, Svahn T, Timberg P, Tingberg A. 2008. Breast tomosynthesis and digital mammography: A comparison of breast cancer visibility and BIRADS classification in a population of cancers with subtle mammographic findings. Eur Radiol 18: 2817–2825.

Berg WA, Blume JD, Cormack JB et al. 2008. Combined screening with ultrasound and mammography versus mammography alone in women at elevated risk of breast cancer. JAMA 299: 2151–2163.

Berry DA, Cronin KA, Plevritis SK et al. 2005. Effect of screening and adjuvant therapy on mortality from breast cancer. N Engl J Med 353: 1784–1792.

Birdwell RL, Bandodkar P, Ikeda DM. 2005. Computer-aided detection with screening mammography in a university hospital setting. Radiology 236: 451–457.

Blanks RG, Wallis MG, Given-Wilson RM. 1999. Observer variability in cancer detection during routine repeat (incident) mammographic screening in a study of two versus one view mammography. J Med Screen 6: 152–158.

Comstock CH, D'Orsi C, Bassett LW et al. 2009. American College of Radiology. ACR Appropriateness Criteria: breast microcalcifications: initial diagnostic workup. http://www.acr.org/(last accessed 1/08/2011).

Crystal P, Strano SD, Shcharynski S, Koretz M. 2003. Using sonography to screen women with mammographically dense breasts. AJR Am J Roentgenol 181: 177–182.

Dean JC, Ilvento CG. 2006. Improved cancer detection using computer-aided detection with diagnostic and screening mammography: Prospective study of 104 cancers. AJR Am J Roentgenol 187: 20–28.

Feig SA, Ehrlich SM. 1990. Estimation of radiation risk from screening mammography: Recent trends and comparison with expected benefits. Radiology 174: 638–647.

Ferlay J, Shin H-R, Bray F, Forman D, Mathers C, Parkin DM. 2010. Estimates of worldwide burden of cancer in 2008: GLOBOCAN 2008. Int J Cancer 127: 2893–2917.

Gilbert FJ, Astley SM, Gillan MG, Agbaje OF, Wallis MG, James J, Boggis CR, Duffy SW, the CADET II Group. 2008. Single reading with computer-aided detection for screening mammography. N Engl J Med 359: 1675–1684.

Holland R, Hendriks JH. 1994. Microcalcifications associated with ductal carcinoma in situ: mammographic–pathologic correlation. Semin Diagn Pathol 11: 181–192.

Hong AS, Rosen EL, Soo MS, Baker JA. 2005. BI-RADS for sonography: Positive and negative predictive values of sonographic features. AJR Am J Roentgenol 184: 1260–1265.

Howlader N, Noone AM, Krapcho M et al. 2011. SEER Cancer Statistics Review, 1975-2008, National Cancer Institute. Bethesda, MD, http://seer.cancer.gov/csr/1975_2008/, based on November 2010 SEER data submission, posted to the SEER web site.

Jansen SA, Newstead GM, Abe H, Shimauchi A, Schmidt RA, Karczmar GS. 2007. Pure ductal carcinoma in situ: Kinetic and morphologic MR characteristics compared with mammographic appearance and nuclear grade. Radiology 245: 684–691.

Jeffries DO, Adler DD. 1990. Mammographic detection of breast cancer in women under the age of 35. Invest Radiol 25: 67–71.

Ko JM, Nicholas MJ, Mendel JB, Slanetz PJ. 2006. Prospective assessment of computer-aided detection in interpretation of screening mammography. AJR Am J Roentgenol 187: 1483–1491.

Kolb TM, Lichy J, Newhouse JH. 2002. Comparison of the performance of screening, mammography, physical examination and breast US and evaluation of factors that influence them: An analysis of 27,825 patient evaluations. Radiology 225: 165–175.

Kopans DB. 2007. Malignant lesions of the lobule. In: Kopans DB, ed. Breast Imaging. pp. 866–870. Baltimore, MD: Lippincott Williams and Wilkins.

Kuhl CK, Mielcareck P, Klaschik S et al. 1999. Dynamic breast MR imaging: Are signal intensity time course data useful for differential diagnosis of enhancing lesions? Radiology 1999 211: 101–110.

Leconte I, Feger C, Galant C et al. 2003. Mammography and subsequent whole-breast sonography of nonpalpable beast cancers: the importance of radiologic breast density. AJR Am J Roentgenol 180: 1675–1679.

Leung JWT, Sickles EA. 2007. Developing asymmetry identified on mammography: Correlation with imaging outcome and pathologic findings. AJR Am J Roentgenol 188: 667–675.

Liberman L, Evans WP 3rd, Dershaw DD, Hann LE, Deutch BM, Abramson AF, Rosen PP. 1994. Radiography of microcalcifications in stereotaxic mammary core biopsy specimens. Radiology 190: 223–225.

Lopez JK, Bassett LW. 2009. Invasive lobular carcinoma of the breast: spectrum of mammographic, US, and MR imaging findings. Radiographics 29: 165–176.

Macura KJ, Ouwerkerk R, Jacobs M, Bluemke D. 2006. Patterns of enhancement on breast MR images: Interpretation and imaging pitfalls. Radiographics 26: 1719–1734.

Meyer JE, Kopans DB, Oot R. 1983. Breast cancer visualized by mammography in patients under 35. Radiology 147: 93–94.

Mossa-Basha M, Fundaro GM, Shah Biren A et al. 2010. Ductal carcinoma in situ of the breast: MR imaging findings with histopathologic correlation. Radiographics 30: 1673–1687.

Morton MJ, Whaley DH, Brandt KR, Amrami KK. 2006. Screening mammograms: Interpretation with computer-aided detection—Prospective evaluation. Radiology 239: 204–212.

Moy L, Elias K, Patel V et al. 2009. Is breast MRI helpful in the evaluation of inconclusive mammographic findings? AJR Am J Roentgenol 193: 986–993.

Newell MS, Birdwell RL, D'Orsi CJ et al. 2010. ACR Appropriateness Criteria® on nonpalpable mammographic findings (excluding calcifications). J Am Coll Radiol 7: 920–930.

Newstead GM, Baute PB, Toth HK. 1992. Invasive lobular and ductal carcinoma: Mammographic findings and stage at diagnosis. Radiology 184: 623–627.

Nunes LW, Schnall MD, Orel SG. 2001. Update of breast MR imaging architectural interpretation model. Radiology 219: 484–494.

O'Flynn EAM, Wilson ARM, Michell J. 2010. Image-guided breast biopsy: State-of-the-art. Clin Radiol 65: 259–270.

Parikh JR, Bassett LW, Mahoney MC et al. 2009. Expert Panel on Breast Imaging. ACR Appropriateness Criteria® Palpable Breast Masses. Reston, VA: American College of Radiology (ACR). [online publication].

Pisano ED, Gatsonis C, Hendrick E et al. 2005. Diagnostic performance of digital versus film mammography for breast-cancer screening. N Engl J Med 353: 1773–1783.

Rafferty EA. 2006. Breast tomosynthesis. Semin Breast Disease 9: 111–118.

Raza S, Vallejo M, Chikarmane S, Birdwell RL. 2008. Pure ductal carcinoma in situ: A range of MRI features. AJR Am J Roentgenol 191: 689–699.

Raza S, Goldkamp AL, Chikaarmane SA, Birdwell RL. 2010. US breast masses categorized as BI-RADS 3, 4, and 5: pictorial review of factors influencing clinical management. Radiographics 1199–1213.

Saslow D, Boetes C, Burke W et al. for the American Cancer Society Breast Cancer Advisory Group. 2007. American Cancer Society guidelines for breast screening with MRI as an adjunct to mammography. CA Cancer J Clin 57: 75–89.

Sickles EA. 1984. Mammographic features of "early" breast cancer. AJR Am J Roentgenol 143: 461–464.

Sickles EA. 1986. Mammographic features of 300 consecutive nonpalpable breast cancers. AJR Am J Roentgenol 146: 661–663.

Schrading S, Kuhl CK. 2008. Mammographic, US, and MR imaging phenotypes of familial breast cancer. Radiology 246: 58–70.

Shaw de Paredes E, Marsteller LP, Eden BV. 1990. Breast cancers in women 35 years of age and younger: Mammographic findings. Radiology 177: 117–111.

Sperber F, Metser U, Gat A, Shalmon A, Yaal-Hahoshen N. 2007. Focal asymmetric density: mammographic, sonographic and pathological correlation in 97 lesions- a call to restrain biopsies. Isr Med Assoc J 9: 720–723

Stomper PC, Geradts J, Edge SB, Levine, EG. 2003. Mammographic predictors of the presence and size of invasive carcinomas associated with malignant microcalcification lesions without a mass. AJR Am J Roentgenol 181: 1679–1684.

Tabar L, Tot T, Dean P. 2004. Breast Cancer-The Art and Science of Early Detection with Mammography. Perception, Interpretation, Histopathologic Correlation. Stuttgart, Germany: Thieme International.

Tabar L, Tot T, Dean P. 2005. The Art and Science of Early Detection with Mammography. In Breast Cancer. Stuttgart, Germany: Thieme International.

Thomassin-Naggara I, Trop I, Cropier J et al. 2011. Nonmasslike enhancement at breast MR imaging: the added value of mammography and US for lesion categorization. Radiology 261: 69–79.

Warner E, Plewes DB, Hill KA et al. 2004. Surveillance of BRCA1 and BRCA2 mutation carriers with magnetic resonance imaging, ultrasound, mammography, and clinical breast examination. JAMA 292: 1317–1325.

Whitman GJ, Lu TJ, Adejolu M, Krishnamurthy S, Sheppard D. 2011. Lymph node sonography. Ultrasound Clin 6: 369–380.

Yang WT, Seun M, Ahuja A, Metreweli C. 1997. In vivo demonstration of microcalcification in breast cancer using high resolution ultrasound. Br J Radiol 70: 685–690.

Yoo JL, Woo OH, Kim YK et al. 2010. Can MR imaging contribute in characterizing well-circumscribed breast carcinomas? Radiographics 30: 1689–1704.

8

Image-Guided Procedures and Surgical Planning

Nehmat Houssami
University of Sydney

Robin Wilson
Royal Marsden NHS Foundation Trust

Jennifer Rusby
Royal Marsden NHS Foundation Trust

8.1 Image-Guided Procedures in Breast Cancer

8.1.1 Introduction

The integration of image-guided procedures in breast assessment provided the critical step of using imaging to obtain a diagnosis of breast abnormalities before excision (surgical) biopsy, paving the way for the current practice of minimizing unnecessary surgical intervention. Over the past three decades, image-guided biopsy has been a key component of "triple testing" (Irwig et al. 2002) and has enabled clinicians to triage women with breast abnormalities to surgery where appropriate, sparing the majority of women with benign breast findings the morbidity from unnecessary surgery. This has been particularly relevant in the screen-detected setting, but applies equally to clinically detected or symptomatic breast lesions. It was also associated with the development of various image-guided localization devices and techniques for accurate localization of lesions requiring surgical excision. Although advances in the technical aspects of image-guide procedures have been made, largely refining biopsy acquisition methods or enhancing the precision of image-directed localization, the emphasis remains that of improving the management of women with breast abnormalities through:

- Discrimination between benign and malignant breast lesions visualized on imaging through percutaneous (nonsurgical) needle biopsy, allowing appropriate selection or triage to surgical management.
- Enhanced quality of specimens retrieved in image-guided needle biopsy, with a shift from analysis of cellular aspirates (cytologic diagnosis) to various core needle devices that provide progressively larger (and generally more intact) tissue samples allowing histologic analysis.
- Improved accuracy and efficiency of image-directed localization, both preoperative and intraoperative techniques.
- Application of image-guided biopsy for axillary staging.

The standard of care for breast diagnosis (see also Chapter 7) is known as the triple test (or triple assessment) (Irwig et al. 2002)—the combination of clinical examination, breast imaging (usually mammography, ultrasound) (Housami et al. 2002) and needle sampling of a significant abnormality (Irwig et al. 2002). This combination of tests is designed to maximize both the sensitivity and specificity of the diagnostic process (Irwig et al. 2002). A large body of evidence exists on the accuracy and role of needle biopsy as part of triple testing: this mostly related to fine needle aspiration biopsy (FNAB), also referred to as fine needle aspiration cytology (FNAC) as the initial (early) application of image-guided breast needle biopsy (Bulgaresi et al. 2006), but increasingly evidence (and practice) has focused on core needle biopsy (Bruening et al. 2010, Wallis et al. 2009, Ciatto et al. 2007, Kettritz et al. 2004). Core needle biopsy (CNB) for tissue sampling and breast diagnosis is now considered superior to FNAB because CNB is more accurate, with less false negative results (hence higher sensitivity) than FNAB, and has fewer inadequate (insufficient) specimens (Ciatto et al. 2007); CNB also provides relatively more of the diagnostic (histological) information required to initiate treatment planning (Provenzano and Pinder 2009). For malignant breast abnormalities, CNB provides tumor histological type and grade, and hormone receptor status and other biomarkers (Provenzano and Pinder 2009), allowing informed discussion of treatment options before surgery.

Worldwide, breast cancer screening has been implemented for approximately two decades, and the great majority of screen-detected abnormalities are impalpable, requiring sampling (and subsequently localization where surgery is performed) under image guidance. Hence, precise targeting of breast abnormalities is essential, and for this reason image-guided breast biopsy is considered a fundamental part of accurate and reliable diagnosis. The application of image-guided needle biopsy has extended to biopsy of palpable breast lesions where it also enhances diagnostic accuracy relative to manual (clinically performed) biopsy methods (Brennan et al. 2011, Houssami et al. 2005)—see also section on CNB accuracy.

It is now generally accepted that where needle biopsy yields a benign diagnosis, and where this is *concordant* with the imaging appearance, then the woman does not require surgical intervention for screen-detected lesions. If the woman has a symptomatic lump, then a benign needle biopsy diagnosis provides her with the option of avoiding surgical intervention.

8.2 Imaging-Based Biopsy

8.2.1 Breast Biopsy Needle Technology

There are four main types of needles used for breast biopsy—fine needle, automated core biopsy, vacuum-assisted core biopsy (often referred to as "mammotomy"), and radiofrequency large core biopsy. FNAB is the simplest technique and involves passing a fine needle (19–23-gauge) multiple times through the target lesion, usually while applying a small amount of suction to the needle. This technique aims to shear off individual and small clumps of cells that are smeared on microscope slides or washed into saline for staining and cytological assessment. This technique is quick to perform and is usually done without local anesthetic. Although the cytopathologist can provide a result within a few minutes, FNAB is not considered as accurate as core needle sampling (but is an acceptable test where it yields adequate samples)—its main limitation is that it is generally more difficult to obtain representative sampling with FNAB, relative to CNB, and FNAB samples may be more difficult to interpret and require dedicated cytopathological expertise. However, FNAB is ideal for axillary lymph node sampling (Ciatto et al. 2007).

Conventional "automated" CNB is carried out using a spring-loaded needle that is made up of two parts. An inner trocar contains a sampling trough, into which a core of breast tissue is collected, and is covered by a hollow needle with a cutting forward tip. This inner trocar is fired through the target lesion, and the cover needle is then fired over this to cut off and trap the cylinder of breast tissue. Automated CNB usually uses 14-gauge diameter needles, and each core sample weighs around 17 mg. The needle needs to be removed to retrieve each sample and then reinserted to collect further tissue samples; two to five (or more) core samples are usually collected. CNB is more invasive than FNAB and requires injection of local anesthesia and a small incision into the skin. It is the technique used for the vast majority of breast biopsies, and usually requires image guidance as the needle must pass through the target abnormality to obtain a representative sample.

Vacuum-assisted core biopsy (VACB), or vacuum-assisted mammotomy (VAM), was developed to facilitate retrieval of larger volumes of tissue, and provides a directional method of sampling. Suction is applied to pull the targeted breast tissue into a sampling chamber at the needle tip, and then a rotating or oscillating hollow inner cutting trocar cuts the sample, which is then retrieved along the needle, again by suction. Most VACB techniques have the advantage of being able to retrieve multiple large samples (approximately 150 to 300 mg/core) without removing the needle from the breast. Various VACB devices with slight variations in specimen acquisition instrumentation, using various needle sizes (11-gauge to 7-gauge) are available. VACB techniques do not require the needle to pass directly through the target lesion as the suction can be used to pull tissue into the sampling chamber. This means that it is particularly useful for sampling abnormalities that are small and subtle and/or lesion located at a site difficult to target. The technique can also be used to entirely remove the abnormality in highly selected cases (Tennant et al. 2008). Radiofrequency large core biopsy is used to retrieve intact single samples 10 to 20 mm in diameter. The needle uses cutting diathermy (electrically induced heat) to collect the sample, and is a single pass technique.

8.2.2 Guided Biopsy Imaging Methods

Breast biopsy may be guided by ultrasound (Chapter 15), x-ray mammography (Chapter 15), and magnetic resonance imaging (MRI; Chapter 16). Selection of the imaging guidance method depends on the type of abnormality. The majority of breast abnormalities that require needle biopsy are mass forming and are visible on ultrasound and, therefore, this is often the method of choice for guiding biopsy. Ultrasound-guided biopsy is the method of first choice as it is accurate, easy to perform, cost-effective, and, most importantly, associated with minimal patient discomfort. The vast majority of breast biopsies can be carried out under ultrasound guidance. However, a significant proportion of breast abnormalities, particularly those detected through screening mammography, are not visible on ultrasound. The commonest of these abnormalities are microcalcifications and parenchymal deformities. These usually require x-ray-guided (stereotactic) biopsy and represent around 5% to 10% percent of breast biopsies. A very small number of abnormalities, usually detected through screening younger women at high familial risk of breast cancer, are only visible on MRI and require MRI-guided biopsy.

Any abnormality undergoing image-guided biopsy that is likely to require subsequent surgical excision, or may be difficult to find again after the needle biopsy, should be marked. There are a variety of different markers available. Some are simple metallic markers and some incorporate ultrasound visible pellets or gel. The most useful markers are those that are visible on both ultrasound and mammography. If the biopsy has required x-ray (stereotactic) guidance, using an ultrasound visible marker will mean that ultrasound can subsequently be used to localize the lesion for surgery. These markers are also used to mark breast cancers before primary (neoadjuvant) chemotherapy so that the site of the cancer can still be found if the tumor shows an apparent complete response. The best markers for this circumstance are those that remain visible on ultrasound for several months.

8.2.3 X-Ray (Mammography)-Guided Biopsy

X-ray guidance is reserved for those abnormalities that are not palpable and are not clearly visible on ultrasound. Microcalcification is the most common abnormality that requires x-ray-guided biopsy. Two-dimensional techniques where the breast is compressed and the target localized using a fenestrated plate or x and y coordinates have been superseded by stereotactic technology that also allows for accurate calculation of lesion depth in the breast. Stereotactic biopsy can be done either using dedicated prone breast biopsy tables or using add-on devices fitted to a standard upright mammogram machine. It is important that the image receptor for stereotactic biopsy has at least the same resolution as the mammography image that demonstrated the abnormality to make sure the target lesion can be imaged for biopsy. Most manufacturers provide upright stereotactic biopsy devices for their mammogram units, and these use the standard digital receptor to obtain the images.

Both upright and prone stereotactic biopsy devices use the same physics principles to calculate the site of the target and to guide the needle to it. The breast is compressed in the biopsy device in a similar way to conventional mammography. A vertical image is obtained on which there are two fixed markers (fiducials) and the target lesion, which is ideally positioned toward the center of the image. This image is used to calculate the x and y coordinates of the target. Two further images are then obtained with the x-ray source angled 15° on either side of vertical. By comparing the position of the target lesion to the fixed position of the fiducials the principle of parallax is used to calculate the depth (z axis) of the target in the compressed breast. Simple software carries out the calculations for the user and transmits the coordinates to the biopsy device. The needle holder is then moved to the correct coordinates to ensure that the needle will target the lesion. The needle can be placed either vertical to the axis of breast compression or parallel using a lateral arm technique. When the vertical plane is used, it is important that the device is aware of the depth of the lower compression plate and image receptor, so that there is a safety mechanism for ensuring that the needle does not pass through the breast and hit the receptor plate.

Most stereotactic devices use Cartesian coordinates and the needle passes parallel to the image plane. This means that it is very unlikely that the needle will pass too close to the chest wall. Some devices use polar coordinates and the needle is angled toward the target and can be used to biopsy lesions that are close to the chest wall or in the axilla. It is important that devices using polar coordinates include fail-safe systems to prevent chest wall damage.

Stereotactic biopsy is used to target lesions that by their nature are likely to be difficult to target and more likely to have borderline pathology that is more difficult to interpret. For this reason, many advocate the use of vacuum-assisted biopsy techniques rather than simple automated CNB for all stereotactic biopsies.

8.2.4 Ultrasound-Guided Needle Procedures

Current ultrasound units operating at high frequency (7.5–15 MHz) have the capability of visualizing the vast majority of clinically palpable masses and the majority of screen-detected masses, providing an ideal and relatively efficient image guidance method for biopsy, or localization, of breast lesions. Several advantages for using ultrasound-guided biopsy (or intervention) relative to other image guidance have been outlined. In particular, standard ultrasound units with linear array probes provide 2-D images that allow real-time feedback throughout intervention. Real-time visualization of the exact position or proximity of needle placement relative to the target lesion, and the biopsy needle advancement toward the target lesion during biopsy acquisition (or localization) helps ensure that the needle passes through and samples the target lesion. Rotating the probe through 90° after firing the needle will confirm that the needle has passed through the target. These features render ultrasound a user-friendly and efficient image-guidance method. In addition, it is generally more comfortable for the patient than stereotactic intervention and does not involve compression to the breast or radiation. In many breast centers around the world, ultrasound-guided biopsy is performed by radiologists or by breast surgeons or other breast clinicians.

Ultrasound-guided "intervention" includes fine needle aspiration of cystic lesions (or lesions likely to be cysts) as well as drainage of abscesses, needle localization, core biopsy, and needle biopsy of axillary nodes. Conventional automated CNB (usually performed using small hand-held devices) is commonly used for ultrasound-guided biopsy; however, VACB may also be used. In automated CNB, the needle is advanced to a short distance proximal to the lesion, or for larger lesions, near the periphery of the lesion. With VACB, the needle is inserted under ultrasound guidance and placed just deep to the lesion to avoid the needle obscuring the biopsy target. Newer ultrasound-guided interventions that use radiofrequency and cryoablative technologies (including localization, excision, and ablative therapy) are practiced in specialized settings, and are not further discussed in this chapter. Three-dimensional ultrasound can be used to potentially enhance targeting, and some manufacturers offer special targeting technology linked to the ultrasound unit to assist the targeting process. However, 2-D ultrasound imaging is all that is required by most radiologists.

Various techniques can be used to perform ultrasound-guided biopsy; the basic technique is similar whether fine or core needle is used, except that a very small incision is made into the skin for CNB insertion. The most effective approach is to enter the skin at the end of the transducer, at a variable distance (usually between 2 and 20 mm) and angle, which are selected based on the depth of the lesion, and then to advance the needle under the long axis of the transducer. As a general guide, the deeper the lesion in the breast, the steeper the angle of approach and the slightly further the skin entry point should be relative to the edge/end of the ultrasound transducer—the operator should have the ability to make real-time changes in needle angle and direction during the intervention, once the needle has been inserted into the breast.

8.2.5 MRI-Guided Biopsy

Most MRI-detected abnormalities prove to be visible on "second-look" breast ultrasound and can be sampled under ultrasound guidance. However, some abnormalities can only been seen with MRI and require MRI-guided biopsy. The technique used is very similar to the fenestrated plate used for the original x-ray-guided biopsy. The breast is compressed in the lateral plane using a grid plate inserted into a made for purpose breast biopsy MR surface coil. The breast is imaged after injection of contrast and the breast is shown with the biopsy grid superimposed. The patient is withdrawn from the magnet. The window in the biopsy grid that corresponds to the lesion on the MR image is chosen and the depth calculated using the image slice thickness. After injection of local anesthetic, a trocar the same length as the biopsy needle, containing MR visible contrast is placed in the breast. The breast is then imaged again to confirm that this trocar is in the correct position, once confirmed the patient is again withdrawn from the MR magnet, the trocar removed and replaced

by the biopsy needle. Most MR-guided biopsies are carried out using VACB to enhance sampling. The breast is then reimaged to confirm that the target lesion has been sampled. To ensure that the biopsy site can be found again, an MR-compatible metal marker is placed at the biopsy site.

8.3 Accuracy and Underestimation of Core Needle Biopsy

8.3.1 Accuracy of Core Needle Biopsy Relative to Surgical Biopsy

Review of the evidence on comparative effectiveness of CNB methods relative to surgical biopsy (Bruening et al. 2010), concluded that CNB was associated with lower risk of complications than surgical biopsy and that women diagnosed with breast cancer by CNB were more likely to be treated with a single surgical procedure than those whose cancer was diagnosed by open surgical biopsy (Bruening et al. 2010). CNB is recommended for these reasons as well as other advantages relating to establishing histological diagnosis (Provenzano and Pinder 2009) before breast surgical intervention (see also Introduction and section on surgical planning). Estimates of the relative accuracy of CNB methods from this comprehensive review are summarized in Table 8.1 (Bruening et al. 2010). It is clear from these estimates, as well as evidence reported from very large series that have examined CNB accuracy (Ciatto et al. 2007, Kettritz et al. 2004), that image-guided CNB sensitivity approximates that of surgical biopsy. The only relative limitation of CNB is that of *underestimation* (Brennan et al. 2011, Houssami et al. 2007)—this can represent a high proportion of subgroups with specific results (such as CNB diagnoses of DCIS) (Bruening et al. 2010, Ciatto et al. 2007, Kettritz et al. 2004) (Table 8.1). It is important to note

TABLE 8.1 Sensitivity of Various Core Needle Biopsy Methods Relative to Surgical Biopsy

| Biopsy Method | Sensitivity (95% CI) | Underestimation Rate[a] (95% CI) | | Number of Missed Cancers Expected for Every 1000 Biopsies[b] |
		Ductal Carcinoma In Situ	Atypical Ductal Hyperplasia	
Open surgical biopsy	98%–99%[c]	Assumed 0%	Assumed 0%	3–6
Stereotactic CNB				
Automated device	97.8% (95.8–98.9)	24.4% (18.0–32.1)	43.5% (35.7–51.7)	3–13
Vacuum-assisted device	99.2% (97.9–99.7)	13.0% (11.1–15.1)	21.7% (17.7–26.4)	1–6
Ultrasound-guided CNB				
Automated device	97.7% (97.2–98.2)	35.5% (27.1, 45.0)	29.2% (23.4, 35.9)	6–9
Vacuum-assisted device	96.5% (81.2, 99.4)	Not calculable	Not calculable	2–56
Freehand CNB (automated device)	85.8% (75.8–92.1)	Not calculable	Not calculable	24–73

Source: Bruening, W., Fontanarosa, J., Tipton, K., Treadwell, J.R., Launders, J., and Schoelles, K. (2010). *Annals of Internal Medicine* 152: 238–246; supplementary data reported in *Comparative Effectiveness of Core Needle and Open Surgical Biopsy for the Diagnosis of Breast Lesions prepared by ECRI Institute Evidence-based Practice Center*. Rockville, MD: Agency for Healthcare Research and Quality (AHRQ); 2009. Available at http://effectivehealthcare.ahrq.gov (accessed August 2010).

[a] Underestimation rate: Lesions diagnosed by CNB as ductal carcinoma in situ (DCIS) that were found to be invasive by the reference standard were considered as DCIS underestimates. Lesions diagnosed by CNB as atypical ductal hyperplasia (ADH) that were found to be invasive by the reference standard were counted as ADH underestimates. The underestimation rate was calculated as the number of underestimates per number of DCIS (or ADH) diagnoses. In the analysis of sensitivity, underestimates were not considered to be missed cancers because current clinical practice is to recommend surgical removal of ADH and DCIS lesions reported on CNB.

[b] For a population of women with a prevalence of breast cancer of 30% (assumes no false-positives).

[c] Reported to be based on limited evidence.

that freehand CNB has lower sensitivity than image-guided CNB, which may partly reflect lack of image guidance but might also be due to factors associated with the type of lesions selected to freehand sampling (Brennan et al. 2011). Studies have also shown that CNB has generally high specificity (Ciatto et al. 2007).

8.3.2 Ductal Carcinoma In Situ on CNB: Factors Associated with Understaging of Invasive Breast Cancer

Evidence on the factors associated with underestimation of CNB, in the context of CNB diagnosis of DCIS (referred to as a "DCIS underestimate," which means understaged invasive breast cancer), was reported recently in a meta-analysis from Brennan et al. (2011). The focus of this meta-analysis of CNB-based DCIS diagnosis was the identification of preoperative factors associated with higher (and conversely lower) proportion of underestimated cases. Using study level data from 7350 subjects with DCIS only on CNB, in whom outcomes were ascertained with excision histology, which identified 1736 invasive breast cancers, the investigators reported a pooled estimate for CNB-related DCIS underestimates of 25.9% (95% CI 22.5%–29.5%) (Brennan et al. 2011). Several preoperative variables were found to be associated with a significantly higher proportion of underestimation (Table 8.2) (Brennan et al. 2011), including the use of 14-gauge automated CNB (vs. 11-gauge vacuum-assisted device), larger lesions on imaging (lesion size on imaging larger than 20 mm, vs. lesions ≤20 mm, used for analytic purposes), Breast Imaging Reporting

TABLE 8.2 Models Examining Potential Association between Variables and Understaged Invasive Breast Cancer in CNB Diagnosis of Ductal Carcinoma In Situ

Variable Examined for Association with Underestimation% in Each Univariate Model	No. with Invasive Cancer on Excision/No. with DCIS on CNB	Underestimation Proportion % (95% CI)	p Value for Association[a]	Odds Ratio (95% CI)
Palpability				
Impalpable	736/3524	23.3 (19.9–27.2)	0.0009	Referent
Palpable	146/242	54.1 (42.4–65.3)		3.87 (2.53–5.93)
Biopsy device				
14-Gauge automated core needle biopsy	669/2479	30.3 (26.1–34.9)	0.0006	Referent
11-Gauge VAB	369/2285	18.9 (14.9–23.6)		0.54 (0.42–0.69)
Imaging size[b]				
Small (<20 mm)	380/2403	20.1 (12.3–31.0)	<.0001	Referent
Large (>20 mm)	300/953	36.4 (24.7–50.0)		2.28 (1.83–2.85)
Mammographic mass				
Absent	862/4036	23.2 (19.1–27.8)	0.0002	Referent
Present	193/585	35.6 (28.7–43.2)		1.83 (1.45–2.32)
Lesion visibility on ultrasound				
Visible	200/506	40.5 (32.7–48.8)	0.17	Referent
Not visible	9/60	13.6 (0.1–95.8)		0.23 (0–38.12)
BI-RADS classification			(p for trend)	
3	11/92	10.7 (4.4–24.0)	0.005	0.33 (0.13–0.80)
4	155/578	26.8 (18.3–37.5)		Referent
5	76/191	34.9 (22.5–49.7)		1.46 (0.90–2.37)
Guidance method				
Stereotactic	705/3795	21.8 (18.8–25.0)	0.007	Referent
Clinical (not image-guided)	43/92	49.2 (31.6–67.0)		3.48 (1.62–7.46)
Ultrasound	224/576	39.9 (31.4–49.0)		2.39 (1.59–3.59)
MRI (based on 1 study only)	4/17	23.5 (5.2–63.4)		1.11 (0.19–6.31)
Grade (based on CNB)				
High grade	321/1030	32.3 (24.6–41.1)	0.0001	Referent
Nonhigh grade (low/intermediate)	243/1142	21.1 (15.4–28.3)		0.56 (0.45–0.71)

Source: Brennan, M.E., Turner, R.M., Ciatto, S., Marinovich, M.L., French, J.R., Macaskill, P., and Houssami N. (2011). *Radiology* 260: 119–128. Details of statistical models are described in the source meta-analysis from the aforementioned source.

[a] p Values refer to p for association except where stated to be p for trend.

[b] Sensitivity analysis for size threshold did not alter association for this variable.

and Data System (BI-RADS) score of 4 or 5 (vs. BI-RADS score of 3), presence of a mammographic mass (vs. calcification only), image guidance method (higher underestimation for ultrasound-guided CNB relative to stereotactic CNB), and lesion palpability.

The importance of establishing a preoperative diagnosis of breast cancer, and in particular diagnosis of invasive status, was highlighted in a national audit from the United Kingdom: Wallis et al. (2009) showed that reoperation rates were much higher in women who did not have a preoperative tissue diagnosis (relative to those that had), with the highest reoperation rates observed in women without a correctly staged diagnosis of invasive disease. Schiller et al. (2008) examined factors potentially associated with a higher likelihood of having negative margins in therapeutic lumpectomy (in 730 women), including diagnostic and imaging factors. In multivariate analysis, CNB with preoperative cancer diagnosis was significantly associated with a higher likelihood of having negative margins (p < 0.0001); however, there was no association with integration of MRI (Schiller et al. 2008).

8.4 Imaging in Guiding Surgical Planning and Preoperative Staging

8.4.1 Background

Breast conservation is associated with better quality of life, body image, and patient satisfaction than mastectomy (Al-Ghazal et al. 2000), but poor cosmesis after breast conserving surgery is associated with the opposite (Al-Ghazal et al. 1999). Historically, large or central tumors, multifocality, and recurrent disease were viewed as indications for mastectomy. However, the use of neoadjuvant chemotherapy and oncoplastic surgical techniques have added a layer of complexity to that decision that now depends on many factors, such as size and position of the tumor, tumor biology and likely response to treatments, and patient preferences. Extensive multifocality, whether identified by mammography, ultrasound or MRI, is still regarded as an indication for mastectomy. The main focus of this section is the role of imaging in planning surgery for early breast cancer. Imaging is crucial when deciding whether breast conservation is possible and when planning the resection, particularly for impalpable disease. The surgeon requires an accurate measurement of tumor size and a description of tumor location including proximity to the nipple.

The modalities used in breast imaging are rarely exactly concordant in measurement of disease extent. Data from a large randomized controlled trial (RCT) (Turnbull et al. 2010) showed that level of agreement in size of the tumor (within 5 mm) between imaging modalities and histopathology was good and very similar for each of mammography (72.4% agreement), ultrasound (71.5%), and MRI (69.5%) when considering size of the index invasive cancer. However, when size measurement incorporated associated DCIS, ultrasound agreement was lower than for either MRI or mammography. DCIS extending beyond an invasive cancer may be undersized by ultrasound, while mammography will demonstrate the extent of calcified but not uncalcified DCIS. The infiltrative growth pattern of invasive

lobular cancer (ILC) without a local fibrotic reaction may make it difficult to measure by mammography or ultrasound; MRI has been shown to be more accurate in this situation in a retrospective study (Mann et al. 2010) and advised for ILC in consensus-based recommendations (Sardanelli et al. 2010). However, an RCT (COMICE) found that although women with ILC were significantly more likely to have a reoperation than women with other invasive breast cancers, the inclusion of preoperative MRI did not reduce the reoperation rate in the subgroup with ILC (Turnbull et al. 2010). If breast density is such that disease extent cannot be determined by either mammography or ultrasound, or if there is significant discrepancy in size assessment between clinical and radiological modalities, then MRI may be used selectively in women considering breast conservation (Sardanelli et al. 2010).

Cochrane et al. (2003), provided evidence of the importance of tumor size relative to the breast volume when planning breast conservation. They estimated the percentage of the volume of the breast excised and compared this with panel assessment of cosmetic outcome and patients' satisfaction with the appearance of their treated breast. They demonstrated that for lateral tumors, panel assessment and patient satisfaction were favorable for excisions of up to 15% of the breast volume, while for medial tumors, only 5% of the breast could be excised before the cosmetic result was adversely affected (Cochrane et al. 2003). These numbers serve as a guide to surgeons toward identifying patients for whom a standard wide local excision is unlikely to give an acceptable cosmetic result. For these patients, neoadjuvant chemotherapy and oncoplastic surgical techniques should be considered with the aim of extending the role of breast conservation. The role of imaging in monitoring response to neoadjuvant therapy, which may include imaging evaluation of early response and/or measuring residual tumor before surgery, is a complex issue, and is beyond the scope of this chapter—however, several imaging modalities are available for this purpose (ultrasound, MRI, and positron emission tomography [PET]).

8.4.2 Mammography-Based Preoperative Localization

Mammographic screening leads to the diagnosis of small breast cancers that are usually impalpable, and wire localization has been the mainstay of radiological guidance for many years. A wire (with a hook or other device to keep it in situ) is placed through the lesion with the tip lying just beyond the lesion. If the lesion is visible on ultrasound, then this is the preferred method of imaging for wire insertion, for the various reasons that have been outlined earlier in this chapter. However, for lesions that are only visible on mammography, stereolocalization is used. Good communication between radiologist and surgeon is essential, as knowledge of the direction and method of wire placement enables the surgeon to visualize the location of the abnormality in three dimensions and to guide the operative approach. Multiple, or bracketing, wires can be used for larger impalpable lesions to assist the surgeon in performing an adequate resection (Liberman et al. 2001) (Figures 8.1, 8.2, and 8.3). This is

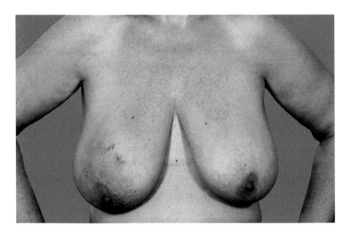

FIGURE 8.1 A woman presenting with a mass in the right breast (bruising from core needle biopsy), before surgical treatment.

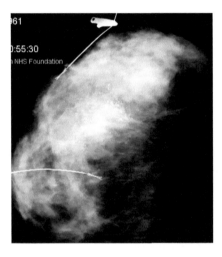

FIGURE 8.2 Wire localization mammogram.

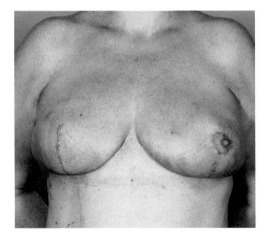

FIGURE 8.3 The same woman 10 days after breast conserving surgery with right therapeutic mammoplasty and symmetrizing left reduction mammoplasty.

particularly useful in women receiving therapeutic mammoplasty where a large volume of breast tissue is excised and extensive remodeling of the remaining breast tissue is undertaken to reshape the breast (Figures 8.1, 8.2, and 8.3), such that reexcision of margins would be complex.

8.4.3 Newer Techniques for Preoperative Localization

Luini et al. (1998), described the first series of patients to undergo radio-guided occult lesion localization (ROLL); 99Tc-labeled colloid was injected preoperatively into the center of the lesion using stereotactic mammography or ultrasound guidance (see also Chapter 17). At operation, a gamma-detecting probe was used to guide the surgeon to the lesion and, after excision, to check the area for residual radioactivity. Similarly, radioactive seed localization involved the placement of a radio-opaque titanium seed containing iodine-125 into the tumor with surgical excision of the lesion guided by a gamma-detecting probe as above. A number of studies of radio-guided localization have been reported, and some of these suggest that this technique may produce lower positive margins rates and fewer reoperations (Lovrics et al. 2011). A systematic review from Lovrics et al. (2011) estimated margin positivity rate and reoperation rate for patients undergoing a therapeutic surgical procedure for breast cancer: the odds ratio for "close or involved" margin status was 0.367 (0.277–0.487) and for reoperation rates was 0.347 (0.25–0.481) in favor of radio-guided localization relative to wire-guided localization (Lovrics et al. 2011). However, there was heterogeneity in the use of isotope, use of specimen radiography and definition of margin involvement, and study sample sizes were generally small limiting definitive conclusions.

8.4.4 Preoperative Staging of the Axilla

Axillary lymph node staging is the single most important prognostic indicator for breast cancer. For decades it has been routine practice to surgically remove axillary lymph nodes for histopathological assessment in all women with invasive disease to determine appropriate treatment options. However, only about one third of women with breast cancer have positive nodes. Therefore, to spare node-negative women unnecessary axillary lymph node clearance (ALNC) and its significant associated morbidity, more limited surgery, in the form of sentinel lymph node biopsy (SLNB), has been developed and is now the operative technique of choice for surgical axillary staging. The surgical management of sentinel node-positive women is evolving and has recently become the subject of some controversy (Giuliano et al. 2011) however, at present, most node-positive women will require completion ALNC.

If axillary metastatic disease can be identified before surgery, node-positive women can avoid SLNB and have ALNC as part of their primary surgery. Clinical examination has limited accuracy and in this era of SLNB, which detects smaller-volume axillary disease, clinical examination is likely to be relatively inaccurate. Several imaging modalities have been used to assess the axilla preoperatively including ultrasound, PET, and MRI (see also Chapters 15–17). Currently, ultrasound (combined with needle biopsy) of axillary nodes is the most commonly used, and has been recommended in some guidelines (Carlson et al. 2009, NICE 2009). Ultrasound examination of the breast can be extended to integrate careful scanning of the axilla in women suspected of having invasive cancer. Ultrasound alone does not provide sufficient accuracy to reliably diagnose node metastases (Houssami et al. 2011), although several node morphologic features can help distinguish nodes likely to contain metastases from nodes that appear enlarged but do not contain metastases (Britton et al. 2009, Houssami et al. 2011). Ultrasound is therefore used to select women with sufficient suspicion on axillary node features to perform ultrasound-guided needle biopsy, using either FNAB or CNB.

Recent meta-analysis (Houssami et al. 2011) reported that ultrasound of axillary nodes (based on 4313 subjects) had a median sensitivity for detection of metastatic nodes of 61.4% (IQR 51.2%–79.4%), and a median specificity of 82.0% (IQR 76.9%–89.0%) (Houssami et al. 2011). In a meta-analysis, pooled estimate for ultrasound-needle biopsy using either FNAB or CNB (2805 needle biopsies in 5981 subjects) excluding insufficient results, had a sensitivity of 79.6% (95% CI 74.1%–84.2%), specificity of 98.3% (95% CI 97.2%–99.0%), and positive predictive value of 97.1% (95% CI 95.2%–98.3%) (Houssami et al. 2011). This work also showed that studies selecting women to axillary needle biopsy based on suspicious node features at ultrasound yielded higher needle biopsy sensitivity (82.2%; 95% CI 76.9%–86.6%) than those selecting to needle biopsy on the basis of visible nodes at ultrasound (70.0%; 95% CI 57.4%–80.1%), $p = 0.037$. Ultrasound sensitivity was shown to be a strong determinant of ultrasound-guided needle biopsy sensitivity, highlighting that ultrasound detection of metastatic nodes needs to be above the (median) 61% ultrasound sensitivity to achieve good accuracy with ultrasound-guided needle biopsy (Houssami et al. 2011).

Of more relevance is confirmation in meta-analysis that ultrasound-guided needle biopsy may be used to triage women to axillary lymph node clearance (ALNC), thus avoiding unnecessary sentinel node biopsy in those with positive needle biopsy results (Houssami et al. 2011). Meta-analysis showed that the median proportion of women triaged directly to ALNC due to positive ultrasound-guided needle biopsy was 19.8% (IQR 11.6%–28.1%) (Houssami et al. 2011). Median proportion of women with metastatic axillary nodes potentially triaged to ALNC was 55.2% (IQR 41.8%–68.2%) based on positive needle biopsy results (Houssami et al. 2011). The median proportion of women with metastatic axillary nodes triaged to ALNC was higher in the subgroup of studies with average tumor size ≥21 mm (Houssami et al. 2011). New imaging techniques, such as contrast-enhanced ultrasound of lymphatics, which may potentially enhance preoperative identification of the sentinel lymph node (SLN), are currently under evaluation.

8.4.5 Surgical Effect of Preoperative MRI of the Breast

The application of preoperative MRI in women with a new breast cancer diagnosis, for local staging of the breast, is an issue of ongoing debate in breast cancer management, with divergent perspectives among experts (Houssami and Solin 2010, Houssami and Hayes 2009, Morrow 2008, Solin et al. 2008, Kuhl et al. 2007). There is consistent evidence from nonrandomized studies that MRI identifies additional disease in the breasts of affected women—this refers to cancer foci occult on conventional imaging in women with an established breast cancer diagnosis (Brennan et al. 2009, Houssami et al. 2008). MRI's additional detection capability, relative to mammography with/out ultrasound, has been summarized as a median proportion of 16% (IQR 11%–24%) for the affected breast (Houssami et al. 2008). Meta-analysis also shows that MRI detects an additional 4.1% (95% CI 2.7%–6.0%) cancers in the contralateral breast (Brennan et al. 2009), which would have also been occult on conventional breast imaging. While there is consensus that MRI has better

TABLE 8.3 Evidence on Surgical Outcomes from Studies Comparing Women Planned for Breast Conservative Surgery and Had Preoperative MRI with Those Who Had Routine Imaging Assessment

Study (First Author)	Surgical Outcome Measured	Did Not Have MRI, No. (%) with Outcome	Had MRI, No. (%) with Outcome	p
Turnbull et al. 2010 (COMICE RCT)	Reoperation/reexcision	156/807 (19%)	153/816 (19%)	0.77
Peters et al. 2011 (MONET RCT)	Reexcision surgery due to positive margins	6/50 (12%)	18/53 (34%)	0.008
Hwang et al. 2009	Reexcision surgery	45/345 (13%)	15/127 (11.8%)	0.5
Bleicher et al. 2009	Positive margins	33/239 (13.8%)	11/51 (21.6%)	0.20
Pengel et al. 2009	Positive margins	35/180 (19.4%)	22/159 (13.8%)	0.17
Schiller et al. 2008	Rate of negative margins	455/553 (82%)	153/177 (86%)	0.21
Mann et al. 2010 (ILC-only)	Reexcision surgery	25/168 (15%)	5/99 (5%)	0.014

Note: RCT, randomized controlled trial (all other studies were nonrandomized); COMICE, Comparative Effectiveness of MRI in Breast Cancer trial; MONET, MR mammography of nonpalpable breast tumors.

sensitivity than conventional imaging for detection of additional disease in affected women, there is little consensus on whether this improves surgical outcomes. From an evidence-based perspective, data have consistently shown that MRI's detection of additional breast cancer foci in newly affected women leads to more extensive breast surgery (Brennan et al. 2009, Houssami et al. 2008).

Meta-analysis has estimated the effect of MRI-only detection on surgical treatment (based on 1908 women from 12 studies) as change to more extensive surgery in 11.3% (95% CI, 6.8–18.3) of women than had been initially planned with conventional imaging assessment; MRI-only detection also caused change to more extensive surgery (mastectomy or wider resection) in 5.5% (95% CI, 3.1–9.5) of women due to false-positive MRI findings (Houssami et al. 2008). The latter is reduced by adopting strict clinical criteria for confirming MRI-detected lesions with CNB. Because RCTs have shown that radical breast surgery in early-stage breast cancer does not confer a survival benefit, MRI-attributable conversion from BCS to more extensive resection or mastectomy may not be beneficial unless it leads to reduced reexcisions, reduced local recurrence, or reduced risk of distant recurrence (Houssami and Hayes 2009): current evidence shows that none of these important clinical end points are reduced from routine application of preoperative MRI in breast cancer. This does not preclude potential benefit from preoperative MRI in highly selected cases, or for use in baseline monitoring in women who will be receiving neoadjuvant therapy.

While initially it was expected that preoperative MRI, through enhanced sensitivity for defining disease extent, would improve surgical precision and reduce reexcision surgeries, this has not been supported by comparative studies that examined surgical outcomes in preoperative MRI—data are summarized in Table 8.3. Six studies (Peters et al. 2011, Mann et al. 2010, Turnbull et al. 2010, Bleicher et al. 2009, Hwang et al. 2009, Pengel et al. 2009, Schiller et al. 2008), including two RCTs, have shown that the inclusion of preoperative MRI in assessment of women planning to receive BCS, relative to those who had conventional imaging only, did not significantly reduce reexcision surgery or the rates of positive tumor margins (Table 8.3). The largest RCT reported the same reexcision rates in both arms of the trial, while the smaller RCT showed that MRI significantly increased the rate of reexcisions due to positive margins (Table 8.3). One retrospective study, restricted to women with ILC, reported that use of MRI in this subgroup was associated with reduced reexcision rates (Mann et al. 2010); however, the RCT of preoperative MRI did not find a reduction in reexcisions in women with ILC. Pre-operative MRI is used in many breast centers around the world, and continues to be an area of applied research to define the role of MRI in preoperative staging.

Acknowledgments

Dr. Houssami dedicates this chapter to the memory of Dr. Stefano Ciatto.

References

Al-Ghazal SK, Fallowfield L, Blamey RW. 1999. Does cosmetic outcome from treatment of primary breast cancer influence psychosocial morbidity? Eur J Surg Oncol 25: 571–573.

Al-Ghazal SK, Fallowfield L, Blamey RW. 2000. Comparison of psychological aspects and patient satisfaction following breast conserving surgery, simple mastectomy and breast reconstruction. Eur J Cancer 36: 1938–1943.

Bleicher RJ, Ciocca RM, Egleston BL, Sesa L, Evers K, Sigurdson ER, Morrow M. 2009. Association of routine pretreatment magnetic resonance imaging with time to surgery, mastectomy rate, and margin status. J Amer Coll Surg 209: 180–187.

Brennan ME, Houssami N, Lord S, Macaskill P, Irwig L, Dixon JM, Warren RM, Ciatto S. 2009. Magnetic resonance imaging screening of the contralateral breast in women with newly diagnosed breast cancer: Systematic review and meta-analysis of incremental cancer detection and impact on surgical management. J Clin Oncol 27: 5640–5649.

Brennan ME, Turner RM, Ciatto S, Marinovich ML, French JR, Macaskill P, Houssami N. 2011. Ductal carcinoma in situ at core-needle biopsy: Meta-analysis of underestimation and predictors of invasive breast cancer. Radiology 260: 119–128.

Britton PD, Goud A, Godward S, Barter S, Freeman A, Gaskarth M, Rajan P et al. 2009. Use of ultrasound-guided axillary node core biopsy in staging of early breast cancer. Eur Radiol 19: 561–569.

Bruening W, Fontanarosa J, Tipton K, Treadwell JR, Launders J, Schoelles K. 2010. Systematic review: Comparative effectiveness of core-needle and open surgical biopsy to diagnose breast lesions. Ann Int Med 152: 238–246.

Bulgaresi P, Cariaggi P, Ciatto S, Houssami N. 2006. Positive predictive value of breast fine needle aspiration cytology (FNAC) in combination with clinical and imaging findings: A series of 2334 subjects with abnormal cytology. Breast Cancer Res Treat 97: 319–321.

Carlson RW, Allred DC, Anderson BO, Burstein HJ, Carter WB, Edge SB, Erban JK et al. 2009. NCCN Breast Cancer Clinical Practice Guidelines Panel. Breast cancer. Clinical practice guidelines in oncology. J Nat Comprehensive Cancer Netw 7: 122–192.

Ciatto S, Brancato B, Risso G, Ambrogetti D, Bulgaresi P, Maddau C, Turco P et al. 2007. Accuracy of fine needle aspiration cytology (FNAC) of axillary lymph nodes as a triage test in breast cancer staging. Breast Cancer Res Treat 103: 85–91.

Ciatto S, Houssami N, Ambrogetti D, Bianchi S, Bonardi R, Brancato B, Catarzi S, Risso GG. 2007. Accuracy and underestimation of malignancy of breast core needle biopsy: The Florence experience of over 4000 consecutive biopsies. Breast Cancer Res Treatm 101: 291–297.

Cochrane RA, Valasiadou P, Wilson AR, Al-Ghazal SK, Macmillan RD. 2003. Cosmesis and satisfaction after breast-conserving surgery correlates with the percentage of breast volume excised. Brit J Surg 90: 1505–1509.

Giuliano AE, Hunt KK, Ballman KV, Beitsch PD, Whitworth PW, Blumencranz PW, Leitch AM et al. 2011. Axillary dissection vs. no axillary dissection in women with invasive breast cancer and sentinel node metastasis: A randomized clinical trial. JAMA 305: 569–575.

Houssami N, Ciatto S, Ambrogetti D, Catarzi S, Risso G, Bonardi R, Irwig L. 2005. Florence–Sydney Breast Biopsy Study: Sensitivity of ultrasound-guided versus freehand fine needle biopsy of palpable breast cancer. Breast Cancer Res Treat 89: 55–59.

Houssami N, Ciatto S, Ellis I, Ambrogetti D. 2007. Underestimation of malignancy of breast core-needle biopsy: Concepts and precise overall and category-specific estimates. Cancer 109: 487–495.

Houssami N, Ciatto S, Macaskill P, Lord SJ, Warren RM, Dixon JM, Irwig L. 2008. Accuracy and surgical impact of magnetic resonance imaging in breast cancer staging: Systematic review and meta-analysis in detection of multifocal and multicentric cancer. J Clin Oncol 26: 3248–3258.

Houssami N, Ciatto S, Turner RM, Cody HS III, Macaskill P. 2011. Preoperative ultrasound-guided needle biopsy of axillary nodes in invasive breast cancer: Meta-analysis of its accuracy and utility in staging the axilla. Ann Surg 254: 243–251.

Houssami N, Hayes DF. 2009. Review of preoperative magnetic resonance imaging (MRI) in breast cancer: Should MRI be performed on all women with newly diagnosed, early stage breast cancer? CA: Cancer J Clin 59: 290–302.

Houssami N, Irwig L, Loy C. 2002. Accuracy of combined breast imaging in young women. Breast 11: 36–40.

Houssami N, Solin LJ. 2010. An appraisal of pre-operative MRI in breast cancer: More effective staging of the breast or much ado about nothing? Maturitas 67: 291–293.

Hwang N, Schiller DE, Crystal P, Maki E, McCready DR. 2009. Magnetic resonance imaging in the planning of initial lumpectomy for invasive breast carcinoma: Its effect on ipsilateral breast tumor recurrence after breast-conservation therapy. Ann Surg Oncol 16: 3000–3009.

Irwig L, Macaskill P, Houssami N. 2002. Evidence relevant to the investigation of breast symptoms: the triple test. Breast 11: 215–220.

Kettritz U, Rotter K, Schreer I, Murauer M, Schulz-Wendtland R, Peter D, Heywang-Kobrunner SH. 2004. Stereotactic vacuum-assisted breast biopsy in 2874 patients: A multicenter study. Cancer 100: 245–251.

Kuhl C, Kuhn W, Braun M, Schild H. 2007. Pre-operative staging of breast cancer with breast MRI: One step forward, two steps back? Breast 16(Suppl 44).

Liberman L, Kaplan J, Van Zee KJ, Morris EA, LaTrenta LR, Abramson AF, Dershaw DD. 2001. Bracketing wires for preoperative breast needle localization. Amer J Roentgenol 177: 565–572.

Lovrics PJ, Cornacchi SD, Vora R, Goldsmith CH, Kahnamoui K. 2011. Systematic review of radioguided surgery for nonpalpable breast cancer. Eur J Surg Oncol 37: 388–397.

Luini A, Zurrida S, Galimberti V, Paganelli G. 1998. Radioguided surgery of occult breast lesions. Eur J Cancer 34: 204–205.

Mann RM, Loo CE, Wobbes T, Bult P, Barentsz JO, Gilhuijs KG, Boetes C. 2010. The impact of preoperative breast MRI on the re-excision rate in invasive lobular carcinoma of the breast. Breast Cancer Res Treat 119: 415–422.

Morrow M. 2008. Magnetic resonance imaging in the breast cancer patient: Curb your enthusiasm. J Clin Oncol 26: 352–353.

National Institute of Clinical Excellence (NICE). 2009. Early and locally advanced breast cancer: Diagnosis and treatment. Developed for NICE by the National Collaborating Center for Cancer. Website: www.nice.org.uk (accessed October 2010).

Pengel KE, Loo CE, Teertstra HJ, Muller SH, Wesseling J, Peterse JL, Bartelink H, Rutgers EJ, Gilhuijs KG. 2009. The impact of preoperative MRI on breast-conserving surgery of invasive cancer: A comparative cohort study. Breast Cancer Res Treat 116: 161–169.

Peters NH, van ES, van den Bosch MA, Storm RK, Plaisier PW, van DT, Diepstraten SC et al. 2011. Preoperative MRI and surgical management in patients with nonpalpable breast cancer: The MONET—Randomised controlled trial. Eur J Cancer 47: 879–886.

Provenzano E, Pinder SE. 2009. Pre-operative diagnosis of breast cancer in screening: Problems and pitfalls. Pathology 41: 3–17.

Sardanelli F, Boetes C, Borisch B, Decker T, Federico M, Gilbert FJ, Helbich T et al. 2010. Magnetic resonance imaging of the breast: Recommendations from the EUSOMA working group. Eur J Cancer 46: 1296–1316.

Schiller DE, Le LW, Cho BC, Youngson BJ, McCready DR. 2008. Factors associated with negative margins of lumpectomy specimen: Potential use in selecting patients for intraoperative radiotherapy. Ann Surg Oncol 15: 833–842.

Solin LJ, Orel SG, Hwang WT, Harris EE, Schnall MD. 2008. Relationship of breast magnetic resonance imaging to outcome after breast-conservation treatment with radiation for women with early-stage invasive breast carcinoma or ductal carcinoma in situ. J Clin Oncol 26: 386–391.

Tennant SL, Evans A, Hamilton LJ, James J, Lee AH, Hodi Z, Ellis IO, Rakha EA, Wilson AR. 2008. Vacuum-assisted excision of breast lesions of uncertain malignant potential (B3)—An alternative to surgery in selected cases. Breast 17: 546–549.

Turnbull L, Brown S, Harvey I, Olivier C, Drew P, Napp V, Hanby A, Brown J. 2010. Comparative effectiveness of MRI in breast cancer (COMICE) trial: A randomised controlled trial. Lancet 375: 563–571.

Wallis MG, Cheung S, Kearins O, Lawrence GM. 2009. Nonoperative diagnosis—Effect on repeat-operation rates in the UK breast screening programme. Eur Radiol 19: 318–323.

9

Displays for Breast Imaging

Aldo Badano
*U.S. Food and
Drug Administration*

9.1 Introduction

A display system needs to represent image data for the human visual system with high fidelity. In particular, in breast imaging, the display characteristics need to ensure that all details and features in the acquired image are conveyed in optimal ways to the human interpreter to help in the accurate diagnosis of the patient. Particularly for mammography, display systems need to be able to convey detail at high spatial frequencies without adding spurious variations that might confuse the reader. At the same time, the display devices need to perform a careful mapping between the values in the image data and the actual luminance levels that constitute the information that the reader uses for the diagnosis. That mapping has to be controlled not only in terms of the calibrated response of the device but also in terms of the variations of the response with angle of view, screen location, time of usage, and for a range of ambient illumination conditions to be found in reading rooms or in rooms where the interpretation of the images take place.

In this chapter, we will review the fundamental aspects of display technology and assessment methodologies that are of relevance for breast imaging systems including major components, technological advances, and emerging challenges. This analysis is presented with a focus on the visual task that is at hand, primarily centered around the essential visual characteristics of breast cancer as seen in a mammographic image including the visualization of mass morphology, the characterization of the distribution in shape and space of microcalcifications, and the rendering of architectural distortions in the breast.

Recent developments in display technology have made possible the introduction of high-quality displays for medical imaging applications based on liquid crystal display (LCD) technology. This now fairly evolved technology can deliver several million pixels for displaying a digital mammogram almost in its entirety with a near 1:1 presentation mode between detector and display pixel. In grayscale displays, general-purpose designs are modified by removal of the color filters to obtain a brighter device. A key advantage of LCDs versus cathode-ray tubes that may prove useful in future breast imaging applications is that, particularly with the availability of brighter backlight technology, by adding color filters to a monochrome design, we obtain a color display without significant degradation in most other areas of performance. However, noise introduced by the variations in LCD panel and by the presence of the subpixels needs to be evaluated and has been, at least to some extent, compensated by noise reduction techniques in some of the products available in the marketplace. These and other techniques to improve image quality become more demanding with large pixel arrays. Current mammographic display devices carry several million pixels with pixel sizes ranging from 100 to 280 μm.

9.2 Performance Characteristics

This section provides a summary description of the most relevant display characteristics that affect image quality in the interpretation of breast images. The relevant parameters are listed in Table 9.1.

9.2.1 Luminance and Contrast

When considering the luminance and contrast characteristics of a display for mammography two aspects need to be considered. The first aspect relates to the consistency in the presentation of images for reviewing breast images using different display devices or workstations. A consistent presentation decreases variability in visual interpretation and diagnostic decisions, and

TABLE 9.1   Key Characteristics of Imaging Displays, Their
Definitions, and Corresponding Relevant References

Parameter	Definition	References
Luminance and contrast	Mapping between pixel intensity in the image to display pixel luminance	Krupinski et al. 2007, Bender et al. 2011, Flynn et al. 1999
Viewing angle	Undesired changes in the luminance mapping due to off-normal viewing	Badano et al. 2003, Fifadara et al. 2004, Badano and Fifadara 2004
Pixel matrix and size	Number and arrangement of display pixels	Siegel et al. 2006
Pixel faults	Malfunctioning pixels	IEC 2010
Spatial resolution	Ability to transfer detail	
Noise	Unwanted variations of luminance	Badano et al. 1999
Veiling glare	Contrast reduction due to scattering processes in the display device	Badano and Flynn 1997, 1998a, 1998b, 2000a, Badano et al. 1999, Badano and Flynn 2000b, Flynn and Badano 1999
Reflections	Added luminance due to ambient illumination sources	IEC 2010
Temporal characteristics	Time-domain resolution of pixel luminance changes in response to stack-mode browsing of images	Liang et al. 2007, Liang and Badano 2007, Liang et al. 2008, Platisa et al. 2011

allows the radiologist to understand how changes in the image data are reflected in variations of luminance in the device screen. To achieve a consistent luminance target response the field has adopted DICOM's Grayscale Standard Display Function (1998) (GSDF), a target model based on Barten's work on perceptually linear visual contrast thresholds (Barten 1989, 1992, 1993, 1999a, 1999b).

It has been reported (Flynn et al. 1999, Siegel et al. 2006, Samei et al. 2005) that, when calibrated to a GSDF model, a luminance ratio (L_{max}/L_{min}) greater than 650 can lead to loss of visibility of image contrast at the edges of the luminance range. The calculation of the luminance ratio should include the contributions from the ambient illumination sources. This immediately points to limitations in the ambient illumination levels. While L_{min} is affected by ambient light, L_{max} is limited by the device output. On the other hand, if the luminance ratio is low (smaller than 250), the range of luminance values to be mapped to image values does not allow for appropriate spacing of the different steps.

A useful way to measure the deviations of the luminance response of a mammographic display with respect to the target model is described in the American Association of Physicists in Medicine (AAPM) Task Group 18 recommendations (Samei et al. 2005). The procedure calls for computing the contrast of each (or a subset of) the gray level steps and by setting application-dependent tolerance levels. For high-quality mammography display, the contrast response of a display should not deviate

from the target values by more than 10%. In practice, this level of accuracy can be achieved with proper calibration but only for a specific viewing direction (Fifadara et al. 2004).

9.2.2 Viewing Angle

Light-modulating devices based on liquid crystals suffer from viewing angle problems. In medical imaging, this effect can be appreciated when viewing different areas of a large display screen (which can reach more than 30 cm on a side). The more severe changes in the luminance presentation curve and in available contrast are likely to happen between the center and the edges or corners of the screen. This effect is exacerbated when more than one subjects are reading an image displayed on one screen, a scenario that happens often at teaching institutions. In this case, the variations can be much larger because of larger angles of viewing.

Display industry laboratories typically report viewing-angle performance with a single metric corresponding to the angle at which the contrast decreases to 10% of the value for normal or perpendicular viewing. This metric does not provide insight into the changes in contrast across the grayscale at different angles. A more useful description (Badano et al. 2003) is the one recommended by the AAPM TG 18 that relies on calculating the normalized contrast ($\Delta L/L$) as a function of the JND index.

Two of the most common methods used to measure the angular response are the goniometric and the conoscopic (or Fourier-optics) methods. In the goniometric approach, a probe is positioned at different angles with respect to the display, and the luminance is measured for a set of gray levels. The goniometric method requires small-spot luminance measurements to be made with a collimated probe. The second method relies on Fourier optics to map luminance intensity to angular luminance using a cooled CCD (Badano et al. 2003). A third approach to measuring viewing angle properties of the display is based on telescopic measurements. In this method, a luminance meter with a narrow acceptance angle is used at a distance from the display and positioned at different viewing angles with respect to the display surface. This approach is not suitable for automatic measurements. Readers interested in the comparison of the conoscopic and goniometric methods with the telescopic method can consult Badano and Fifadara (2004).

In most current mammography systems, the dynamic range of the image acquisition is larger than the dynamic range of the display device in part due to the limited luminance range of current LCD technology and storage and transmission considerations. This is typically compensated by the radiologist by performing some level of time-consuming image manipulation. A compressed or too small luminance range can, on the other hand, lead to the presence of contour artifacts. A combination of 10-bit grayscale resolution with a luminance range greater than 450 seems to be a typical choice (Siegel et al. 2006). With the advent of high-dynamic-range technologies for display devices, and the increased availability of high-dynamic-range detector systems, a high-dynamic-range approach to the display of breast images is under investigation. In this case, the gains of an extended luminance range that accommodates more luminance steps is to be balanced by the potential loss of sensitivity in regions that are away from the adaptation level of the human visual system at any particular instance during the interpretation (Flynn et al. 1999).

How many shades of gray are needed? Evidence seems to suggest that larger than 8-bit palettes are beneficial for image interpretation of breast x-ray images (Krupinski et al. 2007, Bender et al. 2011). Efforts to overcome the current standard driver limitation of 256 gray shades in order to achieve 10- and 11-bit grayscale resolution have so far included subpixel and temporal averaging. The first technique relies on addressing fractional areas of a pixel (corresponding to color subpixel areas) with different pixel driving levels. The second technique is based on rapid modulation of the pixel luminance over many frames. Both approaches can affect other aspects of image quality: the first compromises spatial resolution, the second affects the temporal response of the device.

9.2.3 Pixel Matrix and Size

The matrix size of a breast imaging display systems can often render images in a 1-to-1 presentation, i.e., displaying each image pixel as a display pixel. In most cases, some degree of compression is required, especially with FFDM systems with small pixel size. Although this in principle represents a limitation of the display device, available tools to zoom and pan allow the reader to implement strategies that compensate for this factor. It's clear that the excessive panning needed with small-pixel-array displays, might affect workflow and reader efficiency, and reduce the ability of the reader to contextualize suspicious findings. These requirements have lead to the use of 5-MP devices, although 3-MP displays are being considered as adequate.

9.2.4 Pixel Faults

Pixel faults are display pixels that are malfunctioning and can have a deleterious effect on the visual task of finding small microcalcifications or by masking image details. If a reader knew where the pixel faults were located in the display, he or she could pan the image and see if that feature that was suspicious is indeed a defect. However, this procedure is impractical and pixel faults tend to be problematic when interpreting medical images including and especially, in mammography.

Pixel faults can be evaluated visually or using instrumentation. The visual assessment involves displaying uniform test patters and performing a scan of the image typically using a magnifying glass noting where the faulty pixels are. This procedure requires a trained observer and time to scan the entire display screen consisting of several million pixels. Moreover, this procedure can become more demanding if we consider that pixel faults are of several types. According to the definition in the current International Electrotechnical Commission (IEC) 62563-1 (IEC 2010) a display system can have four types of faulty pixels. Type A faults are defined as subpixels (an addressable part of a grayscale pixel or one of the base color parts of a color pixel) stuck at full luminance state. Type B faults are subpixels stuck at low luminance state. Type C faults are abnormal subpixels not of type A or B, stuck at intermediate states or blinking). Finally, a cluster is defined as a two or more subpixels with faults within a block of 5 × 5 pixels.

9.2.5 Spatial Resolution and Noise

The above recommendation regarding pixel size and pixel array size is related to the spatial resolution capabilities of the monitor. For this purpose, the AAPM TG 18 has recommended that the loss of contrast at high spatial frequencies be at most 35%.

Another significant device characteristic that affects the image interpretation of digital breast images is the unwanted, additional spatial noise that is present in all display devices. Display spatial noise refers to luminance spatial fluctuations that are added by the display system. In general terms, the display noise should be lower than other sources of noise and should not interfere or affect the visual task. However, the frequency content of the display noise and that present in the mammographic image are quite different. Display noise has been shown to be due to two major components: a deterministic component due to pixel design, and a stochastic component due to variations across the screen of the device.

The subpixel structure of LCDs is present in all current designs of medical monochrome and color monitors. It is caused by opaque areas of the pixel layout that correspond to metallic electrodes and thin-film transistor islands. It is possible to increase the aperture ratio of the pixel by reducing the fraction of opaque areas. However, this entails sacrificing other critical performance aspects of the monitor. For example, thinner metal electrodes can generate larger RC delays that can affect the temporal response and pixel grayscale accuracy of high-resolution monitors. We are not aware of any add-on device technology that can achieve a reduction of the subpixel structure. A previous study (Badano et al. 2004) has found that most of the typical monitor noise is due to high-frequency components caused by deterministic subpixel structure.

Recently, an approach to reduce stochastic and pixel-based display noise based on a concept similar to flat-fielding of cameras has been introduced. Display pixels are controlled by thin-film transistors that allow more or less light to be transmitted through the panel, modulating the uniform backlight intensity into a pixel luminance. This same circuitry can be used (through improved gray level-to-luminance tables) to correct for nonuniformities in the screen and reduce unwanted pixel to pixel variations (Badano et al. 2004).

9.2.6 Veiling Glare

Veiling glare has been known to affect different components of the imaging chain including image acquisition (Samei et al. 2005, Seibert et al. 1984, 1985, Zeman et al. 1985), display (Badano and Flynn 1997, 1998a, 1998b, 2000a, 2000b, Badano et al. 1999, Flynn and Badano 1999, Tong, and Prando 1992), and the human visual system (Honda et al. 1993, Paulsson and Sjostrand 1980, Spencer et al. 1995, Stiles 1929). There has been significant research to understand veiling glare effects in image acquisition, though not much is understood at the display and perception level. This is particularly relevant for novel display technologies with a large luminance range (Tisdall et al. 2008). Veiling glare can significantly degrade the quality of an image on different displays as characterized by the veiling glare response function by Flynn and Badano (1999). However, most current breast imaging display systems can achieve a veiling glare ratio of at least 250, which ensures a contrast delivery that is not significantly influenced from bright spots.

9.2.7 Color

Since clinical usage does not require color beyond the monochrome presentation, color in x-ray breast imaging displays is associated with the consistency of the appearance between two monitors in a single workstation or across workstations. This consistency is thought to decrease the variability of mammographic interpretation. Beyond performance, color differences or tints have been associated with discomfort and fatigue.

9.2.8 Reflections and Reading Environment

A critical aspect of display use is related to the level of illumination in the reading room. It is easy to observe that the quality of a displayed image is greatly affected by ambient light. The magnitude of the degradation is associated with the reflection properties of the device, and with the directionality of the light sources and reflectors in the room. In the presence of some diffuse ambient light level (inescapable in practical room designs), the ambient light contribution needs to be taken into account in the luminance calibration of the device, ensuring that contrast in dark areas of the screen is above the luminance level contributed by the ambient light. Otherwise, detail in dark areas is lost.

To minimize this impact, the luminance calibration of the device should always take into account the ambient light reflection, which, according to the TG 18 recommendations, can be achieved by setting the minimum luminance to be at least 2.5 times the luminance caused by the ambient illumination (IEC 2010). In addition, ambient reflections can be minimized by screen designs that rely on antiglare and or antireflection coatings. However, these additional components often affect other aspects of the performance such as spatial resolution and noise.

9.2.9 Temporal Characteristics

The image quality of LCD monitors has been significantly improved over the last few years. However, LCDs have been shown to be inferior to emissive technologies at displaying moving scenes due to their slow response and the hold-type addressing scheme used in their pixel addressing methods (Kurita et al. 2001, Miseli 2004). In breast imaging, the availability of several

modalities where three-dimensional (3-D) data sets are available (i.e., breast CT and breast tomosynthesis), showing a fast sequence of images in stack-mode has been described as a preferred presentation choice. When browsing image data sets in stack mode, slices are displayed in progression and stacked in the display memory buffer. At a given point, only one image is visible on the screen. When a high browsing speed is used, the image is perceived as a 3-D volumetric image instead of a series of 2D images or slices. The refresh rate could reach 30 images per second depending on the application and equipment used (Mathie and Strickland 1997, Reiner et al. 2001).

The slowest intergray level response time of medical LCDs in use can reach 150 milliseconds, with most of the transitions being in the region between 40 and 110 milliseconds. When the refresh rate of stack-mode image reading reaches 30 images/second or higher, the LCD response time is typically far from ideal (Liang et al. 2007, Liang and Badano 2007). The temporal response of LCDs depends on the liquid crystal (LC) cell gap width, on the driving voltage, and on the material properties (i.e., rotational viscosity, dielectric anisotropy, and elastic constants) (den Boer 2005). The mechanical analogy to the slow response is pulling a spring hard, which makes it retract faster when released, illustrating the fact that black-to-white transition times are usually less than inter gray-level transition times for LCDs.

Different techniques can reduce the response time. In addition to the direct design of improved LCD devices by using faster materials and smaller LC gaps, modified addressing schemes including overdrive techniques (Nakamura et al. 2001) have been implemented. These techniques are based on the fact that larger pixel voltages lead to faster transitions between gray levels: if a higher voltage is applied during a fraction of the frame time, the LC molecules move faster and thus the target luminance level might be achieved within one frame time, as desired. However, this technique is not yet used in all medical displays for breast imaging in part because of implementation complexity and added cost.

Liang et al. (2008) described a method for quantifying the effect of slow response of medical LCDs based on the performance of an anthropomorphic model observer for detection in structured backgrounds. In this work, the author relied on an anthropomorphic model for the detection task based on the channelized Hotelling observer (Abbey 2001, Chen et al. 2001, Gallas and Barrett 2003, Park et al. 2010, Platisa et al. 2011) for which significant validation exists, and found that the slow temporal response of displays can degrade the performance of observers in stack-mode readings of volumetric images and showed that the fundamental phenomenon underlying this effect is the reduction of the effective luminance contrast of lesions.

9.3 Routine Checks

The image quality of electronic displays is also susceptible to variations in hardware performance over time. Periodic monitoring of display quality and calibration is essential. Several professional groups including AAPM and ACR have developed guidance documents (although not application specific) for acceptance testing and monitoring of medical imaging displays. The frequency and rigor of the testing is still being discussed in the community. Evidence indicates that, in other areas of the digital imaging department, sporadic and unsystematic quality control leads to significant decreases in performance and ultimately affects diagnostic accuracy and increases inconsistency in the reading. In this sense, new techniques to automate the process of calibration and testing should prove useful for all sizes of healthcare operations where display quality needs to be monitored and maintained.

9.4 Recommendations and Standards

The most up to date guidelines and standard procedures for characterizing display system performance are the TG 18 recommendations (Samei et al. 2005) and the IEC standard on medical image display (IEC 2010). The two documents are for the most part consistent at least in the general sense. The IEC document is an international standard that contains some simplifications to the TG 18 procedures but at the same time, provides a clearer presentation of the methodologies. In addition to the above-mentioned two reference documents, those interested in an expanded set of measurement procedures should consult the new edition of the Flat Panel Display Measurement Standard published by SID (Flat Panel Display Measurements Standard Working Group 2003). Finally, additional standards that are specific to different aspect of display performance can be found to be useful in addressing particular issues. For instance, the International Organization for Standardization (ISO) has published a number of ergonomic requirements standards, ISO 13406-2, that deals with other ways of characterizing display reflections (ISO 1997).

The AAPM Task Group 18 report (Samei et al. 2005), "Assessment of Display Performance for Medical Imaging Systems," provides standard guidelines to practicing medical physicists, engineers, researchers, technologists, and radiologists for the evaluation of display systems intended for medical use. The report consists of sections describing prior standardization efforts, current (as of the time of the report publication) and emerging medical display technologies, and prerequisites for the assessment of display performance. In addition, the report includes a description of required instrumentation and TG 18 test patterns and a description of quantification methods and acceptance levels for key display characteristics limited to grayscale or monochrome display systems.

In December 2009, the IEC published the first international standard on medical image displays. IEC 62563-1 (IEC 2010), "Medical image display systems Part 1: Evaluation methods," describes practical visual and quantitative tests applicable to medical image grayscale display systems using color and grayscale devices. Even though the standard has not specified requirements for acceptance and constancy tests or the frequencies of such tests, it is a useful standard that can be followed to characterize the image quality of a display system. The document contains useful information regarding equipment and tools (luminance and illuminance meter) and introduces methods for characterizing grayscale resolution, luminance response and uniformity, chromaticity, pixel faults, veiling glare, and angular viewing properties. In addition, the standard clearly describes procedures for calibration of display devices in the presence of ambient illumination.

9.5 Emerging Technologies

Several 3-D modalities are being considered for improving the detection of breast cancer, either as adjunct or replacement modalities for full-field digital mammography examinations. Among them, breast tomosynthesis (See Chapter 4) (Lewin and Niklason 2007, Niklason et al. 1997, 1998, Park et al. 2007), a limited-angle tomographic modality, has demonstrated that the additional volumetric information acquired using multiple projections holds promise for improvements in the detection of masses, in part by removing the masking effect of normal anatomical structures that can hide or mimic lesions. Other technologies being currently developed include dedicated breast computed tomography (See Chapter 5) (Boone et al. 2005, Kwan et al. 2005, 2007) and stereo mammography (See Chapter 3) (Andriole et al. 2011). Display systems for these emerging 3-D breast imaging modalities bring new challenges in terms of the characterization methodologies for new criteria that becomes relevant in 3-D displays. The radiologist is now faced with a new reading paradigm that requires browsing over many high-resolution (several million pixels) images (slices from the reconstructed volume). Therefore, in 3-D modalities, the temporal characteristics of the display that were introduced in the previous section become a more relevant characterization topic.

Another approach to visualizing 3-D image sets is the use of 3-D display systems. 3-D systems have recently, in part due to advancements aimed at the consumer markets, become more available and significantly improved. 3-D systems require extending the assessment and measurement methodologies to a third dimension particularly in developing new experimental physical quantities (e.g., stereo-crosstalk and stereo-noise) and relating them to observer performance. Such extension is complicated by the widely different underlying technologies being considered for 3-D displays that include stereoscopic (with active and passive glasses), autostereoscopic, volumetric, and time-sequential implementations, among others. A tentative list of metrics of interest was described by Badano and Cheng (2011) and Badano et al. (2011).

References

American College of Radiology and National Electrical Manufacturers Assoc. 1998. Digital Imaging and Communications in Medicine (DICOM), Part 3.14, Grayscale Standard Display Function. Technical report, ACR/NEMA.

Andriole KP, Wolfe JM, Khorasani R, Treves ST, Getty DJ, Jacobson FL, Steigner ML, Pan JJ, Sitek A, Seltzer SE. 2011. Optimizing analysis, visualization, and navigation of large image data sets: one 5000-section ct scan can ruin your whole day. Radiology 259(2): 346–362 doi: 10.1148/radiol.11091276. URL http://dx.doi.org/10.1148/radiol.11091276.

Badano A and Fifadara DH. 2004. Comparison of Fourier-optics, telescopic, and goniometric methods for measuring angular emissions from medical liquid-crystal displays. Appl Opt 43(26): 4999–5005.

Badano A, Flynn MJ. 1997. Image degradation by glare in radiologic display devices. Proc SPIE 3031: 222–231.

Badano A, Flynn MJ. 1998a. Monte Carlo modeling of glare in cathode-ray tubes for medical imaging. In Proc SID pp. 495–498.

Badano A, Flynn MJ. 1998b. Experimental measurements of glare in cathode-ray tubes. Proc SPIE 3335: 188–196.

Badano A, Flynn MJ. 2000a. Method for measuring veiling glare in high performance display devices. Appl Opt 39(13): 2059–2066.

Badano A. Flynn MJ. 2000b. Method for measuring veiling glare in high-performance display devices. Appl Opt 39(13): 2059–2066.

Badano A, Flynn MJ, Muka E, Compton K, Monsees T. 1999. The veiling glare point-spread function of medical imaging monitors. Proc SPIE 3658: 458–467.

Badano A, Flynn MJ, Martin S, Kanicki J. 2003. Angular dependence of the luminance and contrast in medical monochrome liquid crystal displays. Med Phys 30(10): 2602–2613. URL http://link.aip.org/link/?MPH/30/2602/1.

Badano A, Cheng W-C. 2011. Cutting-edge technology: Part 1: Emerging topics in medical displays. Inf Display 27(4).

Badano A, Gagne RM, Jennings RJ, Drilling SE, Imhoff BR, MukaE. 2004. Noise in flat-panel displays with sub-pixel structure. Med Phys 31(4): 715–723.

Badano A, Cheng W-C, O'Leary BJ, Myers KJ. 2011. Cutting-edge technology: Part 2: Pre-clinical assessment of medical displays for regulatory evaluation. Inf Display 27(4).

Barrett HH, Abbey CK. 2001. Human-and model-observer performance in ramp-spectrum noise: Effects of regularization and object variability. J Opt Soc Am A Opt Image Sci Vis 18(3): 473–488.

Barten PGJ. 1989. Subjective image quality of hdtv pictures. Int Display Res Conf 17: 598.

Barten PGJ. 1992. Physical model for the contrast sensitivity of the human eye. Proc SPIE 1666: 57–72.

Barten PGJ. 1993. Effects of quantization and pixel structure on the image quality of color matrix displays. J SID 1(2): 147–153.

Barten PGJ. 1999a. Contrast Sensitivity of the Human Eye and Its Effects in Image Quality. Bellingham, WA: SPIE Press, PM 72.

Barten PGJ. 1999b. Contrast Sensitivity of the Human Eye and Its Effects on Image Quality. SPIE Press.

Bender S, Lederle K, Wei C, Schoenberg S, Weisser G. 2011. 8-bit or 11-bit monochrome displays. Which image is preferred by the radiologist? Eur Radiol 21: 1088–1096. ISSN 0938-7994. URL http://dx.doi.org/10.1007/s00330-010-2014-1.10.1007/s00330010-2014-1.

Boone JM, Kwan ALC, Seibert JA, Shah N, Lindfors KK, Nelson TR. 2005. Technique factors and their relationship to radiation dose in pendant geometry breast CT. Med Phys 32(12): 3767–3776. doi: 10.1118/1.2128126. URL http://link.aip.org/link/?MPH/32/3767/1.

Chen M, Bowsher JE, Baydush AH, Gilland KL, DeLongand DM, Jaszczak RJ. 2001. Using the hotelling observer on multi-slice and multi-view simulated SPECT myocardial images. Proc Nucl Sci Symp 4: 2258–2262.

den Boer W. 2005. Active Matrix Liquid Crystal Displays: Fundamentals and Applications. Elsevier.

Fifadara DH, Averbukh A, Channin DS, Badano A. 2004. Effect of viewing angle on luminance and contrast for a five-million-pixel monochrome display and a nine-million-pixel color liquid crystal display. J Digit Imag 17(4): 264–270.

Flynn MJ, Badano A. 1999. Image quality degradation by light scattering in display devices. J Digit Imag 12(2): 50–59.

Flynn MJ, Kanicki J, Badano A, Eyler WR. 1999. High-fidelity electronic display of digital radiographs. Radiographics 19(6): 1653–1669.

Gallas BD, Barrett HH. 2003. Validating the use of channels to estimate the ideal linear observer. J Opt Soc Am A 20(9): 1725–1738.

Honda M, Ema T, Kikuchi K. 1993. A technique of scatter-glare correction using a digital filtration. Med Phys 20(1): 59–69.

IEC. 2010. IEC62563-1. Medical electrical equipment—Medical image display systems part 1: Evaluation methods. Technical report, IEC.

ISO. 1997. 13406-2, Ergonomic requirements for visual display terminals units based on flat panels—Part II: Requirements for flat panel displays. Technical report, ISO.

Krupinski EA, Siddiqui K, Siegel E, Shrestha R, Grant E, Roehrig H, Fan J. 2007. Influence of 8-bit vs 11-bit digital displays on observer performance and visual search: A multi-center evaluation. J Soc Inf Display 15: 385–390.

Kurita T, Saito A, Yuyama I. 2001. Consideration on perceived MTF of hold type display for moving images. In Proc of IDW'98.

Kwan ALC, Boone JM, Shah N. 2005. Evaluation of x-ray scatter properties in a dedicated cone-beam breast CT scanner. Med Phys 32(9): 2967–2975.

Kwan ALC, Boone JM, Yang K, Huang S-Y. 2007. Evaluation of the spatial resolution characteristics of a cone-beam breast ct scanner. Med Phys 34(1): 275–281.

Lewin JM, Niklason L. 2007. Advanced applications of digital mammography: Tomosynthesis and contrast-enhanced digital mammography. Semin Roentgenol 42(4): 243–252. doi: 10.1053/j.ro.2007.06.006. URL http://dx.doi.org/10.1053/j.ro.2007.06.006.

Liang H, Park S, Gallas BD, Badano A, Myers KJ. 2007. Assessment of temporal blur reduction methods using a computational observer that predicts human performance. J SID, in press.

Liang H, Badano A. 2007. Temporal response of medical liquid crystal displays. Med Phys 34(2): 639–646.

Liang H, Park S, Gallas BD, Myers KJ, Badano A. 2008. Image browsing in slow medical liquid crystal displays. Acad Radiol 15(3): 370–382. doi: 10.1016/j.acra.2007.10.017. URL http://dx.doi.org/10.1016/j.acra.2007.10.017.

Mathie AG, Strickland NH. 1997. Interpretation of CT scans with PACS image display in stack mode. Radiology 203: 207–209.

Miseli J. 2004. Motion artifacts. Proc SID 86–89.

Nakamura H, Crain J, Sekiya K. 2001. Optimized active-matrix drives for liquid crystal displays. J Appl Phys 90: 2122–2127.

Niklason LT, Christian BT, Niklason LE, Kopans DB, Castle-Berry DE, Opsahl-Ong BH, Landberg CE et al. 1997. Digital tomosynthesis in breast imaging. Radiology 205(2): 399–406.

Niklason LT, Kopans DB, Hamberg LM. 1998. Digital breast imaging: Tomosynthesis and digital subtraction mammography. Breast Dis 10(3–4): 151–164.

Park JM, Franken EA Jr, Garg M, Fajardo LL, Niklason LT. 2007. Breast tomosynthesis: Present considerations and future applications. Radiographics 27(Suppl 1): S231–S240. doi: 10.1148/rg.27si075511. URL http://dx.doi.org/10.1148/rg.27si075511.

Park S, Jennings R, Liu H, Badano A, Myers K. 2010. A statistical, task-based evaluation method for three-dimensional x-ray breast imaging systems using variable-background phantoms. Med Phys 37(12): 6253–6270.

Paulsson L-E, Sjostrand J. 1980. Contrast sensitivity in the presence of a glare light. Investigative Ophtalmology 19(4): 401–406.

Platisa L, Goossens B, Vansteenkiste E, Park S, Gallas BD, Badano A, Philips W. 2011. Channelized hotelling observers for the assessment of volumetric imaging data sets. J Opt Soc Am A Opt Image Sci Vis 28(6): 1145–1163.

Reiner BI, Siegel EL, Hooper FJ, Pomerantz S, Dahlke A, Rallis D. 2001. Radiologists productivity in the interpretation of CT scans: A comparison of PACS with conventional film. Amer J Roentgenol 176: 861–864.

Samei E, Badano A, Chakraborty D, Compton K, Cornelius C, Corrigan K, Flynn MJ et al. 2005. Assessment of display performance for medical imaging systems: executive summary of aapm tg18 report. Med Phys 32(4): 1205–1225.

Seibert JA, Nalcioglu O, Roeck W. 1984. Characterization of the veiling glare PSF in X-ray image intensified fluoroscopy. Med Phys 11(2): 172–179.

Seibert JA, Nalcioglu O, Roeck W. 1985. Removal of image intensifier veiling glare by mathematical deconvolution techniques. Med Phys 12(3): 281–288.

Siegel E, Krupinski E, Samei E, Flynn M, Andriole K, Erickson B, Thomas J, Badano A, Seibert J.A, Pisano ED. 2006. Digital mammography image quality: Image display. J Am Coll Radiol 3(8): 615–627. doi: 10.1016/j.jacr.2006.03.007. URL http://dx.doi.org/10.1016/j.jacr.2006.03.007.

Spencer G, Shirley P, Zimmerman K et al. 1995. Physically-based glare effects for digital images. In Computer Graphics Proceedings, Annual Conference Series SIGGRAPH 95. pp. 325–334.

Stiles WS. 1929. The effect of glare on the brightness difference threshold. Proc Roy Soc Lond B104: 322–351.

Tisdall MD, Damberg G, Wighton P, Nguyen N, Tan Y, Atkins MS, Li H, Seetzen H. 2008. Comparing signal detection between novel high-luminance hdr and standard medical LCD displays. J Display Technol 4(4): 398–409.

Tong HS, Prando G. 1992. Hygroscopic ion-induced antiglare/antistatic coating for CRT applications. In Proc. SID.

Video Electronics Standards Association (VESA). 2003. Flat Panel Display Measurements Standard Working Group. Flat panel display measurements standard, version 2.0. Technical report, VESA.

Zeman HD, Hughes EB, Otis JN et al. 1985. Veiling glare of a linear multichannel Si(Li) detector. Proc SPIE 535: 214–221.

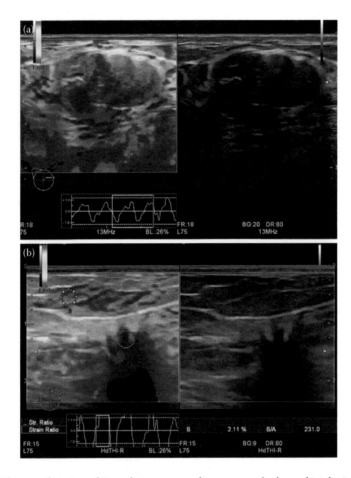

FIGURE 7.7 Elastography. (a) Ultrasound strain and B-mode images on split-screen mode show a histologically proven fibroadenoma. (Right) B-mode image shows a well defined hypoechoic lesion. (Left) Elastogram shows a mosaic pattern consisting predominantly of green areas (high strain) and blue areas (absence of strain), typical for a fibroadenoma. (b) Ultrasound strain and B-mode images on split-screen mode show a breast mass with a histological diagnosis of invasive ductal carcinoma. (Right) B-mode image shows a hypoechoic lesion with angular margins. (Left) Elastogram shows absence of strain information consistent with a "stiff" lesion such as a carcinoma.

FIGURE 12.13 Customized holders for mastectomy tissue specimens, mimicking the uncompressed (left) or compressed (right) breast shape, used for generation of the UMass digital breast phantoms. (Modified from O'Connor, J.M., Das, M., Didier, C., Mah'd, M., and Glick, S. J. (2008). In Krupinski, E.A. (Ed.), *Digital Mammography (IWDM)*. Berlin-Heidelberg. With permission from Springer-Verlag.)

FIGURE 12.20 Photograph of the prototype anthropomorphic physical breast phantom, generated based upon a Penn digital breast phantom of a 450 ml breast with 42% volumetric glandularity. (Reproduced from Carton, A.-K., Bakic, P.R., Ullberg, C., Derand, H., and Maidment, A.D.A. (2011). *Medical Physics*, 38, 891–896. With permission.)

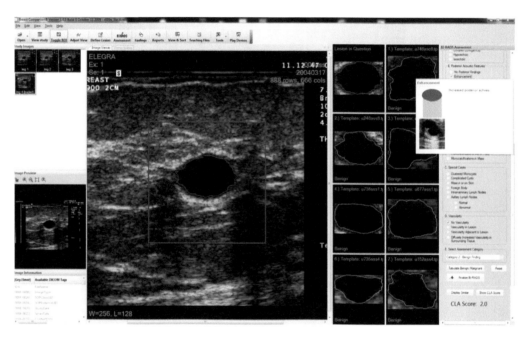

FIGURE 14.3 A screen view of the Breast Companion® CAD system showing analysis of a benign breast cyst (center image with red box) and seven similar masses displayed in the adjacent panel on the right.

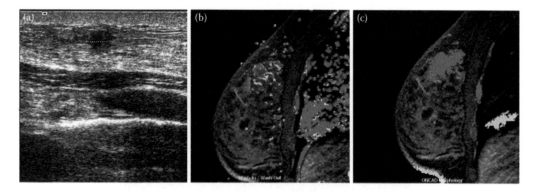

FIGURE 14.5 A 30-year-old woman who presents with a palpable finding in the right breast. (a) An ultrasound of this palpable area revealed an irregular 1.6 cm mass and biopsy confirmed invasive ductal carcinoma. (b) Postcontrast sagittal T1-weighted image of the right breast with color overlay, after CAD implementation, demonstrates the 1.6 cm mass with suspicious enhancing pattern (red color). (c) The automated morphological analysis system ONCAD revealed a larger suspicious area measuring 2.9 cm. Final pathology confirmed 2.4 cm invasive carcinoma with extensive surrounding ductal carcinoma in situ.

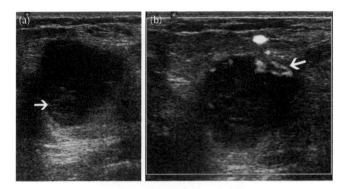

FIGURE 15.9 A 46-year-old woman presented with a palpable mass in the upper outer aspect of the left breast. (a) Antiradial left breast ultrasound in the 2-o'clock region demonstrates a lobular hypoechoic mass with some internal echoes (arrow) and through transmission. (b) Antiradial left breast power Doppler ultrasound shows a peripheral vessel (arrow). Ultrasound-guided core needle biopsy revealed poorly differentiated, high nuclear grade invasive ductal carcinoma (estrogen receptor and progesterone receptor-negative).

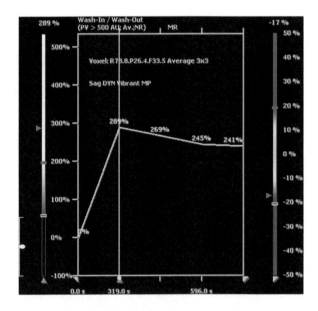

FIGURE 16.2 (b) Signal intensity analysis shows a typically malignant-appearing wash-out curve. Pathology demonstrated invasive ductal carcinoma.

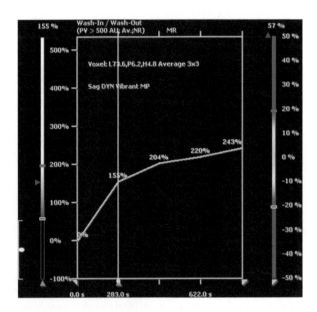

FIGURE 16.3 (b) The signal intensity curve demonstrates persistent kinetics.

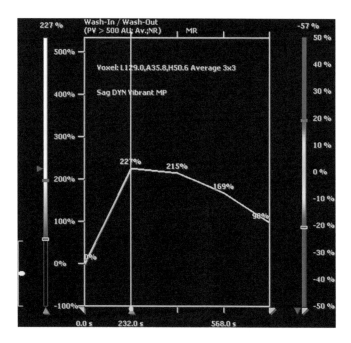

FIGURE 16.4 Early (a) and late (b) sagittal subtraction images demonstrate a lobulated mass with spiculated margins (arrow). The enhancement intensity is greater in the early phase (a) than in the later phase (b), and wash-out of contrast is confirmed with the signal intensity curve (c). Pathology showed invasive ductal carcinoma with associated DCIS.

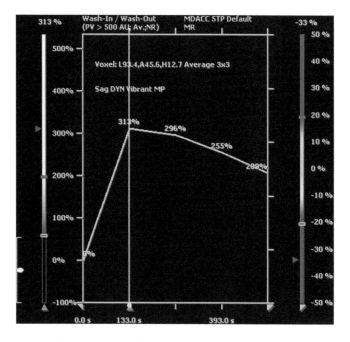

FIGURE 16.6 (b) The signal intensity curve demonstrates rapid wash-in and wash-out kinetics, typical of malignancy. Pathology demonstrated Paget's disease of the nipple with underlying DCIS.

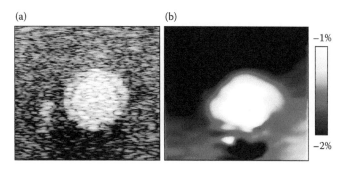

FIGURE 18.5 (a) B-mode ultrasound and (b) axial strain image or "elastogram" of a tissue-mimicking phantom with a stiff inclusion.

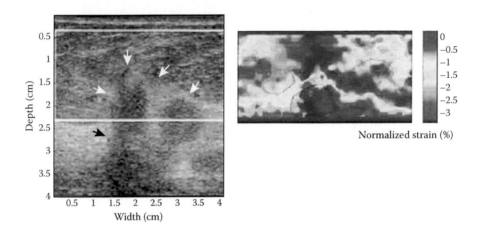

FIGURE 18.8 B-mode ultrasound image (left) and normalized strain image (right) of a ductal carcinoma in a 70-year-old patient. Average normalized strain values inside and outside the lesion were calculated to be 0.25% and 1.28%, respectively. (From Hiltawsky, K.M., Kruger, M., Starke, C., Heuser, L., Ermert, H., Jensen, A. (2001). *Ultrasound Med Biol* 27: 1461–1469. With permission.)

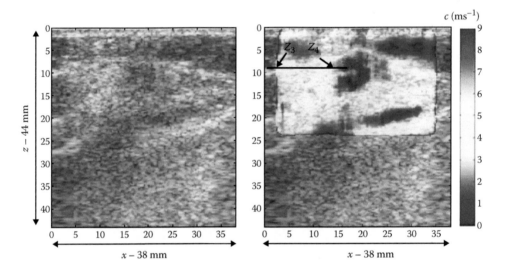

FIGURE 18.10 Comparison between B-mode ultrasound and quantitative elasticity map performed by the SSI mode. Case 1: Infiltrating ductal carcinoma grade III. Hypoechoic lesion with indistinct margins, slightly posterior shadowing classified as BI-RADS category 5. According to the criteria of elastography interpretation, this is a highly suspicious lesion with a high Young's modulus. (From Tanter, M., Bercoff, J., Athanasiou, A., Deffieux, T., Gennisson, J.L., Montaldo, G., Muller, M., Tardivon, A., Fink, M. (2008). *Ultrasound Med Biol* 34: 1373–1386. With permission.)

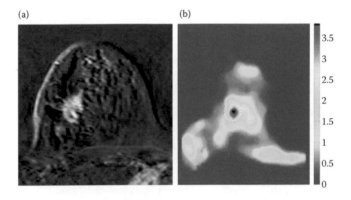

FIGURE 18.11 (a) Gadolinium-DTPA-enhanced substraction image from an MR mammography. (b) In vivo Young's modulus values from a patient suffering from a breast carcinoma. (Adapted from Sinkus, R., Lorenzen, J., Schrader, D., Lorenzen, M., Dargatz, M., Holz, D. (2000). *Phys Med Biol* 45: 1649–1664.)

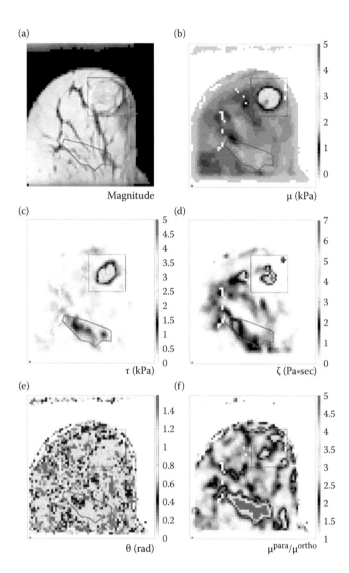

FIGURE 18.12 In vivo results from a patient with a fibroadenoma. (a) The lesion is easily visible in the MR magnitude image (red rectangle) and the corresponding image of the shear modulus (b) shows the tumor well delimited from the surrounding fatty tissue. The MR magnitude image indicates that the lesion is traversed by septae, which probably lead to the enhance values of the anisotropy (c). The core of the lesion exhibits increased viscous properties (d) and the angle of rotation θ indicates a preferred direction of the fibers inside the lesion (e). (f) The ratio μ^{para}/μ^{ortho} shows that parts of the lesion exhibit strong anisotropic properties. (Adapted from Sinkus, R., Tanter, M., Catheline, S., Lorenzen, J., Kuhl, C., Sondermann, E., Fink, M. (2005). *Magn Reson Med* 53: 372–387.)

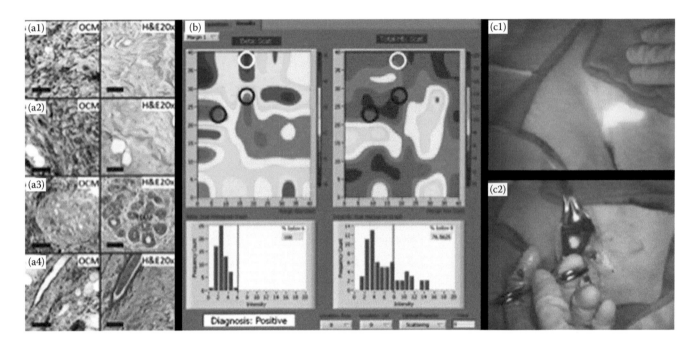

FIGURE 20.2 Representative images of intraoperative optical imaging of breast cancer at microscopic (a), mesoscopic (b), and macroscopic (c) sampling volumes. (a1–4) Microscopic optical coherence microscopy acquired 50 µm below the tissue surface and corresponding histology. Scale bar is 100 µm. (From Zhou C, Cohen DW, Wang Y et al. 2010. *Cancer Res*, 70, 10071–10079. With permission.) (b) Intraoperative breast tumor margin assessment using quantitative diffuse reflectance imaging; image of a pathologically confirmed margin. (From Brown JQ, Bydlon TM, Richards LM et al. 2010. *IEEE Journal of Selected Topics in Quantum Electronics*, 16, 530–544. With permission.) Subcutaneous (c1) and intraoperative (c2) identification of the sentinel lymph node following ICG injection using fluorescence imaging. (From Troyan SL, Kianzad V, Gibbs-Strauss SL et al. 2009. *Annals of Surgical Oncology*, 16, 2943–2952. With permission.)

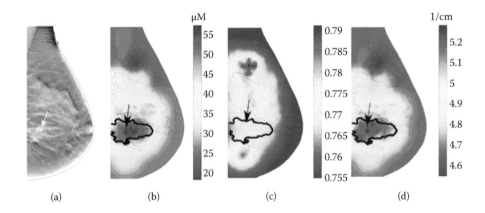

FIGURE 20.6 Results of a reconstruction from an integrated x-ray mammography and optical spectroscopy system at the MGH from a breast with a 2.5-cm invasive ductal carcinoma indicated by the arrow. (a) Tomosynthesis image. (b) Total hemoglobin (micromoles per liter). (c) Oxygen saturation. (d) Scattering coefficient (cm^{-1}). (From Fang Q, Selb J, Carp SA et al. 2011. *Radiology*, 258, 89–97. With permission.)

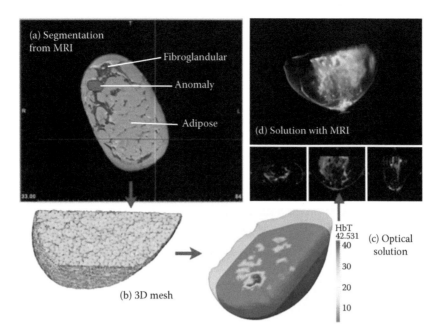

FIGURE 20.8 Mesh creation and inclusion of prior information in MRI-NIR imaging. Tissue types are segmented based on MRI images (a) and a 3-D mesh is created (b) with these assignments preserved. The optical image reconstruction is completed in which the optical parameters in each region are estimated (c). The optical solution is merged with the MRI image (d).

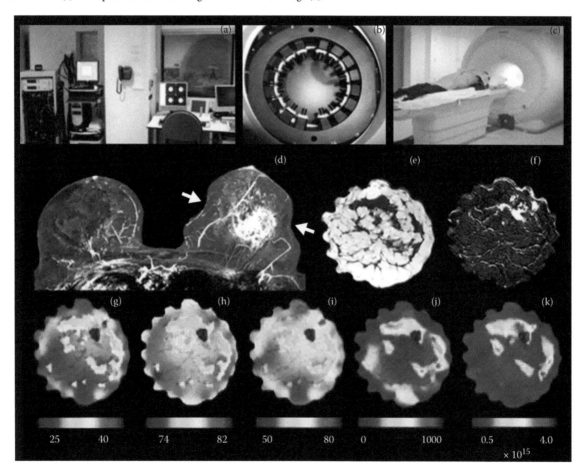

FIGURE 20.9 Clinical example of MRI-NIR imaging. MRI and NIR imaging are performed simultaneously using a clinical scanner (a, c) and circular NIR breast array (b). MRI images (d–f) are used to locate the tumor and assign tissue properties for optical reconstructions of total hemoglobin (g), oxygen saturation (h), percent water (i), scatter amplitude (j), and scatter power (k). (From Carpenter CM, Pogue BW, Jiang SJ et al. 2007. *Optics Letters*, 32, 933–935.)

10

Perception of Mammographic Images

Elizabeth Krupinski
University of Arizona

10.1 What Psychologists Wish Engineers and Physicists Understood About Breast Image Interpretation

The interpretation of mammographic images, as with the rest of radiologic interpretation, is firmly rooted in and dependent upon the science and technology underlying the acquisition, transmission, and display of images. There is however, much more to it than that. The interpretation of any radiologic image is as much an art as it is a science, and the art of interpretation is also firmly rooted and dependent upon the underlying perceptual and cognitive capabilities of the radiologist. Thus, it is important for engineers and physicists involved to understand and appreciate the fact that at the end of the imaging chain are the eyes and brain of a radiologist who must interact with the images presented to them in an efficient and effective manner in order to render a reliable and valid diagnostic interpretation.

Breast imaging technology is actually not as old as many people think, having been available for just over 50 years. The first practical technology dedicated to breast imaging was developed in 1960 (Egan 1960), and the first commercial x-ray units dedicated to breast imaging only became available in the late 1960s (Gold 1990). One of the first mammography studies (Strax et al. 1973) in the 1960s was the Health Insurance Program of New York Project (HIP) with over 60,000 women being enrolled. Subjects had either four consecutive years of screening mammography

or traditional standard care (e.g., physical examination). The women who had yearly screening mammography had a 29% mortality reduction at 9 years and a 23% mortality reduction at 18 years. Screening decreased mortality, and the morbidity associated with a diagnosis of breast cancer was reduced. By 1976 screening mammography became standard practice.

In the past couple of decades, dramatic changes have occurred in the practice of breast imaging, and they have had significant impact on how mammographers interact with the image data and render diagnostic decisions. Mammography was originally film-based and images were interpreted using hardcopy film on dedicated mammographic view boxes. The revolution in technology (see Chapter 1 for a more complete treatment) was digital mammography, which was fully developed in the 1990s, and the first full-field digital mammography (FFDM) system was approved by the Food and Drug Administration (FDA, 2011a) for marketing in 2000. The digitally acquired images were initially still printed to film for interpretation, but softcopy was soon approved and mammographers began interpreting softcopy images using digital displays (see Chapter 9 for a full discussion on the types of displays required for viewing FFDM images). There were a number of technical advantages with FFDM, including a wider dynamic range, lower dose, and the ability to process and manipulate the image data to overcome display limitations (Yaffe 2010a). A key question however, was whether softcopy images could provide the requisite diagnostic information with sufficient quality to render decisions at least as good as the gold standard of film.

Scientific evidence supporting the transition from analog to digital reading was produced in part by a large prospective multi-institutional study done by the American College of Radiology Investigational Network (ACRIN) called the Digital Mammography Imaging Screening Trial (DMIST). This study compared screen-film with FFDM images and found that the overall accuracy of the two modes was equivalent, but mammographers were more sensitive at detecting cancer in certain patient subsets using digital rather than screen film mammography (Pisano et al. 2005). Other studies have shown similar results (Lewin et al. 2001, 2002, Skanne et al. 2003, Skanne and Skjennald 2004, Cole et al. 2004, Del Turco 2007, Lewin 2010, Skaane 2010), and digital mammography use has increased significantly in the 21st century. According to the FDAs MQSA (Mammography Quality Standards Act) Facility Scorecard, as of October 1, 2010, there were 8652 certified FFDM facilities and 12,274 accredited FFDM units in the United States alone (FDA 2011b).

Digital mammography is only one breast imaging modality being used to detect and diagnose breast cancer, each with its advantages and disadvantages (Feig 2011, Stojadinovic et al. 2011). Magnetic resonance imaging (Lehman et al. 2007, Smith 2007, DeMartini et al. 2008, Yeh 2011, Boetes 2011; see Chapter 16 for a discussion of the role of MRI in breast imaging), ultrasound (Yang 2007, Mundinger et al. 2010; see Chapter 15 for a discussion of the role of ultrasound in breast imaging), nuclear medicine (van der Ploeg 2008, Even-Sapir and Inbar 2010; see Chapter 17 for a discussion of the role of nuclear medicine in breast imaging), and molecular imaging (Franc 2007, Oude Munnink et al. 2009) are all part of the arsenal being used to detect breast cancer today. All of these technologies use soft-copy images, which adds another dimension to what must be considered when presenting images to the breast imager—soft-copy images can be manipulated, processed (Karssemeijer and Snoeren 2010) and even "interpreted" by computers (Nishikawa 2010). Chapters 14 and 21 provide a more complete description of the current role of CAD in breast imaging and future directions, respectively. It is becoming increasingly evident that engineers, physicists, and clinicians must consider the fact that in the very near future all of these types of images will need to be available at the same time and data (image, text, decision aids) from a variety of sources must be perceptually and cognitively processed and integrated in order to render an accurate and complete diagnosis. To do this effectively and efficiently, the perceptual and cognitive capabilities of the end user must be considered at all stages in the imaging chain.

10.2 Some Basic Vision Concepts for Reading Mammograms

To appreciate what the human visual system is capable of (and limited by), it is not necessary to fully understand its anatomic and physiological properties. There are, however, a few relevant properties that pertain highly to the display and interpretation of mammographic images. The primary function of the eyes is photoreception or the process by which light from the environment produces changes in the photoreceptors or nerve cells (about 115 million rods and 6.5 million cones) in the retina. Light needs to travel through the pupil, the lens, and the watery vitreous center before it reaches the retina because the retina is located at the back of the eye. The rods sense contrast, brightness, and motion, and are located mostly in the periphery of the retina. Cones are used for fine spatial resolution, spatial resolution, and color vision, and they are located in the foveal and parafoveal regions. As light hits them, pigments in the rods and cones undergo chemical transformations, converting light energy into electrical energy that acts on a variety of nerve cells that connect the eye to the optic nerve and subsequent visual pathways that extend to the visual cortices in the brain.

10.2.1 Spatial Resolution

Spatial resolution is especially important for viewing mammographic images. It is the ability to see fine details (e.g., microcalcifications and fine spiculations extending from masses; see Chapter 7 for more details on the clinical aspects of breast cancer detection and diagnosis), and is highest at the fovea, while declining sharply toward the peripheral retinal regions. The practical result is that in order to see fine details, mammographers must search or move the eyes around the image (more on this later). With digital mammography, both the acquisition and display spatial resolution are controlled by a number of technical factors (Yaffe 2010b, Krupinski and Roehrig 2010; see Chapters 1 and 9 for more details on digital mammography technology and displays, respectively) and these factors need to be optimized in order to maximize the likelihood that microcalcifications and spiculations can be adequately visualized. For example, to improve the limiting spatial resolution of a detector one could use smaller dels, but the tradeoff is reduced SNR (signal-to-noise ratio), which clearly impacts the amount and quality of information available to the breast imager. This type of tradeoff in physical characteristics of imaging systems needs to be understood by those designing the systems so they do not create something that has excellent physical properties in one dimension but negatively impact perception by degrading quality in another dimension.

The question of what resolution a mammography display needs to be can be addressed from a perceptual point of view considering what the human visual system is capable of perceiving. The typical viewing distance for softcopy reading is 30–60 cm and at about 60 cm contrast sensitivity (discussed below) of the eye drops to zero above 2.5 cycles/mm with a peak sensitivity at 0.5 cycles/mm. Based on this fact, display pitch or pixel size should be about 200 μm. Most 3-Mp displays have a pixel pitch of about 207 μm and most 5-Mp displays of about 165 μm. The current recommendation is for 5-Mp displays to be used (Siegel et al. 2006), but there is a strong argument that a 3-Mp display is likely sufficient for softcopy mammography reading (Tabar et al. 2010). If 3 Mp displays are used, however, it is critical that mammographers use zooming as appropriate especially when scanning for microcalcifications as they can be missed (Haygood et al. 2009).

10.2.2 Contrast Resolution

Contrast resolution or the ability to distinguish differences in intensity in an image, thus permitting one to distinguish between objects (masses, see Figure 10.1, and microcalcifications, see Figure 10.2) and background in an image, is another important visual property engineers and physicists should appreciate. There are tests to determine contrast levels perceptible by the human eye. They use a sinusoidal grating pattern (alternating black and white lines where the average luminance remains the same but the contrast between the light and the dark areas differ) and contrast resolution is characterized by the point at which the lines can be discriminated from each other. Grating discrimination is described in terms of cycles per degree or the grating frequency, and contrast sensitivity peaks in the midspatial frequency range around 3 to 5 cycles/degree. This means that low contrast lesions can often go undetected, especially when viewing conditions are not optimal. It is important to recognize that contrast resolution is not a function of either the image or the observer—it depends on both the quality of the image and the visual capabilities of the human observer. Thus, whenever a new type of image, display, or presentation state is developed, it is critical to characterize its contrast resolution using human observers and standardized tests (usually phantom images with test objects that change in size and proximity to each other).

Surprisingly (although they have been proposed; Quaghebeur et al. 1997, Straub et al. 1991), there are no guidelines or requirements for radiologists with respect to having their vision assessed, although it is clear that there is wide variability between radiologists and their visual acuity. For example, Safdar et al. (2009) carried out one study in which they examined visual acuity differences in radiologists over a typical workday. Although they found modest differences in acuity between morning and other times, they concluded that it was unlikely that these differences could impact diagnostic performance. When they surveyed the radiologists regarding when they had their last eye examination, more than a few reported not having had a thorough examination in as many as 15 years. Krupinski et al. (2010) also found that most radiologists had not had an eye examination within the past year.

FIGURE 10.1 Example of a region of interest from a mammogram containing an ill-defined malignant mass (in the center).

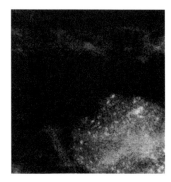

FIGURE 10.2 Example of a region of interest from a mammogram containing a cluster of microcalcifications (lower right).

10.2.3 Depth Perception

It is important to remember that radiographic images are 2-D representations of 3-D structures. Even with advanced imaging technologies that allow us to create volumetric renditions of image data, for the most part, these reconstructions are still displayed using technologies incapable of true 3-D. Therefore, the breast imager generally needs to rely on monocular cues to derive depth information from images. Depth is a critical component of image interpretation because potential lesions must be localized within the breast in order to carry out further imaging, biopsies, and eventual excision or treatment (e.g., radiation) should the finding prove to be malignant. One of the key monocular cues used in interpreting radiographic images is occlusion or interposition. Occlusion is the blocking of one object by another in which the one being blocked is interpreted as being behind the other one, providing some information about relative depth of structures in relation to one another. Although occlusion is clearly beneficial for estimating lesion depth, it is also one of the main obstacles to detecting them in the first place. If the occlusion is too great (e.g., tissues overlap or vessels cross a mass), lesions are easily missed.

Luckily, we are not limited to monocular cues for depth perception. The fact that we have two eyes separated from each other in the horizontal direction accounts for our ability to see depth, or in the case of the radiologist to generate the perception of depth from 2-D images. The separation averages about 65 mm for an adult human, giving each eye a slightly different view of the external world. Stereopsis is the angular horizontal disparity between the corresponding points of a given object in these two different retinal images, and it is what serves as the main basis for stereo vision. The magnitude of the disparity and the relative position of the object compared to the fixation point provide the requisite information to see objects in depth relative to one another. When an object is far away, the disparity between the images is small, and when an object is near the disparity is large, providing the brain with information to perceive depth.

This basic idea of having two images of the same scene with a small angular difference is what forms the basis of stereo imaging. With FFDM systems, it is relatively easy to acquire stereo images by keeping the breast (and detector) stationary and moving the x-ray source by about 6°–8° and acquiring two images. These two images are then presented to the mammographer, but special stereoscopic display systems are required (see Chapter 3 for a more complete discussion of stereo mammography). There are three basic types of display systems. Sequential or active systems typically use a single display and rapidly alternate the presentation of the two images in conjunction with a pair of special glasses that feed the alternating images to the right and left eye sequentially. As long as it is done properly (e.g., the monitor refresh rate is high enough at about 120 Hz), the brain fuses the alternating images and the mammographer perceives a 3-D image. The problem with this system is that the glasses reduce the brightness of the image and certain artifacts (e.g., ghosting) arise.

The second common stereoscopic display system uses simultaneous or passive presentation of the images. This approach gets the images to both eyes simultaneously (still feeding one to the right eye and the other to the left), but typically relies on the use of two displays and special glasses (i.e., using polarized lenses) to present the images to each eye. Again the image brightness is somewhat reduced, the displays must be carefully aligned, and the user has little leeway in changing their position relative to the displays. To date, stereoscopic displays have had little impact in breast imaging. Although studies (Getty et al. 2008) have demonstrated that stereoscopic viewing of mammograms can reduce false-negatives as well as false-positives, the use of stereoscopic breast imaging has not been widely adopted. In all likelihood, the lack of adoption is due to the limitations of the display systems, the need to wear special glasses, and the constraints on mobility imposed by the display systems.

All of this may change, however, in the near future, as true 3-D displays are beginning to hit the commercial entertainment market and are likely to penetrate the medical display market as well. Autostereoscopic systems represent the third option and rely on a variety of technologies related to the displays themselves. There are two core technologies—parallax barrier and lenticular systems. In parallax barrier systems, a layer of material (a grid) is placed in front of the screen. It contains a series of slits that allow each eye to see a slightly different set of pixels, thus allowing the brain to fuse the pixel sets and create the 3-D effect. The limitation is that the viewer needs to stay in the same location in front of the display, sit at a fixed distance, and the grid in front of the display reduces brightness. Lenticular technology uses small lenses on the display to refract the left and right images and send the separately to each eye. Again, the brain automatically fuses these images to create the impression of 3-D. These displays have less reduction in brightness and wider ranges of viewing angles giving them much more potential for widespread use than the parallax barrier displays.

The real question of course is whether the addition of 3-D actually impacts the diagnostic process. Unfortunately, there are very few if any studies directly assessing this question using real radiographic images and radiologists as observers. Nilsson et al. (2007) conducted one study to assess visual–spatial ability and the interpretation of 3-D information in radiographs, but it used nonradiologists (students) as observers. They did, however, find two things worth noting in terms of radiologists using 3-D images. First, they found that the subjects had significant improvements in their ability to interpret 3-D information in radiographs after training on how to use parallax for object depth localization. They also found that the test results were strongly associated with mental rotation test abilities (subjects are asked to compare two 3-D objects rotated slightly differently and determine if they are the same object or mirror images). These results suggest that some people may be better at perceiving depth from images than others, and in general, training seems to improve depth perception abilities. The design of imaging and display systems for 3-D radiography must account for these differences in perceptual abilities and recognize the fact that all users may not benefit equally from the addition of 3-D information.

10.2.4 Motion Perception

Although the majority of radiographic imaging produces static images (including mammography), there are applications in which motion is relevant. The more familiar applications involve the recording of true motion, such as cardiac ultrasound capturing the beating heart or fluoroscopy tracking contrast material as it travels through the functioning alimentary tract. A moving object serves a variety of purposes including drawing the observer's attention to it, segmenting it from the background, aiding in the computation of 3-D shape, computing distance, and helping the observer recognize actions. There is a specific area of the brain (visual area MT) that seems to be the primary area for processing motion information. The neurons in this area are very selective for motion direction and the neural responses are especially well correlated with the perception of motion. The neurons in MT receive their input from the main visual cortex V1, and the visual areas MST and STS also contribute to motion perception.

In medical imaging as with any other type of "filming," the imaging system is capturing a series of still images that when played at the proper frame rate (which determines the temporal resolution) creates the impression of motion. A logical question is what frame rate (frames/sec) is required for the human observer not to see artifacts (i.e., a continuous flow of information). The problem is that no single one-size-fits-all-applications-answer exists—it all depends. It depends on the nature of the stimulus, how much change exists between subsequent frames, the ambient conditions, the device it is displayed on, and the individual observer. This can make the design of imaging and display systems difficult for engineers in some cases.

In traditional mammography where four individual static images are acquired and viewed, motion detection is not an issue. However, some of the newer modalities such as digital breast tomosynthesis (DBT; see Chapter 4 for more details), breast CT (see Chapter 5 for a more complete description), and breast MRI (see Chapter 16) do introduce a certain type of motion by acquiring multiple slices through the breast that then require the breast imager to scroll through the slices (i.e., frames) in order to visualize all the available data. From an engineering perspective, it might make sense to create a single viewing system that scrolls through the images at a set frame rate, but from a perceptual perspective this does not make sense. Again, the nature of the stimulus, the nature of the target/lesion to be detected, the ambient viewing conditions, the display, and the individual observer will all contribute to what frame rate is most appropriate.

As noted above, one function of motion is to capture the attention of the observer and that is one reason why DBT, breast CT, and breast MRI may prove to be very useful in detecting lesions compared to static mammography images. It is true that the primary advantage of the slice approach is to tease apart overlying structures and breast tissue (Diekmann 2010), but the perception of motion created by scrolling through a stack of sequential images could also be an important detection aid. The brain cannot attend to everything in its environment so it needs to select out the relevant objects to attend to. In a stack of individual slices through the breast, the majority of the information is redundant, perfectly normal, and generally flows continuously and evenly from one slice to the next. A mass or cluster of microcalcifications represents an object that does not fit in this flow. As the breast imager cycles through the image slices, the sudden onset of the abnormal object within in the normal background is prone to capture the viewer's attention and direct it to the location of change (Yantis and Jonides 1984). The peripheral visual system, primarily the rods, is particularly sensitive to this type of "abrupt onset" motion, which, from an evolutionary perspective, makes sense—you need to detect the lion approaching from your left before it detects you!

There is some evidence that these newer modalities such as DBT may offer a slight advantage over traditional mammography, but the data to date are not definitive (Spangler et al. 2011, Hakim et al. 2010, Zuley et al. 2010). There is also evidence that viewing times may be longer with DBT, although most new modalities are often associated with longer viewing times in the early adoption phase (Zuley et al. 2010). Developing optimal ways present DBT and similar data sets from CT and MRI need to be explored by engineers, perceptionists, and breast imagers.

10.3 Visual Search

As noted above, high-resolution vision is required to detect and discriminate microcalcifications and mass spiculations in mammographic images. The problem is that this high-resolution foveal vision is restricted to a relatively small (about 5° diameter) "useful visual field" (Figure 10.3). It is true that some abnormalities can be detected with peripheral vision (Kundel and Nodine 1975a), but normal anatomy (or "structured noise") significantly decreases detection the further away lesions are located from the axis of gaze (Kundel 1975b). The result of this limited "useful visual field" is that the mammographer must move her eyes around the image as she searches for signs of abnormalities.

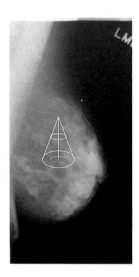

FIGURE 10.3 Schematic of the "useful visual field." High-resolution foveal vision only covers about 5° of visual angle; thus, the mammographer needs to move the eye around the image in order to detect fine details indicative of potential lesions.

The basis for interpreting visual search data in radiology comes from information processing theory (Crowley et al. 2003, Haber 1969, Nodine et al. 1992, Nodine and Kundel 1987). There is evidence that the initial glance (<250 msec) at an image produces a "global impression" or "gist" that includes the processing and recognition of content such as anatomy, symmetry, color, and grayscale. This gist information is compared with information contained in long-term memory contributing to the viewer's cognitive schema (or expectations) of what information is generally in an image. Sometimes, the target of search just "pops out" and the viewer makes a quick decision (Kundel et al. 2007). In most cases, however, abnormalities do not just pop out and the information processed in this global percept is used to guide high-resolution foveal search, whereby extraction of feature details from the complex image backgrounds takes place.

Further evidence to support this holistic model of perception for the detection of cancers on mammograms has been documented (Kundel et al. 2008). This study analyzed data (400 records) from three eye position studies to determine the time required to first fixate a cancer on a mammogram. Time to first fixate was used as an indicator of the initial perception of cancer. The distribution of times was partitioned into two normally distributed components using mixture distribution analysis. It was found that 57% of cancers had a 95% chance of being fixated in the very first second of viewing, while the other 5% took between 1.0 and 15.2 seconds. For most readers, the true-positive fraction was larger for lesions fixated within that first second of search, while lesions that took longer to first fixate had a lower chance of eventually being reported. The initial "gist" is an important component in lesion recognition in mammography. When designing displays and presentation tools for mammography, it is necessary for engineers and physicists to keep in mind the importance of this global impression and the amount of information derived from it as it is clearly impacts diagnostic accuracy.

10.3.1 Visual Search and Errors in Mammographic Interpretation

A very important question in mammography is why are errors made (especially when they are easily detected when viewed a second time)? The miss rate in mammographic screening has been estimated to be about 20%–30% (false-negatives) with a false-positive rate of about 2%–15% (Bird et al. 1992). From a perceptual point of view, false-positives are sometimes easier to understand. They often occur because overlaying anatomic structures/tissues can mimic disease entities, and there is often something in the image (overlapping tissue or dust) that attracts attention, leading to the false perception that a mass or microcalcification is present.

On the other hand, false-negatives are generally more complicated, but more important to understand. With FFDM, technical reasons (e.g., overexposure and underexposure), are virtually eliminated, so errors most likely reside in the reader. The first researchers to suggest that lesions can be missed due to inadequate search of images were Tuddenham and Calvert (1961), who also observed that there was significant variability in search strategies. To classify the types of errors made during search, Kundel et al. (1978) used eye-position recording and developed three categories based on how long missed lesions are dwelled on or fixated, with each category accounting for about one-third of errors.

Search errors constitute the first category, and in this case, the mammographer never fixates the lesion with high-resolution foveal vision—it is a complete miss. The second error is called a recognition error. These lesions are fixated, but not for very long, and the inadequate time reduces the likelihood that relevant lesion features will be detected or recognized. Finally, decision errors occur when the mammographer fixates the lesion for long periods of time, but either does not consciously recognize the features as those of a lesion, or actively dismisses them.

One of the first studies of mammography search was done with hardcopy images (Krupinski 1996), and results have been confirmed in later studies using softcopy images (Nodine et al. 1999, Mello-Thoms et al. 2005). These studies found that true- and false-positive decisions were generally associated with long gaze durations and true-negative decisions with relatively short gaze durations. Some false-negatives were not even looked at (search errors), while the rest were associated with gaze durations falling between the true-positives and true-negatives (recognition and decision errors). Figure 10.4 shows a typical scanning pattern of a mammographer searching a breast image. The dots represent fixations or where the eye directs foveal vision, and the lines represent saccades or the jumps that the eyes make between fixations and reflect the order in which fixations are generated.

10.4 Reader Variability

Even expert mammographers render different decisions on the same images, leading to considerable interobserver (as well as intraobserver) variability rates in mammography (Berg et al. 2000, Ciccone et al. 1992, Elmore et al. 1994, Kerlikowske et al. 1998, Beam et al. 2003). There are numerous factors that contribute to this variability including native ability, training, and reading volume. Although some of these factors such as native ability cannot readily be solved by engineering and physics, there are some aspects of the variability in mammographic interpretation that might be addressed from this perspective.

One factor contributing to reader variability in mammography is sheer volume of reading (Beam et al. 2003, Leung et al. 2007, Miglioretti et al. 2007). For example, in one study (Smith-Bindman et al. 2005), physicians who interpreted a high annual volume (2500–4000) of screening as opposed to diagnostic mammograms, found more cancers with fewer false-positive responses that those read fewer cases. Another study (Esserman et al. 2002) compared British with American radiologists using a standardized test set (PERFORMS 2 = PERsonal PerFORmance in Mammography Screening) designed to provide feedback to radiologists each year they take the test. The 60 U.S. radiologists were grouped as low- (<100 cases/month), medium- (101–300 cases/month), and high-volume radiologist (>301 cases/month), and the U.K. group had 194 high-volume radiologists. Average sensitivity at a specificity of 0.90 was 0.785 and 0.756 for the high-volume U.K. and U.S. readers, respectively, 0.702 for medium- and 0.648 for low-volume radiologists.

Reader volume is clearly an important determinant of mammogram sensitivity and specificity. For mammography in particular, exposure to a large volume of cases is critical because of the relatively low incidence of cancer in the screening environment. Diagnostic radiology residents are required to do a minimum three months rotation in breast imaging. Although some time will be spent in diagnostic sessions, most will be spent

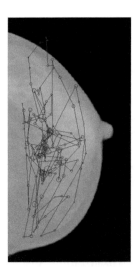

FIGURE 10.4 Example of a typical visual search (scan) pattern of a mammographer searching a breast image for masses and/or microcalcifications. Each circle represents a fixation or location where the eye lands, with the reflecting dwell time (larger circle = longer dwell). The lines between the fixations represent saccades or jumps between fixations and reflect the order in which fixations were generated during search.

in screening. The number and variety of cancers encountered during those few short months is really quite low in most cases and certainly does not expose the resident to the wide variety of lesion appearances likely to be encountered over longer periods of time. Although electronic teaching files exist (e.g., Radiology Education; American College of Radiology), they generally lack the type of educational user interface and associated tools that would make them truly effective for learning (Kim 2004). Engineers could enhance these teaching files by adding tools to provide better, more interactive feedback to the trainee, and even content-based image retrieval (CBIR) systems (Sparks and Madabhushi 2011) to provide residents with more examples of lesions and a way to query a library of images to retrieve similar cases for comparison.

Having prior images, a reliable clinical history, and information from other modalities as with a diagnostic workup all improve accuracy and reduce variability (Berg et al. 2000, Burnside et al. 2002, Houssami et al. 2003, 2004). Optimizing the design of workstations capable of presenting all these sources of information to the breast imager in an efficient and effective manner that does not create information overload, and/or an interface that is simply too complicated to navigate through and use, is a key area of development for software engineers and medical imaging informaticists.

10.5 Hanging Protocols

The way that images are presented to the mammographer is also an important part of the interpretation process that engineers should consider. Again, there is no one-size-fits-all approach, but those developing viewers and workstations need to keep the end user in mind and develop a system interface that presents the necessary information to the mammographer in the most efficient and logical manner. The design must also be flexible enough so that the individual user can readily tailor it to their unique preferences. There is no evidence that the arrangement (i.e., hanging protocol) of images on the display(s) actually affects the perception of the information or diagnostic accuracy, but it is clear that everyone has their own style and preference for hanging protocols that could impact the efficiency and effectiveness of the way and rate with which they sort through and process the information (Zuley 2010).

There are some common hanging protocols with FFDM images. Viewing them sequentially at full resolution (or even 2:1) is one common format, and if presets are used that automatically bring up the relevant pairs (e.g., right and left MLO for current symmetry comparison, then right and left CC, etc.) it can facilitate workflow. This does, however, limit the user to viewing only a pair of images at a time, which could potentially tax short-term memory capacity—moving on to the next pair may make it difficult to remember adequately what was in the previous pair and possibly more importantly where a potential lesion was.

"Fit to viewport" is another common hanging protocol in which a number of images are displayed at the same time for direct comparison. This creates a problem, however, since most FFDM images have more pixels than display so the mammographer cannot view the images at full resolution. This overall presentation of both breasts and both views may have an advantage in that it likely facilitates the gist or global impression. Perception studies suggests that basic patterns and image properties such as symmetry (important in mammography for detecting architectural distortion), gross deviations from normal, and overall context can be detected in the global view (Kundel and Nodine 1975a). Once perturbations in the images are detected in the global view, detailed scanning of the individual images (by clicking on them to bring to full resolution) can ensue.

There is often a technical problem with the "fit-to-viewport" mode. Patients often have mammograms performed on different units from different manufacturers with different acquisition matrix sizes. The result is that the breast images will display at different sizes, making comparisons (especially with regard to symmetry) difficult. Most dedicated mammography workstations now account for different acquisition matrices and scale images appropriately to be the same size, but there is always room for improvement. To a large extent, the issue is solved from an informatics perspective. Workstations can be designed to utilize image information contained in the DICOM (Digital Imaging and Communications in Medicine) header regarding type of image, acquisition size, pixel size, etc., and the profiles developed as part of the IHE (Integrated Healthcare Enterprise) regarding ways that heterogeneous information can be integrated properly tend to display the images correctly (Channin 2001, IHE 2011a).

Briefly, the core IHE profile relevant to the hanging protocol issue is [MAMMO] Mammography Image, and it specifies how mammography images and evidence objects are created, exchanged, used, and displayed. Information from the DICOM Part 10 (Media Storage and File Format for Media Interchange) defines a file format with a preamble that contains the information typically exchanged during association prior to DICOM network communications, and can be used by the MAMMO profile. The MAMMO profile takes FFDM and CR (computed radiography) and creates MG objects that have 40 attributes with specific mandatory requirements to comply with the profile. The IHE Mammography Handbook (IHE 2011b) describes how to use the profile to deal with a variety of digital mammography issues including different sized images, different image formats, and hanging strategies.

10.6 Ergonomics in the Mammography Reading Room

The perception and interpretation of images often depends significantly on the environment in which they are viewed in addition to the quality of the images and displays. There are a number of human factor issues related to the reading room environment that are important, both in terms of creating an environment in which perceptual capabilities are matched to and optimized for that environment, as well in terms of facilitating workflow.

10.6.1 Input Devices

The type of input device the mammographer uses with a workstation is a matter of personal preference, but some common options include keyboards (typically with hot key options), mouse (with or without a scroll wheel and with various number and types of buttons), and track balls. The key to choosing the input device relates to user comfort and task since the mammographer will be using it continuously to interpret cases and, as with any other computer interaction, the task is repetitive and continuous. The risk of carpal tunnel syndrome and other repetitive musculoskeletal injuries is not insignificant (Ruess et al. 2003). Organizations such as the Occupational Safety and Health Administration (OSHA 2011) offer tips and recommendations for choosing an appropriate input device. The advent of the iPad and similar devices may radically change the way users interact with traditional workstations. We may no longer require input devices, but may simply be using our fingers to flick through stacks of images. Whether there are any long-term effects or repetitive motion injuries associated with these types of interfaces has yet to be investigated or reported in the scientific literature, but blogs and other social media outlets do have users complaining about finger injuries from extended iPad use!

10.6.2 Other Sensory Considerations

The visual system is not the only sensory system to consider in the digital reading room. How much heat do workstations produce, how much noise is produced, do ambient lighting, and noise levels impact performance? Computer-related issues are often readily solved. If it produces too much heat just improve airflow around the computer and in the room. If it is too noisy try replacing the cooling fan with a quieter alternative.

It is not clear, however, whether ambient noise levels impact diagnostic accuracy. One interesting study (McEntee et al. 2010) measured noise levels in a variety of hospital locations including radiology reading rooms and found that noise amplitudes rarely exceeded those encountered during normal conversations (maximum mean was 56.1 dB). In addition to simply measuring ambient noise levels, they also carried out a study to study the impact of noise on diagnostic accuracy. They had 26 radiologists view a series of 30 chest images and search for pulmonary nodules. They wore headphones during the study, once with white noise piped through and once with a recording of typical reading room sounds (e.g., general conversation, pulse monitors) being played. Receiver operating characteristic (ROC) analysis revealed no significant differences in performance (area under the curve Az = 0.68 with noise and 0.67 without) as a function of noise. Another study found that playing music generally improved self-rated mood, concentration, perceived diagnostic accuracy, productivity, and work satisfaction, but there was no assessment of impact on diagnostic accuracy (Shapiro 2009).

In terms of ambient lights, it is recommended that 20 lux of ambient light be used since this is generally sufficient to avoid most reflections and still provide sufficient light for the human visual system to adapt to the surrounding environment and the displays (Krupinski et al. 2007). The ambient lighting should be indirect and backlight incandescent lights with dimmer switches rather than fluorescent are recommended. Light-colored clothing and laboratory coats can increase reflections and glare even with today's LCDs (liquid crystal displays) so they should be avoided if possible. The intrinsic minimum luminance of the device should not be smaller than the ambient luminance.

10.6.3 Reader Fatigue

There is increasing evidence that with current demands to read more cases and with the exponential increase in the number of images per case being generated with modalities such as CT and MRI, radiologists are becoming more fatigued in the workplace. As breast imaging becomes more complex with the addition of MRI, DBT, breast CT, and other advanced imaging modalities, the demands are increasing here as well. The question is whether diagnostic accuracy is affected.

One recent study was conducted to address this question (Krupinski et al. 2010). Forty radiologists (20 faculty and 20 residents) viewed a series of 60 skeletal images half with no fracture and half with a single moderate to very subtle fracture. Demographic on the participants included gender, age, months since last eye examination, dominant eye, percentage wearing corrective lenses, type of lenses worn, what time they woke up on the day of the experiment, how many hours sleep they had, how long they had been reading cases that day, the number of cases, what percent had cold/allergies, itchy/watery eyes, or used eye drops that day. Each observer viewed the cases once in the morning (prior to any reading) and once in the late afternoon (after a day of reading). They also completed two surveys that assessed physical, mental, and visual fatigue. Before and after each session, visual accommodation (ability to focus) was measured.

The results revealed a significant drop in detection accuracy for late versus early reading with an average Az of 0.885 for early and 0.852 for late reading (p = 0.049). Total inspection time for interpreting the examinations revealed that each examination took 52.1 seconds on average for early reading and 51.5 seconds for late reading, with no significant difference. In terms of ability to focus, there was more accommodative error after the workday than before (–1.16 diopters late versus –0.72 diopters for early (p < 0.0001)), suggesting that readers were experiencing more visual strain after their workday. The fatigue surveys revealed significantly higher ratings for lack of energy, physical discomfort, sleepiness, and oculomotor strain. These results were obtained after an average of only 8 hours spent reading. Today's radiologist, including mammographers, typically read closer to 12 hours per day, suggesting that the impact of fatigue on diagnostic accuracy may be even higher.

There have been no studies to date examining the impact of fatigue in mammography, but a retrospective analysis of data from four studies that used mammography images and had the majority of readers participating in the experiment once

TABLE 10.1 Reader Az Values for Morning versus Afternoon Reading in Four Different Mammography Studies

	Morning Reading Az	Afternoon Reading Az
Study 1	0.9466	0.9398
	0.9517	0.9563
	0.8850	0.8784
	0.9691	0.9772
	0.9655	0.9603
	0.9843	0.9837
	0.9695	0.9663
Study 2	0.9637	0.9800
	0.9463	0.8704
	0.8948	0.8560
Study 3	0.9172	0.9017
	0.8822	0.8592
	0.9004	0.9120
	0.8783	0.8718
	0.8632	0.9119
	0.9079	0.8544
	0.9175	0.9103
	0.8996	0.8446
Study 4	0.9871	0.9878
	0.8698	0.8301
	0.8927	0.8754
	0.9457	0.8967

in the morning and once in the afternoon does reveal a similar trend in reduced diagnostic accuracy (Krupinski et al. 1999, 2003, Krupinski and Roehrig 2000). Table 10.1 shows the ROC Az values for the readers in the four studies. When examined together, the average Az for morning reading was 0.924 and for late afternoon reading 0.910, which was statistically significant (p = 0.0277). Overall, it also took the readers longer to interpret the cases late in the day (p < 0.0001) compared to early in the day (for screen-film cases 27.01 versus 28.77 sec/case morning versus afternoon, respectively, and for softcopy, 51.38 versus 56.82 sec/case). Although fatigue was not being studied at the time these studies were conducted, it was interesting to find that there were significant drops in diagnostic accuracy interpreting mammograms (detection of both masses and microcalcifications) as a function of whether the cases were read early in the day prior to extended reading or late in the day after having spent a number of hours reading.

10.7 Future Directions

10.7.1 Displays and Perception

Clinicians will continue to rely on images as part of the arsenal with which diseases and other abnormalities are detected, diagnosed, and treated. Currently, only medical-grade high-performance monochrome monitors are recommended for primary interpretation of mammographic images (Siegel et al. 2006). Recently, however, a number of studies have demonstrated

the potential of color displays, both medical-grade (MG) of commercial off-the-shelf (COTS) for use in radiology (Geijer et al. 2007, Krupinski 2009), and there is little reason to believe they would not be appropriate for use in mammography as long as the users understand any limitations they might pose.

COTS and color displays have improved dramatically in recent years and are nearly equivalent to MG displays. A key difference driving the acceptance of color COTS displays in radiology is cost—COTS displays cost far less than MG displays. A disadvantages to COTS displays include the fact that they rarely come preloaded with software to calibrate to the DICOM GSDF (grayscale standard display function) or the necessary hardware and software to carry out (either manually or automatically) the periodic monitoring and resetting of the calibration. From a perceptual perspective, calibrating to the DICOM GSDF is critical, as it has been shown to improve diagnostic accuracy compared to a non-GSDF calibrated display (Krupinski and Roehrig 2000). Calibration issues are readily resolvable, but they require attention on the part of the user, and many clinicians simply do not have the time to carry out the initial calibration or maintain it. The second major difference between COTS and MG displays is lifetime. MG displays typically last significantly longer than COTS displays, although the shorter lifetime COTS displays are more readily replaced due to their lower cost. It is a tradeoff that clinicians need to consider when deciding between the two types of displays.

Color displays also lend themselves readily to "all-purpose" workstations that allow the breast imager to access and integrate multimodality data (MRI, SPECT/PET, CT, FFDM), decision aids (e.g., color coded CAD prompts; see Chapter 21 for more on CAD), as well as image data from other clinical specialties. As breast care becomes more clinically integrated technologically and functionally (Weinstein et al. 2007), breast imagers will be able to access other types of data (e.g., virtual pathology slides from biopsies; see Figure 10.5) to correlate results with their diagnostic interpretations, better appreciate the results of their biopsies, and improve communication with their patients by having access to the entire patient record and clinical history.

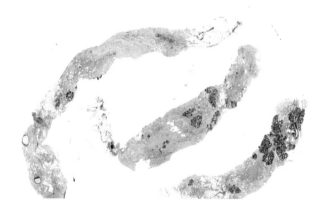

FIGURE 10.5 Typical "virtual" breast biopsy image acquired by scanning a glass specimen slide with a dedicated pathology slide scanner.

Mobile, portable display devices such as the notebook, notepad, phone, and PDA (personal digital assistant) are exploding in their popular use around the world (Shih et al. 2010, Qian et al. 2008). Their use in medicine in general is becoming more prevalent, especially in mobile health (m-health) and telemedicine applications (Johnson and Armstrong 2011, DeCosta et al. 2010, Vezzoni et al. 2011). Radiology has also seen high interest in the potential utility of these portable devices (Applegate 2010, Boonn and Flanders 2005, Flanders et al. 2003) and the number of studies evaluating their potential as well as their limitations is increasing.

One recent study examined the ability of radiologists to diagnose orthopedic injuries (PA wrist images) and intracranial hemorrhage (brain CT) using a PDA and an Apple iPod Touch device (Toomey et al. 2010), comparing these mobile devices to viewing on a secondary-class (review) monitor. The brain study scores were significantly higher with the PDA than with the monitor, but, for the wrist study, there were no statistically significant differences. Neither study revealed significant differences between the iPod Touch and monitor interpretations. It was concluded that handheld devices show promise for emergency teleconsultation of basic orthopedic injuries and intracranial hemorrhage.

Whether these types of mobile devices, at least in the near future, would be appropriate even for secondary review of digital mammograms seems unlikely given the resolution and contrast requirement, but clearly more research needs to be conducted to verify diagnostic accuracy with these portable devices and specific the types of examinations that are and are not appropriate for interpretation. The potential for using these display devices in limited and well-defined circumstances such as emergency and second-opinion consultants is extremely high. A recent study (Chen et al. 2010) did investigate the use of the iPhone for mammography training and found that as long as the device resolution and interface were properly designed, it could be used in some training circumstances. It does, however, require a significant amount of scrolling and zooming to actually access the full image resolution data (making the image only available in small segments/areas), which could be tedious and potentially confusing for more than a few images.

Another area where perception research can facilitate the implementation and use of future displays is with advanced tools and methods to manipulate volumetric data from DBT, CT, and MRI images. There is a considerable amount of research being conducted on the use of virtual reality (VR) and other 3-D rendering techniques to display and view complex volumetric data sets (Adam 2009, Koehring et al. 2008, Berry et al. 2008), although, to date, very little has been done in breast imaging applications. Most of the interest and application so far has been in the area of developing tools for teaching in a VR environment, but large DBT, CT, and MRI data sets will likely lend themselves very well to the creation of 3-D images that can be rendered and manipulated effectively in VR. The real question, however, that has not been addressed to date in any true scientific manner, is whether the use of VR and true 3-D representation of image data

actually impacts the clinician—does it affect diagnostic accuracy and does it reduce viewing time? Does the addition of 3-D really provide more diagnostic information to the human eye–brain system? As noted earlier, the Getty study with stereoscopic presentation of mammographic images does suggest that the third dimension does improve the perception of details in mammographic images, so true 3-D reconstructions and displays (VR and even holographic) could improve sensitivity and specificity even further. More research is needed, however, to confirm these hypotheses and study the impact of 3-D on workflow as well as breast imager (e.g., do 3-D VR environments induce nausea in too many people?).

Eventually, physical displays may not even be necessary at all for viewing mammographic images. Direct retinal imaging (DRI) presents information directly to the retina of the user without any physical display surface by scanning a modulated laser beam across the retina to build up an image. Current prototypes utilize modified glasses equipped with the lasers and optical components. There are obvious portability and mobility advantages with DRI, and the monochromatic laser sources are generally very photon efficient compared with traditional displays, producing collimated light that can be made much brighter at a given power consumption. These displays also have the potential for increased color gamuts compared to traditional displays, extending their utility beyond FFDM to MRI and ultrasound where color overlays are often used. As the system is laser-based, however, there are some safety concerns at this point in time regarding extended use of these systems as would be required if a radiologist or other clinician was to use it for extended reading periods.

10.7.2 Integrating Visual Search into the Clinical Environment

To date there has been very little use of eye-tracking in the clinical environment as the technology is not amenable to wearing for long periods. There are remote systems that do not require the user to wear any glasses or other recording apparatus, but these generally require the user to sit in a rather limited field since the sensors are fixed and cannot track the user as they move outside this field. Why would you want to record eye position during actual clinical reading? One possibility is a perceptual version of computer-aided feedback (Tourassi et al. 2010).

The idea has roots in the series of studies described earlier that explored the types of errors that could be identified using visual search recording methods. It was found that there are three types of errors—search, recognition, and detection, and they differ as a function of how much dwell time they receive during search (Kundel et al. 1978). Subsequent to the classification of false-negative errors based on dwell, another study was done to determine the relationship between dwell and other decisions (true- and false-positives and true-negatives). This study found that the distributions of dwells associated with the various decisions made during search could be characterized by survival analysis, and that overall true- and false-positive decisions tend to be associated with the longest dwell times and true-negatives with the shortest dwell times. False-negatives (the recognition and decision errors that do receive dwell) fall somewhere in between (Kundel et al. 1989). These findings were confirmed for reading of mammograms (Krupinski 1996).

Based on the fact that at least two thirds of false-negatives actually receive substantial visual attention as reflected in the dwell times, a system was designed to provide feedback to radiologists regarding locations where they had looked at for a long time but did not identify any lesions at that location. This perceptual feedback was designed to give radiologists a second chance to recognize missed lesions by using their own perceptual responses to the images (Krupinski et al. 1998). The areas on the image that received prolonged dwell during the initial search of the image were highlighted with a circle immediately afterward. The radiologists reviewed these highlighted areas and had the opportunity to revise their decisions. Figure 10.6 shows how these perceptually based feedback highlights appear on a mammogram. In the control condition, radiologists simply got a second look at the images without the perceptual feedback highlight circles. The alternative free response receiver operating characteristic (AFROC) technique was used to analyze the decision data. In the control condition the initial area under the curve A1 value was 0.540 and in the second look was 0.504, which was not statistically significant. With perceptual feedback however, the initial A1 was 0.495, and with feedback, it increased to 0.618, which was statistically significant (p < 0.001). Perceptual feedback resulted in a 20% increase in true-positive decisions compared to the control, without a significant increase in the false-positives.

These results were very promising for the use of perceptual feedback as a way to improve observer performance. Eye-tracking technology, as noted earlier, is not developed sufficiently to be used on a regular basis in the clinical environment to record eye position in a nonobtrusive and accurate manner to make perceptual feedback a viable decision aid. It may become a viable technology in the future, however, so research needs to continue and that is what Tourassi et al. (2010) is proposing with context-sensitive CADe. Instead of simply developing computer-aided detection and decision tools that operate independently of the radiologist and simply provide "black box" derived indications of where potential lesions might be, the idea is to make CAD more interactive. In context-sensitive CADe, the CADe "opinion" is adapted based on contextual information provided by the radiologists' visual search characteristics and decisions for a given case.

To provide support to move ahead with the idea, a pilot study was carried out in which six radiologists reviewed 20 mammograms while their eye position was recorded. The decisions rendered were correlated with the eye-position data as in previous studies to characterize dwells associated with the decisions. The same cases were then analyzed with a knowledge-based CADe system in two different modes. In the conventional mode, it simply used a globally fixed decision threshold, but in the context-sensitive mode, it incorporated a location-variable decision threshold based on the eye-position data. The conventional CADe system had 85.7% per-image sensitivity (for malignant masses) at a false-positive rate of 3.15 per image. The context-sensitive CADe had 85.7%–100% sensitivity with 0.35–0.40 false-positives/image across the six radiologists, suggesting that the context-sensitive CADe has the potential to reduce perceptual errors in mammography. The system needs to be further refined and tested with a larger set of mammographers and images, but in the near future, CAD could be integrated with and tailored to the individual mammographer. The potential for improved and more efficient training of residents is also quite high.

10.8 Summary

There is little doubt that studying the way that mammographers interpret breast images has improved our understanding of the perceptual mechanisms underlying the interpretation process. Perception studies have also contributed significantly to technology development and validating the impact of new technologies on the way that mammographers interpret images as well as interact with the images presented to them. The future of breast imaging is wide open as new technologies continue to be explored. Breast MRI is becoming much more common as its benefits are being revealed through more clinical research studies (Boetes 2010), although there are still aspects where perception research is important, such as examining the impact of CAD (Shimauchi et al. 2011, Muralidhar et al. 2011), and the interpretation of enhancement patterns (Jansen et al. 2011).

Digital breast tomosynthesis and breast CT are technologies likely to make a significant impact in breast imaging within the next 5 years, but there too are numerous ways that perception research can aid in the optimization and acceptance of these modalities. For example, in DBT, there are still aspects of the acquisition (e.g., mean glandular dose, tube load),

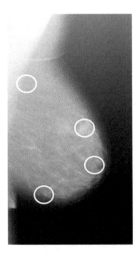

FIGURE 10.6 Example of perceptually based highlight prompts fed back to a mammographer after having viewed the image without any prompts. Each circle represents an area that received prolonged dwell, thus potentially contained a lesion missed during the initial search.

reconstruction, filtering algorithms, and display that need to be optimized. Mathematical models that incorporate characteristics of the human visual system can be used in the early stages of optimization to better understand and narrow down those variables that are most likely going to contribute to observer diagnostic accuracy (Castella et al. 2009).

It is likely that we will also see much more multimodality systems emerging over the next few years, combining structural imaging (mammography, CT, DBT) with functional imaging (MR, SPECT, PET, molecular imaging) to better detect and characterize breast lesions as well as determine and monitor what treatments are most effective at an individual level. At the present time, much of the research in multimodality breast imaging is in the realm of physics and engineering (Williams et al. 2010, Meinel et al. 2010, Cowey et al. 2007), but as these devices reach the point where breast imagers will be validating them and using them in the clinical environment, it will be important again to optimize them with respect to what the human visual system is capable of perceiving and how the integrated data can best be presented to them without causing information overload or errors related to issues in registration, segmentation, and artifacts.

In the end, although technology, engineering and physics are critical components in our fight against breast cancer and finding better ways to detect, diagnose, and treat this disease that affects millions of women every year, it is the breast imager and associated clinicians that must interpret the image and other relevant data and decide on courses of action that will effectively and efficiently treat patients if cancer is found. Until this changes, we must continue to involve mammographers in our research to better understand and appreciate the perceptual and cognitive mechanisms underlying medical image interpretation.

References

Adam A. 2009. Radiology: does it have a sell-by date? Ann Acad Med Sing 38: 1031–1033.

American College of Radiology Mammography Case Review. 2011. http://www.acr.org/SecondaryMainMenuCategories/ACRStore/FeaturedCategories/Education/MultimediaCDs/MamoCaseReview.aspx. Accessed July 13, 2011.

Applegate JS. 2010. The role of mobile electronic devices in radiographer education. Radiol Tech 82: 124–131.

Beam CA, Conant EF, Sickles EA. 2003. Association of volume and volume-independent factors with accuracy in screening mammogram interpretation. JNCI 95: 282–290.

Berg W, Campassi C, Langenberg P et al. 2000. Breast imaging reporting and data system: Inter- and intraobserver variability in feature analysis and final assessment. Am J Roentgenol 174: 1769–1777.

Berry M, Reznick R, Lystig T, Lonn L. 2008. The use of virtual reality for training in carotid artery stenting: A construct validation study. *Acta Radiol* 49: 801–805.

Bird RE, Wallace TW, Yankaskas BC. 1992. Analysis of cancers missed at screening mammography. Radiol 184: 613–617.

Boetes C. 2011. Update on screening breast MRI in high-risk women. Obstet Gyn Clin North Am 38: 149–158.

Boonn WW, Flanders AE. 2005. Survey of personal digital assistant use in radiology. RadioGraphics 25: 537–541.

Burnside E, Sickles E, Sohlich R et al. 2002. Differential value of comparison with previous examinations in diagnostic versus screening mammography. Am J Roentgen 179: 1173–1177.

Castella C, Ruschin M, Eckstein MP et al. 2009. Mass detection in breast tomosynthesis and digital mammography: A model observer study. Proc SPIE Med Imag 7263: 72630O.

Channin D. 2001. Integrating the Healthcare Enterprise: A Primer, Part 2. Seven Brides for Seven Brothers: The IHE Integration Profiles. RadioGraphics 21: 1343–1350.

Chen Y, Turnbull A, James J et al. 2010. Breast screening: Visual search as an aid for digital mammographic interpretation training. Proc SPIE Med Imag 7627: 76270C.

Ciccone G, Vineis P, Frigerio A et al. 1992. Inter-observer and intra-observer variability of mammogram interpretation: A field study. Eur J Cancer 28A: 1054–1058.

Cole E, Pisano E, Brown M et al. 2004. Diagnostic accuracy of Fischer Senoscan digital mammography versus screen-film mammography in a diagnostic mammography population. Acad Radiol 11: 879–886.

Cowey S, Szafran AA, Kappes J et al. 2007. Breast cancer metastasis to bone: Evaluation of bioluminescent imaging and microSPECT/CT for detecting bone metastasis in immunodeficient mice. Clin Exptl Metas 24: 389–401.

Crowley RS, Naus GJ, Stewart J et al. 2003. Development of visual diagnostic expertise in pathology: An information-processing study. J Am Med Inform Assoc 10: 39–51.

Decosta A, Shet A, Kumarasamy N et al. 2010. Design of a randomized trial to evaluate the influence of mobile phone reminders on adherence to first line antiretroviral treatment in South India—The HIVIND study protocol. BMC Med Res Method 10: 25.

Del Turco MR, Mantellini P, Ciatto S et al. 2007. Full-field digital versus screen-film mammography: Comparative accuracy in concurrent screening cohorts. AJR 189: 860–866.

DeMartini W, Lehman C, Partridge S. 2008. Breast MRI for cancer detection and characterization: Review of evidence-based clinical applications. Acad Radiol 15: 408–416.

Diekmann F. 2010. Digital breast tomosynthesis and breast CT. In Bick U, Diekmann F, eds. Digital Mammography. pp. 199–209. New York: Springer.

Egan R. 1960. Experience with mammography in a tumor institution: evaluation of 1,000 cases. Am J Roentgenol 75: 894–900.

Elmore J, Wells C, Lee C et al. 1994. Variability in radiologists' interpretations of mammograms. N Engl J Med 331: 1493–1499.

Esserman L, Cowley H, Carey E et al. 2002. Improving accuracy of mammography: Volume and outcome relationships. J Natl Cancer Inst 94: 369–375.

Even-Sapir E, Inbar M. 2010. PET in women with high risk for breast or ovarian cancer. Lancet Oncol 11: 899–905.

Feig S. 2011. Comparison of costs and benefits of breast cancer screening with mammography, ultrasonography, and MRI. Obstet Gyn Clin North Am 38: 179–196.

Flanders AE, Wiggins RH, Gozum ME. 2003. Handheld computers in radiology. RadioGraphics 23: 1035–1047.

Food and Drug Administration Department of Health and Human Services. 2011a. Mammography. http://www.fda.gov/cdrh/mammography/. Accessed July 12, 2011.

Food and Drug Administration Department of Health and Human Services. 2011b. Mammography http://www.fda.gov/cdrh/mammography/scorecard-statistics.html. Accessed July 12, 2011.

Geijer H, Geijer M, Forsberg L et al. 2007. Comparison of color LCD and medical-grade monochrome LCD displays in diagnostic radiology. J Dig Imag 20: 114–121.

Getty DJ, D'Orsi CJ, Pickett RM. 2008. Stereoscopic digital mammography: Improved accuracy of lesion detection in breast cancer screening. In Krupinski EA, ed. Digital Mammography. pp. 74–79. Berlin: Springer-Verlag.

Gold RH, Bassett LW, Widoff BE. 1990. Highlights from the history of mammography. RadioGraphics 10: 1111–1131.

Haber RN. 1969. Information-Processing Approaches to Visual Perception. New York: Holt, Rinehart and Winston.

Hakim CM, Chough DM, Ganott MA et al. 2010. Digital breast tomosynthesis in the diagnostic environment: A subjective side-by-side review. Am J Roentgen 195: W172–W176.

Haygood TM, Arribas E, Brennan PC et al. 2009. Conspicuity of microcalcifications on digital screening mammograms using varying degrees of monitor zooming. Acad Radiol 16: 1509–1517.

Houssami N, Irwig L, Simpson JM et al. 2003. The contribution of work-up or additional views to the accuracy of diagnostic mammography. The Breast 12: 270–275.

Houssami N, Irwig L, Simpson JM et al. 2004. The influence of clinical information on the accuracy of diagnostic mammography. Br Cancer Res Treatment 85: 223–228.

Integrated Healthcare Enterprise. 2008a. www.ihe.net. Accessed July 13, 2011.

Integrated Healthcare Enterprise Mammography Handbook. 2008b. http://www.ihe.net/Mammo/. Accessed July 13, 2011.

Jansem SA, Lin VC, Giger ML et al. 2011. Normal parenchymal enhancement patterns in women undergoing MR screening of the breast. Eur Radiol 21: 1374–3182.

Johnson MN, Armstrong AW. 2011. Technologies in dermatology: Teledermatology review. Giornale Italiano di Dermatologia e Venereologia 146: 143–153.

Karssemeijer N, Snoeren PR. 2010. Image processing. In Bick U, Diekmann F, eds. Digital Mammography. pp. 69–83. New York: Springer.

Kerlikowske K, Grady D, Barclay J et al. 1998. Variability and accuracy in mammographic interpretation using the American College of Radiology Breast Imaging Reporting and Data System. J Natl Cancer Inst 90: 1801–1809.

Kim Y. 2004. Online education tools. Pub Perf Manag Rev 28: 275–280.

Koehring A, Foo JL, Miyano G et al. 2008. A framework for interactive visualization of digital medical images. J Laproendoscop Adv Surg Tech 18: 697–706.

Krupinski EA. 1996. Visual scanning patterns of radiologists searching mammograms. Acad Radiol 3: 137–144.

Krupinski EA. 2009. Medical grade vs off-the-shelf color displays: Influence on observer performance and visual search. J Dig Imag 22: 363–368.

Krupinski EA, Berbaum KS, Caldwell RT et al. 2010. Long radiology workdays reduce detection and accommodation accuracy. J Am Coll Radiol 7: 698–704.

Krupinski EA, Nodine CF, Kundel HL. 1998. Enhancing the recognition of lesions in radiographic images using perceptual feedback. Opt Eng 37: 813–818.

Krupinski EA, Roehrig H. 2000. The influence of a perceptually linearized display on observer performance and visual search. Acad Radiol 7: 8–13.

Krupinski EA, Roehrig H. 2010. Optimization of display systems. In Samei E, Krupinski E, eds. The Handbook of Medical Image Perception and Techniques. pp. 395–405. New York: Cambridge University Press.

Krupinski EA, Roehrig H, Furukawa T. 1999. Influence of film and monitor luminance on observer performance and visual search. Acad Radiol 6: 411–418.

Krupinski EA, Johnson J, Roehrig H et al. 2003. Using a human visual system model to optimize soft-copy mammography display: Influence of MTF compensation. Acad Radiol 10: 1030–1035.

Krupinski EA, Johnson J, Roehrig H et al. 2005. On-axis and off-axis viewing of images on CRT displays and LCDs: Observer performance and vision model predictions. Acad Radiol 12: 957–964.

Krupinski EA, Williams MB, Andriole K et al. 2007. Digital radiography image quality: Image processing and display. J Am Coll Radiol 4: 389–400.

Kundel HL. 1975b. Peripheral vision, structured noise and film reader error. Radiol 114: 269–273.

Kundel HL, Nodine CF. 1975a. Interpreting chest radiographs without visual search. Radiol 116: 527–532.

Kundel HL, Nodine CF, Carmody DP. 1978. Visual scanning, pattern recognition and decision-making in pulmonary tumor detection. Invest Radiol 13: 175–181.

Kundel HL, Nodine CF, Conant EF et al. 2007. Holistic component of image perception in mammogram interpretation: Gaze-tracking study. Radiol 242: 396–402.

Kundel HL, Nodine CF, Krupinski EA. 1989. Searching for lung nodules: Visual dwell indicates locations of false-positive and false-negative decisions. Acad Radiol 24: 472–478.

Kundel HL, Nodine CF, Krupinski EA et al. 2008. Using gaze-tracking data and mixture distribution analysis to support a holistic model for the detection of cancers on mammograms. Acad Radiol 15: 881–886.

Lehman CD, Gatsonis C, Kuhl CK et al. 2007. MRI evaluation of the contralateral breast in women with recently diagnosed breast cancer. N Engl J Med 356: 1295–1303.

Leung JWT, Margolin FR, Dee KE et al. 2007. Performance parameters for screening and diagnostic mammography in a community practice: Are there differences between specialists and general radiologists? Am J Roentgen 188: 236–241.

Lewin JM. 2010. Digital mammography clinical trials: The North American experience. In Bick U, Diekmann F, eds. Digital Mammography. pp. 145–154. New York: Springer.

Lewin J, Hendrick E, D'Orsi CJ et al. 2001. Comparison of full-field digital mammography with screen-film mammography for cancer detection: Results of 4,945 paired examinations. Radiology 218: 873–880.

Lewin J, Hendrick E, D'Orsi CJ et al. 2002. Comparison of full-field digital mammography and screen-film mammography for detection of breast cancer. Am J Roentgenol 179: 671–677.

McEntee MF, Coffey A, Ryan J et al. 2010. Nuisance levels of noise effects radiologists' performance. Proc SPIE Med Imag 7627: 76270O.

Meinel LA, Abe H, Bergtholdt M et al. 2010. Multi-modality morphological correlation of axillary lymph nodes. Int J Comp Assist Radiol Surg 5: 343–350.

Mello-Thoms C, Hardesty L, Sumkin J et al. 2005. Effects of lesion conspicuity on visual search in mammogram reading. Acad Radiol 12: 830–840.

Miglioretti D, Smith-Bindman R, Abraham L et al. 2007. Radiologist characteristics associated with interpretive performance of diagnostic mammography. J Natl Cancer Inst 99: 1854–1863.

Mundinger A, Wilson ARM, Weismann C et al. 2010. Breast ultrasound—Update. Eur J Cancer Suppl 8: 11–14.

Muralidhar GS, Bovik AC, Sampat MP et al. 2011. Computer-aided diagnosis in breast magnetic resonance imaging. Mt Sinai J Med 78: 280–290.

Nilsson T, Hedman L, Ahlqvist J. 2007. Visual-spatial ability and interpretation of three-dimensional information in radiographs. Dento-Maxillo-Facial Radiol 36: 86–91.

Nishikawa RM. 2010. Computer-aided detection and diagnosis. In Bick U, Diekmann F, eds. Digital Mammography. pp. 85–106. New York: Springer.

Nodine CF, Kundel HL. 1987. Using eye movements to study visual search and to improve tumor detection. RadioGraphics 7: 1241–1250.

Nodine CF, Kundel HL, Mello-Thoms C et al. 1999. How experience and training influence mammography expertise. Acad Radiol 6: 575–585.

Nodine CF, Kundel HL, Toto LC et al. 1992. Recording and analyzing eye-position data using a microcomputer workstation. Behav Res Meth Instrum Comp 24: 475–485.

Occupational Safety and Health Administration. 2011. http://www.osha.gov/SLTC/etools/computerworkstations/index.html. Last accessed July 13, 2011.

Oude Munnink TH, Nagengast WB, Brouwers AH, 2009. Molecular imaging of the breast. Breast 18: S66–S73.

Pisano E, Gatsonis C, Hendrick E et al. 2005. Diagnostic performance of digital versus film mammography for breast-cancer screening. N Engl J Med 353: 1773–1783.

Qian LJ, Zhou M, Xu JR. 2008. An easy and effective approach to manage radiologic portable document format (PDF) files using iTunes. Am J Roentgen 191: 290–291.

Quaghebeur G, Bhattacharya G, Murfitt J. 1997. Radiologists and visual acuity. Eur Radiol 7: 41–43.

RadiologyEducation.com. http://www.radiologyeducation.com/. Accessed July 11, 2011.

Ruess L, O'Connor SC, Cho KH et al. 2003. Carpal tunnel syndrome and cubital tunnel syndrome: Work-related musculoskeletal disorders in four symptomatic radiologists. Am J Roentgen 181: 37–42.

Safdar NM, Siddiqui KM, Qureshi F et al. 2009. Vision and quality in the digital imaging environment: How much does the visual acuity of radiologists vary at an intermediate distance? Am J Roentgenol 192: W335–W340.

Shapiro L. 2009. Baroque music boosts radiologists' mood and cognition. DOTmed News http://www.dotmed.com/news/story/8878/. Accessed July 13, 2001.

Shih G, Lakhani P, Nagy P. 2010. Is Android or iPhone the platform for innovation in imaging informatics. J Dig Imag 23: 2–7.

Shimauchi A, Giger ML, Bhooshan N et al. 2011. Evaluation of clinical breast MR imaging performed with prototype computer-aided diagnosis breast MR imaging workstation: Reader study. Radiol 258: 696–704.

Siegel E, Krupinski E, Samei E et al. 2006. Digital mammography image quality: Image display. J Am Coll Radiol 3: 615–627.

Skaane P. 2010. Digital mammography in European population-based screening programs. In Bick U, Diekmann F, eds. Digital Mammography. pp. 155–173. New York: Springer.

Skanne P, Skjennald A. 2004. Screen-film mammography versus full-field digital mammography with soft-copy reading: randomized trial in a population-based screening program—The Oslo II study. Radiol 232: 197–204.

Skanne P, Young K, Skjennald A. 2003. Population-based mammography screening: Comparison of screen-film and full-field digital mammography with soft-copy reading—Oslo I study. Radiol 229: 877–884.

Smith RA. 2007. The evolving role of MRI in the detection and evaluation of breast cancer. N Engl J Med 356: 1362–1364.

Smith-Bindman R, Chu P, Migloretti DL et al. 2005. Physician predictors of mammographic accuracy. J Natl Cancer Inst 97: 358–367.

Spangler ML, Zuley ML, Sumkin JH et al. 2011. Detection and classification of calcifications on digital breast tomosynthesis and 2D digital mammography: A comparison. Am J Roentgen 196: 320–324.

Sparks R, Madabhushi A. 2011. Content-based image retrieval utilizing explicit shape descriptors: Applications to breast MRI and prostate histopathology. Proc SPIE Med Imag 7962: 79621I.

Stojadinovic A, Summers TA, Eberhardt J et al. 2011. Consensus recommendations for advancing breast cancer: Risk identification and screening in ethnically diverse younger women. J Cancer 2: 210–227.

Straub WH, Gur D, Good BC. 1991. Visual acuity testing of radiologists—Is it time? Am J Roentgen 156: 1107–1108.

Strax P, Venet L, Shapiro S. 1973. Value of mammography in reduction of mortality from breast cancer in mass screening. Am J Roentgenol Radium Ther Nucl Med 117: 686–689.

Tabar LK, Nino AS, Scgreiber E et al. 2010. Three digital mammography display configurations: Observer performance in a pilot ROC study. In Marti J, Oliver A, Freixenet J, Marti R, eds. Digital Mammography. pp. 280–287. Berlin: Springer.

Toomey RJ, Ryan JT, McEntee MF et al. 2010. Diagnostic efficacy of handheld devices for emergency radiologic consultation. Am J Roentgen 194: 469–474.

Tourassi GD, Mazurowski MA, Harrawood BP et al. 2010. Exploring the potential of context-sensitive CADe in screening mammography. Med Phys 37: 5728–5736.

Tuddenham WJ, Calvert WP. 1961. Visual search patterns in roentgen diagnosis. Radiol 76: 255–256.

Vezzoni GM, Guazzelli M, Barachini P. The "phone and mail" system in a teledermatology service for chronic psychiatric patients. Giornale Italiano di Dermatologia e Venereologia 146: 95–101.

Weinstein RS, Lopez AM, Barker GP et al. 2007. The innovative bundling of teleradiology, telepathology, and teleoncology services. IBM Syst J 46: 69–84.

Williams MB, Judy PG, Gong Z et al. 2010. In Marti J, Oliver A, Freixenet J, Marti R, eds. Digital Mammography. pp. 444–451. Berlin: Springer.

Yaffe MJ. 2010a. Basic physics of digital mammography. In Bick U, Diekmann F, eds. Digital Mammography. pp. 1–11. New York: Springer.

Yaffe MJ. 2010b. Detectors for digital mammography. In Bick U, Diekmann F, eds. Digital Mammography. pp. 13–31. New York: Springer.

Yantis S, Jonides J. 1984. Abrupt visual onsets and selective attention: Evidence from visual search. J Exptl Psych: Hum Percep Perf 10: 601–621.

Yeh ED. 2011. Breast magnetic resonance imaging: Current clinical indications. Obstet Gyn Clin North Am 38: 159–177.

Zuley M. 2010. Perceptual issues in reading mammograms. In Samei E, Krupinski E, eds. The Handbook of Medical Image Perception and Techniques. pp. 364–379. New York: Cambridge University Press.

Zuley ML, Bandos AI, Abrams GS et al. 2010. Time to diagnose and performance levels during repeat interpretations of digital breast tomosynthesis: preliminary observations. Acad Radiol 17: 450–455.

Image Analysis
and Modeling

Physical Breast Phantoms

Ernest L. Madsen
University of Wisconsin-Madison

11.1 X-Ray Imaging

11.1.1 Ordinary Mammography

A broad range of tissue-mimicking (TM) materials has been employed in breast phantoms for use in x-ray imaging. (For a discussion of x-ray digital mammography, see Chapter 1.) The degree to which materials mimic breast tissues varies greatly. For conventional projection imaging such as x-ray mammography, early quality assurance (QA) phantoms used acrylic and wax components to mimic normal breast tissues and nylon fibers, alumina particles, and phenolic discs to mimic abnormal structures. One version of such a phantom has been approved for use in satisfying the American College of Radiology (ACR) accreditation requirements, and it is produced by three manufacturers: Gammex, Inc. model 156, CIRS model 015, and Nuclear Associates model 18-220.

A comparison of seven different commercially produced mammography phantoms was done regarding their effectiveness over ranges of x-ray tube potential, focal spot size, and magnification (Faulkner and Law 1994). A scoring system for quantifying the "degree of visibility" of objects in the phantoms is described and all seven phantoms assessed using that system. None of the phantoms was clearly superior to the others in all respects.

The principal component materials of the "Barts" phantom (White and Tucker 1980) were formed from epoxy-based tissue-mimicking TM materials with elemental compositions corresponding to skin, water, adipose tissue, and a mixture of 50% water and 50% adipose tissue. Calcifications were formed from Al_2O_3 and silicon and had spherical, cylindrical, or cubic shapes;

sizes were incremented from small enough to be undetectable to much larger and easily detectable on radiographs. The external shape of the phantom simulated the compressed breast with a half-cylinder shape. The background TM material was the 50%–50% water–adipose material. Cylinders of adipose tissue and water ranged from 0.5 through 10 mm in diameter, and their axes were parallel to the axis of the half-cylinder external breast shape. A diagram of the phantom is shown in Figure 11.1, and images are shown in Figures 11.2 and 11.3.

TM materials aimed at more closely mimicking the x-ray attenuation properties of breast tissue have been assessed in terms of measured linear attenuation coefficients. One study addressed commercially available materials designed to simulate breast fat and glandular tissue (Byng et al. 1998). Close agreement between these materials and breast tissues were found from 18 through 100 keV. In an earlier study involving materials mimicking breast fat and glandular tissue for ultrasound, the TM fat and TM glandular materials were also in good agreement with excised tissues (Burke 1982). Thus, these materials mimic normal breast tissues for both ultrasound imaging and x-ray mammography.

Another parameter to be considered regarding mimicking tissues is scattering. One study determined the differential scattering cross section per unit solid angle at 17.44 keV (K_α-radiation of Mo) for breast tissues and for eight materials used to simulate breast tissues. Results for commercial breast tissue-equivalent materials were found to be similar to adipose tissue, water was similar to glandular tissue.

Attempts to more accurately represent the x-ray properties of breast tissue in phantoms employed preserved excised breast

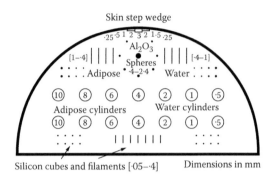

FIGURE 11.1 Breast phantom (test object) with inclusions. Adipose and water cylinders are 5 mm in diameter and range from 0.5 to 10 mm in length. Adipose and water cubes are 0.4 to 2.4 mm. Filaments are 0.4 to 1 mm diameter and 10 mm long. Simulated calcifications (Al_2O_3): spheres are 0.4 and 2.4 mm diameter and cubes are 0.4 to 1 mm on a side. Skin thicknesses are 0.25 to 3 mm thick. Silicon cubes and filaments are also shown near the bottom. (Reproduced from White, D.R., Tucker, A.K. 1980. *British Journal of Radiology* 53: 331–335. With permission.)

tissue. In one study breast tissue fixed in formalin was embedded in an epoxy block (Olsen 1998). The block provided for nine square tiles to be positioned in a 3 × 3 square array. Some of the nine tiles contained simulated masses or calcifications. The phantom was employed in a receiver operating characteristic (ROC) study for comparing results at five different laboratories.

Another phantom contained preserved breast tissue embedded in a paraffin block with a 2-mm-thick acrylic test plate and 25 holes in which were placed egg shell fragments to simulate calcifications (Obenaur 2003). ROC analyses were done using phantom images made with conventional screen-film mammography (SFM) and beam quality parameters selected automatically by the SFM system. ROC analyses were also done using a full-field digital mammography (FFDM) system with the same beam quality parameters. The average glandular dose (AGD) was the same for both systems. Two other sets of beam quality parameters were also employed with the FFDM system with resulting lower AGD and no decrease in detectability of the egg shell fragments.

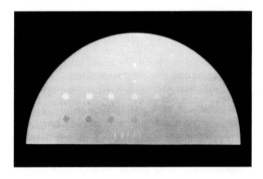

FIGURE 11.2 Radiograph of test object with Mo target with screened film. (Reproduced from White, D.R., Tucker, A.K. 1980. *British Journal of Radiology* 53: 331–335. With permission.)

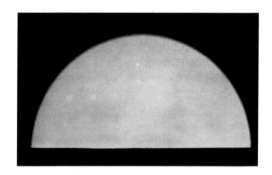

FIGURE 11.3 Radiograph of test object with W target with non-screened film. (Reproduced from White, D.R., Tucker, A.K. 1980. *British Journal of Radiology* 53: 331–335. With permission.)

11.2 3-D X-Ray Breast Imaging

In recent years, 3-D breast imaging systems such as digital breast tomosynthesis (DBT) (see Chapter 4 for a description of breast tomosynthesis) and computed tomography (CT) (see Chapter 5 for a description of breast CT) have been gaining interest in academia and industry. Thus, phantoms that mimic the physical and geometric properties of breast tissues are of interest. Perhaps the first 3-D anthropomorphic phantom to meet these specifications was one developed primarily for research in ultrasound breast imaging (Madsen et al. 1982b). (See Chapter 15 regarding ultrasound imaging of the breast.) The adipose and glandular tissue parts of the phantom were made with materials with x-ray attenuation properties described by Burke et al. (1982). That phantom contained simulated masses and calcifications. These materials were also used in more recent phantoms (Madsen et al. 2006a).

Another 3-D anthropomorphic phantom designed specifically for 3-D x-ray imaging was made following a computer model and "rapid prototyping" which produced successive layers from four forms of "acrylic-based photopolymer" (Carton 2011).

A compressible 3-D phantom was made with polyvinyl alcohol and water where published values of Young's moduli for glandular and tumor were replicated. Adipose tissue was not represented (Price 2011).

An anthropomorphic breast phantom for simulating x-ray and MR properties simultaneously has also been reported (Freed 2011a). Both adipose and glandular mimicking components are included. The materials used were egg whites and lard.

11.3 Ultrasound Imaging

11.3.1 Tissue-Mimicking Materials

Considerable work has been done at the University of Wisconsin-Madison developing tissue-mimicking (TM) materials. The basic material is a water-based gel with microscopic solid or liquid particles to produce tissue-mimicking frequency-dependent attenuation coefficients and backscatter coefficients; any water-based gel will suffice, including gelatin or agar; a detergent can be used to facilitate production of microscopic oil droplets (Madsen et al. 1981, 1982a, 1991b). Nonfat TM materials use solid particles; the concentration of solid particles with mean diameters less than 10 μm allows production of a broad range of the frequency-dependent attenuation coefficients, and larger particles (diameters of tens of microns) allows production of a broad range of backscatter coefficients (Madsen et al. 1991b, 1999). TM fat

(adipose tissue) is made with the oil-in-gel material. Propagation speed in the TM materials can be raised with addition of various water-soluble agents; examples are *n*-propanol, propylene glycol and glycerol (Madsen et al. 2006a, 2006b).

Various preservatives have been used in making these materials. Currently a commercial version used in preserving cosmetics is employed, viz., Liquid Germall Plus® (Madsen et al. 2006b).

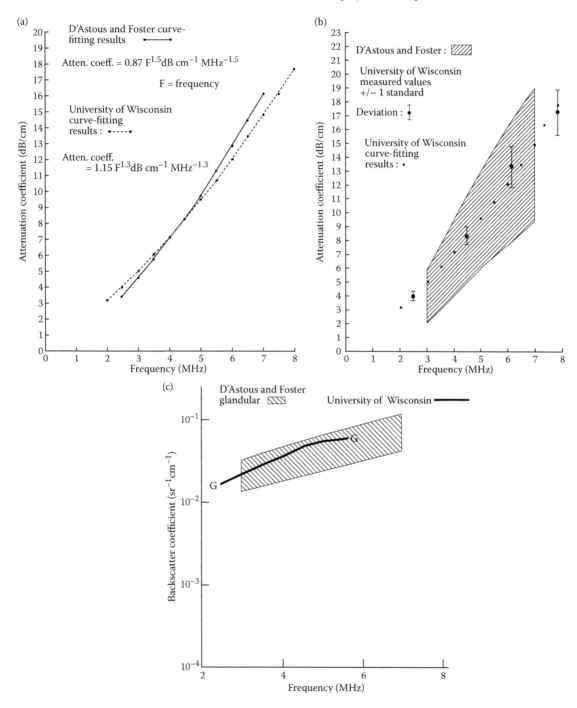

FIGURE 11.4 (a) Results of curve fitting to results of measurements of attenuation coefficients for the TM glandular material at 22°C and for measurements on breast glandular tissue at 37°C. (b) Four measured values of attenuation coefficient for the TM glandular material (closed circles) and the range of measured values for breast glandular tissue (cross-hatched area). Also shown are the curve-fitted values for the TM material. (c) Range of measured values of backscatter coefficients for breast glandular tissue (cross-hatched area) and values measured on the TM material.

A recent article describes versions of the above materials (Cannon 2011). Nonfat materials consist of agar, Al_2O_3 particles to adjust attenuation, larger SiC particles to adjust backscatter and glucose to adjust propagation speed. Adipose tissue consists of a dispersion of olive oil in agar using a detergent to facilitate creation of a dispersion of microscopic droplets.

An alternative TM material for mimicking glandular parenchyma has been reported by Madsen (1995). The material consists of 50% by volume graphite-in-gelatin spheres with diameter range 0.6 to 1.4 mm randomly distributed in an oil-in-gelatin dispersion. The spheres were produced via "shot tower" (Chin 1990). The material mimics (at 22°C) the frequency dependencies of both the attenuation coefficient and backscatter coefficient of in vitro breast glandular (at 37°C) tissue reported by D'Astous (1986). A comparison of the TM material measurements (Madsen et al. 1995) with breast glandular measurements (D'Astous1986) is shown in Figure 11.4.

11.3.2 Anthropomorphic Breast Phantoms

11.3.2.1 Freely Suspended Breast

In the early 1980s, clinical compound B-scans and were done on the breast freely suspended in water. Also, research ultrasound computerized tomography (UCT) was being done. Recently, there has been a resurgence of UCT imaging with much more sophisticated systems. Thus, we describe phantoms mimicking the freely suspended breast.

Four versions of anthropomorphic breast phantoms were reported in 1982, and these are referred to as phantoms 1, 2, 3, and 4 in this section. These phantoms possessed a realistic undistorted breast shape making the phantoms particularly useful for assessing breast computerized ultrasound tomography (CUT); with the recent resurgence of more sophisticated CT, the freely suspended breast shape is again of interest to researchers (Duric 2005, 2007, Glide 2007, Li 2009, Simonetti 2009). Representations include subcutaneous fat, retromammary fat, fat globules, glandular parenchyma, subareolar tissue, cysts, and malignant and benign tumors. The melting point of the materials is above 70°C because of cross-linking with formaldehyde.

Phantom 1 (Madsen et al. 1982b) has (nonfat) glandular parenchyma with main constituents being water, gelatin, graphite, or talc powder with diameters between 1 and 50 μm, The TM

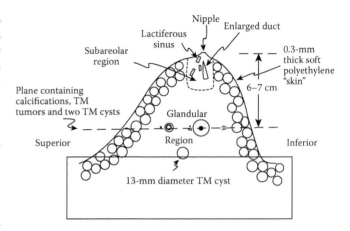

FIGURE 11.5 Side view of phantom 1. (From Madsen, E.L., Zagzebski, J.A., Frank, G.R., Greenleaf, J.F., Carson, P.L. 1982b. *Journal of Clinical Ultrasound* 10: 67–75. Copyright Wiley-VCH Verlag GmbH & Co. KGaA. Reproduced with permission.)

fat is an oil-in-gelatin dispersion where the oil is a solution of 50% olive oil and 50% kerosene by volume. Cystic TM material contains no liquid or solid particles. Table 11.1 shows ultrasonic properties of the various components in phantom 1. Values are given at 22°C and 34°C to show temperature dependencies. Figures 11.5 and 11.6 show depictions of the internal structure of the phantom. Note that the subcutaneous fat is approximated by a collection of 13-mm diameter TM fat spheres beneath the 0.3-mm-thick polyethylene "skin." Five spherical inclusions and Al_2O_3 simulated calcifications are also present. The degree of beam distortion due to refraction at the TM fat sphere boundaries is more than expected for a real breast, but the refraction effect is well demonstrated. Figure 11.7 shows images made with three different single transducers with beam axes at different angles to the phantom; the deleterious affect of refraction is demonstrated. Note that the fat and glandular materials are tissue-mimicking for x-rays in the mammographic range of photon energies (Burke et al. 1982). Figure 11.8 shows an x-ray CT scan of phantom 1.

Phantom 2 (Madsen et al. 1982c) differs from phantom 1 in that a more realistic subcutaneous fat layer is simulated and the glandular parenchyma is a dispersion of olive oil droplets

TABLE 11.1 Ultrasonic Properties of Materials in First Breast Phantom

Tissue-Mimicking Material	Speed of Sound (m/sec)		Attenuation Coefficients—Frequency dB/cm/MHz		Density (g/cm³)	Scatter Strength
	22°C	34°C	22°C	34°C		
Glandular tissue	1568	1583	0.07	0.65	1.06	Intermediate
Subareolar tissue	1565	1582	0.88	0.85	1.07	Intermediate
Cysts, lactiferous sinuses, enlarge duct	1568	1583	0.11	0.1	1.02	No scatter
Tumors	1557	1574	1.13	1.1	1.08	Low
Fat	1459	1437	0.48	0.39	0.94	Low

Source: Madsen EL, Zagzebski JA, Frank GR, Greenleaf JF, Carson PL. 1982b. *Journal of Clinical Ultrasound* 10: 67–75. With permission from Wiley & Sons.

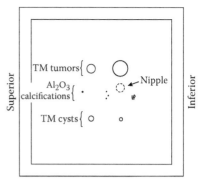

FIGURE 11.6 Arrangement of calcification, TM tumors, and TM cysts in phantom 1. (From Madsen, E.L., Zagzebski, J.A., Frank, G.R., Greenleaf, J.F., Carson, P.L. 1982b. *Journal of Clinical Ultrasound.* 10: 67–75. Copyright Wiley-VCH Verlag GmbH & Co. KGaA. Reproduced with permission.)

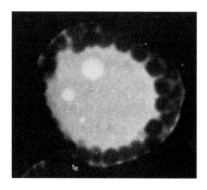

FIGURE 11.8 X-ray CT scan of phantom 1 corresponding to the diagram in Figure 11.6. (From Madsen, E.L., Zagzebski, J.A., Frank, G.R., Greenleaf, J.F., Carson, P.L. 1982b. *Journal of Clinical Ultrasound* 10: 67–75. Copyright Wiley-VCH Verlag GmbH & Co. KGaA. Reproduced with permission.)

in gelatin resulting in a lower propagation speed of 1519 m/s at 22°C; also, Cooper's ligaments and a spiculated tumor are represented. Figures 11.9 and 11.10 show diagrams of the phantom, and Figure 11.11 shows an image made with the more decreased refraction effect due to the subcutaneous fat. Figure 11.12 shows propagation speed and attenuation reconstructions via ultrasound computed tomography (UCT). Perhaps tests with anthropomorphic phantoms with known ultrasound properties throughout resulted in decreased pursuit of UCT for decades.

Phantom 3 (Madsen et al. 1982b) is similar to phantom 1 except that the irregularity of the subcutaneous fat-to-glandular parenchyma is reduced. Nevertheless, UCT scans of this phantom were very poor compared to those of phantom 2. Also, B-scans and UCT images of this phantom demonstrated considerable degradation due the realistic refraction effects.

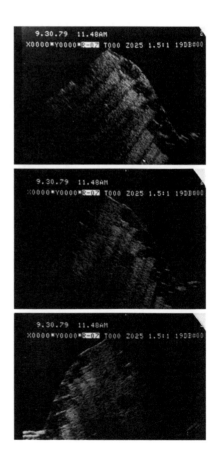

FIGURE 11.7 Demonstration in the top image of the affect of refraction at fat sphere boundaries on the distal mass. (From Madsen, E.L., Zagzebski, J.A., Frank, G.R., Greenleaf, J.F., Carson, P.L. 1982b. *Journal of Clinical Ultrasound* 10: 67–75. Copyright Wiley-VCH Verlag GmbH & Co. KGaA. Reproduced with permission.)

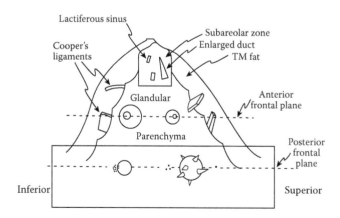

FIGURE 11.9 Side-view diagram of phantom 2. (From Madsen, E.L., Zagzebski, J.A., Frank, G.R., Greenleaf, J.F., Carson, P.L. 1982c. *Journal of Clinical Ultrasound* 10: 91–100. Copyright Wiley-VCH Verlag GmbH & Co. KGaA. Reproduced with permission.)

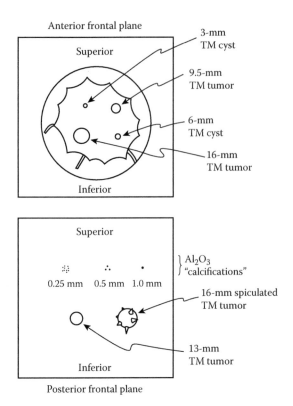

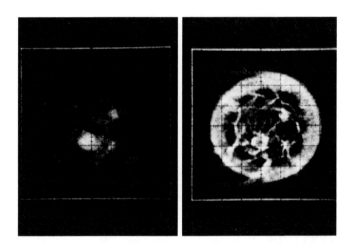

FIGURE 11.12 Ultrasound CT reconstructions of propagation speed (left) and attenuation (right) of the anterior frontal plane of phantom 2 (see Figure 10 diagram.) (From Madsen, E.L., Zagzebski, J.A., Frank, G.R., Greenleaf, J.F., Carson, P.L. 1982c. *Journal of Clinical Ultrasound* 10: 91–100. Copyright Wiley-VCH Verlag GmbH & Co. KGaA. Reproduced with permission.)

FIGURE 11.10 Diagrams of inclusions in two coronal planes for phantom 2. (From Madsen, E.L., Zagzebski, J.A., Frank, G.R., Greenleaf, J.F., Carson, P.L. 1982c. *Journal of Clinical Ultrasound* 10: 91–100. Copyright Wiley-VCH Verlag GmbH & Co. KGaA. Reproduced with permission.)

Another phantom representing the freely suspended breast was recently made for researchers developing a sophisticated UCT system. Since accurate backscatter properties are important, the TM glandular parenchyma was made with the material described at the end of Section 2a. The phantom has been used in the development of UCT system and publications have resulted with images of the

Phantom 4 also has the shape of the freely suspended breast and is a modified version of phantom 3 mainly regarding the inclusion of a range of fat spheres in the glandular parenchyma to assess the effects of distributed fat clumps on image quality. A layer of TM skin is also included.

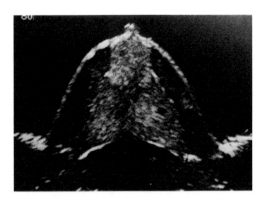

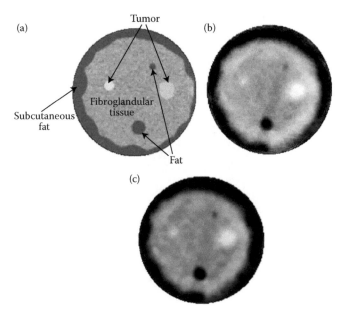

FIGURE 11.11 Image of phantom 2 showing decreased refraction effect since the fat-to-glandular interface presents angles of incidence closer to right angles than in the case of phantom 1. (From Madsen, E.L., Zagzebski, J.A., Frank, G.R., Greenleaf, J.F., Carson, P.L. 1982c. *Journal of Clinical Ultrasound* 10: 91–100. Copyright Wiley-VCH Verlag GmbH & Co. KGaA. Reproduced with permission.)

FIGURE 11.13 Tomography images of breast phantom. (a) X-ray CT scan; (b) total variation bent-ray ultrasound tomogram; (c) Tikhonov bent-ray ultrasound tomogram. (Reprinted from *Ultrasound in Medicine and Biology*, 35, Li, C., Duric, N., Littrup, P., Huang, L., 1615–1628, Copyright 2009, with permission from Elsevier.)

TABLE 11.2 Comparison of Calculated and Known Sound Speeds for Breast Phantom (see Figure 11.13)

Material	Diameter (mm)	Known Sound Speed (m/s)	TV-Calculated Sound Speed (m/s)	Tikhonov-Calculated Sound Speed (m/s)
Big fat sphere	12	1470	1468	1472
Small fat sphere	6	1470	1480	1488
Subcutaneous fat		1470	1468	1470
Glandular		1515	1515	1508
Small tumor	9	1549	1544	1539
Irregular tumor	5	1559	1558	1552

phantom (Li et al. 2009, Simonetti et al. 2009). Three coronal images of the phantom are shown in Figure 11.13. Figure 11.13a is an x-ray CT scan through two simulated tumors, the larger of which has an irregular boundary. Two TM fat spheres are also depicted. Figure 11.13b and c show reconstructed propagation speed (speed of sound) images using two different reconstruction methods. Table 11.2 shows a comparison of the known (measured) speeds in the phantom materials and the speeds calculated using the two reconstruction methods.

11.3.2.2 The Breast Compressed toward the Chest Wall

Since the advent of linear arrays, the usual method for breast imaging is with the breast compressed to some extent toward the chest wall. Therefore, breast phantoms corresponding to that geometry have been made using the TM materials described in Section 12.2.1. Two versions of such phantoms were reported by Madsen (1988). Diagrams of the first phantom (phantom 1) are shown in Figure 11.14. An example of imaging of 6 mm diameter

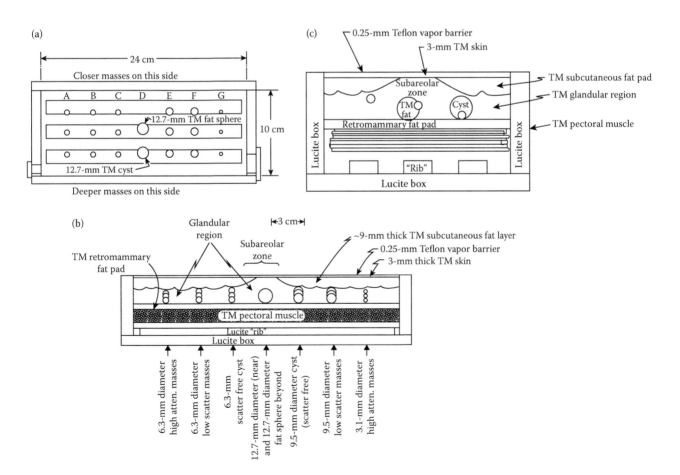

FIGURE 11.14 (a) Top view of phantom 1. (b) Side view of phantom 1. (c) End view of phantom 1. (Reprinted from *Ultrasound in Medicine and Biology*, 14, Madsen, E.L., Kelly-Fry, E., Frank, G.R., 183–201, Copyright 1988, with permission from Elsevier.)

low scatter masses at different depths in the glandular paren-chyma is shown in Figure 11.15. Figure 11.15a has white markers indicating where the masses are (markers on the right side of the masses) and Figure 11.15b is a corresponding image without the markers. The other phantom (phantom 2) is diagrammed in Figure 11.16. In this phantom there are clusters of simulated

calcifications, an irregularly shaped cancer with a necrotic core, and two cysts. An example image of the cancer where the image slice passes through the necrotic core is shown in Figure 11.17.

A recent version of a compressed breast phantom was produced for qualifying investigators for participation in the American

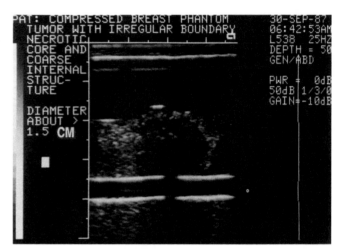

FIGURE 11.15 Images of phantom 1 showing three 6-mm diam-eter low scatter lesions simulating fibroadenomas at different depths in the glandular parenchyma. The image quality degrades with depth. (Reprinted from *Ultrasound in Medicine and Biology*, 14, Madsen, E.L., Kelly-Fry, E., Frank, G.R., 183–201, Copyright 1988, with permission from Elsevier.)

FIGURE 11.17 Image of phantom 2 with scan slice through the irregularly shaped cancer including the necrotic (very high attenua-tion) necrotic core. The strong shadowing due to the necrotic core is seen passing through the horizontal bright specular reflections at the surfaces of the retromammary fat pad. (Reprinted from *Ultrasound in Medicine and Biology*, 14, Madsen, E.L., Kelly-Fry, E., Frank, G.R., 183–201, Copyright 1988, with permission from Elsevier.)

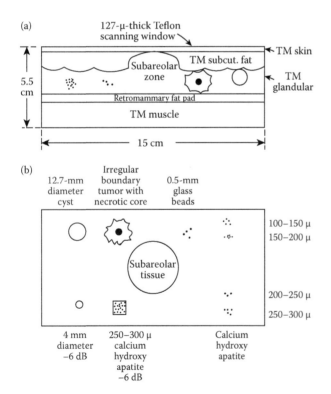

FIGURE 11.16 (a) Side view diagram of phantom 2. (b) Top view dia-gram of phantom 2. (Reprinted from Madsen, E.L., Kelly-Fry, E., Frank, G.R., *Ultrasound in Medicine and Biology*, 14, 1988, 183–201, with per-mission from Elsevier.)

FIGURE 11.18 Ultrasound image shows 3- and 10-mm subtle hypoechoic lesions with attenuation equal to that of surrounding glandular region (i.e., no distal shadows). The 3-mm lesion is on the right (solid arrow), and the 10-mm lesion is on the left (open arrow). (Reproduced from Madsen, E.L., Berg, W.A., Mendelson, E.B., Frank, G.R. 2006a. *Radiology* 239: 869–874. With permission.)

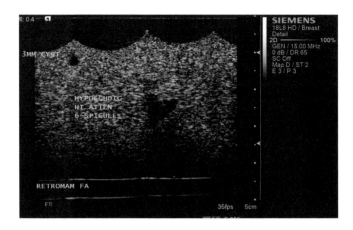

FIGURE 11.19 A recent (2009) 15-MHz linear array image of an ACRIN-type phantom showing the undulating interface between the TM subcutaneous fat (top) and TM glandular parenchyma. A well-defined 3-mm diameter spherical simulated cyst is shown, and a slice through a TM-spiculated, low-echo, high-attenuation cancer; the number of spicules is 6. The 5-mm thick TM retromammary fat pad is below the TM glandular parenchyma.

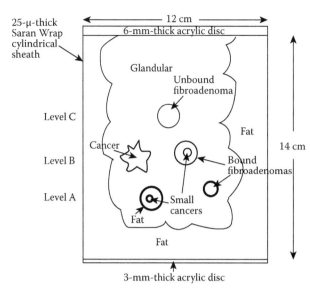

FIGURE 11.20 Side view of the MRE breast phantom with polyvinylidine chloride and acrylic vapor barrier. Three lesions at level A have very bold outlines, three at level B have less bold outlines, and the single inclusion at level C has the lightest outline. The purpose of the different outlines is to facilitate inclusion identification in Figure 11.21. (Reproduced from Madsen, E.L., Berg, W.A., Mendelson, E.B., Frank, G.R. 2006a. *Radiology* 239: 869–874. With permission.)

College of Radiology Imaging Network (ACRIN) protocol 6666. Seven distinct masses are represented in the phantom, all being spherical except for one small low-echo double-ended cone, the latter for testing the user's ability to detect the mass, determine the orientation of the cone axis of the cones, and distinguish between the two cones. In Figure 11.18 is an image showing hypoechoic masses with attenuation equal to that of the surrounding TM glandular parenchyma. A retromammary fat pad is imaged and beneath that a layer of simulated pectoral muscle.

A replica of the ACRIN phantom—differing in that cancers with spiculations are present—has been produced and may become commercially available. The phantom should be useful for training sonographers. In Figure 11.19 is an image at 15 MHz showing a 3-mm diameter spherical TM cyst and a low-echo, high-attenuation spiculated tumor.

11.4 Elastography

11.4.1 Ultrasound and MR Elastography Phantoms

One publication reports on an anthropomorphic breast phantom for use in testing both ultrasound and MR elastography of the breast (Madsen et al. 2006b). (See Chapter 18 for a description of breast elastography.) Two identical phantoms were reported, one for use in quasistatic ultrasound elastography (USE) and one for use in MR elastography (MRE). The difference is that the USE version was kept in an oil bath to allow quasistatic compressions, and the MRE version was enclosed in Saran Wrap and acrylic to facilitate MR scanning. Diagrams of the phantom are shown in Figures 11.20 and 11.21. Complex Young's moduli and ultrasound and MR properties

are tabulated for the various components. A large number of elastograms are provided and only few are shown here. Figure 11.22 shows a 14-mm diameter TM bound fibroadenoma, and for comparison, a 14-mm diameter unbound fibroadenoma is shown in Figure 11.23; a halo is seen around the unbound fibroadenoma. Figure 11.24 shows a simulated cancer with an

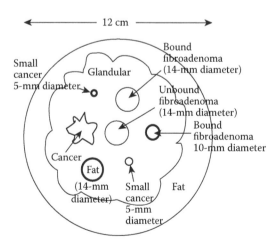

FIGURE 11.21 Coronal view of the MRE breast phantom. (Reproduced from Madsen, E.L., Berg, W.A., Mendelson, E.B., Frank, G.R. 2006a. *Radiology* 239: 869–874. With permission.)

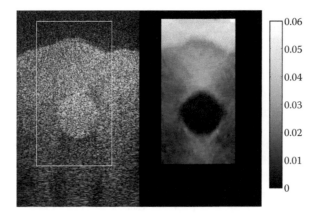

FIGURE 11.22 B-mode image (left) and elastogram (right) with the Siemens Antares of the 14-mm diameter TM bound fibroadenoma. Increased strain areas appear in the TM glandular region just above and below the harder TM fibroadenoma. Note the shadowing distal to the lateral surfaces of the TM fibroadenoma on the B-mode. (Reproduced from Madsen, E.L., Berg, W.A., Mendelson, E.B., Frank, G.R. 2006a. *Radiology* 239: 869–874. With permission.)

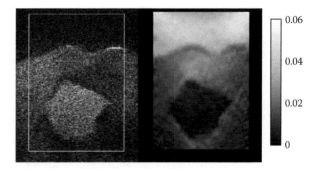

FIGURE 11.24 B-mode image (left) and elastogram (right) with the Siemens Antares of the TM cancer with irregular boundary. The outline of the TM cancer in the elastogram agrees rather well with that on the B-mode image. (Reproduced from Madsen, E.L., Berg, W.A., Mendelson, E.B., Frank, G.R. 2006a. *Radiology* 239: 869–874. With permission.)

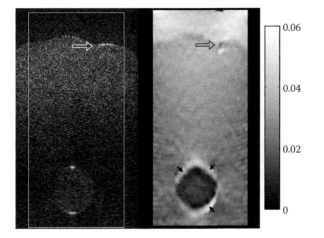

FIGURE 11.23 B-mode image (left) and elastogram (right) with the Siemens Antares of the 14-mm diameter TM unbound fibroadenoma. Slipping at the boundary of the lesion gives rise to the approximately 2-mm wide, bright ring around it in the elastogram (filled black arrows). In the upper right part of the elastogram (open black arrow) at the boundary between the TM subcutaneous fat and the TM glandular region, there is an artifact related to the specular reflection seen on the B-mode image (open white arrow). (Reproduced from Madsen, E.L., Berg, W.A., Mendelson, E.B., Frank, G.R. 2006a. *Radiology* 239: 869–874. With permission.)

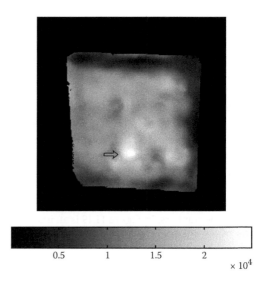

FIGURE 11.25 Elastogram consisting of a gray-scale mapping of the shear modulus (scale values are in pascals). The stiffer 14-mm diameter TM fibroadenoma (arrow) is seen as the brighter (stiffer) object; the TM subcutaneous fat is seen as the darker edges on the top, bottom, and right side. (Reproduced from Madsen, E.L., Berg, W.A., Mendelson, E.B., Frank, G.R. 2006a. *Radiology* 239: 869–874. With permission.)

irregular shape and bound to its surrounding glandular parenchyma. The shape in the elastogram is nearly identical to that in the B-scan to its left. One MRE and corresponding MRI are shown in Figures 11.25 and 11.26.

After acceptance of the manuscript images of the MRE phantom were obtained at the Mayo Clinic laboratory of Richard L.

Ehman in 2006. Corresponding MR and MRE images at two levels in the phantom are shown in the Figures 11.27 and 11.28. The MR image is on the left); a direct inversion (DI) (Oliphant 2001) MRE image is in the middle; and a local frequency estimator (LFE) (Knutsson et al. 1994, Manduca et al. 1996) is on the right.

FIGURE 11.26 MR image of a slice of the MR breast phantom through the 14-mm diameter TM bound fibroadenoma (open arrow). Evidence of the TM unbound fibroadenoma is also seen (black arrow) as is a grazing incidence of the TM irregular tumor (white arrow). (Reproduced from Madsen, E.L., Berg, W.A., Mendelson, E.B., Frank, G.R. 2006a. *Radiology* 239: 869–874. With permission.)

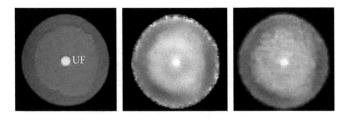

FIGURE 11.27 Images from a slice through the unbound 14-mm diameter spherical fibroadenoma (level C in Figure 11.20). The MR image is on the left; a direct inversion (DI) (Reproduce from Oliphant, T.E., Manduca, A., Ehman, R.L., Greenleaf, J.F. 2001. *Magnetic Resonance in Medicine* 45: 229–310. With permission.) MRE image is in the middle; and a local frequency estimator (LFE) (Reproduce from Knutsson et al. 1994. *Proceedings IEEE International Conference on Image Processing*: 36–40; Manduca, A. et al. 1996. *SPIE Medical Imaging* 2710: 616–623. With permission.

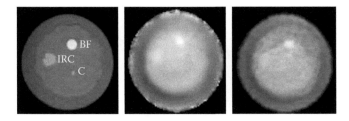

FIGURE 11.28 Images from a slice through level B in Figure 11.20 showing the irregularly shaped cancer (IRC), a 14-mm diameter bound fibroadenoma (BF), and a 5-mm diameter spherical cancer.

11.4.2 Higher Loss Modulus MR Tissue-Mimicking Material

Soft tissues have a significant shear loss modulus with shear loss modulus/shear storage modulus as high as 0.3, unlike most

present TM materials. A new TM material consisting of gelatin, sucrose, and a preservative has been reported by Doyley (2010). At a dry weight sucrose concentration of 20%, the ratio is 0.12 compared to 0.04 with no sucrose.

11.5 Optical Imaging

Optical imaging of the breast is described in Chapter 20. Anthropomorphic breast phantoms made from paraffin wax with wax color pigments have been reported for testing a laser trans-illumination tomographic system (Srinivasan and Singh 2004). The concentrations of the pigments allow scattering and attenuation properties of normal and abnormal breast tissues to be mimicked. The outer shape of the breast phantoms approximates a truncated parabola of revolution, 8 cm diameter at the base and 8 cm in height. (The authors refer to the shapes as conical.) The bulk of each phantom mimics normal breast tissue, and parallel cylindrical inclusions are embedded in that. The diameters of the inclusions range from 3 to 15 mm.

A breast phantom for assessing fluorescence-enhanced breast imaging in the near infrared frequency range was used to assess depth limits of imaging (Ge et al. 2009). A 1% Liposyn solution with dimensions $10 \times 10 \times 6.5$ cm and with or without some concentration of indocyanine green (ICG) represented normal breast tissue, and clear plastic spherical shells filled with a solution of 1% Liposyn and various concentrations of ICG.

Materials for creating a broad range of absorbing and scattering properties in the near infrared have been reported (Vernon et al. 1999). TiO_2 particles are used for determination of scattering and ProJet 900P (Zeneca Specialties, Manchester, U.K.) is used to determine absorption properties. A diagram of a phantom containing eight cylindrical inclusions is shown in Figure 11.29, and the optical and geometrical properties are given in Table 11.3. A time-resolved transmission image of the phantom is shown in Figure 11.30. To appreciate the degree to which the materials mimic breast tissues, see the valuable report by Durduran (2002).

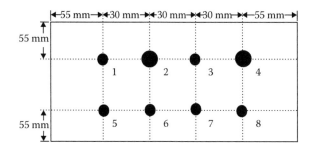

FIGURE 11.29 Diagram of optical phantom. (Reproduced from Vernon, M.L., Frechette, J., Painchaud, Y., Caron, S., Beaudry, P. 1999. *Applied Optics* 38: 4247–4251. With permission.)

TABLE 11.3   Physical and Optical Parameters of Inclusions Used in the Phantom Inclusion

Number	μ_s (mm^{-1})	μ_a (mm^{-1})	Diameter (mm)	Thickness (mm)
1	1.14	0.0023	8.0	10.0
2	1.34	0.0023	10.0	8.0
3	1.43	0.0023	7.0	7.0
4	1.43	0.0023	10.0	8.0
5	1.06	0.008	7.0	7.0
6	1.06	0.008	8.0	10.0
7	1.06	0.011	8.0	10.0
8	1.00	0.014	7.0	7.0

Source: Vernon ML, Frechette J, Painchaud Y, Caron S, Beaudry P. 1999. *Applied Optics* 38: 4247–4251.

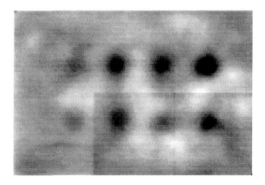

FIGURE 11.30   Scanned image of the phantom from time integration of the first 0.8 ns of the time-resolved transmission. (Reproduced from Vernon, M.L., Frechette, J., Painchaud, Y., Caron, S., Beaudry, P. 1999. *Applied Optics* 38: 4247–4251. With permission.)

11.6 Acousto-Optical Imaging

Somewhat realistic heterogeneous phantoms have been reported (Ephrat et al. 2010, Jaeger et al. 2011). In one type of phantom, the background TM material is composed of an aqueous mixture of 1% agarose and 1% Intralipid® (Ephrat et al. 2010). Two types of simulated spherical lesions were produced. A homogeneous type was a mixture of 1% agarose and a variety of whole blood concentrations, viz., 75%, 50%, 25%, and 12.5%; the remaining weight% was water. Four spherical lesion sizes were made for each blood concentration, viz., 9, 6, 3, and 1.5 mm. Blood vessel-mimicking objects were made to achieve a 1% agarose and 50% whole blood. Vessels with a diameter of 0.5 mm were enclosed in a 6-mm diameter sphere surrounded by the background material (1% agarose and 50% whole blood). The larger diameter vessels were wrapped around a 6-mm diameter sphere of the background material. A set of breast phantoms was then made, each of which contained one of the simulated lesions.

The other type of phantom normal breast tissue was simulated with a 20% gelatin solution, about 1 g/L of TiO$_2$ (powder), about 0.03 g/L India ink, and about 80 g/L of flour (presumably ordinary white cooking flour) (Jaeger et al. 2011). The flour provides acoustic attenuation and some optical scattering, the India ink provides optical absorption, and TiO$_2$ provides additional (probably most) optical scattering. Inclusions consisted of clusters of spheres with diameters of approximately 1 mm with the interstices filled with the background (normal breast) material. These inclusions were then suspended in the bulk phantom consisting of the normal breast tissue mimicking material. The sphere material contained no flour and a much larger concentration of India ink than the background material, viz., 2 mL/L. The external shape of the breast phantom simulates that of a real breast (see Figure 11.31a). An ultrasound image showing two of the inclusions is shown in Figure 11.31b.

(a)   Irradiation profile converter

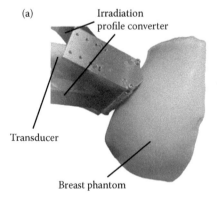

Transducer

Breast phantom

(b)

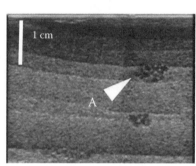

1 cm

A

FIGURE 11.31   (a) Breast phantom home-built from gelatine, flour, TiO², and ink, covered by latex skin. Transducer head combining a commercial ultrasound linear transducer with fiber bundle profile converters for laser irradiation. (b) Typical echo ultrasound image obtained from a breast phantom, showing the speckle pattern caused by the flour in the bulk material, and the small optically absorbing, but acoustically transparent, insets. The layer structure of the bulk material resulted from the casting process. The larger inset is marked "A" with a white arrow. (Reproduced from Jaeger, M., Preisser, S., Kitz, M., Ferrara, D., Senegas, S., Schweizer, D., Frenz, M. 2011. *Physics in Medicine and Biology* 56: 5889–5901. With permission.)

11.7 MR Imaging

See Chapter 16 for a description of breast MRI. The primary strength of MR breast imaging has been its use in assessing the rate of uptake of MR contrast, such as the chelate Gadolinium-diethylenetriamine penta-acetic acid (Gd-DTPA), for distinguishing between malignant and benign tumors. Mimicking this perfusion process in a phantom does not seem to have been accomplished based on a reasonably thorough literature search. An anthropomorphic breast phantom has been reported where adipose tissue is represented with lard and glandular tissue with preserved egg white, but no perfusion mimicking is proposed (Freed et al. 2011b).

Most phantoms produced for use in MRI are not used for assessment of MRI per se for diagnosis of disease. The majority of publications involve phantoms to aid in MRI-guided biopsies. One breast phantom used to test a breast biopsy system employed materials having glandular- and tumor-mimicking T_1's and T_2's with long-term stability of those values as well as of the size and shape of the simulated spherical tumors. The materials were enclosed in a thin polyethylene skin. MRI images of the phantom are shown in Figure 11.32 from an article by Smith (2008). Long-term stability of these agar/gelatin materials was demonstrated previously (Madsen et al. 1991a).

Other rather simple biopsy phantoms have been employed. Two other biopsy phantoms used a piece of chicken breast to simulate human breast. One has an implanted solid simulated mass made of gelatin, water, and a vitamin E capsule (Lehman

and Aikawa 2004), and the other simulated a 3.2-mm diameter lesion with polyvinyl alcohol. In another case, agarose gel represented breast tissue and the targeted simulated lesions were peas (garden vegetable) (Werner et al. 2006). In another study, lard was used to represent breast tissue and gelatin doped with gadopentetate dimeglumine (Magnevist®) represented 6- to 7-mm diameter lesions (Daniel et al. 1997).

11.8 Summary and Discussion

The reader is cautioned that the author of this chapter has published mostly on ultrasound and elastography phantoms, with some in MR and x-ray imaging (including CT). Phantoms for use in x-ray mammography have been made from materials that mimic approximately the mass density and attenuation properties of breast tissues using acrylic and wax materials. Other materials mimic the mass density and linear attenuation coefficients of breast adipose and glandular tissue more closely (White and Tucker 1980) without attempting to mimic scattering properties of breast tissues. Phantoms based on excised and preserved actual breast tissues likely are closest in x-ray properties to real breast tissues and internal geometries. However, in the breast tissue case, detailed knowledge of internal geometric distribution of the x-ray properties is difficult to determine, although much information can be gleaned from x-ray CT.

Regarding MR, materials have been developed for mimicking relaxation times of breast tissues and complex geometries can be

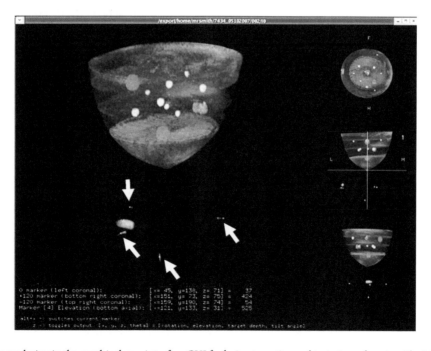

FIGURE 11.32 Volume rendering in the graphical user interface GUI for lesion targeting and trajectory planning. The 3-D volume can be rotated freely and orthographic maximum intensity projections MIP are displayed right column to facilitate trajectory planning. Reference cursors are placed on the device's fiducial markers white arrows and the targeted lesion to register the device geometry to the imaging field of view FOV. Once registration is complete, the trajectory positioning output generated by the planning algorithm is displayed for allowing access to the target. (Reproduced from Smith, M., Zhai, X., Harter, R., Sisney, G., Elezaby, M., Fain, S. 2008. *Medical Physics* 35: 3779–3786. With permission.)

produced. However, currently the strength of MR for breast cancer diagnosis involves assessing the rate of uptake of MR contrast via the vascular system, and mimicking that requirement has not yet been adequately addressed.

Anthropomorphic phantoms for use in ultrasound grayscale imaging are perhaps the best developed in terms of mimicking ultrasound properties of mass density, propagation speed, attenuation coefficients, and backscatter coefficients. The phantoms are water-based, and desiccation can occur over months or years; however, rejuvenation via "transfusion" restores the phantoms. The same type of materials for ultrasound has been used for elastography phantoms.

11.9 Comparison of Physical and Digital Phantoms

The major strength of digital phantoms is that any values of most physical properties (such as MR T1 and T2, ultrasound propagation speeds, and attenuation and backscatter coefficients, and x-ray linear attenuation coefficients) can be easily introduced. Also, once representative geometric distributions of various normal and abnormal tissues have been produced, many simulations are feasible.

There are disadvantages for digital phantoms, however. A major disadvantage is that they are digital, and testing performance of a hardware imager depends on how adequately the hardware imaging process can be mimicked with software. The case of ultrasound grayscale imaging is worth mentioning. For example, a great many configuration options are available in ultrasound imaging, e.g., focus, nominal frequency, time-gain compensation, size and shape of the image, etc. Once those configuration parameters are specified, there is the problem of software generation of the spatial and time dependencies of the pulses; then refraction and diffraction effects—and also attenuation effects—on the beam as it propagates through the phantom pose a daunting task for current computer power and programming times.

Regarding elastography, mimicking the 3-D time-dependent changes in shapes of various component materials may be very challenging.

Physical phantoms, although limited in the number of feasible geometric configurations of internal components, can be used in the clinic with the real hardware imagers using any imaging parameters chosen to compare performances of different imager makes and models. Training of imaging personnel seems best done using real imagers and tangible phantoms.

References

Burke TM, Madsen EL, Zagzebski JA. 1982. X-ray linear attenuation coefficients in the mammographic range for ultrasonic breast phantom materials. Radiology 142: 755–757.

Byng JW, Mainprize JG, Jaffe MJ. 1998. X-ray characterization of breast phantom materials. Phys Med Biol 43: 1367–1377.

Cannon LM, Fagan AJ, Browne JE. 2011. Novel tissue-mimicking materials for high frequency breast ultrasound phantoms. Ultrasound Med Biol 37: 122–135.

Carton A-K, Bakic P, Ullberg C, Derand H. 2011. Development of a physical 3D anthropomorphic breast phantom. Med Phys 38: 891–896.

Chin RB, Madsen EL, Zagzebski JA, Jadvar H, Wu XK, Frank GR. 1990. A reusable perfusion supporting tissue-mimicking material for ultrasound hyperthermia phantoms. Med Phys 17: 380–390.

Daniel BL, Birdwell RL, Black JW, Ikeda DM, Glover GH, Herfkens RJ. 1997. Interactive MR-guided, 14 gauge core-needle biopsy of enhancing lesions in a breast phantom model. Acad Radiol 4: 508–512.

D'Astous FT, Foster FS. 1986. Frequency dependence of ultrasound attenuation and backscatter in breast tissue. Ultrasound Med Biol 12: 795–808.

Doyley MM, Perreard I, Patterson AJ, Weaver JB, Paulsen KM. 2009. The performance of steady-state harmonic magnetic resonance elastography when applied to viscoelastic materials. Med Phys 37: 3970–3979.

Durduran T, Choe R, Culver JP, Zubkov L, Holboke MJ, Giammarco J, Chance B, Yodh AG. 2002. Bulk optical properties of healthy female breast tissue. Phys Med Biol 47: 2847–2861.

Duric N, Littrup P, Babkin A, Chambers D, Azevedo S, Pevzner R, Tokarev M, Holsapple E, Rama O, Duncan R. 2005. Development of ultrasound tomography for breast imaging: Technical assessment. Med Phys 32: 1375–1386.

Duric N, Littrup P, Poulo L, Babkin A, Pevzner R, Holsapple E, Rama O, Glide C. 2007. Detection of breast cancer with ultrasound tomography: First results with the computed ultrasound risk evaluation (CURE) prototype. Med Phys 34: 773–785.

Ephrat P, Albert GC, Roumeliotis MB, Belton M, Prato FS, Carson JJL. 2010. Localization of spherical lesions in tumor-mimicking phantoms by 3D sparse array photoacoustic imaging. Med Phys 37: 1619–1628.

Faulkner K, Law J. 1999. A comparison of mammographic phantoms. Br J Radiol 67: 174–180.

Freed M, Badal A, Jennings RJ, de las Heras H, Myers KJ, Badano A. 2011a. X-ray properties of an anthropomorphic breast phantom for MRI and x-ray imaging. Phys Med Biol 56: 3513–3533.

Freed M, de Zwart JA, Loud JT, El Khouli RH, Myers KJ, Greene MH, Duyn JH, Badano. 2011b. An anthropomorphic phantom for quantitative evaluation of breast MRI. Med Phys 38: 743–753.

Ge J, Erickson SJ, Godavarty A. 2009. Fluorescence tomographic imaging using a handheld-probe-based optical imager: Extensive phantom studies. Appl Opt 48: 6408–6416.

Glide C, Duric N, Littrup P. 2007. Novel approach to evaluating breast density utilizing ultrasound tomography. Med Phys 32: 744–753.

Jaeger M, Preisser S, Kitz M, Ferrara D, Senegas S, Schweizer D, Frenz M. 2011. Improved contrast deep optoacoustic imaging using displacement-compensated averaging: Breast tumour phantom studies. Phys Med Biol 56: 5889–5901.

Knutsson et al. 1994. Proc IEEE Intl Conf on Image Processing: 36–40.

Lehman CD, Aikawa T. 2004. MR-guide vacuum-assisted breast biopsy: Accuracy of targeting and success in sampling in a phantom model. Radiology 232: 911–914.

Li C, Duric N, Littrup P, Huang L. 2009. In vivo breast sound speed imaging with ultrasound tomography. Ultrasound Med Biol 35: 1615–1628.

Madsen EL, Zagzebski JA, Frank GR. 1982a. Oil-in-gelatin dispersions for use as ultrasonically tissue-mimicking materials. Ultrasound Med Biol 8: 277–287.

Madsen EL, Zagzebski JA, Frank GR, Greenleaf JF, Carson PL. 1982b. Anthropomorphic breast phantoms for assessing ultrasonic imaging system performance and for training ultrasonographers: Part I. J Clin Ultrasound 10: 67–75.

Madsen EL, Zagzebski JA, Frank GR, Greenleaf JF, Carson PL. 1982c. Anthropomorphic breast phantoms for assessing ultrasonic imaging system performance and for training ultrasonographers: Part II. J Clin Ultrasound 10: 91–100.

Madsen EL, Kelly-Fry E, Frank GR. 1988. Anthropomorphic phantoms for assessing systems used in ultrasound imaging of the compressed breast. Ultrasound Med Biol 14(Suppl 1): 183–201.

Madsen EL, Blechinger JC, Frank GR. 1991a. Low-contrast focal lesion detectability phantom for 1H MR imaging. Med Phys 18: 549–554.

Madsen EL, Zagzebski JA, Macdonald MC, Frank GR. 1991b. Ultrasound focal lesion detectability phantoms. Med Phys 18: 1171–1180.

Madsen EL, Frank GR, Lorriane PW, Dong F. 1995. Anthropomorphic US breast phantom with a realistic heterogeneous glandular region. Radiology 197 (Suppl): 190–191.

Madsen EL, Dong F, Frank GR et al. 1999. Interlaboratory comparison of ultrasonic backscatter, attenuation, and speed measurements. J Ultrasound Med 18: 615–631.

Madsen EL, Berg WA, Mendelson EB, Frank GR. 2006a. Anthropomorphic breast phantoms for qualification of Investigators for ACRIN Protocol 6666. Radiology 239: 869–874.

Madsen EL, Hobson MA, Frank GR, Shi H, Jiang J, Hall TJ, Varghese T, Doyley MM, Weaver JB. 2006b. Anthropomorphic breast phantoms for testing elastography systems. Ultrasound Med Biol 32: 857–874.

Manduca A et al. 1996. SPIE Med Image 2710: 616–623.

Obenaur S, Hermann K-P andf Grabbe E. 2003. Dose reduction in full-field digital mammography: An anthropomorphic breast phantom study. Br J Radiol 76: 478–482.

Oliphant TE, Manduca A, Ehman RL, Greenleaf JF. 2001. Complex-valued stiffness reconstruction for magnetic resonance elastography by algebraic inversion of the differential equation. Mag Reson Med 45: 299–310.

Olsen JB, Sketting A. 1998. Detectability of simulated masses and calcifications in mammography. Acta Radiologica 398: 501–506.

Price BD, Gibson AP, Tan LT, Royle GJ. 2011. An elastically compressible phantom material with mechanical and x-ray attenuation properties equivalent to breast tissue. Phys Med Biol 55: 1177–1188.

Simonetti F, Huang L, Duric N, Littrup P. 2009. Diffraction and coherence in breast tomography: A study with a toroidal array. Med Phys 36: 2955–2965.

Smith M, Zhai X, Harter R, Sisney G, Elezaby M, Fain S. 2008. A novel MR-guided interventional device for 3D circumferential access to breast tissue. Med Phys 35: 3779–3786.

Srinivasan R, Singh M. 2004. Multislice tomographic imaging and analysis of human breast-equivalent phantoms and biological tissues. IEEE Trans Biomed Eng 51: 1830–1837.

Vernon ML, Frechette J, Painchaud Y, Caron S, Beaudry P. 1999. Fabrication and characterization of a solid polyurethane phantom for optical imaging through scattering media. Appl Opt 38: 4247–4251.

Werner R, Krueger S, Winkel A, Albrecht C, Schaeffter T, Heller M, Frahm C. 2006. MR-guided breast biopsy using an active marker: A phantom study. J Magn Reson Imaging 24: 235–241.

White DR, Tucker AK. 1980. A test object for assessing image quality in mammography. Br J Radiol 53: 331–335.

Digital Phantoms for Breast Imaging

Predrag R. Bakic
The University of Pennsylvania

Andrew D. A. Maidment
The University of Pennsylvania

12.1 Motivation for Developing Digital Phantoms

Due to the complexity of clinical breast cancer imaging systems, their validation and optimization are challenging tasks, which require both preclinical and clinical studies. This is particularly true for systems used in breast cancer screening due to the low prevalence of disease in a screening population. Validation for screening requires clinical trials involving very large numbers of volunteers and repeated imaging using different acquisition conditions. This results in delayed dissemination of new technology due to the prohibitive duration and cost of such trials, increased radiation risk in the case of imaging systems utilizing ionizing radiation, and strict limitations on the number of test conditions.

Preclinical system evaluations are typically performed with simple geometric phantoms designed to stress specific aspects of the imaging system or to measure specific image properties, such as spatial resolution or noise characteristics. However, improved technical performance with simple phantoms does not necessarily predict improved clinical performance. Preclinical testing with anthropomorphic breast phantoms may offer a valuable alternative that can reduce the burden of clinical trials. Anthropomorphic phantom studies can be used to identify the most promising systems or system parameters for detailed clinical validation.

Digital anthropomorphic breast phantoms (also referred to as software breast phantoms) offer distinct advantages in preclinical testing, providing the flexibility to simulate anatomic variations and providing ground truth of the simulated tissues. The availability of absolute ground truth can be used for quantitative validation. By contrast, clinical studies require a biopsy for each positive finding and long-term follow-up to confirm negative findings. Furthermore, clinical studies require manual segmentation of the clinical images by radiologists, which are subjective. Even knowledge of the presence or absence of a lesion might be insufficient for validation of certain imaging tasks. For example, dual-energy imaging requires ground truth information about tissue composition and contrast-enhanced imaging requires knowledge of the contrast agent distribution (see Chapter 2 for overview of contrast-enhanced mammography).

Modern clinical breast imaging is largely a multimodality endeavor. Digital anthropomorphic phantoms are ideally suited to the validation of multimodality imaging because the establishment of absolute ground truth between modalities is particularly challenging in the clinical setting. For a given anthropomorphic phantom, individual imaging modalities can be simulated by associating modality-specific physical properties (e.g., linear x-ray attenuation coefficient or MRI relaxation times) to each simulated tissue type. Ideally, support of multimodality imaging requires simulation of the mechanical deformation of the breast under different acquisition geometries.

Synthetic phantom images can be assessed by either human (Chapter 10) or mathematical observers (Chapter 13). Mathematical observer models have been developed to mimic the clinical decision process performed by radiologists (Barrett and Myers 2004). An example of digital breast phantom applications using observer models is the work of Young et al. (2009) on the estimation of lesion detectability in digital breast tomosynthesis (DBT) (see Chapter 4 for overview of DBT). Synthetic images of digital phantoms also have application in the assessment of computer image processing methods and computer image analysis tools (Chapter 21). Examples include assessment of accuracy in mammographic image registration (Richard et al. 2006), testing a wavelet-based noise reduction method (Tischenko et al. 2005), comparison of breast density estimates from phantom mammograms and ultrasound tomography images (Bakic et al. 2011a), and assessment of the effect of acquisition parameters on texture estimation in DBT (Kontos et al. 2008).

Additional applications of digital breast phantom images include radiation dosimetry (Hoeschen et al. 2005, Dance et al. 2005, Hunt et al. 2005, Chen et al. 2011), radiation therapy modeling (Mundy et al. 2006, Sztejnberg Goncalves-Carralves and Jevremovic 2007), and breast surgery simulation. Recent advances in the simulation of breast anatomy with very small voxel size (a few micrometers; Pokrajac et al. 2011) allow simulation and analysis down to the cellular level.

Finally, a digital breast phantom has been used to design a novel anthropomorphic physical breast phantom (Carton et al. 2011, Pokrajac et al. 2012). Images of the physical phantom acquired using clinical systems can be compared with ground truth in the form of synthetic images of the corresponding digital phantom.

12.2 Critical Issues in Phantom Design

The greatest challenge in the design of digital anthropomorphic phantoms for breast imaging is to accommodate the complexity and variability of the normal breast anatomy. Breast images include tissue structures organized uniquely for each woman, yet visualized differently at various scales. For example, in mammography, dense fibroglandular tissue is visualized as large-scale radio-opaque regions, while, in fact, fibroglandular tissue is composed of a complex spatial arrangement of Cooper's ligaments, fibrous connective tissue, glandular epithelium, and other nonadipose structures (vascular, lymphatic, etc.) (Chapter 1). In addition, hormonal influences on breast tissue cause significant variations in breast composition both short-term and throughout a woman's life. Furthermore, the elasticity of breast tissue results in complex deformations due to breast positioning and compression during clinical examinations, so that no two images of the same breast are identical. The degree of detail required in simulation depends on both the imaging tasks and the imaging modalities. An optimal breast phantom should simulate all of the variability observed in the breast.

An important issue of digital breast phantoms is the ability to realistically simulate normal tissue structures that contribute to the image background texture (sometimes called the parenchymal pattern or anatomical noise). The complex arrangement of breast texture, as evident by the parenchymal pattern seen in projection mammography, adversely affects breast cancer detection either by masking existing abnormalities or by introducing false ones. Prior work has shown that the parenchymal pattern

can be simulated mathematically based on statistical properties in 2-D. This work is based upon the experimental observation that the parenchymal pattern exhibits a power spectrum with $1/f^{\beta}$ ($\beta \approx 3$) proportionality to the spatial frequency, f (Burgess 1999). A number of methods, including clustered lumpy backgrounds (Bochud et al. 1999, Castella et al. 2008) and self-similar filtering of a random field (Heine et al. 1999), can successfully match some statistical properties of clinical mammograms. Unfortunately, 2-D tissue simulations cannot be used to model the breast for 3-D imaging modalities. 3-D digital breast phantoms, on the other hand, directly simulate the breast anatomical structures that make up the parenchymal pattern observed in images.

In addition to normal anatomy, digital breast phantoms should also provide the ability to simulate breast abnormalities, including masses, clusters of microcalcifications, architectural distortion, or bilateral breast asymmetry. As with the normal anatomy, the appearance of abnormalities is scale-dependent. For example, individual microcalcifications seen in mammograms are an admixture of proteinaceous and calcific materials (Lanyi 1988). Optimal simulation of the various types of abnormalities represents a substantial challenge for breast phantom designers, and remains an open problem. Current methods for simulation include the use of abnormalities segmented from clinical images and the use of mathematical models. Digital phantoms offer the advantage that it is possible to control the properties of the simulated abnormalities directly. For example, it is possible to generate phantoms with and without abnormalities, both simulated with the same normal background anatomy. Furthermore, by varying the size and/or the contrast of the simulated abnormality, one can test the visibility in synthetic images by human or model observers.

The majority of digital breast phantoms today are voxel-based. The choice of phantom voxel size and image pixel size is related to both the scale of the simulated breast anatomy and the image modality. The optimal scale of the simulated anatomical details depends upon the intended use. Ideally, both the voxel size and the detector element size should be chosen to avoid aliasing artifacts. In practice, constraints exist, which are based upon the scale and complexity of the tissue simulation, and the memory and computational limitations of the simulation platform.

Generating phantoms with smaller voxels requires both longer computation times and larger memory storage. The time needed to generate a phantom also depends upon the computational complexity of the simulation method. Reducing computational complexity is a current research topic and a key area for optimization in phantom design. One solution is a phantom design that is scalable, i.e., generate a phantom at higher resolution based upon a lower resolution version. Innovations in the storage of simulation data are also of current research interest. Generally, voxel-based phantoms tend to be less scalable; each step up in phantom resolution dramatically increases the number of voxels. A more efficient approach would be to use an object-based phantom representation; for example, octrees (Meagher 1981) or marching cubes (Lorensen and Cline 1987) would more easily transfer between different phantom scales. Object-based methods also promise to accelerate breast deformation modeling and image simulation.

Multimodality breast imaging is a key motivation for digital phantom development. There are a number of critical issues related to the simulation of multimodality breast imaging. The models of tissue physical properties, breast compression, orientation, and positioning are all modality specific. The simulated features vary depending upon whether anatomic or function imaging is simulated (e.g., blood flow and contrast agents are required for contrast-enhanced imaging). The dimensionality of the images also varies for different modalities: 2-D in mammography, 3-D in DBT or ultrasound (Chapter 15, Ultrasound Imaging of the Breast), and 4-D (i.e., 3-D + time) in contrast-enhanced DBT or dynamic contrast enhanced (DCE)-MRI. An outstanding problem is the lack of measurements of breast vascular dynamics, particularly when the breast is placed under compression for imaging.

Finally, a decision must be made with regard to the necessary level of phantom realism. Traditionally, it has been argued that one should strive to achieve as high a level of realism as possible. Objectively, it is more appropriate to simulate only to the level of realism that is necessary. In the simulation of a clinical imaging study, the phantoms should be able to mimic clinical performance of the simulated imaging systems. For example, depending upon the purpose of the simulation, it may be sufficient to correctly rank the performance of the imaging systems. In that instance, the simplest phantom that could guarantee the correct ranking would suffice; further efforts to improve phantom realism are not justified. This assertion is supported by the extensive use of simple geometric phantoms (physical and digital) for optimization and quality control of projection mammography. Such phantoms are sufficient for the task. However, 3-D imaging, quantitative imaging (e.g., the quantification of breast density or parenchymal pattern), and more complex imaging tasks (e.g., multimodality imaging) require the use of appropriately realistic anthropomorphic phantoms.

Phantom realism can be ascertained either directly by comparing the simulated anatomy with anatomic specimens or indirectly by comparing simulated images with clinical images. Direct validation of phantom realism is challenging, largely due to the fact that there are very few accurate studies of breast anatomy conducted over multiple scales. Clinical evaluation of surgical specimens is limited to analyses relevant to clinical management and is typically performed with small specimens. Detailed visualization of whole breast histology sections, which would allow direct validation of anthropomorphic phantoms, is relatively rare (Wellings et al. 1975, Ohtake et al. 1995, Moffat and Going 1996, Ohtake et al. 2001, Going and Moffat 2004, Clarke et al. 2006, Clarke et al. 2011). The seminal work by Sir Astley Cooper (Cooper 1840) still represents one of the most detailed treatises of breast anatomy. Only a few direct validation studies have been conducted, including a comparison of simulated ductal branching patterns with published drawings of breast sections (Taylor and Owens 1998).

Indirect testing of phantom realism is more common. Examples of indirect validation include comparison of texture properties between phantom and clinical images (Bakic et al. 2002b, Bliznakova et al. 2010), power spectral analysis of phantom images (Bakic et al. 2002b, Lau et al. 2010, Bakic et al. 2010a, Bliznakova et al. 2003, Chen et al. 2011, Li et al. 2009b, Bliznakova et al. 2010, O'Connor et al. 2010), the assessment of simulated ductal trees (Bakic et al. 2003), and comparison of methods for tissue segmentation in breast CT for the purpose of generating digital phantoms from clinical data sets (O'Connor et al. 2010) (see Chapter 5 for overview of breast CT).

12.3 Existing Phantom Designs

There have been several attempts to develop realistic digital breast phantoms by simulating the 3-D anatomy of the breast. These simulation methods can be divided into two major categories: methods based upon rules for generating anatomical structures in the breast (Bakic et al. 2002a, Bakic et al. 2011b, Taylor and Owens 1998, Bliznakova et al. 2003, Ma et al. 2009, Bliznakova et al. 2010, Reiser and Nishikawa 2010, Chen et al. 2011) and methods based upon the use of individual clinical 3-D breast images (Hoeschen et al. 2005, O'Connor et al. 2008, Li et al. 2009b). These two methods are complementary; the first offers more flexibility in simulating all possible variations in breast anatomy, while the second provides the potential for greater realism since they are based on clinical data.

12.3.1 Rule-Based Phantoms

The philosophy of developing phantoms based upon rules for simulating anatomical structures forms the basis for the authors' work at the University of Pennsylvania (Penn). Started in 1996, the Penn phantom has undergone substantial refinement. The initial breast anatomy simulation used simple geometric primitives—spheres and spherical shells for adipose compartments and Cooper's ligaments and a ramified tree for the breast ductal network (Bakic et al. 2002a). The phantom was created with the breast undeformed, i.e., before positioning or compression for a specific imaging examination. A simple mechanical deformation model was then applied to simulate clinical mammographic compression.

The initial design of the Penn digital phantom was validated indirectly based upon the analysis of parenchymal texture features. Good agreement was observed between synthetic mammographic projections through the phantom and clinical mammograms, in terms of the mean texture features and the range of features seen over a large group of images (Bakic et al. 2002b). This initial design was limited by the sharp, overly geometric boundary between the large-scale regions of predominantly adipose and predominantly fibroglandular tissue and the uniform spherical shape of the adipose compartments. These features noticeably reduced the realism of the resulting synthetic images. However, as described in Sections 12.4.3 through 12.4.5, phantoms based on this design have been extensively used.

The Penn digital anthropomorphic phantom design was subsequently updated to include a more realistic simulation of the adipose compartments and Cooper's ligaments using a region growing approach (Zhang et al. 2008, Bakic et al. 2011b). Figure 12.1 compares the structure of the region-growing-based Penn digital phantom with breast anatomical features illustrated in anatomical and pathological sections. The updated phantom allows for great flexibility in simulating different breasts. Figure 12.2 illustrates cross sections of phantoms simulated with

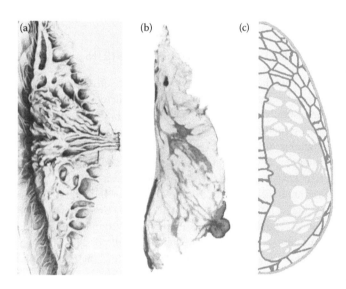

FIGURE 12.1 Comparison of clinical and simulated breast anatomy. (a) Drawings of an anatomical breast section from Sir Astley Cooper's collection. (From Cooper A. (1840). *On the Anatomy of the Breast*, London: Longman.) (b) Subgross breast pathological section. (From Wellings, S.R., Jensen, H.M., and Marcum, R.G. (1975). *Journal of the National Cancer Institute*, 55, 231–273.) (c) Section through the region-growing based Penn anthropomorphic breast model.

different glandularity (i.e., the amount of simulated dense, fibroglandular tissue) and various simulated adipose compartment sizes, thus demonstrating the ability to model women with different breast composition.

Synthetic x-ray images of the region-growing based Penn phantom are generated by simulating the breast deformation during the mammographic compression using a finite element model proposed by Ruiter et al. (2006, 2008), followed by a model of the x-ray projections of the compressed phantom, typically assuming monoenergetic x-rays without scatter. The acquisition geometry and the detector size and resolution are selected so as to correspond to the clinical mammography or DBT system to be simulated. Figure 12.3 shows examples of synthetic DBT projections and reconstruction images for a 450-ml phantom.

The Penn phantom also allows for the simulation of breast abnormalities, including masses and clusters of microcalcifications, as illustrated by the synthetic mammograms and DBT images shown in Figure 12.4 (Nishikawa et al. 2011). The microcalcifications are simulated using a database of clinical images in which calcification clusters were extracted from clinical stereo mammographic images of biopsy specimens (Maidment et al. 1996, 1998). Spiculated lesions are simulated by morphological dilation of a selected intersection of simulated Cooper's ligaments. Continuous refinement and validation of the simulations are ongoing.

The most current version of the Penn anthropomorphic phantom is generated using an octree-based recursive partitioning approach (Pokrajac et al. 2011). It can be shown that such an approach results in anatomy simulation that is equivalent to the region-growing method, while allowing for a substantial speed up of the phantom generation process. The simulation times for an octree-based phantom are inversely proportional to the square of the voxel size (as compared to the fourth order for the original region growing simulation method). With a parallelized implementation of the octree method (Chui et al.

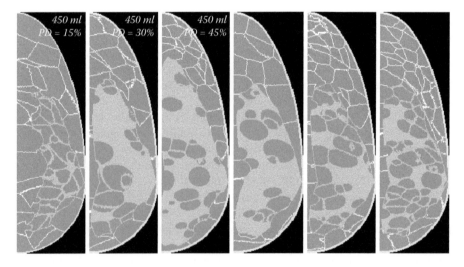

FIGURE 12.2 Sections through the Penn phantom simulating different breast glandularity (left) or different size of simulated adipose compartments (right). (Reproduced from Bakic, P.R., Zhang, C., and Maidment, A.D.A. (2011b). *Medical Physics*, 38, 3165–3176.)

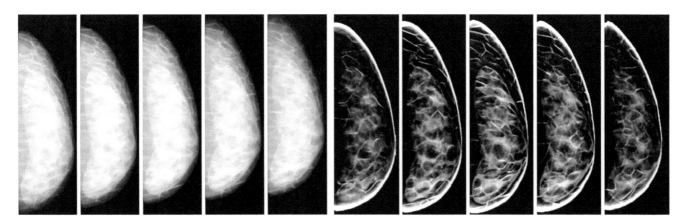

FIGURE 12.3 Synthetic DBT projections (left) generated using the region-growing based Penn digital breast phantom and slices through the phantom (right) reconstructed from the synthetic projections DBT slices. (Reproduced from Bakic, P.R., Zhang, C., and Maidment, A.D.A. (2011b). *Medical Physics*, 38, 3165–3176. Reconstructed DBT images provided by Real-Time Tomography, LLC.)

2012), simulation of very small voxel size phantom is possible. Figure 12.5 shows orthogonal cross sections of phantoms with the same internal arrangement simulated at voxel sizes of 400, 100, and 25 μm³. The magnified regions of interest illustrate the level of detail achievable at different spatial resolutions. The new simulation method also provides improved user control of the thickness of the simulated Cooper's ligaments and skin. Figure 12.6 shows synthetic mammograms of phantoms with simulated ligament thickness of 1200, 800, and 400 μm. The development of a direct validation method for the octree-based Penn phantom is ongoing, based upon the analysis of ellipsoids fitted to the adipose compartments (Contijoch et al. 2012).

Further improvement of the phantom realism is achievable by simulating small-scale variations in the breast tissue. The distribution of the dense fibroglandular breast tissue in the current octree-based Penn digital phantom is modeled by filling selected adipose compartments with simulated dense tissue. Researchers from the University of Chicago have recently extended the Penn breast phantom to simulate small-scale tissue variations. This extension is based upon a random-phase noise model with power-law spectrum by Reiser and Nishikawa (2010). The inverse Fourier transform of the noise is binarized to create an admixture of adipose and fibroglandular tissue. The threshold value is varied iteratively until the proportion of voxels above the

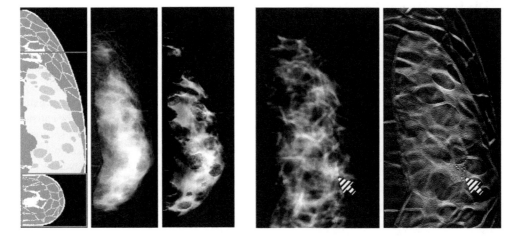

FIGURE 12.4 Examples of simulated abnormalities in the region-growing-based Penn digital breast phantom. (Left) Orthogonal sections through a phantom, a synthetic mammographic projection, and a reconstructed DBT image with simulated speculated lesion. (Right) A synthetic mammographic projection and a reconstructed DBT image of a phantom with simulated cluster of microcalcifications. In addition to the improved realism and flexibility of the updated design, the region-growing-based Penn digital phantom was used for developing the first prototype 3-D anthropomorphic physical phantom. More details about the physical phantom are given in Section 12.4.6. (From Carton, A.-K., Bakic, P.R., Ullberg, C., Derand, H., and Maidment, A.D.A. (2011). *Medical Physics*, 38, 891–896.)

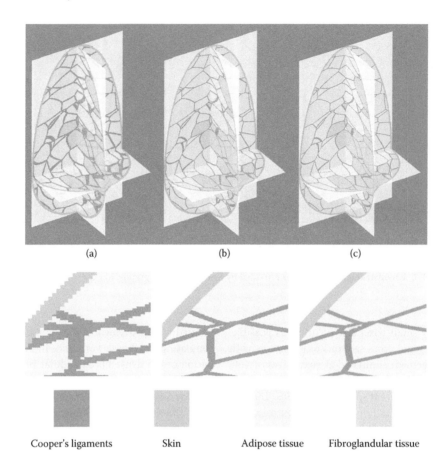

Cooper's ligaments   Skin   Adipose tissue   Fibroglandular tissue

FIGURE 12.5 Cross sections (upper row) and details (middle row) of three Penn digital breast phantoms simulated using identical adipose compartment seed positions with voxel size of (a) 400, (b) 100, and (c) 25 µm. (From Pokrajac, D.D., Maidment, A.D.A., and Bakic, P.R. (2012). *Medical Physics*, 39, 2290–2302. With permission.)

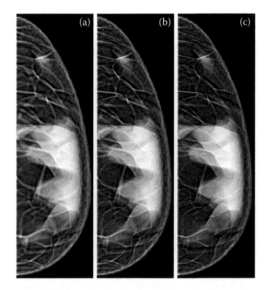

FIGURE 12.6 Synthetic mammographic projections through the compressed phantoms with voxel size of 200 µm and simulated Cooper's ligaments of thickness (a) 1200, (b) 800, and (c) 400 µm. (From Pokrajac, D.D., Maidment, A.D.A., and Bakic, P.R. (2012). *Medical Physics*, 39, 2290–2302. With permission.)

threshold equals the required breast glandular fraction. Once the threshold value is determined, attenuation values corresponding to fibroglandular and adipose tissues are assigned to the voxels above and below the threshold, respectively. Figure 12.7 shows slices through a volume of the binarized noise model, as well as the corresponding tomosynthesis projections and reconstructed slices. Figure 12.8 shows a section and a synthetic projection through a Penn digital breast phantom extended by inclusion of binarized noise. A validation of the extended Penn phantom based upon the analysis of the power spectrum is ongoing.

Another early effort in breast anatomy simulation was performed by Paul Taylor from the University of Western Australia. The simulation featured an admixture of adipose and glandular tissue. Of note, the breast ductal network was simulated using a fractal description of duct lengths, diameters, and branching angles (Taylor and Owens 1998). The shape of the simulated breast was derived from images acquired with structured light (Taylor et al. 2000).

An alternative rule-based design of digital breast phantoms has been developed at the University of Patras, Greece. Bliznakova et al. (2003, 2010) have created a hybrid method that combines features based upon geometric primitives with breast

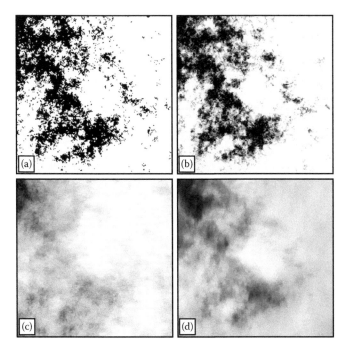

FIGURE 12.7 (a) A thin (80 μm) and (b) a thick (0.8 mm) slice through the binarized noise volume; (c) central projection and (d) a reconstructed tomosynthesis slice. ROI size is 2.5 cm². (Reproduced from Reiser, I., and Nishikawa, R.M. (2010). *Medical Physics*, 37, 1591–1600. With permission.)

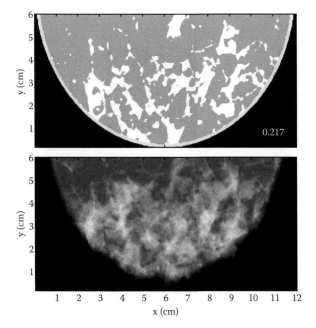

FIGURE 12.8 A single (central) slice through the Penn digital phantom modified by inclusion of the binarized noise model of small-scale tissue variations (top) and parallel projection of the phantom (bottom). (Courtesy of Beverly Lau, University of Chicago.)

anatomical noise approximated with a clustered lumpy background (Bochud et al. 1999), as illustrated in Figure 12.9. A similar phantom design is shared by Ma et al. (2009). The geometric primitives are used to model the large- and medium-scale anatomical structures, including the breast outline and skin, ductal network, Cooper's ligaments, the pectoralis muscle, and blood and lymph vessels (Bliznakova et al. 2010).

The small-scale tissue variations, making up the background mammographic texture, are simulated in a series of 3-D statistical filtering steps (Figure 12.10). Initially, a 3-D texture matrix is filled using an algorithm based upon a fractional Brownian motion, which allows long-range correlations and scaling properties similar to mammograms (Bliznakova et al. 2010). The resulting matrix may be morphologically dilated and/or low-pass and Gaussian filtered to improve the realism of the simulated 3-D breast tissue distribution; these steps are optional. Finally, the texture matrix voxel values (ranging from 0 to 1) are converted to linear x-ray attenuation coefficients for the specified photon energy. The resulting texture voxels are combined with the simulated duct system, Cooper's ligaments, optional abnormalities, and lymphatic and blood vessels, the muscles, and skin to form the final phantom matrix (Bliznakova et al. 2010).

Projections of the Patras phantom have been generated assuming a monochromatic x-ray beam, an ideal photon counting detector with 100-μm resolution, and a simulated Poisson quantum noise (Bliznakova et al. 2010). Phantom deformation due to mammographic breast compression has been performed by modeling the interaction of small tissue subvolumes based upon spring equations (Zyganitidis et al. 2007); it is, however, not routinely simulated. Figure 12.11 shows examples of breast images synthesized using the Patras digital phantom.

Synthetic images of the Patras digital breast phantom have been validated indirectly based upon image texture properties (Bliznakova et al. 2010). It has been reported that calculated values of skewness, kurtosis, exponential parameter of the power law spectrum (β), and fractal dimension show good correlation between clinical and phantom images.

A rule-based phantom design has also been developed at Duke University (Chen et al. 2011). The phantom includes an outer layer of skin and subcutaneous fat and an admixture of glandular and adipose tissue modeled as a power law distribution. The breast ductal network is generated stochastically, based upon self-similarity models of arterial blood vessels. The phantom also allows for simulation of microcalcifications and masses. The breast is simulated assuming an ellipsoidal compressed shape; other compression levels of the same phantom are simulated by reciprocal stretching in two orthogonal directions, assuming incompressible tissue properties.

Synthetic phantom images have been generated using a Penelope-based Monte-Carlo simulation (Saunders et al. 2009), assuming a half-cone beam x-ray source at 28 kVp. A Selenium direct flat-panel detector was modeled with 85-μm resolution, assuming only quantum noise. The Monte-Carlo code has been used to record the energy deposited in each phantom voxel and hence generate 3-D dose maps (Chen et al. 2011).

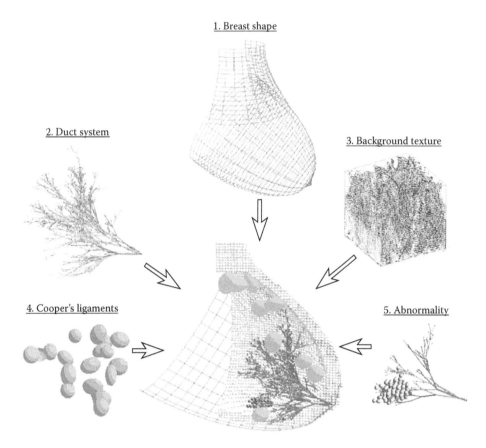

FIGURE 12.9 The composition of the Patras digital breast phantom, combining geometric primitives and anatomical noise simulated by a clustered lumpy background model. (Reproduced from Bliznakova, K., Bliznakov, Z., Bravou, V., Kolitsi, Z., and Pallikarakis, N. (2003). *Physics in Medicine and Biology*, 48, 3699–3719. With permission.)

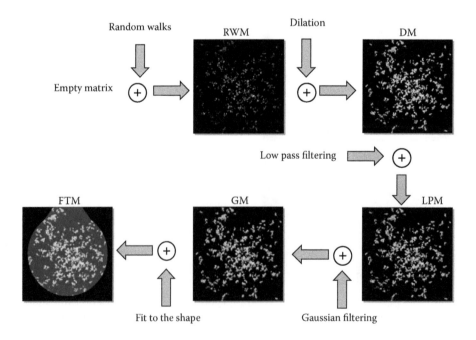

FIGURE 12.10 Illustration of the main steps for simulating the breast image texture in the Patras digital phantom. Shown are slices extracted from a phantom with size of 75 mm in each direction and voxel dimension of 150 μm. (Reproduced from Bliznakova, K., Suryanarayanan, S., Karellas, A., and Paiilikarakis, N. (2010). *Medical Physics*, 37, 5604–5617. With permission.)

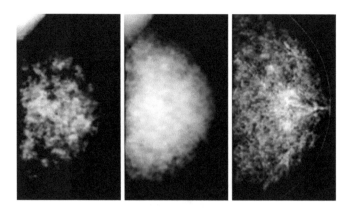

FIGURE 12.11 Examples of mammograms synthesized using the Patras digital breast phantom. (Reproduced from Bliznakova, K., Suryanarayanan, S., Karellas, A., and Paiilikarakis, N. (2010). *Medical Physics*, 37, 5604–5617. With permission.)

12.3.2 Phantoms Based upon Individuals

The alternative is to create digital phantoms based on specific clinical images. The earliest report on using individual clinical breast images to create digital phantoms was published by Hoeschen et al. (2005) from the German National Research Center for Environment and Health (GSF). High resolution CT images of a mastectomy specimen were acquired, segmented, and postprocessed to separate skin, adipose, and fibroglandular tissue. Figure 12.12 illustrates the GSF digital phantom. The phantom has been used for estimating the dose in mammography,

based upon a Monte-Carlo radiation transport simulation (Zankl et al. 2005).

The concept of using mastectomy specimens for digital breast phantom design has been further developed at the University of Massachusetts (UMass). O'Connor et al. (2008) have used a dedicated, bench-top, breast CT prototype for acquiring 3-D images of mastectomy specimens. Using an IRB-approved protocol, fresh mastectomy specimens were obtained immediately after surgery and prior to tissue gross pathology. Each specimen was placed in an appropriately sized holder, mimicking the compressed or uncompressed breast shape, corresponding to digital breast tomosynthesis or breast CT, respectively, as illustrated in Figure 12.13.

The specimens were imaged at approximately 10 times the exposure typically given in a clinical breast CT study. The specimen CT images were reconstructed using a Feldkamp filtered back-projection algorithm. The reconstructed images were then corrected for slice nonuniformity and smoothed by anisotropic diffusion filtering. The postprocessed images were converted into a digital breast phantom either by segmentation into voxels corresponding to adipose or glandular tissue or by scaling reconstructed values into percentage of glandular tissue contained within each voxel, as illustrated in Figure 12.14.

Synthetic mammograms of the UMass digital breast phantoms were generated using a mastectomy specimen holder to mimic mammographic compression, as shown in Figure 12.15. Synthetic CT images of the UMass digital breast phantoms were simulated using a ray-tracing method with a realistic model of signal and noise propagation through the detector (Vedula et al. 2003), and compared with the original specimen CT images

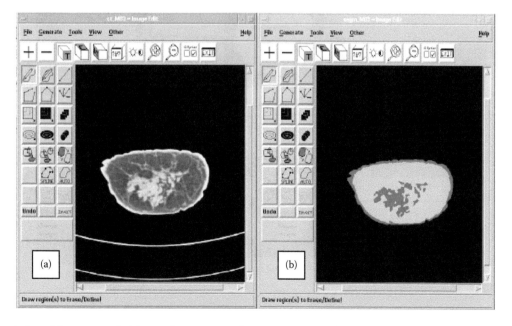

FIGURE 12.12 A slice of the segmented CT-scan of a mastectomy breast specimen, used as the GSF digital breast phantom (a), indicating different tissue distribution (b). (Reproduced from Hoeschen, C., Fill, U., Zankl, M., Panzer, W., Regulla, D., and Dohring, W. (2005). *Radiation Protection Dosimetry*, 114, 406–409. With permission from Oxford University Press.)

FIGURE 12.13 **(See color insert.)** Customized holders for mastectomy tissue specimens, mimicking the uncompressed (left) or compressed (right) breast shape, used for generation of the UMass digital breast phantoms. (Modified from O'Connor, J.M., Das, M., Didier, C., Mah'd, M., and Glick, S. J. (2008). In Krupinski, E.A. (Ed.), *Digital Mammography (IWDM)*. Berlin-Heidelberg. With permission from Springer-Verlag.)

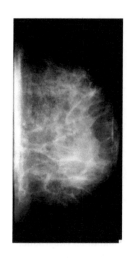

FIGURE 12.15 A synthetic mammogram generated using the UMass digital breast phantom. (Courtesy of Dr. Stephen J. Glick, University of Massachusetts.)

(O'Connor et al. 2010). The power-law exponents β showed very close agreement between the originally and simulated images; slightly better agreement was observed for phantoms generated based upon the segmented tissue method. Both types of phantoms showed very good agreement to the original images in terms of the mean contrast between the regions of adipose and glandular tissue.

An alternative phantom design based upon simulating individual clinical cases has been developed at Duke University (Li et al. 2009b) using a postprocessed clinical data set acquired by a dedicated breast CT scanner (Lindfors et al. 2008). The reconstructed breast CT images are postprocessed to suppress noise and scatter and segmented using a multistep algorithm into different tissue types. Figure 12.16 shows a slice through the

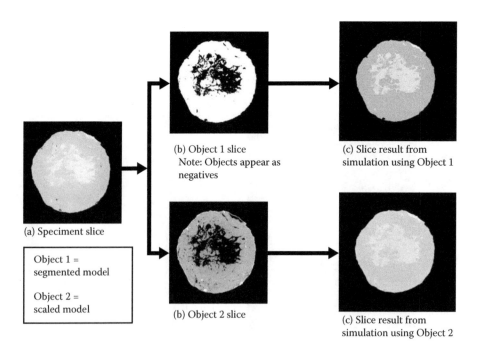

FIGURE 12.14 Illustration of two methods for generating the UMass digital breast phantoms. CT images of mastectomy specimens are either segmented into voxels corresponding to the adipose or glandular tissue (top) or the reconstructed values denote the percentage of glandular tissue (bottom). (Reproduced from O'Connor, J.M., Das, M., Didier, C., Mah'd, M., and Glick, S. J. (2008). In Krupinski, E.A. (Ed.), *Digital Mammography (IWDM)*. Berlin-Heidelberg. With permission from Springer-Verlag.)

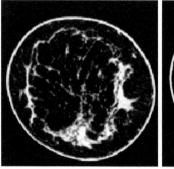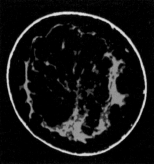

FIGURE 12.16 Illustration of the Duke digital breast phantom design. (Left) The original reconstructed breast CT image. (Right) The corresponding section of the digital breast phantom generated by denoising and segmentation of the CT data. (Reprinted from Li, C.M., Segars, W.P., Tourassi, G.D., Boone, J.M., and Dobbins III, J.T. (2009b). *Medical Physics*, 36, 3122–3131. With permission.)

original reconstructed breast CT volume along with the corresponding phantom slice. The mammographic compression was originally approximated by simplified geometric deformation rules (Li et al. 2009b); this has recently been replaced by a finite element compression model employing a high density mesh (Hsu et al. 2011). The x-ray image acquisition uses a polyenergetic x-ray spectrum to simulate a breast tomosynthesis geometry with 250 μm pixels. Figure 12.17 illustrates a synthetic breast image generated using the Duke digital phantom. Synthetic images have been validated indirectly by comparing their fractal dimensions with reported values from clinical and other simulated data (Li et al. 2009b).

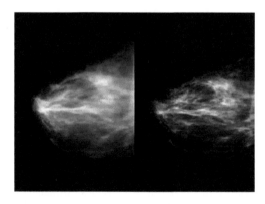

FIGURE 12.17 Synthetic mammograms generated using the Duke digital breast phantom. The image on the right was generated with artificially enhanced attenuation coefficients (compared to the image on the left) to illustrate the high resolution tissue detail. (Reprinted from Li, C.M., Segars, W.P., Tourassi, G.D., Boone, J.M., and Dobbins III, J.T. (2009b). *Medical Physics*, 36, 3122–3131. With permission.)

12.4 Phantom Applications

The main motivation for developing digital breast phantoms has been the validation and optimization of breast imaging systems and breast image analysis methods. Perhaps the most valuable contribution of digital phantom studies is expected in preclinical validation of novel image acquisition systems, as preclinical phantom studies can guide the design of actual clinical studies. Phantom studies have value in breast image analysis, particular for problems in which clinical ground truth is unavailable, such as breast image registration. Digital anthropomorphic phantoms also have value in applications where traditional geometric phantoms have found utility, such as image quality assessment and radiation dosimetry. Table 12.1 lists the mostly reported applications of digital breast phantoms. The following section describes some of these applications in more detail.

12.4.1 Optimization and Validation of DBT Systems

The Penn digital breast phantom has been extensively used for validation and optimization of DBT imaging systems, including optimization of the DBT reconstructions (Bakic et al. 2010b) and assessment of the DBT reconstruction geometric accuracy (Bakic et al. 2010c). Synthetic images of the phantom have been used in an analysis of the effects of DBT acquisition parameters on reconstructed image texture (Kontos et al. 2008), as well as comparison of power spectral properties in DM and

TABLE 12.1 Reported Applications of Digital Breast Phantoms

X-ray imaging	
Mammography	Dance et al. 2005, Chen et al. 2011, Li et al. 2009, Zankl et al. 2005, Bakic et al. 2011
DBT	Kontos et al. 2008, Ruiter et al. 2008, Saunders et al. 2009, Young et al. 2009, Bakic et al. 2010a, Bakic et al. 2010b, Diaz et al. 2012
BCT	O'Connor et al. 2010
CE-DBT	Carton et al. 2011
Nonionizing imaging	
Ultrasound tomography	Bakic et al. 2011
MRI	Nishikawa et al. 2011
Image analysis	
Mammogram registration	Richard et al. 2006
Breast ductal tracing	Bakic et al. 2003
Dense tissue segmentation	Bakic et al. 2011
Image compression	Hoeschen et al. 2005b
Dosimetry	Hoeschen et al. 2005a, Dance et al. 2005, Zankl et al. 2005, Saunders et al. 2009, Chen et al. 2011
Radiation therapy	Mundy et al. 2006, Sztejnberg Goncalves-Carralves and Jevremovic 2007

Sources: See table for multiple data references.

DBT (Lau et al. 2010, Bakic et al. 2010a) and scatter field estimation (Diaz et al. 2012). Synthetic images of the Penn phantom are also being used in an ongoing observer study of tumor visibility as a function of the DBT acquisition parameters (Young et al. 2009).

As an example of phantom-based DBT validation, Figures 12.18 and 12.19 illustrate the assessment of geometric accuracy in a DBT reconstruction (Bakic et al. 2010c). Figure 12.18 shows a phantom section (left) with simulated fiducial markers (small attenuating objects with a size of one 200 μm voxel), along with a DBT image reconstructed at the position of the fiducial markers (right). The images were reconstructed with a pixel size of 115 μm. The phantom contains three sets of fiducial markers (each having four markers) positioned at different depths within the phantom.

To measure geometric accuracy, a series of images was reconstructed with subpixel shifts within the plane of reconstruction. These images were combined to form an image supersampled 10-fold in the y-axis (scanning direction). The position of each fiducial marker was identified as the center of mass of the marker in the reconstructed plane in which the spatial extent of the marker was smallest). Based upon the ground truth marker positions, available from the phantom, the geometric accuracy was assessed using the Euclidean distance between the estimated and the true marker position. Figure 12.19 shows the error in the estimated marker positions (Ep), as a function of the reconstructed plane depth. Also shown are the errors calculated separately for each of the three marker coordinates (x, y, z). The error averaged over all 12 fiducial markers was 0.105 ± 0.086 mm (average ± standard deviation). The interslice (z direction)

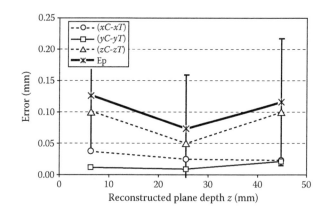

FIGURE 12.19 Average error in the measured 3-D position of reconstructed fiducial markers (Ep) as a function of the reconstructed plane depth. (Error bars represent one standard deviation.) Shown also are the errors calculated separately for each marker coordinate. (Modified from Bakic, P.R., Ng, S., Ringer, P., Carton, A.-K., Conant, E.F., and Maidment, A.D.A. (2010b). In Samei, E., and Pelc, N. (Eds.), *SPIE Medical Imaging: Physics of Medical Imaging.* San Diego, CA: SPIE.)

error is substantially larger than the in-plane (x, y) errors, due to the limited angular range of the DBT acquisition geometry. Both the in-plane and interslice errors are substantially smaller than the reconstructed image voxel size, validating the reconstruction method. However, the more important result is that this work demonstrated that DBT is capable of super-resolution (Acciavatti and Maidment 2011); this illustrates the essential role of digital phantoms in preclinical assessment of imaging systems.

12.4.2 Breast Density Measurement

Breast density is a measure of the fractional area or volume of fibroglandular (dense) tissue in a breast image. Breast density has been shown to correlate with the lifetime risk of breast cancer. (Martin and Boyd 2008) The Penn phantom has been used for validation of 3-D ultrasound tomography (UST) of the breast, including the measurement of breast density by UST (Li et al. 2009a). In the phantom study (Bakic et al. 2011a), the performance of UST and DM in the task of breast density estimation was simulated. The UST-based volumetric breast density (VBD) was calculated by thresholding the reconstructed UST images. Percent density (PD) values from DM images were estimated interactively by a clinical breast radiologist using Cumulus software (Byng et al. 1994). The VBD values estimated from UST showed a high correlation with the ground truth volumetric breast density. The PD estimates were also correlated with the ground truth; however, the standard error of DM estimates was 1.5 times higher than the UST estimates, potentially due to the use of global thresholding and the variability arising from human observers in the PD method.

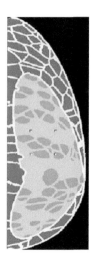 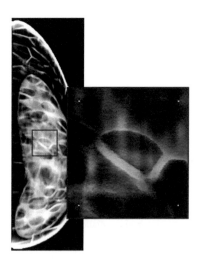

FIGURE 12.18 A section through the Penn digital phantom (left) with simulated fiducial markers, along with a DBT reconstructed image (right) at the position of the fiducial markers. The insert shows a magnified region containing the markers. (Modified from Bakic, P.R., Ng, S., Ringer, P., Carton, A.-K., Conant, E.F., and Maidment, A.D.A. (2010b). In Samei, E., and Pelc, N. (Eds.), *SPIE Medical Imaging: Physics of Medical Imaging.* San Diego, CA: SPIE.)

12.4.3 Validation of Mammogram Registration

The Penn digital breast phantom (Bakic et al. 2002a) has been used to validate a nonrigid mammogram registration method (Richard et al. 2003, 2006). In that study, images of the same phantom were produced with various amounts of simulated mammographic compression. Images produced with different compressions were registered pairwise, and positional errors of simulated fiducial markers were calculated. An average displacement error of 1.6 mm was observed for synthetic mammograms with compressed thickness differences of 1–3 cm. Digital phantom studies have value in registration problems because the ground truth of the tissue displacement exists. As with other uses of digital phantoms, the validity of the results does depend upon the realism of the model, including the simulated deformation or positioning of the breast.

12.4.4 Dosimetry

Digital breast phantoms have also been used to estimate the radiation dose received in mammography. The availability of ground truth information about the phantom composition allows for tissue-specific dosimetry; the received dose can be calculated separately for different simulated tissues. Using the Penn phantom with a Monte-Carlo model of mammography (Dance et al. 2005) demonstrated that significant differences (up to 48%) occurred in comparison to tabulated values of incident air kerma to mean glandular dose conversion coefficients. These differences arise from variations in the spatial distribution of glandular tissue within the breast. The observed results demonstrate the limitations of the standard breast dosimetry approach, which does not account for patient specific breast anatomy. Similar limitations were observed in a study using the GSF digital breast phantom (Zankl et al. 2005). In that study, better agreement with tabulated conversion coefficients was observed when the phantom was positioned so that the majority of the glandular tissue was located close to the entrance surface of the breast. More recently, the Duke digital phantom has also been used to generate 3-D dose maps in simulated breasts under different levels of mammographic compression (Saunders et al. 2009).

12.4.5 Radiation Therapy Optimization

Researchers from Purdue University have used the Penn digital breast phantom to explore the feasibility of a novel radiation targeted therapy for HER-2-positive breast cancers (Mundy et al. 2006, Sztejnberg Goncalves-Carralves and Jevremovic 2007). In order to conduct the study, the phantom was extended to include a deep-seated tumor model, simulated by a cylindrical volume in which the tumor tissue was 10% denser than the surrounding fibroglandular tissue. The dose distributions within the simulated tumor and the surrounding healthy tissue were calculated using Monte-Carlo methods (MCNP5). Based on the success of this preclinical trial, more complex patient-specific simulations were then undertaken by the authors.

12.4.6 Fabrication of a Physical Anthropomorphic Phantom

The Penn digital phantom was used as the basis of a 3-D physical anthropomorphic phantom of the breast (Carton et al. 2010, 2011). The physical phantom was produced in a series of steps. First, the digital phantom was separated into several sections up to 1 cm thick. The simulated nonadipose tissue (i.e., the fibroglandular tissue, skin, and Cooper's ligaments) in each section was then fabricated with rapid prototyping using a single material having 50% glandular tissue equivalence. Next, the adipose compartments were filled manually with an epoxy-based resin with 100% adipose equivalence. Finally, the phantom sections were stacked to form the physical anthropomorphic phantom. Figure 12.20 shows slices of the physical anthropomorphic phantom fabricated based upon a digital breast phantom corresponding to a 450 ml breast with 45% dense tissue, deformed to a 5-cm compressed thickness. A clinical mammographic projection image and a reconstructed tomosynthesis slice of the physical phantom are shown in Figure 12.21.

Rapid prototyping of a monolithic physical breast phantom is challenging since no rapid prototyping materials that accurately simulate the x-ray properties of both adipose and dense breast tissues currently exist. The stepwise approach described above was successful because the Penn digital phantom design is based upon a region growing algorithm that ensures that the regions of dense tissue are contiguous (i.e., no isolated dense regions exist), and thus can be fabricated as a single object. A superior method for fabricating 3-D anthropomorphic phantoms requires development of suitable rapid prototyping materials to ensure accurate representation of x-ray properties of the breast.

FIGURE 12.20 (See color insert.) Photograph of the prototype anthropomorphic physical breast phantom, generated based upon a Penn digital breast phantom of a 450 ml breast with 42% volumetric glandularity. (Reproduced from Carton, A.-K., Bakic, P.R., Ullberg, C., Derand, H., and Maidment, A.D.A. (2011). *Medical Physics*, 38, 891–896.)

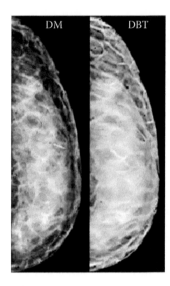

FIGURE 12.21 Clinically acquired mammographic projection (left) and a reconstructed tomosynthesis image (right) of the Penn anthropomorphic physical breast phantom from Figure 12.20. (Reproduced from Carton, A.-K., Bakic, P.R., Ullberg, C., Derand, H., and Maidment, A.D.A. (2011). *Medical Physics*, 38, 891–896.)

12.5 Future Directions of Digital Breast Phantoms Research

Given that computational power and algorithm efficiency are likely to continue to improve for the foreseeable future, we can reasonably presume that the demand for and complexity of breast simulations will increase with time. Thus, the role of digital breast phantoms will only increase in importance. There are a number of phantom design issues that will need to be addressed in the near future; these include further acceleration of phantom generation methods, concomitant improvements in compression and image simulation methods, further reductions in voxel or feature size to model the breast at finer scales, inclusion of temporal dynamics, and the development of a standard ontology for the creation, search, and exchange of simulated breast data. In addition, the role of digital phantoms needs to be established either alone or in combination with conventional geometric test objects and physical versions of digital anthropomorphic phantoms.

The octree-based Penn phantom features an optimized computational complexity, which allows fast generation of a large number of phantoms with small voxel size. The improvement in synthetic image quality comes with a cost of increased storage and transfer times to accommodate the large number of voxels. An alternative approach to reduce quantization artifacts, without an increase in the number of voxels, is simulation of partial volumes during the phantom generation process (Chen et al. 2012). Phantom generation methods, which replace the voxel-based representation of the simulated anatomical structures by an object-based representation, as, e.g., octrees, may also help to accelerate the simulation of breast deformation or image acquisition; of particular interest are optimized implementations of object-based phantoms for use with Monte-Carlo methods of image synthesis and further development of GPU-based simulation systems.

Future applications of digital breast phantoms will likely widen the range of simulated imaging modalities, including breast DCE-MRI (Chapter 16), PET/SPECT (Chapter 17), or optical imaging (Chapter 20). Current digital phantoms have limited ability to simulate dynamical physiological processes in the breast, e.g., vascular kinetics. Resolving this limitation requires an extension of the phantom design to include temporal changes in simulated vasculature anatomy or in simulated perfusion properties of the breast tissue. The necessary level of realism would depend upon the imaging task, similar to the realism requirements for the simulation of anatomical variations (as discussed in Section 12.2).

A key limitation in the synthesis of phantom images is related to the level of realism achieved in modeling the clinical examination setup, i.e., the breast positioning and compression. The majority of digital breast phantoms discussed in this chapter includes a compression model, simulated using either a finite element approach or simplified geometric rules for calculating the compressed breast shape. Improvements in the simulation of breast deformation are needed, including improved support of the various breast imaging positions (prone, supine, erect, etc.) and deformation directions (mediolateral, craniocaudal, etc.). An improved simulation of breast positioning and deformation could also support the simulation of interventional imaging procedures, e.g., presurgical and intrasurgical imaging, radiation therapy, and radiation or chemotherapy follow-up.

The ability to efficiently generate a large number of digital phantoms opens a question about the optimal representation of the phantom- and simulation-related parameters. Medical data standards (e.g., DICOM, HL7) regulate the exchange of the information related to the image acquisition procedure, as well as patient-related information. In current phantom studies, many of those information fields are used informally (as "dummy" variables) simply to ensure compatibility with the standard data formats. As phantom designs become more sophisticated, we envision that these variables will be used formally; for example, simulations relevant to specific target patient populations or simulations modeling the age-specific changes to a given breast. This illustrates the need to develop an appropriate ontology for the representation of the simulation data.

When viewed together, the development of a suitable ontology coupled with improvements to computational speeds may form a complementary solution to the problems of increased data size associated with reduced feature size. Specifically, the current repositories of prefabricated digital phantoms might be replaced by databases of optimized phantom descriptors, for which we have coined the phrase "phantom DNA." Using an appropriate ontology to describe the phantom design, it should be possible to create, group, and query phantoms based on their "DNA."

Predicting future advances in phantom designs require that we better define the roles for these digital phantoms. Intrinsically, one would expect that as the realism of the phantom increases, the number of uses of the phantoms will increase. Examples of applications already in planning include the use of digital phantoms in virtual clinical trials designed to precede more focused clinical trials, and simulation at mesoscopic and microscopic scales to explore radiology–pathology correlation. Advancements in the fabrication of physical versions of anthropomorphic digital phantoms introduce additional uses.

With increased realism, comes the requirement for improved methods of both direct and indirect validation; digital phantoms will only be as good as our knowledge of the breast. This introduces one of the greatest limitations of current phantoms—the lack of good morphological data at various scales. It is likely that to further improve breast phantoms will require that the study of breast pathology be revisited with modern computational methods. Similarly, use of digital breast phantoms in physiological simulations at cellular or molecular level will need interdisciplinary efforts and may introduce novel requirements to the phantom design.

In conclusion, the use of digital breast phantoms has grown significantly in the last decade. The contemporaneous advancements in computational methods, computational hardware, clinical data, and the demand for imaging simulations have all fueled this growth. As we move forward, we expect that the principles used in developing digital breast phantoms can be applied to the simulation of other organs or tissues in the human body. In specific applications, phantoms based upon individual clinical images have already been used for quite some time. We believe that heuristic rules for simulating anatomical structures need to be developed for other tissues, allowing the inherent benefits associated with the knowledge of the ground truth and the flexibility of such designs to be applied more widely in medical research.

References

Acciavatti RJ, Maidment ADA. 2011. Investigating the potential for super-resolution in digital breast tomosynthesis. In Physics of Medical Imaging. San Diego, CA: SPIE.

Bakic PR, Albert M, Brzakovic D, Maidment, ADA. 2002a. Mammogram synthesis using a 3D simulation. I. Breast tissue model and image acquisition simulation. Med Phys 29: 2131–2139.

Bakic PR, Albert M, Brzakovic D, Maidment ADA. 2002b. Mammogram synthesis using a 3D simulation. II. Evaluation of synthetic mammogram texture. Med Phys 29: 2140–2151.

Bakic PR, Albert M, Brzakovic D, Maidment ADA. 2003. Mammogram synthesis using a three-dimensional simulation. III. Modeling and evaluation of the breast ductal network. Med Phys 30: 1914–1925.

Bakic PR, Lau B, Carton A-K, Reiser I, Maidment ADA, Nishikawa RM. 2010a. An anthropomorphic software breast phantom for tomosynthesis simulation: Power spectrum analysis of phantom projections. In International Workshop on Digital Mammography. Girona, Spain: Springer.

Bakic PR, Li C, West E, Sak M, Gavenonis SC, Duric N, Maidment ADA. 2011a. Comparison of 3D and 2D breast density estimation from synthetic ultrasound tomography images and digital mammograms of anthropomorphic software breast phantoms. In Pelc NJ, Samei E, Nishikawa RM, eds. SPIE Medical Imaging: Physics of Medical Imaging. Lake Buena Vista, FL.

Bakic PR, Ng S, Ringer P, Carton A-K, Conant EF, Maidment ADA. 2010b. Validation and optimization of digital breast tomosynthesis reconstruction using an anthropomorphic software breast phantom. In Samei E, Pelc N, eds. SPIE Medical Imaging: Physics of Medical Imaging. San Diego, CA.

Bakic PR, Ringer P, Kuo J, Ng S, Maidment ADA. 2010c. Analysis of geometric accuracy in digital breast tomosynthesis reconstruction. In International Workshop on Digital Mammography. Girona, Spain: Springer.

Bakic PR, Zhang C, Maidment ADA. 2011b. Development and characterization of an anthropomorphic breast software phantom based upon region-growing algorithm. Med Phys 38: 3165–3176.

Barrett HH, Myers KJ. 2004. Foundations of Image Science, New York: John Wiley & Sons, Inc.

Bliznakova K, Bliznakov Z, Bravou V, Kolitsi Z, Pallikarakis N. 2003. A three-dimensional breast software phantom for mammography simulation. Phys Med Biol 48: 3699–3719.

Bliznakova K, Suryanarayanan S, Karellas A, Paiilikarakis N. 2010. Evaluation of an improved algorithm for producing realistic 3D breast software phantoms: Application for mammography. Med Phys 37: 5604–5617.

Bochud FO, Abbey CK, Eckstein MP. 1999. Statistical texture synthesis of mammographic images with clustered lumpy backgrounds. Opt Exp 4.

Burgess AE. 1999. Mammographic structure: Data preparation and spatial statistics analysis. In Hanson KM, ed. SPIE Medical Imaging: Image Processing. San Diego, CA.

Byng JW, Boyd NF, Fishell E, Jong RA, Yaffe MJ. 1994. The quantitative analysis of mammographic densities. Phys Med Biol 39: 1629–1638.

Carton A-K, Bakic PR, Ullberg C, Derand H, Maidment ADA. 2011. Development of a physical 3D anthropomorphic breast phantom. Med Phys 38: 891–896.

Carton A-K, Bakic PR, Ullberg C, Maidment ADA. 2010. Development of a 3D high-resolution physical anthropomorphic breast phantom. In Samei E, Pelc NJ, eds. SPIE Medical Imaging: Physics of Medical Imaging. San Diego, CA.

Castella C, Kinkel K, Descombes F, Eckstein MP, Sottas P-E, Verdun FR, Bochud FO. 2008. Mammographic texture synthesis: Second-generation clustered lumpy backgrounds using a genetic algorithm. Opt Exp 16: 7595–7607.

Chen B, Shorey J, Saunders RSJ, Richard S, Thompson J, Nolte LW, Samei E. 2011. An anthropomorphic breast model for breast imaging simulation and optimization. Acad Radiol 18: 536–546.

Chen F, Pokrajac DD, Shi X, Liu F, Maidment ADA, Bakic PR. 2012. Partial volume effect simulation in software breast phantoms. In Physics of Medical Imaging. San Diego, CA: SPIE.

Chui JH, Pokrajac DD, Maidment ADA, Bakic PR. 2012. Roadmap for efficient parallelization of breast anatomy simulation. In Physics of Medical Imaging. San Diego, CA: SPIE.

Clarke GM, Peressotti C, Constantinou P, Hosseinzadeh D, Martel A, Yaffe MJ. 2011. Increasing specimen coverage using digital whole-mount breast pathology: Implementation, clinical feasibility and application in research. Comput Med Imag Graph 35.

Clarke GM, Peressotti C, Mawdesley GE, Yaffe MJ. 2006. Design and characterization of a digital image acquisition system for whole-specimen breast histopathology. Phys Med Biol 51: 5089–5103.

Contijoch F, Lynch JM, Pokrajac DD, Maidment ADA, Bakic PR. 2012. Shape analysis of simulated breast anatomical structures. In Physics of Medical Imaging. San Diego, CA: SPIE.

Cooper A. 1840. On the Anatomy of the Breast, London: Longman.

Dance DR, Hunt RA, Bakic PR, Maidment ADA, Sandborg M, Ullman G, Carlsson GA. 2005. Breast dosimetry using high-resolution voxel phantoms. Radiat Protect Dosimetry 114: 359–363.

Diaz O, Dance DR, Young KC, Elangovan P, Bakic PR, Wells K. 2012. A fast scatter field estimator for Digital Breast Tomosynthesis. In Physics of Medical Imaging. San Diego, CA: SPIE.

Going JJ, Moffat DF. 2004. Escaping from Flatland: Clinical and biological aspects of human mammary duct anatomy in three dimensions. J Pathol 203: 538–544.

Heine JJ, Deans SR, Velthuizen RP, Clarke LP. 1999. On the statistical nature of mammograms. Med Phys 26: 2254–2265.

Hoeschen C, Fill U, Zankl M, Panzer W, Regulla D, Dohring W. 2005. A high resolution voxel phantom of the breast for dose calculations in mammography. Radiat Protect Dosimetry 114: 406–409.

Hsu CM, Palmeri ML, Segars WP, Veress AI, Dobbins III JT. 2011. An analysis of the mechanical parameters used for finite element compression of a high-resolution 3D breast phantom. Med Phys 38: 5756–5770.

Hunt RA, Dance DR, Bakic PR, Maidment ADA, Sandborg M, Ullman G, Carlsson GA. 2005. Calculation of the properties of digital mammograms using a computer simulation. Radiat Protect Dosimetry 114: 395–398.

Kontos D, Zhang C, Ruiter NV, Bakic PR, Maidment ADA. 2008. Evaluating the effect of tomosynthesis acquisition parameters on image texture: A study based on an anthropomorphic breast tissue software model. In Krupinski EA, ed. 9th International Workshop on Digital Mammography. Tucson, AZ: Springer.

Lanyi M. 1988. Diagnosis and Differential Diagnosis of Breast Calcifications. Berlin, Germany: Springer-Verlag.

Lau AB, Bakic PR, Reiser I, Carton A-K, Maidment ADA, Nishikawa RM. 2010. An anthropomorphic software breast phantom for tomosynthesis simulation: Power spectrum analysis of phantom reconstructions. Med Phys 37: 3473.

Li C, Duric N, Littrup P, Huang L. 2009a. In vivo breast sound-speed imaging with ultrasound tomography. Ultrasound Med Biol 35: 1615–1628.

Li CM, Segars WP, Tourassi GD, Boone JM, Dobbins III JT. 2009b. Methodology for generating a 3D computerized breast phantom from empirical data. Med Phys 36: 3122–3131.

Lindfors KK, Boone JM, Nelson TR, Yang K, Kwan ALC, Miller DF. 2008. Dedicated breast CT: Initial clinical experience. Radiology 246: 725–733.

Lorensen WE, Cline HE. 1987. Marching cubes: A high resolution 3D surface construction algorithm. Comput Graph 21: 163–169.

Ma AKW, Gunn S. Darambara DG. 2009. Introducing DeBRa: A detailed breast model for radiological studies. Phys Med Biol 54: 4533–4545.

Maidment ADA, Albert M, Conant EF. 1998. Three-dimensional imaging of breast calcifications. In Selander JM, ed. 26th AIPR Workshop: Exploiting New Image Sources and Sensors. Washington, DC: SPIE.

Maidment ADA, Albert M, Feig SA. 1996. 3-D mammary calcification reconstruction from a limited number of views. Physics of Medical Imaging. San Diego: SPIE.

Martin LJ, Boyd NF. 2008. Mammographic density. Potential mechanisms of breast cancer risk associated with mammographic density: Hypotheses based on epidemiological evidence. Breast Cancer Res 10: 201–214.

Meagher D. 1981. Geometric modeling using octree encoding. Comput Graph Image Process 19: 129–147.

Moffat DF, Going JJ. 1996. Three dimensional anatomy of complete duct systems in human breast: Pathological and developmental implications. J Clin Pathol 49: 48–52.

Mundy DW, Herb W, Jevremovic T. 2006. Radiation binary targeted thearpy for HER-2 positive breast cancers: Asumptions, theoretical assessment and future directions. Phys Med Biol 51: 1377–1391.

Nishikawa RM, Glick SJ, Bakic PR, Reiser I. 2011. 3D breast models. Med Phys 38: 3706.

O'Connor JM, Das M, Didier C, Mah'd M, Glick SJ. 2008. Comparison of two methods to develop breast models for simulation of breast tomosynthesis and CT. In Krupinski EA, ed. Digital Mammography (IWDM). Berlin-Heidelberg: Springer-Verlag.

O'Connor JM, Das M, Didier C, Mah'd M, Glick SJ. 2010. Development of an ensemble of digital breast object models. In Marti J, ed. Digital Mammography (IWDM). Berlin-Heidelberg: Springer-Verlag.

Ohtake T, Abe R, Kimijima I, Fukushima T, Tsuchiya A, Hoshi K, Wakasa H. 1995. Intraductal extension of primary invasive breast carcinoma treated by breast-conservative surgery. Computer graphic three-dimensional reconstruction of the mammary duct-lobular systems. Cancer 76: 32–45.

Ohtake T, Kimijima I, Fukushima T, Tyasuda M, Sekikawa K, Takenoshita S, Abe R. 2001. Computer-assisted complete three-dimensional reconstruction of the mammary ductal/lobular systems: Implications of ductal anastomoses for breast-conserving surgery. Cancer 91: 2263–2272.

Pokrajac DD, Maidment ADA, Bakic PR. 2012. Optimized generation of high resolution breast anthropomorphic software phantoms. Med Phys 39: 2290–2302.

Pokrajac DD, Maidment ADA, Bakic PR. 2011. A method for fast generation of high resolution software breast phantoms. Med Phys 38: 3431.

Reiser I, Nishikawa RM. 2010. Task-based assessment of breast tomosynthesis: Effect of acquisition parameters and quantum noise. Med Phys 37: 1591–1600.

Richard FJP, Bakic PR, Maidment ADA. 2003. Non-rigid registration of mammograms obtained with variable breast compression: A phantom study. In Gee JC. et al. ed. Biomedical Image Registration. Berlin, Germany: Springer-Verlag.

Richard FJP, Bakic PR, Maidment ADA. 2006. Mammogram registration: A phantom-based evaluation of compressed breast thickness variation effects. IEEE Trans Med Imag 25: 188–197.

Ruiter NV, Mueller TO, Gemmeke H, Reichenbach JR, Kaiser WA. 2006. Model-based registration of x-ray mammograms and MR images of the female breast. IEEE Trans Nucl Sci 53: 204–211.

Ruiter NV, Zhang C, Bakic PR, Carton A-K, Kuo J, Maidment ADA. 2008. Simulation of tomosynthesis images based on an anthropomorphic software breast tissue phantom. In Miga MI, Cleary KR, eds. SPIE Medical Imaging: Visualization, Image-guided Procedures, and Modeling. San Diego, CA.

Saunders RSJ, Samei E, Lo JY, Baker JA. 2009. Can compression be reduced for breast tomosynthesis? Monte Carlo study on mass and microcalcification conspicuity in tomosynthesis. Radiology 251: 673–682.

Sztejnberg Goncalves-Carralves ML, Jevremovic T. 2007. Numerical assessment of radiation binary targeted therapy for HER-2 positive breast cancers: Advanced calculations and radiation dosimetry. Phys Med Biol 52: 4245–4264.

Taylor P, Owens R. 1998. Simulated mammography using synthetic 3D breasts. In Karssemeijer N, ed. 4th International Workshop on Digital Mammography. Nejmegen, Netherlands: Kluwer.

Taylor P, Owens R, Ingram D. 2000. 3D fractal modelling of breast growth. In Yaffe MJ, ed. 5th International Workshop on Digital Mammography. Toronto, Canada: Medical Physics Publishing.

Tischenko O, Hoeschen C, Dance DR, Hunt RA, Maidment ADA, Bakic PR. 2005. Evaluation of a novel method of noise reduction using computer-simulated mammograms. Radiat Protect Dosimetry 114: 81–84.

Vedula AA, Glick SJ, Gong X. 2003. Computer simulation of CT mammography using a flat-panel imager. In Physics of Medical Imaging. San Diego, CA.

Wellings SR, Jensen HM, Marcum RG. 1975. An atlas of the subgross pathology of the human breast. J Natl Cancer Inst 55: 231–273.

Young S, Park S, Anderson K, Badano A, Myers KJ, Bakic PR. 2009. Estimating DBT performance in detection tasks with variable-background phantoms. In Samei E, Hsieh J, eds. SPIE Medical Imaging: Physics of Medical Imaging. Lake Buena Vista, FL.

Zankl M, Fill U, Hoeschen C, Panzer W, Regulla D. 2005. Average glandular dose conversion coefficients for segmented breast voxel models. Radiat Protect Dosimetry 114: 410–414.

Zhang C, Bakic PR, Maidment ADA. 2008. Development of an anthropomorphic breast software phantom based on region growing algorithm. In Sonka M, Manduca A, eds. SPIE Medical Imaging. San Diego, CA: SPIE.

Zyganitidis C, Bliznakova K, Pallikarakis N. 2007. A novel simulation algorithm for soft tissue compression. Med Biol Eng Comput 45: 661–669.

13

Observer Models for Breast Imaging

François Bochud
Lausanne University Hospital

13.1 Image Quality and Observer Models

13.1.1 Objective Assessment of Image Quality

A medical image is produced for some specific diagnostic purpose, and the most meaningful measure of its quality is how well it fulfills that purpose (Barrett 1990). The usefulness of an image can, however, be determined at different levels (Wagner 2007), from its pure technical properties to its impact on society (see Table 13.1). Depending on the point of view, each of these levels could be valid, but in the context of this chapter, we leave aside the technical parameters of level 1 such as contrast, resolution, noise, DQE, NEQ and we focus on level 2, which is where medical physicists' and radiologists' efforts meet. At this level, we are interested in quantifying how the information available in the image can be efficiently used by the radiologist for making a correct diagnostic decision.

While subjective assessment methods of image quality have been proposed in the literature, we will follow Barrett and Myers (2004) and only consider objective methods that take into account the four following points. (1) The task should be well defined. In this chapter, we will essentially deal with the detection task, which is a particular case of the classification task. (2) The properties of signals and backgrounds have to be defined. For example, the evaluation of mammographic imaging systems for the task of breast lesion detection requires the characterization of normal breast tissues and breast lesions in terms of the full probability density function of the pathology present or absent. These functions are difficult to quantify but can often be estimated by limited order statistics. (3) Given a task and a set of backgrounds and signals, the next step is to specify which observer is performing the task. The observer might be a human or a model observer. (4) Finally, a performance metric should be precisely defined and quantified. This could be done with a percentage of correct hits, a point on the ROC curve or the full ROC curve.

13.1.2 Psychophysics and Image Quality

Although randomized clinical studies are ultimately best suited for a task-based estimation of image quality, they are not always practical when many confounding factors have to be taken into account. This is especially the case when several observers, imaging units, imaging protocols, diagnostic centers, and lesion types are involved. For these reasons, a growing interest in psychophysical studies has been shown since Tanner and Birdsall (1958) estimated human efficiency. Knowing the reference classification of each image (the "gold standard") and controlling the viewing conditions, one can vary the properties of the task

TABLE 13.1 Levels of Assessment of Image Quality

Level of Assessment	Example of Quantities Used to Perform the Assessment
1. Technical efficacy	Physical performance measurements of imaging system characteristics like MTF, NPS, DQE; preclinical standalone and bench tests
2. Diagnostic accuracy	Sensitivity, specificity, ROC curve, and their summary measures
3. Diagnostic thinking	Difference in clinicians' subjective estimates of correct diagnostic probabilities, pretest to posttest
4. Therapeutic efficacy	Effect of diagnostic imaging or test on therapeutic management of patients
5. Patient outcome	Expected value of test information in terms of gains in quality-adjusted life years (QALYs); also, cost per QALY gained
6. Societal efficacy	Cost-effectiveness and/or cost–benefit analysis from the societal viewpoint

Source: Wagner RF, Metz CE, Campbell G. 2007. *Academic Radiology* 14: 723–748.

along a limited number of physical properties. The strength of the psychophysics framework is that it significantly reduces the amount of uncontrolled variables and allows isolating how different image acquisition or display variables affect diagnostic performance (Castella 2009a). Depending on the experimenters' goals, two characteristic psychophysical studies paradigms have been developed and used: receiver operating characteristic (ROC) (Barrett and Myers 2004, ICRU 2008) and multiple alternative forced choice (MAFC) (Barrett and Myers 2004, Burgess 1995, Eckstein 2000, Gallas 2007, Castella 2009a).

13.1.3 Observer Models as a Way to Assess Image Quality

Obtaining information with human psychophysical studies is often costly and time consuming, and generalizing the conclusions to other experimental conditions is usually complex (see Chapter 10, Perception of Mammographic Images). Moreover, human observers are known to be prone to variability, coming from internal or external sources. This is why model observers, called *anthropomorphic model observers*, have been developed to mimic the human behavior in a reproducible way (Eckstein 1998). Such models will never replace the radiologists in the reading room, but they allow one to automatically perform repetitive psychophysical studies and explore a large space of image parameters. Alternatively, the *ideal observer* defines an upper bound of the performance of any observer for a given task with the information contained in the image (Barrett 1995). This model is especially useful when the image acquisition parameters have to be tuned.

13.1.4 Goal of This Chapter

This chapter focuses on image quality defined at the radiologists' level. Although this may involve very sophisticated

actions, the assessment of their interpretive performance is achieved by analyzing parameters like sensitivity, specificity, recall rate, positive predictive value and cancer detection rate (Carney 2010), which can ultimately be linked to a detection task. Therefore, we consider the simple detection task as the first step for defining image quality. This task has been widely used in the literature and both anthropomorphic and ideal model observers have been developed in the framework of breast imaging.

This chapter starts with a brief description of the signals and backgrounds in breast images. Then, we present the major characteristics of the human observers that are of interest when we model them. The following sections are more mathematical and present the main model observers and their application in the context of mammography. We finish this chapter with the limitations of the models and some potential developments, in particular their generalization to 3-D imaging.

13.2 Signals and Backgrounds Used by Model Observers in Mammography

13.2.1 Clinical and Visual Aspects of Signals of Interest in Mammography

The main question raised during the radiographic examination of the breast is to determine if a breast carcinoma is present or not. The primary sign of breast carcinoma is the presence of a mass, generally described as spiculated, lobulated, or smooth (Vyborny 1989). While a large spiculated mass on a fatty background should be easy to detect, it is much more demanding for smaller masses on the order of 5 mm diameter. Identifying the subtle spiculated borders or the filamented tentacles of neoplastic spread requires high resolution and contrast as well as low noise. The secondary signs of breast carcinoma are architectural distortion, asymmetric density, dilated duct, skin thickening, vascular asymmetry and clustered microcalcifications. Both primary and secondary signs have to be analyzed together to establish a diagnosis. However, in the simplified context of the model observers, researchers have concentrated on the detection of masses and microcalcifications.

Psychophysics experiments with masses have often been performed with smooth masses approximated by a simple projected sphere. More realistic simulations have been proposed by Saunders et al. (2006) who developed an algorithm capable of generating benign or malignant breast masses based on the analysis of real masses' characteristics. Many simulations of microcalcification have been proposed. For a recent publication on the subject, we propose to refer to Zanca et al. (2008).

13.2.2 Mammographic Backgrounds

The noise power spectrum (NPS) is widely used in the framework of the technical efficacy of an imaging system (see Table 13.1), and it has also been used as a descriptor of the mammographic backgrounds in psychophysics experiments. One important result of the NPS analyses is that they can be well fitted by a power-law defined by $NPS(f) = k/f^\beta$, where f is the radial spatial frequency and k a constant. The exponent β can vary between 1.5 and 4.0, with an average value of about 3.0 (Burgess 2001).

Other parameters, such as gray-level histogram, gray-level co-occurrence matrices, primitive matrices, fractal analysis, or neighborhood gray-tone difference matrix have also been proposed as summary descriptors (Castella 2007a). However, these second order statistics assume that the image is stationary, which is not the case at all scales, as can be seen in Figure 13.1.

Early psychophysical experiments were performed with simple geometric patterns on uncorrelated noise (Burgess 1981, 1984). Researchers then explored correlated noise textures, which were meant to better represent the influence of projected anatomical structures in radiographic images (Myers 1985, Bochud 1999). These studies confirmed the fact that anatomical structure could outweigh the effect of the detector noise in real images. This has been well investigated in mammography (Burgess 2001, Huda 2006).

Full 3-D phantoms are presented in details in Chapter 12 of the present book. They allow the generation of synthetic images that can be used in lieu of real 2-D mammography. However, more simple solutions have been proposed in order to generate synthetic mammographic 2-D backgrounds that maximize both visual and statistical realism. Based on the lumpy backgrounds (Rolland 1992), the clustered lumpy backgrounds (CLB) are produced by the random superposition of small structures ("blobs") (Bochud 1999b, Castella 2008).

Whether they are synthetic or extracted from real asymptomatic mammograms, these backgrounds can then be combined (or not) with a signal (mass or microcalcification) in order to produce a set of images with known statistical properties and a well-defined gold standard.

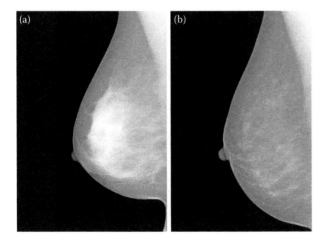

FIGURE 13.1 Typical example of mammographic backgrounds.

13.3 Human Observer

13.3.1 Characteristics of the Human Observer

This section summarizes the main characteristics of human observers that can be useful in the development of model observers perceiving gray-scale static images. It has been greatly inspired by Chapter 14.2 of Barrett and Myers (2004).

With about 7 million cones and 75–150 million rods, the human eye contains about 10 times more photoreceptors than a current digital camera. Although it is tempting to model the eye as a camera, its functioning is much more complex and the optical part is only the beginning of a process performed by bipolar and amacrine cells that transfer the signal from the photoreceptors to the retinal ganglion cells. The signal is then transmitted to the lateral geniculate nucleus of the thalamus, which relays the signal to the visual cortex in the posterior part of the brain, where it is finally processed to produce our visual perception of the world.

The first understanding of the mammal visual system was obtained with cats and monkeys by electrophysiological studies in which electrodes were placed into single cells of the visual pathway. Nowadays, similar experiments can be performed noninvasively with fMRI or positron emission tomography on human subjects. One of the main results of these experiments is that the visual response can be linked to specific orientation and location patterns (bars, edges, and spots). This leads to the assumption that visual information is perceived through channels that are independent processors tuned to different narrow ranges of spatial or temporal frequencies.

One of the most important image process performed at the retinal level is lateral inhibition. This comes from the fact that the output of a retinal ganglion is not only defined by its associated photoreceptors but also by nearby ganglions: A ganglion cell can be inhibited when a surrounding ganglion is excited. This implies that the signal ultimately transmitted to the visual cortex actually is more about edges than about the light intensity at each part of the retina.

Another basic aspect of the human visual system is our inability to detect very small and very large objects: we could not easily see an unmoving mosquito from 10 meters away and it would also be hard to recognize an elephant if one's nose touched its skin. The limitation on small objects (i.e., high spatial frequency) comes from the limited size of our photoreceptors. On the other hand, our inefficiency to perceive large objects (i.e., low spatial frequency) can be explained by lateral inhibition. This observation, together with the fact that an object with a high contrast is easier to detect than an object with a low contrast, defines the human eye contrast sensitivity function (CSF), which is the overall sensitivity of the visual system as a function of spatial frequency. Figure 13.2 attempts to visualize this concept on an image with increasing frequencies and decreasing contrast. Many estimations of the CSF have been performed, and the result is that it has a band-pass shape with a maximum at around 4 cycles per degree. This can be visualized by noting that our thumb (as well as the sun) covers about 0.5° when our arm is

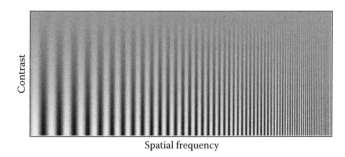

Contrast

Spatial frequency

FIGURE 13.2 Example of a sinusoid with increasing frequency and a decreasing contrast from the bottom to the top of the image. Depending on the distance you are looking at this image from, the sinusoid should be easier to perceive in the middle frequency range.

extended. This means that we are most sensitive to frequencies corresponding to about two cycles in the angle defined with our thumb. We remark, however, that the CSF varies among individuals and is also dependent on the mean luminance level, noise level, color, accommodation, eccentricity, and image size.

The presence of a pattern in the image can make the signal less visible. This property, known as masking, is frequent in medical imaging where the normal anatomical backgrounds with high contrast can prevent the detection of pathology.

Repeatedly asking a human observer to detect a given signal on a given image does not necessarily lead to the same outcome. This means that the human observer possesses some internal noise. Internal noise has been shown to have two components: one that is additive and independent of the image luminance and one that is image-dependent and is proportional to the variance of the image noise (Burgess 1988).

Human observers are also dependent on their a priori knowledge when they perform a task. Depending on the prior information we have on the imaging system we tend to perform better or worse. This is one of the reasons that explain why senior radiologists perform better than their less-seasoned colleagues: the more experience they have, the larger the database they can rely in their memory. In experiments performed with human observers, this has to be taken into account by training them until they reach a stable performance level and by defining the task as unambiguously as possible. The easiest to consider is the signal known exactly (SKE) paradigm, but more clinically realistic experiments can be performed with signal known statistically (SKS). In this latter case, the experimenter has to make sure that the observer has a precise idea about what the signal could look like. It has been shown that when the task is simple enough and the observer is well trained with sufficient information, no difference can be seen between radiologists and (trained) naïve observers (Brettle 2007).

13.3.2 Performance Metrics of the Human Observer

Having specified the task (a detection), the type of signal (a mass or a microcalcification), the backgrounds (synthetic or

real mammogram), and the main characteristics of the human observer (CSF, channel analysis and a priori knowledge), we need to describe which kind of experiments can be performed in order to produce a figure of merit that quantifies the performance. Ideally, one would like to insert an electrode in the brain of the observers and directly measure their certainty of the presence of the signal: the higher the impulse, the higher the probability of presence of the signal. While this is obviously not feasible, several psychophysical experiments try to extract this parameter usually denoted as t, like test statistics, through the observer responses. As shown in Figure 13.3, if we manage to estimate the probability density function of the response when the signal is absent $p(t|H_0)$ and when the signal is present $p(t|H_1)$, setting a decision threshold t_c defines a true and a false-negative fraction (TNF and FNF, below t_c) as well as a false- and a true-positive fraction (FPF and TPF above t_c). Varying the decision threshold allows plotting the TPF (also known as sensitivity) versus 1 minus TNF (also known as specificity). This graph is called the receiver operating characteristic curve or ROC curve (see Figure 13.4). It is often summarized by its area under the curve (AUC). A value of 0.5 corresponds to an informationless observer for which the two distributions are superimposed. On the other hand, AUC = 1 corresponds to the supreme observer for which the two distributions are perfectly separated.

Several scalar parameters have been developed in order to synthesize the separateness of the two distributions (Barrett and Myers 2004). The most straightforward is the signal-to-noise ratio defined as the difference between the mean of each distribution normalized by the quadratic mean of the standard deviations:

$$\mathrm{SNR}_t = \frac{\langle t \rangle_1 - \langle t \rangle_0}{\sqrt{\frac{1}{2}\left(\sigma_0^2 + \sigma_1^2\right)}}. \qquad (13.1)$$

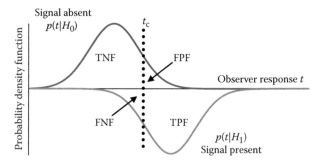

FIGURE 13.3 Probability density functions of the observer response for signal-absent and signal-present images. Setting a threshold response t_c defines false- and true-positive fractions (FPF and TPF), as well as true- and false-negative fractions (TNF and FNF). Note that the scale of the "signal present" probability density function has been inverted for reading clarity with the "signal absent" function.

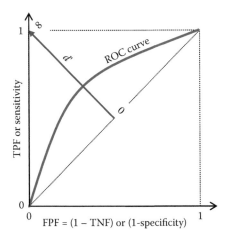

FIGURE 13.4 Example of ROC curve and its link with the detectability d'. When the "signal present" and "signal absent" responses are Gaussian with the same variances, d' is defined at the crossing point of the ROC curve and the minor diagonal. Detectability $d' = 0$ when the ROC curve is a straight line and $d' = 1$ when the ROC curve is a square passing by the top-left corner (TPF = TNF = 1).

SNR_t is useful when the two distributions are close to Gaussians, but becomes problematic when the tails are larger. When the two distributions are Gaussian with same variances, SNR_t takes the name of detectability index d'. In this case, the ROC curve is symmetric and d' can be directly plotted on the minor diagonal of the ROC curve with the value 0 at the center of the square at coordinate (0.5, 0.5) and an infinite value at the top left corner (see Figure 13.4).

If both distributions are Gaussian (but not necessarily of equal variances), the AUC can be derived from the signal-to-noise ratio, assuming that $\mathrm{erf}\ (z) = \dfrac{2}{\sqrt{\pi}} \displaystyle\int_0^z e^{-y^2}\, \mathrm{d}y$:

$$\mathrm{AUC} = \frac{1}{2} + \frac{1}{2}\mathrm{erf}\left(SNR_t/2\right). \tag{13.2}$$

With no information about the distributions but with an estimate of the AUC, one could compute a signal-to-noise ratio as if the distributions were Gaussian and express an effective detectability index d_A as:

$$d_A = 2\ \mathrm{erf}^{-1}\ [2\mathrm{AUC} - 1]. \tag{13.3}$$

13.3.3 Reading Experiments with Human Observers

Mainly two types of detection experiments have been conducted in the field of mammography: the rating (or ROC) and the multiple alternative forced-choice experiments (MAFC). The fact that they are described in a report from the International Commission on Radiation Units and Measurements (ICRU 2008) is a clear sign that these methods have become mainstream and are used by the medical community and the industry.

The rating experiment consists of presenting a series of images and asking the observers to notify their level of certainty, for instance on a 5-level scale (from 1 "signal definitely present" to 5 "signal definitively absent"). This allows the experimenter to sample four points on each response function of Figure 13.3 and therefore four points of the ROC curve. In many cases, the observer response t is assumed to be Gaussian and the ROC curve can be plotted as a fit of the responses (Metz 2006).

In an MAFC experiment, the observers are shown M images and are forced to make their decision knowing that only one image is containing the signal. This experiment has the advantage to be more robust because the decision is performed relative to other images in each trial. Contrary to the rating experiment, it does not require the observer to maintain a decision scale separate from the images. The figure of merit provided by the MAFC experiment is a percentage of correct answers (PC), which can be linked to the AUC (for instance, in a 2AFC PC = AUC). This usually leads to the derivation of the detectability index d_A which can be expressed as d_{MAFC} (Bochud 2000).

13.4 Observer Models

13.4.1 The Task

We can formalize the detection task assigned to an observer as being the observation of an M-pixels image **g** written as a column vector of M elements (an M × 1 vector). The image is either in the "signal absent" (H_0) or "signal present" (H_1) state. The observer has to make a decision: D_0 for "signal absent" and D_1 otherwise.

13.4.2 Ideal Observer

The ideal observer (IO) has no internal noise, it has full knowledge of the statistical information of the image, and uses it optimally (Barrett and Myers 2004). At least five optimization rules have been proposed:

1. Maximization of the diagnostic accuracy as measured by the AUC.
2. Minimization of the mean Bayesian cost associated to each combination of decisions D_i and states H_j.
3. Minimization of both types of errors $P(D_1|H_0)$ and $P(D_0|H_1)$.
4. Neyman–Pearson criterion (sensitivity above a certain level and maximization of the sensitivity).
5. Choice of the state H_j that has the highest likelihood when observing an image **g** $p(\mathbf{g}|H_j)$.

Surprisingly, each of these rules leads to the same strategy with the observer response t to an image $\mathbf{g}$ defined as a scalar by:

$$t(\mathbf{g}) = \frac{p(\mathbf{g}|H_1)}{p(\mathbf{g}|H_0)} \begin{array}{c} D_1 \\ > \\ < \\ D_0 \end{array} t_c, \tag{13.4}$$

where t_c is the response threshold above which decision D_1 is taken and decision D_0 otherwise. The likelihood ratio on the left-hand side of Equation 13.4 defines the ROC curve and the threshold t_c its operation point. The only difference between the five optimization rules is the actual value of t_c.

Any reversible transformation of $\mathbf{g}$ has no effect on the performance of the IO. Alternatively, irreversible transformations reduce its performance. Unintuitively, this is in particular the case for actions beneficial to the human observer like edge enhancement or noise reduction.

The problem with Equation 13.4 is that we usually do not know the likelihoods sufficiently in order to compute the IO performance. A substantial exception is the case where the image background $\mathbf{b}$ is Gaussian noise and the task is to detect a super-imposed signal $\mathbf{s}$. In this case, the IO becomes linear and can be calculated, assuming we have enough computing power and a sufficient number of images to estimate its statistics:

$$t(\mathbf{g}) = \mathbf{s}^t \mathbf{K}_b^{-1} \mathbf{g}, \tag{13.5}$$

where $\mathbf{K}_\chi$ is the covariance matrix of the image background χ in which the diagonal elements are the variances of each pixel and the nondiagonal elements are the covariances between each couple of pixels of the image backgrounds χ. For an image containing M pixels, $\mathbf{K}_\chi$ is therefore an M × M matrix.

In this case, the IO is expressed linearly in term of $\mathbf{g}$ and the background statistical information taken into account is simply the second order noise statistics. This form of the IO is often called pre-whitening–matched filter (PWMF or PW) because the application of the inverse of the covariance matrix in Equation 13.5 realizes a decorrelation of the noise contained in $\mathbf{g}$ $\left(\mathbf{K}_b^{-\frac{1}{2}} \mathbf{g} \right)$ and filters it by the searched signal as seen by the same decorrelation process $\left(\mathbf{K}_b^{-\frac{1}{2}} \mathbf{s} \right)^t$.

Equation 13.5 is also useful because it expresses the optimal linear observer when the task is the detection of a signal $\mathbf{s}$ superimposed on a noise $\mathbf{n}$.* Indeed, if we force the observer to be linear:

$$t(\mathbf{g}) = \mathbf{w}^t \mathbf{g}, \tag{13.6}$$

---

* And also assuming that the signal amplitude is sufficiently weak in order not to change significantly the background correlation whether the signal is present or not.

where $\mathbf{w}$ is a vector of the same dimension as $\mathbf{g}$ and usually called the observer linear template, it can be shown that $\mathbf{w} = \mathbf{s}^t \mathbf{K}_b^{-1}$ is optimal in the sense that it maximizes the observer's SNR (Barrett and Myers 2004). This ideal linear observer is known as the Hotelling observer (HO). When the noise is Gaussian-distributed, the HO is the IO.

13.4.3 Anthropomorphic Pixel Observers

The objective assessment of image quality also involves sub-optimal model observers that aim at mimicking human observers. As mentioned above, human observers have internal noise and they are not able to assess all statistical properties of the image, let alone take them into account for making their decisions. Furthermore, they have access to the image through their CSF and the information delivered to the visual cortex is already processed in a way that appears like channels.

These considerations have led to the development of different anthropomorphic observers derived from the IO. Starting from Equation 13.5, we can assume that the human observers are processing the image as the HO, except that they cannot decorrelate the noise. This produces the non-pre-whitening (NPW) observer (Wagner 1979):

$$t(\mathbf{g}) = \mathbf{s}^t \mathbf{g}, \tag{13.7}$$

which can be seen as the filtering of the image by the searched signal. In other words, the NPW computes the scalar product of the image $\mathbf{g}$ by the signal $\mathbf{s}$: each pixel of the image is summed and weighted by a large amount where the signal is expected and by zero where it is not. This observer usually performs poorer than the human observer.

In order to take into account the CSF of the human observer, Equation 13.7 can be modified by convolving the image $\mathbf{g}$ and the searched signal $\mathbf{s}$ by the CSF expressed as a matrix $\mathbf{E}$. This gives the non-pre-whitening–matched filter with an eye-filter-(NPWE) observer (Burgess 1994):

$$t(\mathbf{g}) = (\mathbf{Es})^t \mathbf{Eg}. \tag{13.8}$$

The convolution of $\mathbf{E}$ in the NPWE acts as an edge-enhancement factor and greatly improves the performance of the NPW in mammography. Without the addition of internal noise, this observer provides often better performance than the human observer.

While both NPW and NPWE observers do not depend on the background statistics, this is not the case of the human linear template (HLT) (Beard and Ahumada 1998, Abbey and Eckstein 2002, Castella 2007b). The HLT is based on the assumption that the human observers process the image linearly as in Equation 13.6 and that it is possible to estimate their templates by analyzing their responses in present/absent or 2AFC experiments. The basic idea behind the HLT is to make use of all the trials performed by the human observers in which they believe they detected the searched signal. Using a large amount of images (typically several thousands), the clever averaging of many noise structures adds up and allow the estimation of the HLT.

13.4.4 Channelized Hotelling Observers

The channelized Hotelling observers (CHO) also derive from Equation 13.5, but keep the prewhitening effect while reducing the dimensionality of the image. Instead of directly processing the M pixels of image **g**, the CHO starts by preprocessing the image by P channels $\mathbf{u}_i$ (M × 1 vectors) on **g** as shown in Figure 13.5. The P scalar outputs v_i of each channel i can be written as an M × 1 vector **v** on which the HO strategy defined in Equation 13.5 is applied (Myers and Barrett 1987). Mathematically, this can be written as:

$$t(\mathbf{v}) = \left(\mathbf{U}^t \mathbf{s}\right)^t \mathbf{K}_v^{-1} \mathbf{v}, \quad \mathbf{v} = \mathbf{U}^t \mathbf{g} \text{ and } \mathbf{K}_v = \mathbf{U}^t \mathbf{K}_b \mathbf{U}, \quad (13.9)$$

where U is an M × P matrix whose columns are the channel profiles. The first practical advantage of the CHO is its mathematical tractability. While $\mathbf{K}_b$ is an M × M matrix that is difficult to compute and even more to inverse, $\mathbf{K}_v$ is a P × P matrix that can be much more simple as the number of channels P can be lowered down to less than a dozen in many cases and no more than about 50 in most cases.

The second, and probably more fundamental, advantage of CHOs is the plasticity of their channels. They can be chosen in order to lose as little information as possible in order to stay close to the performance of the HO (and therefore the IO under Gaussian assumption). On the other hand, the channels can integrate human characteristics, like the CSF, in order to elaborate an anthropomorphic observer. Examples of channels for IO estimations are Laguerre–Gauss functions (Gallas and Barrett 2003) or the singular-value decomposition (Park 2009). For anthropomorphic observers, channels like Gabor filters (Eckstein 1998) or difference-of-Gaussian and square channels (Abbey and Barrett 2001) have been proposed.

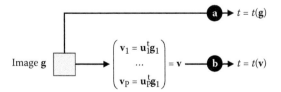

FIGURE 13.5 Main difference between pixel and channelized observers. (a) The pixel observer directly processes the M pixels of the image **g**. (b) The channelized observer only has access to the P scalars that have been preprocessed by the P channels. **v** is a P × 1 vector.

13.5 Practical Observer Models Used in Mammography

13.5.1 Developments of Anthropomorphic Observer Models

Many psychophysical experiments have been made in order to better understand the characteristics of the human visual system and to develop anthropomorphic model observers. Ultimately, one would like to have a single model usable for any detection task. It should be tunable for any kind of pathological signal (nodule, fibrosis, microcalcification, masses, etc.) whatever its size and for any kind of background (uniform noise from Plexiglas, lung, liver, breast, etc.). Although we are still far from this goal, much research has been done in the field of breast imaging, because it is a demanding type of examination widely used in medical diagnoses.

After the first developments of model observers on white noise, filtered white noise, and lumpy backgrounds, the obvious question was to understand the effect of the breast anatomy on the detection ability of the human observer. For this, model observers have been used in conjunction with 2AFC human experiments. The main result was that the role of the anatomy depends on the scale at which we are looking at the mammograms. At the small scale used for microcalcification detection, the breast anatomy acts as pure noise. At the larger scale of the masses (above 1 mm), the anatomy has both noise and masking components (Bochud 1999b).

The role of the anatomy at the mass scale was treated in details by Burgess et al. (2001), who considered 2AFC experiments performed on real mammograms and randomly synthesized backgrounds that had the same power law NPS as the real mammograms. Apart from human observers, they also computed the PW observer (which is equal to the IO in the case of the synthesized backgrounds) and the CHO with Laguerre-Gauss channels (which is a suboptimal approximation of the PW). They observed that PW models performed better than humans, which means that humans are not able to fully decorrelate image noise. Furthermore, they observed that humans are much closer to PW models on real mammograms than on synthesized backgrounds. This confirmed that humans probably perceive some anatomy as a mask that is taken into account in the detection process, whereas the PW cannot. A surprising result that might be counterintuitive at first was that the contrast detection threshold increased with larger signal sizes (for both real and synthesized mammograms): above 1 mm size, larger signals were harder to detect. This can be explained by the fact that mammographic masses have blurred edges and are surrounded by a background whose fluctuations increase at lower spatial frequencies. This means that for a given contrast, the SNR at the signal location is decreasing for larger signals. This was also coherent with the predictions of the PW models.

Burgess et al. (2007) came back to this problem by considering power law backgrounds. They realized different sets of images by varying the exponent β between 1.5 and 3.5, and as in 2001, they computed the IO (equal to the PW observer in this case) and the CHO with Laguerre–Gauss channels and performed 2AFC detection experiments with human observers. They started by computing the models responses for different signal sizes as a function of the exponent β and showed that it was more difficult to detect larger signals with exponents β > 2. For exponents β < 2, it was the opposite: larger signals were easier to detect. One explanation is that our natural visual world is surrounded by power law noise with exponents around β = 2 and that one of the first goals of our visual system could be the removal of redundant information. Burgess et al. (2007) went even further and computed the slope *m* of the contrast detail diagram (contrast threshold versus object size) for each exponent β with the IO ($m = (β – 2)/2$). The agreement between the IO and the human observers was very good. However, and as expected, in terms of the detection performance, the human observers were less effective than models (between 40% and 60% in terms of the ratio of the square of the detectabilities).

Apart from the detection experiment, Burgess et al. (2001) also conducted search experiments with the two types of backgrounds. In this case, the performance was the same for the real and the synthesized backgrounds. This suggests that for search, and contrary to detection, mammogram structure can be regarded as pure noise.

Burgess et al. (2001) also computed the performance of the NPWE observer. They noticed that it was close to the human observer, but less efficient. Therefore they concluded that the NPWE was not realistic and must be used with care. Reiser and Nishikawa (2006) were, however, more enthusiastic with the NPWE observer. They were able to fit the human results of detection and discrimination experiments on white noise and mammographic background. However, they had to explicitly include location uncertainty in the model.

The HLT was computed for mammographic backgrounds (Castella 2007b) with 1400 real images and 4000 CLB images presented to five human observers that were asked to detect a synthetic yet realistic mass in a 2AFC experiment. The HLT represented in the frequency domain oscillates around zero, with nodes very close to those of the signal profile. The main difference with the signal profile is that the HLT is high-pass in the low-frequency range and, therefore, shows an inhibition process. This is interesting because it is similar to what is obtained with the NPWE observer and CHOs with appropriate channels.

Castella et al. (2009b) also investigated the effect of signal shape on different kinds of realistic mammographic backgrounds (real and CLB). For detections performed in the SKE or SKS conditions, human observers had the same performance when the signal size was kept constant. However, when the signal size was variable, the performance dropped. Several model observers were computed, and excellent agreement was found with the human observers for the NPWE, the HLT, and the CHO with Gabor channels. For each of these three model observers, the agreement was obtained once an adequate quantity of internal noise was included. This was not the case for the NPW observer and the CHO with different other channels (SRQ, SDOG, and DDOG): even before the introduction of internal noise, their performance was lower than the humans.

13.5.2 Use of Observer Models to Assess Image Quality or to Avoid Repetitive Tasks

Because the IO has full knowledge of the statistical information of the image and uses it optimally, it is best suited to tune a hardware imaging system. Indeed, at this level, what is important is the amount of information available before any image processing. In other words, the IO is able to guide the engineer to obtain as much information as possible from the image acquisition system. The way this information will be made available and usable to the observer is another problem that can be dealt with by an anthropomorphic observer.

An example of this approach was attempted by Veldkamp et al. (2003). Using images of a CDMAM phantom of different thicknesses obtained with a mammographic unit with and without antiscatter grid at different kilovoltages and target–filter combinations, they built the contrast detail curve with the IO, which was looking for gold disks of different sizes. One of the conclusions was that breasts of thicknesses below about 5 cm could be imaged without antiscatter grid, "at lower dose to the patient without losing image quality."

This kind of statement emphasizes the need to clearly define the meaning of image quality and the four aspects mentioned in the first section of this chapter. (1) The task is a detection, which is fine for this purpose; (2) the signal is a gold disk, which contrast is similar to that of microcalcifications but too high to properly simulate masses; furthermore, the background is uniform and Gaussian with a white spectrum, which is very different from what can be encountered on a real breast image; (3) the observer was supposed to be the IO but was indeed the NPW observer, which is equal to IO only when the noise is white and Gaussian; and (4) the metric is a contrast-detail plot, which is also fine. This example shows the difficulty of extrapolating an observation in a very narrow set of conditions into a very different application without further justification. Despite these limitations, the European Reference Organisation for Quality Assured Breast Screening and Diagnostic Services (EUREF) will probably be using the NPW observer on CDMAM phantom images for quality assurance as described by Suryanarayanan et al. (2002). The main advantage of this approach is that it automates otherwise fastidious human experiments and uniforms the qualification throughout Europe.

Gagne et al. (2006) used two model observers (NPW and PW) in order to replace the painstaking use of human observers for selecting between two kinds of image detectors in mammography with the ACR/MAP phantom. They also used these models in order to replace the human reading of the EUREF images of the CDMAM phantoms. In both cases, they observed a better precision of the model observer versus the variability in the scores from human observers. More recently, and also in the EUREF context, the NPWE observer has been proposed as an alternative to the NPW observer because of its closer link to human behavior (Marshall 2011, Monnin 2010, 2011).

One of the ultimate purposes of model observers is to be able to use them for optimizing imaging devices. This has a great potential because nowadays imaging devices do not only vary in terms of hardware but also in terms of image processing. This highly increases the number of possibilities to adjust the parameters and renders almost impossible the determination of image quality with human observers. In this context, the use of model observers could be more useful because it could at least test for the main combinations before submitting the first choices to a panel of human readers. But for this, we need to be confident that the performance of the model can really be used as a surrogate for a human observer. A nice example has been proposed by Segui and Zhao (2006) in which they tested several digital imaging modalities with a NPWE observer as well as with expert radiologists. The results were compatible.

Chen and Barrett (2005) went one step further and trusted the model observer at the design step of a new digital imaging system. They numerically simulated the whole imaging process of a digital mammographic system with different signal locations and different parameters of the lens that couples a fluorescent screen to a CCD detector. They used a CHO for detecting a lesion in a CLB mimicking breast tissue. This allowed them to directly compare the lens parameters in a task-based manner through the CHO.

Chawla et al. (2007) used two model observers to investigate the potential dose reduction in mammographic screening procedures: CHO with Laguerre–Gauss channels (approximation of the linear IO) or with Gabor channels (approximation of the visual human response). Different kinds of signals (masses and calcifications) were superimposed on real mammographic backgrounds and noise was added in order to simulate different dose levels. The results indicated room for optimization and suggested that a 50% dose reduction was reachable. However, the authors were careful to mention that their "findings need[ed] additional confirmation by rigorous clinical trials and human observer studies before being implemented clinically."

13.6 Perspectives of the Observer Models in Mammography

Model observers are becoming widespread in mammography, but a good understanding of their limitations is necessary in order to use their full potential. A good example is the idea of using the NPW observer in the European screening program, which avoids repeating detection experiments with human observers. The main limitations of this approach are that it is performed with an unrealistic signal (a gold disc) on a uniform background and with a different strategy than the human observer (the NPW observer does not decorrelate the noise and looks at the absolute value of the signal). A better use of the model observers in such a context would be either to use the IO with a realistic signal in order to qualify the available information on the raw data or to use an anthropomorphic model observer on the processed data in order to estimate image quality at the radiologist's level. The importance of understanding the difference between the potential information contained in the raw data and the actual information usable by the radiologists has become crucial. Modern digital images are now processed with many nonstationary and nonreversible "black box filters" that are applied based on the local image statistics.

This leads to the comment that anthropomorphic model observers in mammography are required to perform realistic tasks on realistic backgrounds. Phantoms available for detection experiments are currently too simple and more sophisticated backgrounds and more difficult tasks should be used to stress the imaging system (Gagne 2006).

The NPWE or the CHO with appropriate channels (like Gabor for instance) have been used as anthropomorphic model observers. Although the CHO can be made indistinguishable from NPWE, it is probably closer to the real human observer because we know that the channel mechanism is how the brain is actually working. The NPWE, however, still remains popular, because it usually gives good practical results and is easy to compute. However, the CHO has a greater potential, as it can be more easily generalized to other tasks or imaging conditions.

Apart from the simplified detection experiments described in this chapter, many other factors influence the radiologist's ability to correctly detect masses or microcalcifications on mammograms: the search space is not limited to a defined square, the signal location has much uncertainty, and the prevalence is extremely low (of the order of 7 per 1000 cases). Moreover, real clinical strategies also include comparison with prior images and the contralateral breast, and the context of the global breast architecture, which are absolutely not taken into account by the current model observers.

Finally, all the models presented here are designed for 2-D images, while most modern medical modalities are becoming 3-D. Breast imaging is no exception with breast CT (see Chapter 5 or Boon et al. 2001) or tomosynthesis (see Chapter 4 or Dobbins and Godfrey 2003). Some preliminary 3-D aspects have already been incorporated (Liang 2008, Diaz 2011), but much needs to be done before they can be used for practical purposes. A recent paper from Platiša et al. (2011) proposed several ways of implementing 3-D linear CHOs in medical imaging: by considering a single slice (for instance, the center slice of an object), by doing the same for all the slices or a restricted number of them, or directly on the whole 3-D image. Figure 13.6 schematically presents these different options of deriving a scalar response t from a 3-D image. The practical choice will depend both on the clinical viewing conditions and on the knowledge that we have about the observer characteristics.

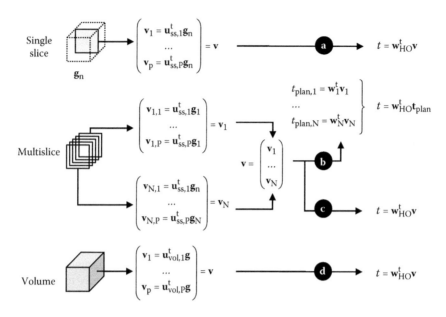

FIGURE 13.6 Single-slice, multislice, and volumetric implementation of the linear CHO in 3-D medical images. The 3-D image **g** contains M voxels distributed in N slices of Q voxels each. (a) The single slice method consists in applying P single-slice channels $\mathbf{u}_{ss,i}$ ($Q \times 1$ vectors) on one image slice $\mathbf{g}_n$ in order to produce one channel output **v** (a $P \times 1$ vector). The Hotelling response t is computed by applying the relations of Equation 13.9. The multislice method starts by computing the single-slice channel outputs $\mathbf{v}_i$ on each of the N slices; each output being a $P \times 1$ vector. From them, one can (b) compute N planar responses $t_{plan,i}$ and then compute the Hotelling response t. (c) An alternative is to directly compute the Hotelling response t from the channel outputs considered as an $N \times P$ column vector. (d) Finally, one can directly apply P volume channels $\mathbf{u}_{vol,i}$ ($M \times 1$ vectors) on the M voxels of the whole 3-D image and obtain one channel output **v**. Again, the Hotelling response t is computed by applying the relations of Equation 13.9. Figure inspired from Platiša et al. (2011).

References

Abbey CK, Barrett HH. 2001. Human- and model-observer performance in ramp-spectrum noise: Effects of regularization and object variability. J Opt Soc Am A 18: 473–488.

Abbey CK, Eckstein MP. 2002. Optimal shifted estimates of human-observer templates in two-alternative forced choice experiments. IEEE Trans Med Imag 21: 429–440.

Barrett HH. 1990. Objective assessment of image quality: Effects of quantum noise and object variability. J Opt Soc Am A 7: 1266–1278.

Barrett HH, Denny JL, Wagner RF et al. 1995. Objective assessment of image quality. II. Fisher information, Fourier crosstalk, and figures of merit for task performance. J Opt Soc Am A 12: 834–852.

Barrett HH, Myers KJ. 2004. Foundations of Image Science. Hoboken, NJ: Wiley.

Beard BL, Ahumada AJ Jr. 1998. Technique to extract relevant image features for visual tasks. In Rogowitz BE, Pappas TN, eds. Human Vision and Electronic Imaging III SPIE Proceedings 3299 Paper 10. Presented at the 1998 IS&T/SPIE Electronic Imaging Symposium, January 24–30, San Jose, CA.

Bochud FO, Valley J-F, Verdun FR et al. 1999a. Estimation of the noisy component of anatomical backgrounds. Med Phys 26: 1365–1370.

Bochud FO, Abbey CK, Eckstein MP. 1999b. Statistical texture synthesis of mammographic images with clustered lumpy backgrounds. Opt Exp 4: 33–43.

Bochud FO, Abbey CK, Eckstein MP. 2000. Visual signal detection in structured backgrounds. III. Calculation of figures of merit for model observers in statistically nonstationary backgrounds. J Opt Soc Am A 17: 193–205.

Bochud FO, Abbey CK, Eckstein MP. 2004. Search for lesions in mammograms: non-Gaussian observer response. Med Phys 31: 24–36.

Boone JM, Nelson TR, Lindfors KK et al. 2001. Dedicated breast CT: Radiation dose and image quality evaluation. Radiology 221: 657–667.

Brettle DS, Berry E, Smith MA. 2007. The effect of experience on detectability in local area anatomical noise. Brit J Radiol 80: 186–193.

Burgess AE, Wagner RF, Jennings RJ et al. 1981. Efficiency of human visual signal discrimination. Science 214: 93–84.

Burgess AE, Ghandeharian H. 1984. Visual signal detection. I. Ability to use phase information. J Opt Soc Am A 1: 900–905.

Burgess AE, Colborne B. 1988. Visual signal detection. IV. Observer inconsistency. J Opt Soc Am A 5: 617–627.

Burgess AE. 1994. Statistically defined backgrounds: Performance of a modified nonprewhitening model observer. J Opt Soc Am A 11: 1237–1242.

Burgess AE. 1995. Comparison of receiver operating characteristic and forced choice observer performance measurement methods. Med Phys 22: 643–655.

Burgess AE, Jacobson FL, Judy PF. 2001. Human observer detection experiments with mammograms and power-law noise. Med Phys 28: 419–437.

Burgess AE, Judy PF. 2007. Signal detection in power-law noise: Effect of spectrum exponents. J Opt Soc Am A 24: B52–B60.

Carney PA, Sickles EA, Monsees BS et al. 2010. Identifying minimally acceptable interpretive performance criteria for screening mammography. Radiology 255: 354–361.

Castella C, Kinkel K, Eckstein MP et al. 2007a. Automatic mammographic parenchymal patterns classification using multiple statistical features. Acad Radiol 14: 1486–1419.

Castella C, Abbey CK, Eckstein MP et al. 2007b. Human linear template with mammographic backgrounds estimated with a genetic algorithm. J Opt Soc Am A 24: B1–B12.

Castella C, Kinkel K, Descombes F et al. 2008. Mammographic texture synthesis: Second-generation clustered lumpy backgrounds using a genetic algorithm. Opt Exp 16: 7595–7607.

Castella C. 2009a. Breast texture synthesis and estimation of the role of the anatomy and tumor shape in the radiological detection process: From digital mammography to breast tomosynthesis. Ph.D. Dissertation, École Polytechnique Fédérale de Lausanne.

Castella C, Eckstein MP, Abbey CK et al. 2009b. Mass detection on mammograms: Influence of signal shape uncertainty on human and model observers. J Opt Soc Am A 26: 425–436.

Chawla AS, Samei E, Saunders R et al. 2007. Effect of dose reduction on the detection of mammographic lesions: A mathematical model observer analysis. Med Phys 34: 3385–3398.

Chen L, Barrett HH. 2005. Task-based lens design with application to digital mammography. J Opt Soc Am A 22: 148–167.

Diaz I, Timberg P, Zhang S et al. 2011. Development of model observers applied to 3D breast tomosynthesis microcalcifications and masses. In Manning DJ, Abbey CK, eds. Proceeding SPIE 7966. Medical Imaging 2011: Image Perception, Observer Performance, and Technology Assessment.

Dobbins JT, Godfrey DJ. 2003. Digital x-ray tomosynthesis: current state of the art and clinical potential. Phys Med Biol 48: R65–R106.

Eckstein MP, Abbey CK, Whiting JS. 1998. Human vs. model observers in anatomic backgrounds. Proc SPIE 3340: 16–26.

Eckstein M, Abbey CK, Bochud FO. 2000. A practical guide to model observers for visual detection in synthetic and natural noisy images. In Handbook of Medical Imaging. Physics and Psychophysics, Vol. 1. SPIE Press.

Gagne RM, Gallas BD, Myers KJ. 2006. Toward objective and quantitative evaluation of imaging systems using images of phantoms. Med Phys 33: 83–95.

Gallas BD, Barrett HH. 2003. Validating the use of channels to estimate the ideal linear observer. J Opt Soc Am A 20: 1725–1738.

Gallas B, Pennello GA, Myers KJ. 2007. Multireader multicase variance analysis for binary data. J Opt Soc Am A 24: B70–B80.

Huda W, Odgen KM, Scalzetti EM et al. 2006. How do lesion size and random noise affect detection performance in digital mammography? Acad Radiol 13: 1355–1366.

ICRU. 2008. ICRU report 79: Receiver operating characteristic analysis in medical imaging. J ICRU 8.

Liang H, Park S, Gallas BD et al. 2008. Image browsing in slow medical liquid crystal displays. Acad Radiol 15: 370–82.

Marshall NW, Monnin P, Bosmans H et al. 2011. Image quality assessment in digital mammography: Part I. Technical characterization of the systems. Phys Med Biol 56: 4201–4220.

Metz CE. 2006. Receiver operating characteristic (ROC) analysis: A tool for quantitative evaluation of observer performance and imaging systems. J Am Coll Radiol 3: 413–422.

Monnin P, Bochud FO, Verdun FR. 2010. Using a NPWE Model observer to assess suitable image quality for a digital mammography quality assurance programme. Radiat Protect Dosimetry 139(1–3): 459–462.

Monnin P, Marshall NW, Bosmans H et al. 2011. Image quality assessment in digital mammography: Part II. NPWE as a validated alternative for contrast detail analysis. Phys Med Biol 56: 4221–4238.

Myers KJ, Barrett HH, Borgstrom MC et al. 1985. Effect of noise correlation on detectability of disk signals in medical imaging. J Opt Soc Am A 2: 1752–1759.

Myers KJ, Barrett HH. 1987. Addition of a channel mechanism to the ideal-model observer. J Opt Soc Am A 4: 2447–2457.

Park S, Clarkson E. 2009. Efficient estimation of ideal-observer performance in classification tasks involving high-dimensional complex backgrounds. J Opt Soc Am A 26: B59–B71.

Platiša L, Goossens B, Vansteenkiste E et al. 2011. Channelized Hotelling observers for the assessment of volumetric imaging data sets. J Opt Soc Am A 28: 1145–1163.

Reiser I, Nishikawa RM. 2006. Identification of simulated microcalcifications in white noise and mammographic backgrounds. Med Phys 33: 2905–2911.

Rolland JP, Barrett HH. 1992. Effect of random background inhomogeneity on observer detection performance. J Opt Soc Am A 9: 649–658.

Suryanarayanan S, Karellas A, Vedantham S et al. 2002. Flat-panel digital mammography system: Contrast-detail comparison between screen-film radiographs and hard-copy images. Radiology 225: 801–807.

Saunders R, Samei E, Baker J et al. 2006. Simulation of mammographic lesions. Acad Radiol 13: 860–870.

Segui JA, Zhao W. 2006. Amorphous selenium flat panel detectors for digital mammography: Validation of a NPWE model observer with CDMAM observer performance experiments. Med Phys 33: 3711–3722.

Tanner WP, Birdsall TG. 1958. Definitions of d' and eta as psychophysical measures. J Acoust Soc Am 30(10): 922–928.

Veldkamp WJ, Thijssen MA, Karssemeijer N. 2003. The value of scatter removal by a grid in full field digital mammography. Med Phys 30: 1712–1718.

Vyborny CJ, Schmidt RA. 1989. Mammography as radiographic examination: An overview. Radiographics 9: 723–764.

Wagner RF, Brown DG, Pastel MS. 1979. Application of information theory to the assessment of computed tomography, Medical Physics 8(2): 83–94.

Wagner RF, Metz CE, Campbell G. 2007. Assessment of medical imaging systems and computer aids: A tutorial review. Acad Radiol 14: 723–748.

Zanca F, Chakraborty DP, Van Ongeval C et al. 2008. An improved method for simulating microcalcifications in digital mammograms. Med Phys 35: 4012–4018.

Current Status of Computer-Aided Detection of Breast Cancer

Huong (Carisa) Le-Petross
The University of Texas MD Anderson Cancer Center

Tanya W. Stephens
The University of Texas MD Anderson Cancer Center

14.1 Introduction

Computer-assisted detection (CAD), which was introduced in the 1960s, is being increasingly incorporated into diagnostic imaging practice. Several commercial CAD systems designed for different imaging modalities and disease sites are now available to help radiologists detect cancer, perform diagnostic work-ups of suspicious findings, and stage newly diagnosed cancers. These CAD systems reduce false-negative rates by identifying subtle findings that would otherwise be missed by even experienced radiologists. In addition, most CAD systems can perform repetitive pattern matching and mark suspicious areas, which may help reduce detection errors during the analysis of high-volume workloads and enable radiologists to spend more time performing complex tasks such as tissue biopsies.

One area in which CAD is being used with increasing frequency is breast imaging. Breast cancer remains the leading cause of new cancer cases and the second most common cause of cancer death among women in the United States, and thus, many Western countries have instituted screening programs to help detect early-stage breast cancer (Chapter 7). This chapter will focus on the role of CAD in mammography, breast sonography, and breast MRI.

14.2 Mammography CAD

Screening mammography detects only 3–10 cancers per 1000 women (American Cancer Society 2011) but has been reported to contribute to the observed reduction in breast cancer mortality in recent years (Tabar et al. 2003) (Chapter 1). However, screening errors can make interpreting a large volume of mammograms difficult. Two common types of screening errors are perceptual errors and interpretation errors. Perceptual errors are failures

in the detection of an abnormality in film reading. This type of error is common to visual perceptual tasks and can be related to multiple factors such as the level of observer alertness, observer fatigue, duration of the observation task, distracting factors, the subtlety of the radiographic finding(s), or the poor conspicuity of the finding. Interpretation error is defined as correctly identifying an abnormality, but failure to correctly categorize the abnormality. One patient-related physiological feature that can affect a radiologist's perception of an image and cause a cancer to be missed is heterogeneously or extremely dense background breast parenchyma. On mammograms, breast cancer usually appears as a white mass, and fatty tissue appears as a dark background. However, glandular tissue in the breast is also white on mammograms. Heterogeneously or extremely dense breast parenchyma has more glandular tissue than fatty tissue and, thus, appears as a white background on mammograms, making the detection of a white-appearing lesion difficult. Perceptual errors may be compounded by the level of radiologist's fatigue at the time of image interpretation, the complexity of the breast parenchyma or the density of the breast tissue, and the subtle appearance of certain cancers such as invasive lobular carcinoma. CAD systems have been suggested to help reduce these perceptual errors or deficits in perception in interpreting screening mammograms. Reduction in these potential errors are beneficial, especially since malpractice in radiology has been reported to have increased in the last several years, with reported rate increase from 8% in 2002 to 10% in 2006, and the radiologist being the most common defendant (Dick et al. 2009).

Mammography CAD first became available in 1998. In the past, when only film-screen mammography was available, analog screening mammography images had to be converted to digital images with a digitizer before CAD could be applied.

However, many breast imaging and radiology centers have converted from film-screen mammography to digital mammography, obviating the need for a digitizer. Now, CAD systems can process and apply markings directly on the primary digital images. However, a radiologist must review the mammograms with the CAD markings and interpret the examination.

The efficacy of mammography CAD varies. Factors that may influence the efficacy of mammography CAD include the setting in which mammography is performed (i.e., academic versus private practice) and the interpreting radiologist's level of experience (Freer and Ulissey 2001, Gur et al. 2004, Gilbert et al. 2006). However, the CAD system used does not seem to influence CAD efficacy, as the performance of various systems does not differ substantially (Leon et al. 2009, Lechner et al. 2002). Commercially available CAD systems for mammography include the Image Checker M1000 (R2 Technology, Los Altos, CA), Mammoreader (Intelligent Systems Software, Clearwater, FL), and iCAD Second Look (CADx Medical Systems, Montreal, Quebec, Canada). All these systems have similar major limitations such as the large number of marks made per mammography examination. These CAD-generated markings on the mammography examination can distract the interpreting radiologist by redirecting the radiologist's focus onto the CAD markings instead of on the mammography examination. The high false-positive rate associated with the CAD-generated markings resulted in the implementation of a limit to the number of marks per patient that commercially available systems can make on an examination. In one retrospective review, 172 subtle cancers depicted in 169 mammograms to which CAD was applied were reviewed by five radiologists who were not blinded to the CAD markings and final pathology reports. The study revealed that the majority of radiologists would not have selected patients in whom CAD marked the mammographic findings for recall or biopsy (Ikeda et al. 2004). Today, an average of 3.9–4.5 marks is noted with the available commercial systems (Ikeda et al. 2004, Destounis et al. 2004).

14.2.1 Detection Accuracy

The radiologists' goal in using CAD to help interpret screening mammograms is to increase the rate of cancer detection by reducing the rate of false-negatives (Figure 14.1). In multicenter retrospective studies of screening mammography from cases of breast cancer, retrospective application of CAD on the prior screening mammogram found that CAD correctly marked the missed cancers in 65%–77% of cases (Warren Burhenne et al. 2000, Brem et al. 2003, Birdwell et al. 2001). Similarly, a single-institution study found that CAD correctly marked 71% (37/52) of the findings that had been previously interpreted as negative but should have prompted additional evaluation (Destounis et al. 2004). In that study, CAD reduced the false-negative rate from 31% (98/318) to 19% (61/318). Prospective clinical studies have shown that CAD increases cancer detection rates 4.7%–19.5% at the cost of increasing recall rates 8.1%–18.5% (Freer and Ulissey 2001, Birdwell et al. 2005, Morton et al. 2006, Dean and Ilvento 2006, Ko et al. 2006). One study found that the implementation of CAD in clinical practice increased radiologists' cancer detection rate by 16.1% and recall rate by 8.1% (Cupples et al. 2005).

Not all studies of mammography CAD have shown that it improves breast cancer detection rates. A clinical radiology practice of 24 radiologists in an academic setting reported no difference in the cancer detection rates or in recall rates for the entire group before and after implementation of CAD (Gur et al. 2004). Another large community-based clinical trial that evaluated screening mammography practices at 43 facilities in three states found that CAD not only failed to significantly improve breast cancer detection rates (Fenton et al. 2007) but it also may have actually been harmful: its use significantly reduced the specificity of screening mammography, reduced the positive predictive value of screening mammography, and resulted in higher recall and biopsy rates. However, this observational study was heavily criticized for its design flaws and for the reported results

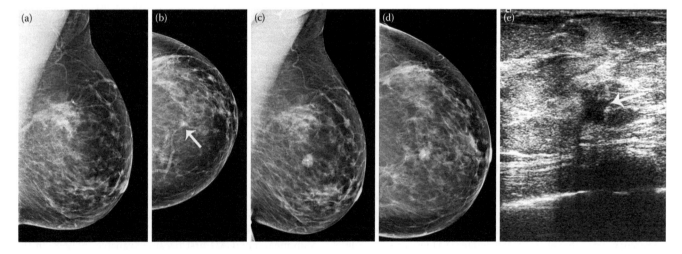

FIGURE 14.1 A 53-year-old woman who had a screening mammogram (a and b) where CAD marked a density in the central breast (white arrow) seen only on the craniocaudal view. One year later (c and d), the density has developed into a lobulated mass that is now visible on both views. Ultrasound of the left breast revealed a 6-mm mass (e), and ultrasound-guided biopsy confirmed invasive ductal carcinoma.

that contradicted those of prior prospective clinical studies. For example, several studies have reported that CAD improved cancer detection (Birdwell 2009) as well as the sensitivity, which was not demonstrated in the study by Fenton et al. (2007).

Other investigators looked at the performance of CAD for evaluating specific mammographic findings such as breast masses, architectural distortion, focal asymmetries, or microcalcifications. Some radiologists find CAD more helpful in detecting microcalcifications than in detecting masses, perhaps because the sensitivity of CAD for detecting calcifications (98%) is higher than that for detecting masses (86%) (Castellino et al. 2000). One prospective analysis of CAD markings on screening mammography at 75 sites where cancer later developed revealed that CAD correctly marked 94% (16/17) of the microcalcifications, 52% (11/21) of the masses, and 88% (7/8) of masses with microcalcifications (Markey et al. 2002). In a prospective study of 8682 women who underwent screening mammography, the percentage of microcalcifications representing biopsy-proven cancers detected with CAD (38%) was significantly higher than that detected without CAD (3%); meanwhile, the rate of invasive breast cancer decreased by 12% (Birdwell et al. 2005) (Figure 14.2).

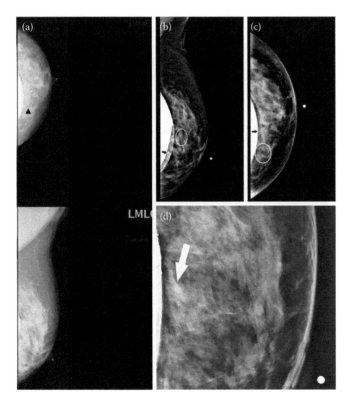

FIGURE 14.2 A 65-year-old woman who had a screening mammogram where CAD marked some faint calcifications in the left breast only on the craniocaudal view (a). The calcifications appear suspicious to the radiologist on both craniocaudal and mediolateral oblique views (b and c—white circles). CAD did not mark other calcifications in the left breast or along the implant capsule, which have benign appearances (black arrows). Additional magnification view (d) confirmed suspicious microcalcifications, and biopsy was appropriately recommended.

14.2.2 Single Reading with CAD versus Double Reading without CAD

Double reading, in which two radiologists independently review each screening mammogram, is one alternative to the use of CAD to improve the sensitivity of mammography in detecting breast cancer and maintain the consistency of screening mammography interpretation. Some studies have suggested that double reading improves breast cancer detection rates by 3%–15% (Gilbert et al. 2006, Gromet 2008, Karssemeijer et al. 2003). However, because of a shortage of radiologists and the fact that financial reimbursement is not given for having a second radiologist read the same radiology examination, double reading is often not achievable or practical in many settings in the United States. Thus, double reading is not currently a standard of care in the United States, even though it is widely used in other countries.

Studies conducted in several countries have compared the efficacy of double reading without CAD to that of single reading with CAD. In the United Kingdom, a retrospective analysis of 10,267 mammography cases reported that a single reading with CAD resulted in a significantly higher recall rate than a double reading without CAD (8.6% and 6.5%, respectively; p < 0.001) and that a single reading with CAD detected 6.5% more cancers than double reading without CAD (Gilbert et al. 2006). In another retrospective study, Gromet (2008) reviewed 231,221 screening mammograms interpreted by experienced radiologists at community-based practices in the United States and found that the sensitivity of double reading without CAD (88%) and the sensitivity of a single reading with CAD (90%) were higher than that of a single reading without CAD (81%). Although the sensitivity of a single reading with CAD was not significantly different than that of double reading without CAD, the sensitivity of a single reading with CAD was significantly higher than that of a single reading without CAD (p < 0.0001). However, the higher sensitivity came at a cost of higher recall rates. Whereas a single reading without CAD resulted in a recall rate of 10.2%, a single reading with CAD and double reading without CAD resulted in recall rates of 10.6% and 11.9%, respectively (Cho et al. 2010). In a study conducted in the Netherlands, Karssemeijer et al. (2003) reviewed 500 mammography cases from the Dutch Breast Cancer Screening Program (250 positive cases [125 screening-detected cancers and 125 interval cancers] matched with 250 negative screening cases performed during the same time interval and in the same geographic regions) and found that compared to a single reading without CAD, double reading offered a significantly higher increase in sensitivity than did a single reading with CAD (10.5% and 7%, respectively; p < 0.001). From these studies findings, it would seem that although double reading without CAD yields a higher sensitivity than a single reading without CAD, a single reading with CAD is less expensive than and provides an acceptable alternative to double reading without CAD.

Another study that compared the efficacy of single reading with CAD to that of double reading without CAD was the prospective randomized multicenter Computer Aided Detection Evaluation Trial II (CADET II) (James et al. 2010). In this trial, 227 cancers were detected in 28,204 women (a detection rate of about 8 cancers per 1000 women). There was no difference in the detection rate between single reading with CAD (7.02 per 1000) and double reading without CAD (7.02 per 1000). However, the types of cases being recalled and the types of lesions being detected in each study arm were different. Parenchymal deformities or architectural distortion were recalled more frequently in the double reading without CAD arm than in the single reading with CAD arm, whereas asymmetric densities were recalled more frequently in the single reading with CAD arm than in the double reading without CAD arm. The study also found that CAD was more effective in detecting microcalcifications than in detecting other mammographic features (James et al. 2010).

Given the cost constraints of the healthcare systems of the United States and other Western countries, the addition of CAD to a single reading seems preferable to hiring another radiologist so that a double reading without CAD can be performed. CAD offers cost savings over hiring a second radiologist; in contrast, double reading requires the nonbillable time of an additional radiologist. In addition, the cost of CAD has decreased from approximately $16.00 in 2006 to approximately $11.55 in 2010. The cost of CAD consists of a technical fee and a professional fee. The professional fee—the cost of having a radiologist interpret CAD findings—is only about $3.00, which makes it hard to justify paying the higher equivalent salary fee of a second radiologist.

In summary, the addition of CAD to a single reading improves sensitivity regardless of lesion type. CAD appears to be more effective in detecting microcalcifications than in detecting lesions (including architectural distortion and asymmetries suggestive of lesions), which may be related to the fact that the interpretation of mammographic densities or mass lesions is more complex than the interpretation of mammographic microcalcifications, and radiologists find it more difficult to use the CAD prompts for mass lesions than the CAD prompts for microcalcifications.

14.2.3 Advantages and Limitations

CAD technology is now available worldwide, and its increased use throughout the United States has made CAD a standard of practice for breast cancer screening with mammography. Several currently published trials have shown that CAD does assist the radiologist in detecting additional screen-detected cancers with a reasonable increase in the recall rate, as long as the radiologist makes the final decision. Thus, a radiologist would not recall a patient based on CAD markings only when no suspicious lesion is detected by the radiologist. Unlike the human eye, software system such as CAD can perform repetitive tasks without being vulnerable to fatigue, environmental or emotional distractions. In a busy clinical practice or a large volume mammography screening practice, the chance of fatigue and/or distractions is common. Therefore, imaging tools such as CAD serve a similar purpose as a spell check program by alerting the reader to findings that are typical of cancer. This process helps the radiologist avoid overlooking a cancer, making CAD very valuable in today's medical practice.

Although CAD can help decrease perceptual or detection errors, it has less potential to reduce interpretation errors. Several factors that can affect the interpretation of CAD markings include multiple adjacent or superimposed benign findings, a slow-growing lesion that remains stable over several years, cancers with features suggestive of benign lesions, and the subtlety of suspicious features. Another limitation of CAD is the large number of false-positive findings it generates (Brenner et al. 2006). Although CAD has been reported to increase the sensitivity of mammography in detecting cancer, one must also consider the balance between CAD marks that radiologists overlook and the false-positive results radiologists identify and act upon. If a radiologist acts upon every false-positive CAD markings or findings, then unnecessary recall would lead to additional, costly testing and additional anxiety for patients. Such unnecessary recall practice would thereby decrease the cost-effectiveness of CAD compared with double reading without CAD. The interpreting radiologists' level of experience correlates with the affect CAD has on screening mammography. For example, a prospective study of 303 patients with benign and/or malignant mammography findings revealed that CAD increased the reading time of less experienced medical students and residents but not that of experienced radiologists (Sohns et al. 2010).

14.3 Sonography CAD

Breast sonography is not a primary screening tool for breast cancer (Chapter 15). Instead, breast sonography is used predominantly as a diagnostic tool or as an adjuvant to mammography in the evaluation of questionable mammographic findings and physical examination in the evaluation of palpable findings. Therefore, breast sonography CAD is not as widely used as mammography CAD or MRI CAD. Currently, breast sonography CAD is used as an image-viewing and report-generating tool to help analyze sonographic images of breast masses (Figure 14.3). Since sonography CAD systems are considered image processing and visualization tools, they are regulated similarly to devices such as Picture Archiving and Communication Systems (PACS), i.e., U.S. FDA class 2 devices, as opposed to mammographic CAD systems, which are considered to be medical image analyzing tools that require more stringent review, i.e., U.S. FDA class 3 devices. However, despite the way that available systems have been approved by the U.S. regulatory authorities, the term *computer-aided detection*, or CAD, is widely used to refer not only to systems that aid in detection but also to systems that assist more generally with lesion visualization and characterization. Thus, for the purpose of this chapter, the term CAD is used throughout even though sonography CAD systems actually aid in the characterization of breast lesions and not in the detection of lesions (Starvos 2009).

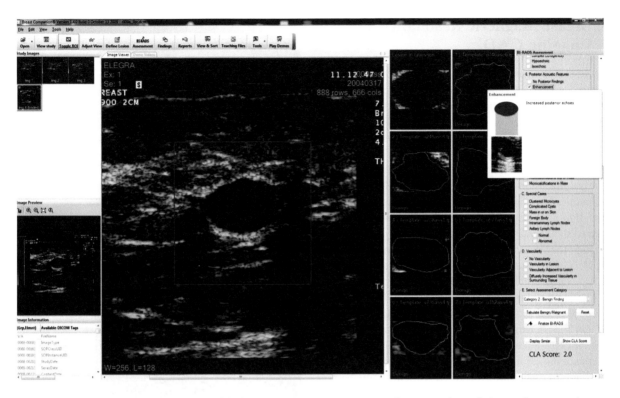

FIGURE 14.3 (See color insert.) A screen view of the Breast Companion® CAD system showing analysis of a benign breast cyst (center image with red box) and seven similar masses displayed in the adjacent panel on the right.

Commercially available CAD systems for breast sonography include the B-CAD (Medipattern, Toronto, Ontario, Canada) and BI-RADS Companion (Almen Laboratories, Vista, CA). The BI-RADS Companion compares an area of interest to known imaging findings in a database. The radiologist selects image(s) from a case and applies a user-selected threshold or outlines the lesion manually with a cursor to segment the lesion from the background. Some sonography CAD systems display images of similar, biopsy-proven masses from the database and computes an assessment based on the American College of Radiology BI-RADS Lexicon to generate a standardized report. However, the radiologist's final interpretation of the sonography CAD assessment is still necessary. Unlike screening mammography CAD which aids in lesion detection, the goal of sonography CAD is to aid in lesion characterization and thus requires the radiologist to either manually or use software to automatically outline the borders of an area of interest before CAD can be applied.

Many sonography CAD systems use techniques such as image texture extraction, morphologic feature extraction, and/or decision-making algorithms to characterize lesions in order to comprise a CAD result. Textural features are characteristics of the gray-level distribution in an image. However, the texture analysis is system-dependent and can only be performed well using one specific ultrasonography system. Recently, researchers have used sonography CAD mainly to homogenize the textural features of images generated using different sonography systems before the images are processed (Yang et al. 2008). Other sonography CAD systems analyze the morphologic features of lesions to diagnose solid breast tumors. These systems have the advantage of functioning nearly independently of the settings of different ultrasonography systems and machines. Sonography CAD systems may also assess the acoustic features of a lesion on the basis of sound velocity and tissue attenuation. In the future, the use of sonography CAD systems that analyze the acoustic, textural, and morphological features of breast lesions will likely improve the performance of sonography CAD in detecting and diagnosing breast cancer.

Studies investigating the sensitivity and specificity of sonography CAD in detecting breast cancer and differentiating benign from malignant masses have been limited to a few single-institution trials with small sample sizes. One trial that evaluated the ability of CAD or radiologists to differentiate malignant tumors from benign lesions on sonograms from 508 patients found that CAD had sensitivity and specificity rates of 100% and 30%, respectively, while manual interpretation by radiologists had sensitivity and specificity rates of 100% and 77%, respectively (Drukker et al. 2008). This would suggest that radiologists with expertise in breast imaging do not find sonography CAD useful. However, because the study included only a small number of cases and because these cases were the same as those in the library of known cases or the comparison database, the study's findings are not a true reflection of clinical practice and thus provide no basis for definitive conclusions about the role of CAD in ultrasonography systems.

14.4 Breast Magnetic Resonance Imaging CAD

Breast MRI is more sensitive than mammography or sonography in the detection of breast cancer (Kuhl 2007) (Chapter 16). Breast MRI also provides information about the morphologic, kinetic, and functional features of breast lesions. However, because benign and malignant breast lesions share some morphologic and kinetic features, visually estimating a lesion's kinesis on a grayscale reading station is not an accurate method of determining whether the lesion is malignant or benign. Therefore, several companies have developed computer systems to aid in the interpretation of breast MRI. In this chapter, all such systems are referred to as CAD systems in keeping with general usage, even though they do not perform CAD in the sense that mammographic CAD systems do. Commercially available breast MRI CAD systems include the DynaCAD (Invivo, Gainesville, FL), CADstream (Merge Healthcare, Chicago IL), and Aegis (Sentinelle Medical Inc., Toronto, Ontario, Canada).

MRI CAD performs several functions that simultaneously improve the detection of an enhancing breast lesion and provide the kinetics of the lesion to make image interpretation more efficient and improve workflow. The kinetics or enhancing pattern of a breast lesion is analyzed by measuring the pixel intensity of the lesion of interest during the precontrast and postcontrast sequences and generating a kinetic curve (Figure 14.4). A minimum of three measurements taken at different times, after the administration of the intravenous contrast agent, is required to generate a kinetic curve. At many institutions, radiologists perform more than three measurements or time points. An arbitrary enhancement threshold ranging from 50% to 100% is set to enable the system to determine which pixels to code as the threshold level. The peak enhancement is usually observed 60–120 seconds after intravenous contrast has been injected into the patient. The sequence acquired at this time is expected to contain the peak pixel intensity level used by the system to identify the chosen threshold level. When pixel intensity increases above the predetermined threshold, the CAD system generates a colorized overlay of the breast MR image or colorized mapping. Each pixel is assigned a color indicating the pattern of enhancement. Three common enhancing pattern descriptions are persistent enhancement (usually blue or green), plateau enhancement (yellow), and washout enhancement (red). Most MRI CAD systems also perform segmentation of the region of interest to determine the lesion volume and quantify the percentage of volume each kinetic pattern constitutes.

Most MRI CAD systems include a tool for motion correction, also known as image registration correction. Motion correction is very important for studies in which active fat suppression is not performed and subtraction images (postcontrast images minus precontrast images), which demonstrate only enhancing structures, are necessary. Patient motion affects the quality of these postprocessing subtraction images; the MRI CAD system can provide motion correction by improving the alignment of

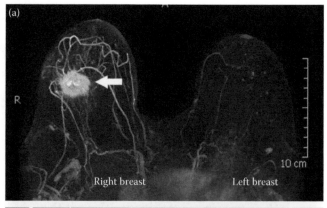

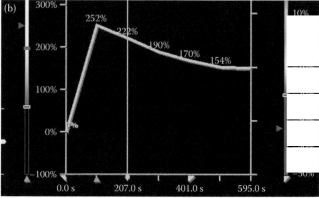

FIGURE 14.4 A 65-year-old woman with newly diagnosed right breast invasive ductal carcinoma. (a) The axial maximum intensity projection image of the bilateral breast MRI examination demonstrates an irregular 3.6 cm right breast mass (white arrow), compatible with the biopsy proven cancer. (b) The kinetic assessment using MRI CAD system revealed initial rapid enhancement with delayed washout curve, which is consistent with malignant lesion.

the postcontrast images with the precontrast images. This postprocessing tool helps to reduce misregistration, the subtraction artifact, which can result in a false-positive finding and unnecessary biopsy. Newer MRI CAD systems can perform registration correction in 3-D imaging.

Only a few studies have aimed to validate the ability of MRI CAD to distinguish benign breast lesions from malignant breast lesions. One retrospective study of 154 lesions in 125 women reported that setting the enhancing threshold level for MRI CAD at 100% reduced the false-positive rate by 23% (Williams et al. 2007). Another retrospective analysis of 42 lesions in 36 patients who had indeterminate mammographic and/or sonography results or high-risk screening and who were scanned using a 3.0-T MRI system found that MRI CAD (CADStream) significantly improved the specificity of manual interpretation by two experienced radiologists from 68.8% to 81.3% (p < 0.05) and markedly improved the sensitivity from 84.6% to 90.4% (Meeuwis et al. 2010). More recently, a meta-analysis of 10 studies revealed that MRI CAD (CADstream or DynaCAD) had little influence on the sensitivity or specificity rates of manual

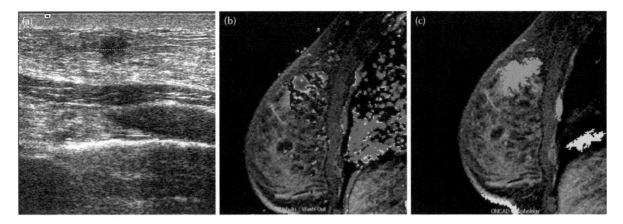

FIGURE 14.5 (**See color insert.**) A 30-year-old woman who presents with a palpable finding in the right breast. (a) An ultrasound of this palpable area revealed an irregular 1.6 cm mass and biopsy confirmed invasive ductal carcinoma. (b) Postcontrast sagittal T1-weighted image of the right breast with color overlay, after CAD implementation, demonstrates the 1.6 cm mass with suspicious enhancing pattern (red color). (c) The automated morphological analysis system ONCAD revealed a larger suspicious area measuring 2.9 cm. Final pathology confirmed 2.4 cm invasive carcinoma with extensive surrounding ductal carcinoma in situ.

interpretation by an experienced radiologist (Dorrius et al. 2011) but improved those of trainees, residents, and inexperienced radiologists. The specificity of breast MRI increased when the threshold level of enhancement for the first postcontrast image was set at 100% than when the threshold level was set at a lower value. Only one study has investigated the use of CAD as an aid to breast MRI in assessing chemotherapy response; only 15 patients were enrolled in the study, and too many false-positive results were noted to justify its use in this capacity (DeMartini et al. 2005).

Studies investigating the use of MRI CAD for detecting specific breast cancer subtypes are also sparse. One study found that although MRI can be used to identify mammographically occult ductal carcinoma in situ (DCIS), a large portion of false-negative findings on MRI are later found to be DCIS (Kuerer et al. 2009). Another study found that retrospectively implementing an MRI CAD prototype did not improve the sensitivity of MRI in detecting DCIS in a high-risk population (Arazi-Kleinman et al. 2009). Thus, additional, larger studies are needed to determine whether MRI CAD could improve the detection rate of mammographically occult DCIS. In addition, because DCIS commonly presents as a mix of DCIS and invasive carcinoma, which may affect the sensitivity and specificity of currently available CAD systems, more advanced CAD algorithms may be needed for the evaluation of nonmass-like enhancement on MR images (Newell et al. 2010). DCIS commonly presents as a mix of DCIS and invasive carcinoma, which may affect the calculations of sensitivity and specificity of the CAD systems.

Because of the limitations of breast MRI and MRI CAD, additional imaging-enhancing products have been developed to complement the tools available in MRI CAD systems. For example, Z3D Contrast Acuity Software (Clario Medical Imaging, Inc., Seattle, WA), which has received marketing clearance from U.S. Food and Drug Administration, is used in conjunction with DynaCAD to closely evaluate small (5–15 mm in diameter) enhancing lesions in the breast by displaying 2-D images in three dimensions and enhancing the visualization of signal intensity changes. ONCAD, an automated morphological analysis system, uses patented algorithms to detect morphologically suspicious lesions, which can then be easily identified by a color overlay (Figure 14.5). When ONCAD is combined with DynaCAD, kinetics and morphology analyses can be performed simultaneously.

14.5 Conclusion

The literature on this topic is confusing due to the variable study models and validation methods used. Few definitive conclusions can be made about the utility of mammography, sonography, or MRI CAD. Nevertheless, CAD can be a valuable addition to diagnostic practices and has influenced how breast radiologists detect cancer, perform diagnostic workups of suspicious findings, and stage newly diagnosed cancers. Although a double reading has largely been shown to be more effective than CAD, the hiring of a second radiologist is not always feasible. Very busy radiology practices with large case volumes may find CAD useful in preventing radiologist fatigue and in reducing large quantities of images needing assessment. In all types of practices, the use of CAD may lead to increased efficiency and productivity, which will allow radiologists to perform complex tasks such as biopsies. In the meantime, further study is needed to determine the circumstances under which CAD systems can be optimally used.

References

American Cancer Society. 2011. Leading Sites of New Cancer Cases and Deaths—2011 Estimates. Accessed on 07/06/2011. http://www.cancer.org/acs/groups/content/@epidemiology surveilance/documents/document/acspc-029821.pdf.

Arazi-Kleinman T, Causer PA, Jong RA, Hill K, Warner E. 2009. Can breast MRI computer-aided detection (CAD) improve radiologist accuracy for lesions detected at MRI screening and recommended for biopsy in a high-risk population? Clin. Radiology 64: 1166–1174.

Birdwell RL, Bandodkar P, Ikeda DM. 2005. Computer-aided detection with screening mammography in a university hospital setting. Radiology 236: 451–457.

Birdwell RL, Ikeda DM, O'Shaughnessy KF, Sickles EA. 2001. Mammographic characteristics of 115 missed cancers later detected with screening mammography and the potential utility of computer-aided detection. Radiology 219: 192–202.

Birdwell RL. 2009. The preponderance of evidence supports computer-aided detection for screening mammography. Radiology 253: 9–16.

Brem RF, Baum J, Lechner M et al. 2003. Improvement in sensitivity of screening mammography with computer-aided detection: a multiinstitutional trial. Am J Roentgenol 181: 687–693.

Brenner RJ, Ulissey MJ, Wilt RM. 2006. Computer-aided detection as evidence in the courtroom: Potential implications of an Appellate Court's ruling. Am J Roentgenol 186: 48–51.

Castellino RA, Roehrig J, Zhang W. 2000. Improved computer-aided detection (CAD) algorithms for screening mammography (abstr). Radiology 217: 400.

Cho N, Kim SJ, Choi HY, Lyou CY, Moon WK. 2010. Features of prospectively overlooked CAD marks on prior screening digital mammograms in women with breast cancer. Am J Roentgenol 195: 1276–1282.

Cupples TE, Cunningham JE, Reynolds JC. 2005. Impact of computer-aided detection in a regional screening mammography program. Am J Roentgenol 185: 944–950.

Dean JC, Ilvento CC. 2006. Improved cancer detection using computer-aided detection with diagnostic and screening mammography: Prospective study of 104 cancers. Am J Roentgenol 187: 20–28.

DeMartini WB, Lehman CD, Peacock S, Russell MT. 2005. Computer-aided detection applied to breast MRI: Assessment of CAD-generated enhancement and tumor sizes in breast cancers before and after neoadjuvant chemotherapy. Acad Radiol 12: 806–814.

Destounis SV, DiNitto P, Logan-Young W, Bonaccio E, Zuley ML, Willison KM. 2004. Can computer-aided detection with double reading of screening mammograms help decrease the false-negative rate? Initial experience. Radiology 232: 578–584.

Dick III JF, Gallagher TH, Brenner RJ et al. 2009. Predictors of radiologists' perceived risk of malpractice lawsuits in breast imaging. Am J Roentgenol 192: 327–333.

Dorrius MD, Jansen-van der Weide MC, van Ooijen PMA, Pijnappel RM, Oudkerk M. 2011. Computer-aided detection in breast MRI: A systematic review and meta-analysis. Eur Radiol 21: 1600–1608.

Drukker K, Gruszauskas NP, Sennett CA, Giger ML. 2008. Breast US computer-aided diagnosis workstation: Performance with a large clinical diagnostic population. Radiology 248: 392–397.

Fenton JJ, Taplin SH, Carney PA et al. 2007. Influence of computer-aided detection on performance of screening mammography. New Engl J Med 356: 1399–1409.

Freer TW, Ulissey MJ. 2001. Screening mammography with computer-aided detection: Prospective study of 12,860 patients in a community breast center. Radiology 220: 781–786.

Gilbert FJ, Astley SM, McGee MA et al. 2006. Single reading with computer-aided detection and double reading of screening mammograms in the United Kingdom National Breast Screening Program. Radiology 241: 47–53.

Gromet M. 2008. Comparison of computer-aided detection to double reading of screening mammograms: Review of 231,221 mammograms. Am J Roentgenol 190: 854–859.

Gur D, Sumkin JH, Rockette HE et al. 2004. Changes in breast cancer detection and mammography recall rates after the introduction of a computer-aided detection system. J Natl Cancer Inst 96: 185–190.

Ikeda DM, Birdwell RL, O'Shaughnessy KF, Sickles EA, Brenner RJ. 2004. Computer-aided detection output on 172 subtle findings on normal mammograms previously obtained in women with breast cancer detected at follow-up screening mammography. Radiology 230: 811–819.

James JJ, Gilbert FJ, Wallis MG et al. 2010. Mammographic features of breast cancers at single reading with computer-aided detection and at double reading in a large multicenter prospective trial of computer-aided detection: CADET II. Radiology 256: 379–386.

Karssemeijer N, Otten JD, Verbeek AL et al. 2003. Computer-aided detection versus independent double reading of masses on mammograms. Radiology 227: 192–200.

Ko JM, Nicholas MJ, Mendel JB, Slanetz PJ. 2006. Prospective assessment of computer-aided detection in interpretation of screening mammography. Am J Roentgenol 187: 1483–1491.

Kuerer HM, Albarracin CT, Yang WT et al. 2009. Ductal carcinoma in situ: State of the science and roadmap to advance the field. J Clin Oncol 27: 279–288.

Kuhl C. 2007. The current status of breast MR imaging. Part I. Choice of technique, image interpretation, diagnostic accuracy, and transfer to clinical practice. Radiology 244: 356–378.

Lechner M, Nelson M, Elvecrog E. 2002. Comparison of two commercially available computer-aided detection (CAD) systems. Appl Radiol 31: 31–35.

Leon S, Brateman L, Honeyman-Buck J, Marshall J. 2009. Comparison of two commercial CAD systems for digital mammography. J Digit Imag 22: 421–423.

Markey MK, Lo JY, Floyd Jr CE. 2002. Differences between computer-aided diagnosis of breast masses and that of calcifications. Radiology 223: 489–493.

Meeuwis C, van de Ven SM, Stapper G et al. 2010. Computer-aided detection (CAD) for breast MRI: Evaluation of efficacy at 3.0 T. Eur Radiol 20: 522–528.

Morton MJ, Whaley DH, Brandt KR, Amrami KK. 2006. Screening mammograms: Interpretation with computer- aided detection—Prospective evaluation. Radiology 239: 375–383.

Newell D, Nie K, Chen J-H et al. 2010. Selection of diagnostic features on breast MRI to differentiate between malignant and benign lesions using computer-aided diagnosis: Differences in lesions presenting as mass and non-mass-like enhancement. Eur Radiol 20: 771–781.

Sohns C, Angic BC, Sossalla S, Konietschke F, Obenauer S. 2010. CAD in full-field digital mammography-influence of reader experience and application of CAD on interpretation of time. Clin Imag 34: 418–424.

Starvos T. 2009. New advances in breast ultrasound: Computer-aided detection. Ultrasound Clin 4: 285–290.

Tabar L, Yen MF, Vitak B et al. 2003. Mammography service screening and mortality in breast cancer patients: 20-year follow-up before and after introduction of screening. Lancet 361: 1405–1410.

Warren Burhenne LJ, Wood SA, D'Orsi CJ et al. 2000. Potential contribution of computer-aided detection to the sensitivity of screening mammography. Radiology 215: 554–562.

Williams TC, DeMartini WB, Partridge SC, Peacock S, Lehman CD. 2007. Breast MR imaging: computer-aided evaluation program for discriminating benign from malignant lesions. Radiology 244: 94–103.

Yang H-C, Chang H-C, Huang S-W, Chou Y-H, Li P-C. 2008. Correlations among acoustic, texture and morphological features for breast ultrasound CAD. Ultrason Imag 30: 228–236.

IV

Complementary Approaches

15

Breast Ultrasound

Gary J. Whitman
The University of Texas MD Anderson Cancer Center

Raunak Khisty
The University of Texas MD Anderson Cancer Center

R. Jason Stafford
The University of Texas MD Anderson Cancer Center

15.1 Introduction

Advances in ultrasound technology have dramatically impacted ultrasound's role in breast imaging. Ultrasound is an inexpensive, convenient, and portable modality used regularly in breast imaging, especially in the evaluation of breast masses. Breast ultrasound has undergone tremendous changes since 1953, when the first breast ultrasound was performed (Woo 2008). While ultrasound has commonly been used for differentiating cysts from solid breast masses, ultrasound is now used to characterize benign and malignant breast masses. Breast masses are usually biopsied with sonographic guidance, and ultrasound is being considered as a screening modality in selected patients (Thompson 2007). In addition, sonography plays a major role in the evaluation of palpable breast abnormalities. Today, ultrasound is an important component of nearly all breast imaging centers.

Ultrasound is especially useful for evaluating palpable abnormalities (Figure 15.1). Gumus et al. (2011) reported a study of 251 women with palpable abnormalities and negative ultrasound and mammographic examinations. In all 251 women, clinically guided biopsy was performed. Of 251 biopsies, there were three (1.2%) malignancies, and two were invasive tumors. All three of the malignancies were classified as clinically suspicious or malignant. The remaining 248 lesions were categorized as benign or probably benign on clinical examination. Thus, in patients with palpable abnormalities and negative ultrasounds and mammograms, the negative predictive value is very high (98.8%) and the likelihood of malignancy is very low (1.2%). Also, if the clinical examination reveals suspicious or malignant-appearing findings, then biopsy should be performed, even if mammography and sonography are negative?

Ultrasound also plays an important role in clarifying mammographic and sonographic abnormalities. Following mammography, sonography is utilized to determine if the targeted finding is a real lesion or due to superimposition of normal structures (Figure 15.2). If the finding is thought to be real on sonography, ultrasound is important in determining if the lesion is cystic or solid, and if solid, in determining if the lesion is likely benign or malignant. When performing breast sonography, it is important that the imaging parameters are optimized (Figure 15.3) and that the descriptors are accurate and based on real-time evaluation. Findings on breast ultrasound should be assessed according to the American College of Radiology Breast Imaging–Reporting and Data System (BI-RADS) lexicon (American College of Radiology 2003). Stavros et al. (1995) prospectively classified 750 solid breast masses as benign, indeterminate, or malignant. Two masses out of 424 prospectively classified as benign were found to be malignant at biopsy. Thus, the classification scheme used by Stavros et al. (1995) (before the publication of the first BI-RADS lexicon) had a negative predictive value of 99.5%. Of 125 malignant masses, 123 were correctly classified as indeterminate or malignant, resulting in a sensitivity of 98.4% (Stavros 1995).

Ultrasound also plays an important role in further evaluation of lesions identified on magnetic resonance imaging (MRI). Following breast MRI, it is common for targeted or second-look ultrasound to be performed to verify the presence of the MRI-detected abnormality, to help characterize it, and to facilitate

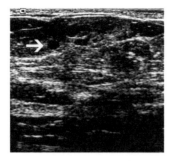

FIGURE 15.1   A 72-year-old woman presented with a palpable abnormality in the right breast 9–10-o'clock position. Transverse ultrasound in the region of the palpable abnormality in the right breast 9–10-o'clock region demonstrates an area of increased echogenicity with round (arrow) and oval hypoechoic masses, consistent with oil cysts.

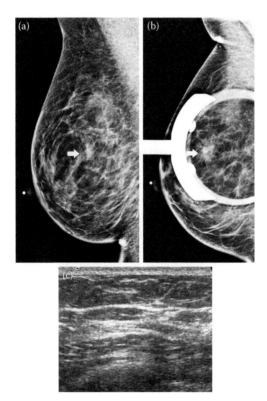

FIGURE 15.2   A 43-year-old woman presented for screening mammography. (a) A focal asymmetry (arrow) was noted in the right breast at 1 o'clock, as seen on the mediolateral oblique (MLO) view. (b) Spot compression right breast MLO view showed the persistent asymmetry (arrow). (c) Transverse ultrasound in the 1-o'clock region revealed no suspicious findings. The asymmetry noted on mammography was due to overlap of normal structures.

percutaneous biopsy. Abe et al. (2010) reviewed the records of 158 consecutive patients with 202 breast abnormalities initially identified on MRI. All of the lesions were identified on MRI as enhancing findings, and all of the lesions were subsequently evaluated with ultrasound. MRI–ultrasound correlation was established for 115 (57%) of the 202 abnormalities identified on MRI, including

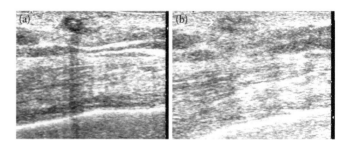

FIGURE 15.3   A 66-year-old woman presented with a palpable abnormality in the right breast 8-o'clock region. (a) Transverse ultrasound in the right breast 8-o'clock region demonstrates a hypoechoic mass (arrow) in the dermis, consistent with a sebaceous cyst, at the level of the focal zone (triangle on right). (b) Transverse ultrasound in the right breast 8-o'clock region obtained with the focal zone (triangle on right) positioned in the deep part of the breast shows the hypoechoic mass (arrow). The sebaceous cyst was seen more clearly in (a) rather than in (b) due to appropriate positioning of the focal zone in (a) and the focal zone positioned too deep relative to the lesion in (b).

33 malignant lesions (Figure 15.4) and 82 benign lesions. The remaining 87 lesions identified on MRI, including 11 malignant lesions and 76 benign lesions, did not have sonographic correlates. Mass lesions noted on MRI were more likely to have ultrasound correlates than nonmasslike lesions. Sixty-five percent of the mass lesions and 12% of the nonmasslike lesions had ultrasound correlates. The malignant lesions with ultrasound correlates tended to be subtle on sonography.

Over the last decade, increased attention has been directed to breast ultrasound as a supplemental screening modality in addition to mammography. While digital mammography performs

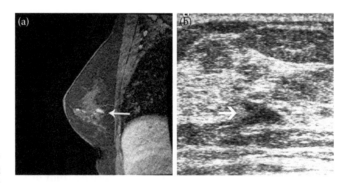

FIGURE 15.4   A 51-year-old woman presented for screening breast MRI. The patient was at high risk for the development of breast cancer, and the patient's mother was diagnosed with breast cancer at age 79 years. (a) Contrast-enhanced dynamic sagittal T1-weighted MRI shows an irregular enhancing mass with jagged margins (arrow) in the posterior aspect of the right breast in the 8–9-o'clock position. (b) Transverse right breast ultrasound in the 8-o'clock region demonstrates an irregular hypoechoic mass (arrow), correlating with the finding noted on MRI. Ultrasound-guided fine needle aspiration was performed, showing evidence of a papillary neoplasm. The patient then underwent right mastectomy, which showed evidence of intermediate grade DCIS with cribriform, solid, and papillary features.

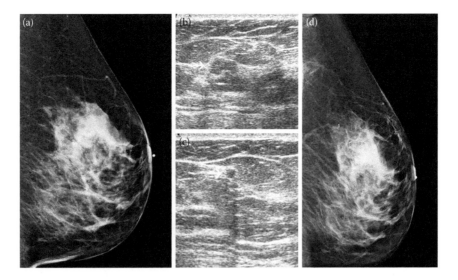

FIGURE 15.5 A 49-year-old woman with a history of right breast invasive ductal carcinoma presented for screening mammography. (a) Left lateromedial mammogram demonstrates dense tissue in the upper breast, not significantly changed when compared to the mammograms from the previous year. (b) Left breast longitudinal ultrasound in the 11-o'clock position demonstrates a hypoechoic lobular mass (arrow). Ultrasound-guided core needle biopsy was performed, revealing evidence of invasive ductal carcinoma, intermediate grade, with apocrine features. (c) Left breast longitudinal ultrasound in the 11-o'clock position shows a hypoechoic round mass (arrow). Ultrasound-guided fine needle aspiration was performed, and cytology showed evidence of ductal breast carcinoma with apocrine features. Clips were placed with sonographic guidance at both biopsy sites. (d) Postprocedural left lateromedial mammogram shows both clip markers.

with increased sensitivity in women with dense breasts compared to screen-film mammography, digital mammography does not eliminate one of mammography's major limitations—that noncalcified breast cancers may be obscured on mammography due to overlying dense parenchyma (Figure 15.5). Berg et al. (2008), reporting the results of the American College of Radiology Imaging Network (ACRIN) 6666 trial, noted that the supplemental yield of screening sonography was 4.2 cancers per 1000 women screened. In the ACRIN 6666 trial, there were 12 malignancies identified only with breast ultrasound. Eleven of the 12 cancers were invasive, with a median size of 1 cm (range, 0.5–4.0 cm). Eight of nine of the cancers were associated with negative lymph nodes. The positive predictive value of biopsy recommendation after a full diagnostic workup (PPV2) was 22.6% for mammography, 8.9% for ultrasound, and 11.2% for combined mammography and ultrasound (Berg et al. 2008).

15.2 Grayscale Ultrasound

Ultrasound is usually defined as the sound frequencies above the audible range of 20–20,000 Hz (Lieu 2010). Medical ultrasound devices operate in the range of approximately 1–15 MHz, with higher frequencies being preferable, as higher frequencies are associated with increased resolution. Ultrasound systems have scanheads composed of piezoelectric crystals that act as reciprocal energy transducers in that they vibrate to generate an ultrasound wave when an electrical current is applied to them and similarly produce an electric current when ultrasound waves are incident upon them. Ultrasound is a longitudinal mechanical wave that propagates through tissue. As the wave propagates, it interacts

with the tissue via scattering and absorption, based on the acoustic properties of the tissue. Ultrasound imaging utilizes the part of the wave that is backscattered to the transducer. Two-dimensional images are generated by recording this backscattered signal over time across an array of piezoelectric transducing elements (Lieu 2010). The assumed speed of sound in tissue (1540 m/s) is used to scale the arrival time of the echoes to the depth in the tissue as well as to determine timing delays between firing elements to "focus" each beam to achieve higher lateral resolution. The differences in the intensities of the received echoes are displayed in grayscale. Grayscale imaging is often referred to as brightness mode or B-mode imaging. The lateral field of view in B-mode imaging is determined primarily by the geometry of the scanhead. Linear transducers generate a field of view approximately the length of their footprint, while curvilinear arrays generally have a radially expanding field with depth that provides a large field of view at depth. Breast imaging primarily relies on high-frequency, high-resolution linear scanheads operating at >10 MHz.

The ability to electronically steer and focus the beam facilitates the ability to generate a trapezoidal acquisition if a larger field of view at depth is required. The speed and the real-time processing capabilities of modern ultrasound units facilitate a method of "extending" the lateral field of view of a linear array beyond the footprint of the scanhead. This is accomplished by pulling the scanhead along the direction of the lateral field of view manually. The images are captured, registered, and stitched together in real-time to form an extended field of view. Extended field of view imaging, also referred to as panoramic field of view imaging, has been incorporated into breast ultrasound (Weng 1997) to aid in the visualization of large structures and to facilitate measuring distances between structures in the breast (Fornage et al. 2000), including the nodal basins; large masses, scars, and seromas; and multifocal and multicentric malignant processes. At The University of Texas MD Anderson Cancer Center (UTMDACC), extended field-of-view imaging is routinely employed to document the distance between a breast lesion and the nipple. The information provided by extended field of view imaging is useful for surgical planning. As the real-time acquisition and processing capabilities have evolved, many modern systems additionally allow advanced contrast modes to be acquired in the extended field of view mode, such as tissue harmonic imaging and spatial compounding, as well as duplex modes, such as power Doppler imaging.

15.3 Speckle Reduction

Contrast in B-mode imaging is primarily generated by the differences in backscatter intensity from various tissues, not from coherent reflections at tissue interfaces. This backscatter is a function of the acoustic properties of the tissue and the waves. Speckle is a granular pattern caused in an ultrasonic image by the constructive and destructive coherent interference of backscattered echoes from scatterers in the tissue, small in comparison to the beam wavelength. Speckle results in a "salt-and-pepper"-type modulation of the image in tissue with homogeneous acoustic parameters that would have otherwise generated a uniform level of signal from the constant backscatter amplitude. Speckle degrades image contrast and hampers the ability to resolve structures, such as small lesions, against the background. Because of this, speckle reduction techniques are extremely important in breast ultrasound as they aid in delineating lesion boundaries and improving accuracy in detecting and measuring breast cancer volumes (Su 2010).

Speckle reduction is achieved primarily by two different approaches, compounding and real-time filtering, that are often used together to achieve the maximum amount of speckle reduction while minimizing tradeoffs with respect to image frame rates, resolution, and penetration. Speckle is not simply random noise that can be reduced effectively via simple averaging of multiple acquisitions. In body ultrasound, there can be enough motion between acquisitions to slightly alter and decorrelate the speckle patterns, making temporal compounding (averaging) of images somewhat useful in reducing speckle. However, in breast ultrasound, there tends to be less tissue movement between subsequent acquisitions and temporal compounding is limited. Engineers have thus tapped into the technology behind electronically steerable broadband transducers to exploit other approaches (Quistgaard 1997, Li 1994).

Frequency compounding uses the wide bandwidth of modern transducers and real-time digital signal processing capabilities to simultaneously generate images with separate center frequencies and combine them. Because coherence depends strongly on the center frequency, the speckle patterns become less coherent, and speckle signal is attenuated. However, with the center frequencies far apart, there is a difference in contrast, penetration, and resolution between the two images, which can be very undesirable.

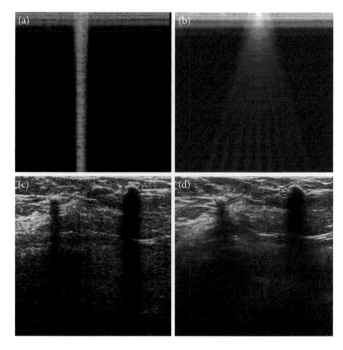

FIGURE 15.6 Spatial compounding reduces speckle by introducing decoherence via the use of multiple beams electronically steered in different directions and compounded together to form an image. Reverberations from a strong ultrasound reflector placed near the scanhead can be used to illustrate the differences in beamforming for a standard acquisition (a) versus compound imaging (b). Clinical breast ultrasound images depicting superficial fat necrosis without (c) and with (d) aggressive spatial compounding with real-time speckle reduction filtering demonstrate differences in speckle as well as differences in distal shadowing.

The most often used compounding technique for speckle reduction is spatial compounding (Figure 15.6). This technique uses the real-time electronic beam steering capabilities of modern ultrasound scanheads to form images from several different angles. Since coherence depends strongly on the scatterers in the path, this approach also attenuates speckle. Additionally, since it takes time to acquire each separate beam steered line, temporal compounding is effectively built into this approach as well. Usually, three to nine beams are incorporated into the image, and real-time filtering is incorporated to optimize the speckle reduction. While effective for enhancing the contrast of lesion borders, it is important to note that this approach (which averages beams at different angles) can also obscure critical image characteristics, such as distal enhancement or shadowing. It is important to be cognizant of the impact of a particular vendor's compounding approach on the ultrasound image when incorporating spatial compounding into breast imaging, particularly screening protocols.

15.4 Phase Aberration Correction

Focusing in ultrasound is achieved using assumptions regarding the speed of sound in tissue to estimate the needed time delays of the array elements. Phase aberrations are focusing errors caused by heterogeneous tissue velocity along the path of the beam that violate these assumptions. These focusing errors result in degradation of the quality of ultrasound images in the form of loss of signal, contrast, and resolution. Because breast tissue generally consists of a high percentage of adipose tissue, which generally has a slower speed of sound (1450 m/s) compared to that of soft tissue, phase aberrations do impact image quality in breast ultrasound. While several techniques have been proposed to quantify the speed of sound for corrections, currently many vendors employ a user selectable or fixed speed of sound correction option for imaging in fatty tissue.

15.5 Harmonic Imaging

Ultrasound imaging is based on the transmission of ultrasound pulses and the receipt of reflected sound echoes. In conventional ultrasonography, the frequency of the received echoes is usually taken to be the same as the transmitted frequency (Athanasiou 2009). However, tissue represents a nonlinear environment for

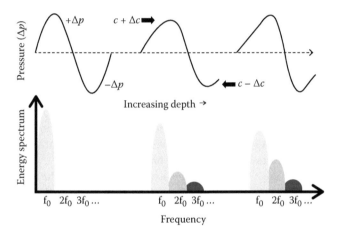

FIGURE 15.7 The ultrasound wave becomes distorted as it propagates in viscoelastic material, such as tissue. The ultrasound wave propagates at a higher speed of sound in the presence of excess pressure and at a slower speed of sound during rarefaction. More energy is transferred from the center frequency (f_0) to the harmonic frequencies, and the relative amplitude of the harmonic frequency increases with depth.

ultrasound pulse propagation, and the wave becomes increasingly distorted into a shock wave as a function of intensity, depth, and the nonlinear acoustic properties of the tissue. This results in the presence of integer multiples of the original center frequency (Figure 15.7) in echoes received from viscoelastic material, such as tissues, which are lower in amplitude than the primary frequency echo signals but can only be generated when sufficient intensity is incident on the nonlinear responding material.

Modern ultrasound scanners using broadband scanheads have the bandwidth to be able to effectively receive both frequencies, and techniques for isolating the harmonic band use real-time pulse inversion as well as novel pulse encoding, beam forming, and frequency band filtering techniques. Because tissue harmonic imaging (THI) relies on high incident intensity, THI tends to reduce reverberation and clutter from weak grating or side lobes as well as reduce the width of the primary lobe, effectively providing increased resolution. Additionally, fluids generate weak harmonic signal with respect to tissue and, thus, THI tends to improve the detection of complicated breast cysts (Figure 15.8). Similarly, fatty tissue tends to have a stronger harmonic response than soft tissue, and the application of harmonics may render isoechoic solid lesions more conspicuous against a fatty background. In a study by Rosen and Soo (2001) 73 breast masses were studied by both conventional sonography and THI. THI was significantly preferred for lesion conspicuity, margin analysis, and overall quality by the reviewing radiologists for both solid masses and cystic lesions (Rosen 2001). Szopinski and colleagues reported that THI used as an adjunct to conventional breast sonography may improve lesion detectability and characterization (Szopinski et al. 2003).

15.6 Color and Power Doppler Ultrasound

Color Doppler imaging is a technique that can quantitatively depict the direction and the speed of blood flow by detecting the motion induced shift in frequency (Kim 2011). In contrast to color Doppler, power Doppler imaging is a much more flow sensitive technique, but without directional information

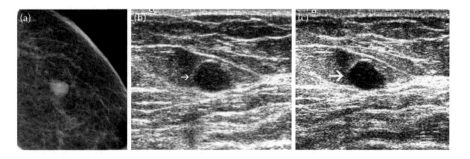

FIGURE 15.8 A 65-year-old woman presented for screening mammography. (a) Left breast craniocaudal spot compression view shows an oval mass in the outer left breast. (b) Transverse left breast ultrasound in the 2-o'clock region shows a hypoechoic mass with internal echoes and through transmission. (c) Transverse left breast ultrasound with harmonic imaging in the 2-o'clock region demonstrates a hypoechoic mass (arrow) with fewer internal echoes. The findings were consistent with an inspissated cyst.

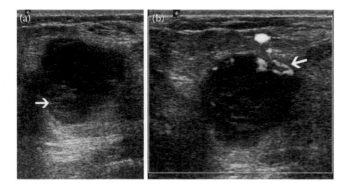

FIGURE 15.9   (See color insert.) A 46-year-old woman presented with a palpable mass in the upper outer aspect of the left breast. (a) Antiradial left breast ultrasound in the 2-o'clock region demonstrates a lobular hypoechoic mass with some internal echoes (arrow) and through transmission. (b) Antiradial left breast power Doppler ultrasound shows a peripheral vessel (arrow). Ultrasound-guided core needle biopsy revealed poorly differentiated, high nuclear grade invasive ductal carcinoma (estrogen receptor and progesterone receptor-negative).

(Kim 2011). Color and power Doppler imaging (Figure 15.9) are important tools in breast cancer imaging because neovascularization has been proven to be an essential event in the growth and development of malignant tumors (Srivastava 1989). Kook et al. (1999) noted that the typical color Doppler signs of malignancy are central, penetrating, branching, and disordered intratumoral vessels. Milz et al. (2001) demonstrated a significant increase in intratumoral blood flow as compared to the blood flow in normal breast tissue using color Doppler. Several studies have shown that color Doppler and power Doppler sonography can be helpful techniques in distinguishing between benign and malignant breast lesions (Kubek 1996, Birdwell 1997, Raza 1997). Malignant tumors tend to show color flow signals at a higher rate than benign tumors (Kubek 1996, Birdwell 1997, Raza 1997). It has also been reported that vascularity in breast tumors as noted on power or color Doppler sonography correlates with the rate of lymph node involvement (Holcombe 1995, Mehta 1997). Hypervascular breast cancers detected by color Doppler ultrasound tend to be of a higher histological grade (Holcombe 1995).

15.7   Three-Dimensional Breast Ultrasound

Three-dimensional (3-D) breast ultrasound has been developed. 3-D ultrasound is thought to add value for staging (and showing the relationship between a satellite mass and a dominant tumor), evaluating intraductal lesions, and for guiding procedures. 3-D sonography can allow for improved visualization of breast structures (Lin 2011). In addition, 3-D ultrasound can lead to decreased patient wait times and decreased examination times (as an image volume is acquired, rather than multiple individual scans) (McDonald 2011).

When performing 3-D ultrasound, the images can be obtained with manual methods or in an automated manner. With the manual method, a two-dimensional (2-D) transducer is used and images are obtained. The resultant images are a set of 2-D image frames. These frames are later optimized and processed to display a 3-D image (McDonald 2011). With automated methods, robotic assistance is used to image the breast. The data are acquired and then reconstructed in three planes (Kelly and Richwald 2011). Image processing involves enhancing the images, segmenting the regions of interest, and fusing multimodality images (McDonald 2011). The final ultrasound image can be shown in several ways including multiplanar, multislice, and shaded surface rendered displays (McDonald 2011).

15.8   Automated Breast Ultrasound

Mammography has been the gold standard for breast cancer screening. However, mammograms are summation images, with regions of overlapping breast tissue (Shin 2011). Overlapping normal tissue structures result in recalls on mammography and decreased cancer detection in patients with dense breast tissue (Shin 2011). Ultrasound can serve as a helpful supplemental screening modality in addition to mammography, especially in women with mammographically dense breasts. However, handheld ultrasound requires training and expertise. Ultrasound is an operator-dependent modality. Automated breast ultrasound (ABUS) eliminates some of the disadvantages of standard, handheld ultrasound, and ABUS can be a useful supplement to mammography. ABUS uses sound waves like a traditional ultrasound and generates 3-D images of the breast tissue.

There are three basic approaches to automating the breast ultrasound process. With the first approach, the patient lies supine and, using robotic assistance, a standard ultrasound probe is guided over the breast (Kelly and Richwald 2011). Using the second method, the patient lies supine and, using robotic assistance, a large transducer is guided over the breast (Kelly and Richwald 2011). With the third approach, the patient lies prone, and the breast is immersed in an acoustically conductive material (e.g., water). The breast tissue is then scanned with a customized transducer, and the resulting images are reconstructed (Kelly and Richwald 2011).

The main advantage of ABUS is that it can be used by personnel with less training than sonographers, it is less operator dependent than handheld ultrasound, and physicians can interpret the images efficiently on a workstation (Shin 2011). ABUS also provides ultrasound in data in the coronal plane, which can be helpful for evaluating ductal processes and the spread of tumors. In a study by Lin et al. (2011), ABUS exhibited 100% sensitivity and 95% specificity in diagnosing breast lesions. In addition, ABUS had a higher diagnostic accuracy (97.1%) in evaluating breast tumors when compared to handheld ultrasound (91.4%) (Lin 2011). Kelly and Richwald (2011) reported that ABUS improved overall breast cancer detection by a factor of two and tripled the detection rate for 1 cm or less invasive cancers in women with mammographically dense breasts when

ABUS was compared to mammography. In the study by Kelly and Richwald (2011), mammography has a significantly greater detection rate for ducal carcinoma in situ (DCIS) than ABUS.

15.9 Computer-Aided Diagnosis in Breast Ultrasound

Computer-aided detection (CADe) and computer-aided diagnosis (CADx) involve the use of computerized image data to enhance detection (in CADe) and facilitate diagnosis (in CADx) (Masala 2006). In cases in which radiologists are not confident, it is expected that the final decision can be improved by use of computer outputs (Doi 2007). With CADe and CADx, the potential gain is due to the synergistic effects of the radiologist's competence and the computer's capability (Doi 2007). Fujita et al. (2008) described a CADe system utilizing an automatic whole breast scanner. The scanner was an automated device, scanning the whole breast in sweeps. 3-D volumetric whole breast data were reconstructed from the original scans, and the data were then viewed on a workstation. Fujita et al. (2008) reported a sensitivity of 80.5% for breast mass detection with this CADe system.

Giger and colleagues utilized a linear discriminant analysis (LDA) model to differentiate between benign and malignant lesions using computer-extracted features including shape, margin, texture, and posterior acoustic attenuation (Giger 1999, Ayer 2010). Receiver operating characteristic (ROC) analysis yielded an area under the curve (AUC) of 0.94 for the entire database and 0.87 for the database that included only proven cases (Giger 1999, Ayer 2010). The authors concluded that their study showed that computerized analysis could improve the specificity of breast sonography (Giger 1999, Ayer 2010).

15.10 Contrast-Enhanced Breast Ultrasound

Some limitations of conventional B-mode color and power Doppler ultrasound can be overcome by using contrast-enhanced ultrasound (CE-US). Ultrasound contrast agents are generally microbubbles containing a gas that responds nonlinearly with the acoustic pressure from ultrasound, generating a strong harmonic signal that can be often be separated from the tissue signal. With contrast agents, it is possible to examine the parenchymal microcirculation (Albrecht 2004). Conventional color Doppler imaging along with grayscale ultrasound is sensitive only to vessels larger than about 200 mm in diameter (Santamaría et al. 2005, Yang et al. 2000, Cha et al. 2005). With the use of ultrasound contrast agents, the morphology and flow of microvessels (7–10 mm in diameter) can be shown (Liu 2008). CE-US is particularly useful in distinguishing between benign and malignant breast lesions. Caproni et al. (2010) demonstrated that the enhancement of mammary lesions on CE-US can provide parameters useful in distinguishing between malignant and benign lesions.

Wan et al. (2011) used CE-US to demonstrate that contrast enhancement originated from the peripheral region and developed centripetally in 84.62% of malignant breast lesions. Liu et al. (2008) demonstrated that the peripheral enhancement pattern was suggestive of malignancy, with a specificity of 98.3% and a PPV of 94.4%. Wan et al. (2011) showed that centrifugal and diffuse enhancement were associated with benign lesions. Malignant tumors demonstrated heterogeneous vessels. Thus, it is thought that CE-US will be a useful tool in differentiating between benign and malignant breast lesions. Contrast enhancement may help in providing more accurate measurements of breast masses. In addition, CE-US appears as a promising technique in sentinel lymph node evaluation in patients with breast cancer (Goldberg et al. 2011, Sever 2011, Yang and Goldberg 2011).

15.11 Future Developments

In the future, there will be continual advancements in breast sonography, as adjunctive techniques are improved and streamlined. These developments will result in more sophisticated images with multiple parameters being produced in a shorter time period. In addition, there will be continued efforts to make sonography less operator dependent (Berg et al. 2006). Nevertheless, those performing breast sonography must practice with excellent scanning and biopsy skills. Optimal machine and transducer settings are also critical (Stavros 1995). Those interpreting breast ultrasound examinations should be familiar with the BI-RADS lexicon and the criteria for benign and malignant lesions. Meticulous scanning and careful interpretation along with refinements in established techniques and the emergence of new techniques, such as shear wave elastography, should result in improved specificity (Berg et al. 2012) and increased reproducibility (Cosgrove 2011).

References

Abe H, Schmidt RA, Shah RN et al. 2010. MR-directed ("second-look") ultrasound examination for breast lesions detected initially on MRI: MR and sonographic findings. AJR Am J Roentgenol 194: 370–377.

Albrecht T, Blomley M, Bolondi L et al. 2004. Guidelines for the use of contrast agents in ultrasound. Ultraschall Med 25: 249–256.

American College of Radiology. 2003. BIRADS: Ultrasound, 1st ed: Breast Imaging Reporting and Data System: BIRADS Atlas. 4th ed. Reston, VA: American College of Radiology.

Athanasiou A, Tardivon A, Ollivier L et al. 2009. How to optimize breast ultrasound. Eur J Radiol 69: 6–13.

Ayer T. 2010. Computer-aided diagnostic models in breast cancer screening. Imaging Med 2: 313–323.

Berg WA, Blume JD, Cormack JB, Mendelson EB. 2006. Operator dependence of physician-performed whole-breast US: Lesion detection and characterization. Radiology 241: 355–365.

Berg WA, Blume JD, Cormack JB et al. 2008. Combined screening with ultrasound and mammography compared to mammography alone in women at elevated risk of breast cancer: Results of the first-year screen in ACRIN 6666. JAMA 299: 2151–2163.

Berg WA, Cosgrove DO, Dore CJ et al. 2012. Shear-wave elastography improves the specificity of breast US: The BE1 multinational study of 939 masses. Radiology 262: 435–449.

Birdwell RL, Ikeda DM, Jeffrey SS, Jeffrey RB. 1997. Preliminary experience with power Doppler imaging of solid breast masses. AJR Am J Roentgenol 169: 703–707.

Caproni N, Marchisio F, Pecchi A et al. 2010. Contrast-enhanced ultrasound in the characterization of breast masses: Utility of quantitative analysis in comparison with MRI. Eur Radiol 20: 1384–1395.

Cha JH, Moon WK, Cho N et al. 2005. Differentiation of benign from malignant solid breast masses: Conventional US versus spatial compound imaging. Radiology 237: 841–846.

Cosgrove DO, Berg WA, Dore CJ et al. 2012. Shear wave elastography for breast masses is highly reproducible. Eur Radiol 22: 1023–1032.

Doi K. 2007. Computer-aided diagnosis in medical imaging: Historical review, current status and future potential. Comput Med Imag Grap 31: 198–211.

Fornage BD, Atkinson EN, Nock LF, Jones PH. 2000. US with extended field of view: Phantom-tested accuracy of distance measurements. Radiology 214: 579–584.

Fujita H, Uchiyama Y, Nagakawa T et al. 2008. Computer-aided diagnosis: The emerging of three CAD systems induced by Japanese health care needs. Comput Methods Programs Biomed 92: 238–248.

Giger M, Al-Hallaq H, Huo Z et al. 1999. Computerized analysis of lesions in US images of the breast. Acad Radiol 6: 665–674.

Goldberg BB, Merton DA, Liu JB et al. 2011. Contrast-enhanced ultrasound imaging of sentinel lymph nodes after peritumoral administration of Sonazoid in a melanoma tumor animal model. J Ultrasound Med 30: 441–453.

Gumus H, Gumus M, Mills P et al. 2012. Clinically palpable breast abnormalities with normal imaging: Is clinically guided biopsy still required? Clin Radiol 67: 437–440.

Holcombe C, Pugh N, Lyons K et al. 1995. Blood flow in breast cancer and fibroadenoma estimated by colour Doppler ultrasonography. Br J Surg 82: 787–788.

Kelly KM, Richwald GA. 2011. Automated whole-breast ultrasound: Advancing the performance of breast cancer screening. Semin Ultrasound CT MRI 32: 273–280.

Kim MJ, Kim YJ, Yoon JH et al. 2011. How to find an isoechoic lesion with breast US. Radiographics 31: 663–676.

Kook S, Park HW, Lee YR et al. 1999. Evaluation of solid breast lesions with power Doppler sonography. J Clin Ultrasound 27: 231–237.

Kubek KA, Chan L, Frazier TG. 1996. Color Doppler flow as an indicator of nodal metastasis in solid masses. J Ultrasound Med 15: 835–841.

Li P-C, O'Donnell M. 1994. Elevational spatial compounding. Ultrason Imag 16: 176–189.

Lieu D. 2010. Ultrasound physics and instrumentation for pathologists. Arch Pathol Lab Med 134:1541–1556.

Lin X, Wang J, Han F et al. 2012. Analysis of eighty-one cases with breast lesions using automated breast volume scanner and comparison with handheld ultrasound. Eur J Radiol 81:873–878.

Liu H, Jiang YX, Liu JB et al. 2008. Evaluation of breast lesions with contrast-enhanced ultrasound using the microvascular imaging technique: initial observations. Breast 17: 532–539.

Masala GL. 2006. Computer aided detection on mammography. World Acad Sci Eng Technol 15: 1–6.

McDonald DN. 2011. 3-Dimensional breast ultrasonography: What have we been missing? Ultrasound Clin 6: 381–406.

Mehta TS, Raza S. 1997. Power Doppler sonography of breast cancer: Does vascularity correlate with node status or lymphatic vascular invasion? AJR Am J Roentgenol 173: 303–307.

Milz P, Lienemann A, Kessler M, Reiser M. 2001. Evaluation of breast lesions by power Doppler sonography. Eur Radiol 11: 547–554.

Quistgaard J. 1997. Signal acquisition and processing in medical diagnostics ultrasound. IEEE Signal Proc Mag 14: 67–74.

Raza S, Baum JK. 1997. Solid breast lesions: Evaluation with power Doppler US. Radiology 203: 164–168.

Rosen EL, Soo MS. 2001. Tissue harmonic imaging sonography of breast lesions: Improved margin analysis, conspicuity and image quality compared to conventional ultrasound. Clin Imag 25: 379–384.

Santamaría G, Velasco M, Farré X et al. 2005. Power Doppler sonography of invasive breast carcinoma: Does tumor vascularization contribute to prediction of axillary status? Radiology 234: 374–380.

Sever AR, Mills P, Jones SE et al. 2011. Preoperative sentinel node identification with ultrasound using microbubbles in patients with breast cancer. AJR Am J Roentgenol 196: 251–256.

Shin HJ, Kim HH, Cha JH et al. 2011. Automated ultrasound of the breast for diagnosis: Interobserver agreement on lesion detection and characterization. AJR Am J Roentgenol 197: 747–754.

Srivastava A, Hughes LE, Woodcock JP, Laidler P. 1989. Vascularity in cutaneous melanoma detected by Doppler sonography and histology: Correlation with tumour behavior. Br J Cancer 59: 89–91.

Stavros AT, Thickman D, Rapp CL, Dennis MA, Parker SH, Sisney GA. 1995. Solid breast nodules: Use of sonography to distinguish between benign and malignant lesions. Radiology 196: 123–134.

Su Y, Wang H, Wang Y et al. 2010. Speckle reduction approach for breast ultrasound image and its application to breast cancer diagnosis. Eur J Radiol 75: e136–e141.

Szopinski KT, Pajk AM, Wysocki M et al. 2003. Tissue harmonic imaging: Utility in breast sonography. J Ultrasound Med 22: 479–487.

Thompson M, Klimberg VS. 2007. Use of ultrasound in breast surgery. Surg Clin North Am 87: 469–84.

Wan C, Du J, Fang H et al. 2012. Evaluation of breast lesions by contrast enhanced ultrasound: Qualitative and quantitative analysis. Eur J Radiol 81: e444–e450.

Wang X, Xu P, Grant EG. 2011. Contrast-enhanced ultrasonographic findings of different histopathologic types of breast cancer. Acta Radiol 52: 248–255.

Weng L, Tirumalai AP et al. 1997. US extended-field-of-view imaging technology. Radiology 203: 877–880.

Woo J. 2008. A short history of the development of ultrasound in obstetrics and gynecology. Last revised March 2008, http://www.ob-ultrasound.net/history1.html.

Yang WT, Chang J, Metreweli C. 2000. Patients with breast cancer: Differences in color Doppler flow and gray-scale US features of benign and malignant axillary lymph nodes. Radiology 215: 568–573.

Yang WT, Goldberg BB. 2011. Microbubble contrast-enhanced ultrasound for sentinel lymph node detection: Ready for prime time? AJR Am J Roentgenol 196: 249–250.

16

Breast MRI

Deanna L. Lane
*The University of Texas MD
Anderson Cancer Center*

R. Jason Stafford
*The University of Texas MD
Anderson Cancer Center*

Gary J. Whitman
*The University of Texas MD
Anderson Cancer Center*

16.1 Introduction

Mammography has been the mainstay of breast imaging since the 1980s (see Chapter 1 for fundamentals of digital mammography). Because of widespread mammographic screening, as well as improved treatments for breast cancer, the death rate from breast cancer has decreased by approximately 30% since 1990 (American Cancer Society 2009). That said, mammography is far from perfect and its limitations are well known. Cancers may not be detected on mammography, especially in women with dense breasts. The reported sensitivity of mammography ranges in the literature from 63% to 98%; in dense breasts, cancers may be obscured by surrounding dense tissue and sensitivity in these women may be as low 30% to 48% (Berg et al. 2004).

In recent years, the field of breast imaging has evolved to include newer technologies that, when used appropriately, can aid in the management of patients. Arguably one of the most important of these technologies is breast MRI. The strength of MRI is its extremely high sensitivity for detecting breast cancers, some of which would otherwise go undetected on mammography and ultrasound (Chapter 15). The sensitivity of MRI for invasive breast cancer is reported between 89% and 100% and is higher in all studies than the sensitivity of mammography (Kuhl 2007). If a breast MRI shows no abnormality, the likelihood that

an invasive breast cancer is present is extremely low. This chapter will discuss the important role that breast MRI now plays in clinical practice, as well as the challenges and controversies associated with breast MRI.

16.2 Equipment

MRI uses a strong, static magnetic field (B_0) to polarize the magnetic moments of nuclei in the body in order to generate the net magnetization needed for imaging. The amount of magnetization induced is proportional to the field strength used. Because the net magnetization induced is small, nuclei with a high natural abundance and a large magnetic moment, such as protons from water associated with soft tissue, are generally used for clinical MRI. These nuclear magnetic moments precess around the longitudinal axis of the magnetic field with a frequency (Larmor frequency) given by the product of their gyromagnetic ratio (42.58 MHz/T) and the field strength and falls in the radio-frequency (RF) range. For imaging, an oscillating magnetic field (B_1) applied at this resonance frequency is used to rotate the net magnetization vector of the polarized spin population into a plane transverse to the field. This oscillating magnetization can be measured by coils tuned to the resonance frequency since an oscillating flux through a conducting loop will induce a current in the loop (Faraday's law of induction). This is the primary source of the signal detected in MRI and is the reason why RF coils used for signal reception are designed to conform as well as possible to the anatomy of interest. Imaging is performed by systematically varying the Larmor frequency across the volume in order to encode the spatial frequency distribution. The inverse Fourier transform of this distribution is the image.

Modern clinical MRI systems used for breast imaging are generally superconducting cylindrical bore magnets operating at 1.5 T. However, 3.0 T system use is steadily rising as the increased signal-to-noise ratio (SNR) afforded by these systems may be used to increase the spatial or the temporal resolution of the acquisition as well as to facilitate higher image quality in functional acquisitions, such as diffusion-weighted imaging (DWI) or magnetic resonance spectroscopy (MRS). One of the tradeoffs in moving to 3 T for breast imaging lies in the challenge of consistently obtaining homogeneous B_0 and B_1 fields across the entire field of view for a variety of patient shapes and sizes during simultaneous bilateral breast imaging. These issues are being addressed in the latest generation of equipment via more optimal approaches to shimming the static field as well as modifying the transmit properties of the built-in body coil. Dynamic acquisitions of contrast uptake in bilateral imaging are the most important and challenging acquisitions with respect to spatiotemporal needs. A high performance gradient subsystem with a high bandwidth receiver will go a long way in facilitating shorter echo-times (TE) and pulse repetition times (TR) for faster acquisition times as well as higher spatial resolution acquisitions. Modern cylindrical bore systems tend to have gradient strengths in excess of 25 mT/m with slew rates greater than 140 mT/m/ms operating at 100% duty cycle with usable fields of view over 45 cm.

The RF coils used to receive the signal in breast imaging are designed to conform as closely as possible to the patient's anatomy while accommodating a wide variety of breast shapes and sizes. As MRI systems increase the number of independent receiver elements that can be simultaneously active (currently ≥16), breast MRI coils have increased the number of elements used in order to increase the SNR. An additional benefit of a higher number of independent receivers across the breast volume is the opportunity to optimize the design of the coil for parallel imaging across the volume. Parallel imaging is an acquisition acceleration technique that essentially reduces the number of views needed to reconstruct the image by undersampling k-space. The spatial information inherently embedded in the receive coils is used to remove the aliasing artifact that appears in the phase-encoding direction from this technique. It is worth noting that given that some suspicious lesions visualized using MRI cannot be seen on mammography or ultrasound to guide a biopsy needle to the site of interest, the RF coils must also be designed to accommodate access to the breast in order to perform percutaneous biopsy procedures with MRI guidance (see Chapter 8). While most RF coils for diagnostic breast imaging support biopsy, as the channel density is increased in the evolution of diagnostic coils, one might expect a divergence between coils used for diagnosis versus those used for biopsy, where a lower channel density coil design, which provides better access to the sites of interest in the breast could be more optimal.

16.3 Technique and Image Acquisition

Patients lie prone with their chest supported by the dedicated breast coil and the breasts and tissue from the axillary portion of the chest brought into the active area of the phased array coils. The patient's arms are placed above them, out of the imaging field of view. It should be noted that while every effort is made to minimize discomfort, this positioning can only be tolerated for a limited time for many patients and therefore the procedure time is a crucial element in breast MRI. Since the primary diagnosis is made from contrast-enhanced imaging, it is important to minimize the amount of time the patient must stay on the table before and during contrast injection.

The breast imaging examination typically consists of a fast acquisition used to localize the anatomy for prescribing subsequent acquisitions. Typically, bilateral examinations are performed unless the patient is status post-mastectomy or imaging of a specific breast is warranted. Bilateral coverage may be acquired in either the sagittal or the axial planes; however, in either case, the frequency encoding direction is kept anterior to posterior in order to minimize ghosting artifacts from heart motion. The decision for a particular orientation is not trivial as the optimal sequence parameters and system hardware performance greatly impact the achievable spatiotemporal resolution and the image quality that can be obtained. Typically T2-weighted fast spin-echo images are acquired prior to contrast administration. Cystic masses appear bright on T2-weighted images, and T2-weighted images can be useful in characterizing lesions seen on the dynamic contrast-enhanced images. These acquisitions tend to use the same imaging plane as the dynamic contrast enhancement acquisition. Adipose tissue also appears bright on T2-weighted fast spin-echo images. Because of the substantial adipose tissue in the breast, the signals from lipid protons are suppressed. Often, this is performed using a chemical shift selective saturation (CHESS) technique, which applies an RF pulse centered on lipid (−3.4 ppm from water) and spoils the signal prior to the normal acquisition. However, the shape of the breast tissue–air interface tends to result in field inhomogeneities that result in suboptimal fat saturation, which can be exacerbated at higher fields. An alternative technique, which uses a short tau inversion recovery (STIR) time to suppress the lipid signal can be used, but typically suffers from both reduced SNR and prolonged acquisition times. Fast Dixon techniques, which use multiple echo times to modulate the phase between water and lipid magnetization and then algorithmically separate them, are more immune to field inhomogeneities and may provide an alternative to chemical saturation in the future if acquisition times can be decreased. A special case arises when there is a desire to image silicone from an implant specifically to assess leakage. Silicone is spectrally further from water than lipid (−5 ppm) and is hyperintense on T2 images. To visualize silicone, CHESS is tuned to saturate the water signal and a STIR approach is used to simultaneously null the lipid signal from adipose tissue. The remaining high-intensity signal in the breast is from the silicone from the implant.

In addition to T2-weighted imaging, T1-weighted image may be acquired. Recently, DWI has been performed to aid in increasing specificity. Tissue with abnormally high cellularity, such as cancer, tends to exhibit more restricted water diffusion, which can be characterized with DWI. DWI is usually acquired bilaterally using a spin-echo echo-planar imaging (EPI) sequence with one or more diffusion weightings (*b* values) that are typically less than 1000 s/mm². An estimate of the apparent diffusion coefficient (ADC) of the tissue water can be made using two or more *b* values. Again, fat suppression across the large field of view is extremely challenging, but important, in that lipid signal shifts substantially on EPI images and is not attenuated substantially by diffusion weighting. Because DWI EPI acquisitions are typically very fast, the time penalty of using STIR is not prohibitive and the SNR penalty can be mitigated with an increased number of signal averages (NSA).

The primary diagnostic sequence is the simultaneous bilateral MRI acquisition acquired dynamically to illustrate the pattern of contrast uptake in suspected lesions. This entails acquisition of a volume prior to administration of a gadolinium based contrast agent (typically 0.01 mmol/kg body weight), followed by dynamic acquisitions of the volume after injection to illustrate the rapid wash-in and wash-out kinetics of the lesions. Because the spatial distribution of enhancement is extremely important in evaluating lesion morphology, high spatial resolution of the acquired volume tends to take precedence over temporal resolution, which is usually on the order of 2 minutes or less per acquired volume. On current systems, volume acquisitions are typically performed using a high-resolution 3-D gradient echo volume acquisition with radiofrequency spoiling of the transverse steady state in order to emphasize the T1 weighting. The American College of Radiology's *Practice Guideline for the Performance of Contrast-Enhanced MRI of the Breast* recommends that, regardless of acquisition orientation, the slice thickness remain less than 3 mm and the in-plane resolution remain less than 1 mm in order to maximize sensitivity to small lesions. Additionally, timing between subsequent volume acquisitions for kinetic evaluation should be less than 3 minutes.

Because the lipid signal from the adipose tissue appears hyperintense, like the contrast enhancement, fat suppression techniques are usually employed. Because the quality of fat suppression is dependent on field homogeneity, but absolutely crucial to image quality, special RF pulses that incorporate both chemical shift-selective saturation as well as T1 relaxation based nulling of the lipid signal are used. Additionally, some vendors adjust their shimming protocols to accommodate two shim volumes, one over each breast, for more optimal fat suppression. Image subtraction to illustrate enhancement is also useful in removing the bright lipid signal. However, this technique is easily contaminated by patient motion during the examination and therefore is not recommended as the sole method for suppression of lipid signal from the images.

In order to provide coverage over the large volume with these exacting spatiotemporal resolution and fat suppression standards, parallel acquisition techniques are employed to speed up the acquisition at the expense of SNR and potential artifacts. Because of this SNR loss and the enhanced appearance of the contrast agent against background tissue, higher field systems are advantageous. Parallel acquisitions use the spatial localization of the RF receivers in order to unalias the undersampled images. MRI systems coming on the market with higher channel counts may result in much higher quality dynamic bilateral contrast enhanced imaging in the future.

16.4 Interpretation of Breast MRI

Dynamic contrast enhanced breast MRI is based on the fact that invasive breast malignancies undergo neoangiogenesis. The enhancement of a breast cancer is influenced by its level of perfusion, the density of the blood vessels within it, and the permeability of those vessels. Newly formed blood vessels within a malignancy are abnormally leaky, allowing accumulation of contrast material within the cancer. After the administration of low molecular weight gadolinium chelate contrast agents, the gadolinium circulates in the bloodstream until it leaks through the vasculature into the extracellular or interstitial space of the tumor. Wash-out typically occurs when the permeability of vessels is high or as a result of arteriovenous shunting, allowing contrast to pass quickly back into the bloodstream (Padhani and Khan 2010, O'Flynn and deSouza 2011). Gadolinium is eventually excreted by the kidneys.

The new vascular networks formed by a cancer do not function as normal blood vessels would. Their excess permeability and shunting result in the typical enhancement pattern of a breast malignancy: rapid wash-in of contrast followed by rapid wash-out. Benign breast parenchyma, on the other hand, is perfused with normal vessels and enhances more slowly and continuously. This fundamental difference in enhancement pattern means that most cancers can best be visualized in the early post-contrast phase, usually 60–120 seconds after the administration of gadolinium (Kuhl 2007). As surrounding normal tissue gradually but progressively enhances, enhanced normal tissue can obscure a rapidly washing out malignancy in the later phases. Thus, the ability to obtain images quickly after contrast administration is of substantial importance; there is a great emphasis on temporal resolution. The same image stack is acquired before and 4–5 times after contrast is given. To achieve the required temporal resolution, each acquisition should take no more than 60–120 seconds (Kuhl 2007). Dynamic imaging yields kinetic information; time–signal intensity curves can be created to aid in differential diagnosis.

Interpretation of mammography is highly dependent on morphologic assessment; analysis of a lesion's shape and margins will determine its level of suspicion. The same can be said for MRI, where morphology is no less important in suggesting the correct diagnosis. High spatial resolution allows better morphologic analysis, but can be acquired only with added time and at the expense of temporal resolution (Kuhl et al. 2005). There are, therefore, competing demands to achieve both temporal and spatial resolution, and protocols must be designed to maximize both. Recent improvements in MRI systems and advanced sequences have allowed us to achieve better temporal and spatial resolution.

The accumulation of contrast material within a cancer allows us to visualize it on T1-weighted images, especially if fat suppression is used or subtraction images are generated. A lesion's morphology is best assessed on early postcontrast images, before the lesion washes out or its margins become obscured by surrounding enhancing breast tissue. Kinetic information can be visually assessed on source or subtraction images or can be evaluated using signal intensity curves generated by placing small regions of interest over the most intensely enhancing portion of the lesion on the first postcontrast image (Kuhl 2007). Kinetic curves yield information regarding both the wash-in and the wash-out of contrast from a lesion. The first part of the kinetic curve represents the early phase (approximately the first 2 minutes after injection) and classifies the initial enhancement rate as slow, intermediate, or fast. The wash-in enhancement rate may be quantified and expressed as a percentage increase in signal intensity over baseline (Kuhl 2005). The greater the slope of the first part of the kinetic curve, the greater the degree of enhancement immediately after injection.

The second part of the kinetic curve describes the signal intensity in the later phases, and curves are typically described according to their appearance on these later phases. Two minutes after contrast injection, the signal intensity may continue to increase (persistent), stay the same (plateau), or decrease (washout) (Figure 16.1) (Kuhl 2005). In general, persistent (type 1) curves are usually seen in benign lesions, and wash-out (type 3) curves are suspicious for malignancy (Figure 16.2). Plateau (type 2) curves are indeterminate. It is critical to keep in mind,

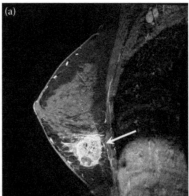

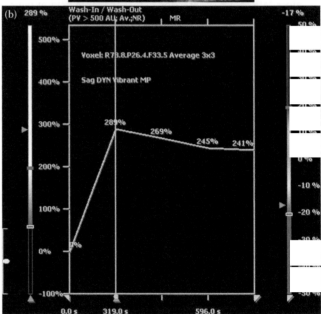

FIGURE 16.2 **(See color insert.)** (a) Sagittal postcontrast T1-weighted MRI with fat suppression demonstrates a 3.5 cm lobulated mass with irregular margins (arrow). There is heterogeneous internal enhancement. (b) Signal intensity analysis shows a typically malignant-appearing wash-out curve. Pathology demonstrated invasive ductal carcinoma.

however, that some malignancies may display a benign-appearing persistent curve (Figure 16.3), and therefore, a lesion that appears suspicious based on morphology should be biopsied regardless of its kinetic curve. On the same note, some benign lesions such as intramammary nodes may demonstrate malignant-appearing wash-out type kinetics. It is therefore imperative to use both morphologic and kinetic information in conjunction with each other and not rely on one source of information instead of the other.

Several commercially available CAD systems also aid in kinetic analysis (see Chapter 14). These systems display color in lesions that enhance above certain threshold levels. This may allow for easier detection of lesions, decreased variability among radiologists, and fewer false-positive examinations (Levman et al. 2009, Williams et al. 2007). The color that is displayed depends upon the later phase of the kinetic curve; lesions with

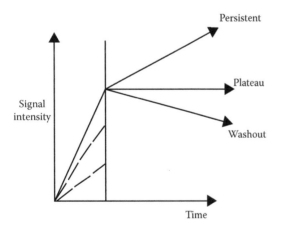

FIGURE 16.1   Signal intensity plotted over time. The initial upstroke of the kinetic curve describes contrast uptake within the first 2 minutes as slow, intermediate, or rapid. After 2 minutes or in the delayed phase, signal intensity may continue to increase (persistent curve), remain unchanged (plateau curve), or decrease (wash-out curve).

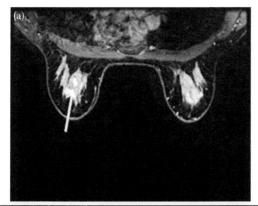

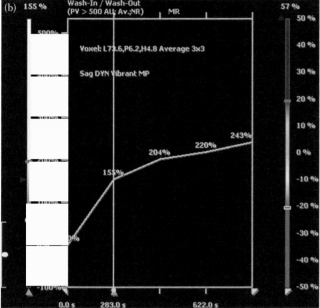

FIGURE 16.3 (See color insert.) (a) Axial T1-weighted postcontrast MRI with fat suppression demonstrates an oval mass with smooth margins and homogeneous enhancement (arrow). Gradual and progressive enhancement allowed this benign fibroadenoma to be best visualized on delayed series. (b) The signal intensity curve demonstrates persistent kinetics.

persistent enhancement will typically be displayed in blue and lesions with wash-out curves will be displayed in red.

At many institutions, clinicians' reliance on breast MRI has increased in recent years, perhaps due in part to increased radiologist comfort with interpreting these studies in clinically relevant terms. In 2003, the American College of Radiology (ACR) expanded the Breast Imaging-Reporting and Data System (BI-RADS) lexicon to include standardized descriptions of breast MRI findings; updates to further refine reporting and interpretation are expected soon (American College of Radiology 2003). It is recommended that every breast MRI report contain a statement describing the degree of background enhancement of the breast parenchyma. Increasing levels of background enhancement can decrease the sensitivity of MRI for detecting cancers. This recommendation will likely be incorporated into the new edition of the MRI BI-RADS lexicon.

According to the BI-RADS lexicon, an enhancing lesion seen on MRI must be classified as one of three things: a mass, a focus, or an area of nonmass-like enhancement. A mass is characterized as having definable margins and a ball-like 3-D structure; it may be seen on precontrast T1-weighted images. Its level of suspicion is based on morphologic analysis of size, shape, and margins, as well as its kinetics. A focus is a small spot of enhancement, typically less than 5 mm in size, which has no space occupying features and no corresponding abnormality on precontrast images. Management of a focus depends on the clinical context in which it is seen. In simplistic terms, nonmass-like enhancement is anything that cannot be accurately characterized as a mass or a focus. Nonmass-like enhancement may occur over small or large areas and often has normally enhancing breast tissue or fat interspersed among areas of abnormal enhancement. No abnormality is typically seen on precontrast images (American College of Radiology 2003).

Areas of nonmass-like enhancement are further characterized based on the pattern of enhancement within them and their distribution. An example of nonmass-like enhancement is the clumped or stippled enhancement pattern, which can be seen in ductal carcinoma in situ (DCIS) (American College of Radiology 2003). It is important to note that kinetic information may not be reliable in areas of nonmass-like enhancement and should be used with caution, if at all. Both DCIS and lobular carcinomas can present on MRI as areas of nonmass-like enhancement; both these entities may not exhibit the increased levels of angiogenesis needed to produce typically malignant appearing kinetics (Kuhl 2007). Bilaterally symmetric areas of nonmass-like enhancement may be benign.

Invasive cancers can present as masses or nonmass-like areas of enhancement. In a study by Bartella et al. (2006), malignant masses usually had irregular margins and heterogeneous enhancement; interestingly, the number that displayed plateau type kinetics was equal to the number that had wash-out kinetics. As stated earlier, even a persistent curve cannot exclude malignancy in the presence of suspicious morphologic features. In the study by Bartella et al. (2006), the most common MRI features of invasive cancer presenting as nonmass-like enhancement were focal clumped enhancement and plateau kinetics. Seventy-three percent of invasive lobular cancers presented as an area of nonmass-like enhancement rather than as a mass. Although increased signal on the T2-weighted images is typically thought of as a benign feature, some cancers may have high signal on the T2-weighted images (Le-Petross and Lane 2011).

Approximately 20% of mammographically detected breast cancers are DCIS (Ernster et al. 2002). The sensitivity of mammography for detecting DCIS has been reported at 86%, which is higher than its reported sensitivity for invasive cancers (Ernster et al. 2002). While the large majority of DCIS manifests as calcifications on mammography, some noncalcified DCIS remains occult on conventional imaging. Although the high sensitivity of MRI for invasive cancers was established, early studies regarding MRI sensitivity for detecting DCIS reported variable results

(Menell 2005). The literature suggested that the MRI sensitivity for DCIS was not adequate and possibly lower than the sensitivity of mammography for DCIS. More recently, however, higher MRI sensitivities for DCIS have been reported, which are similar to or greater than those of mammography. This increase is likely multifactorial, but could be due in part to improved imaging techniques, more experienced radiologists, and a better understanding of the MRI features of DCIS. The sensitivity of MRI for DCIS reported in the recent literature now approaches its sensitivity for invasive cancers and is therefore extremely high (Warner et al. 2011, Kuhl et al. 2007).

MRI can detect both calcified and noncalcified DCIS. DCIS most commonly presents as clumped nonmass-like enhancement, with a ductal or a segmental distribution. As stated earlier, kinetic evaluation of DCIS is not reliable and should be interpreted with caution; all three types of kinetic curves may be seen in DCIS. Again, a persistent curve is not proof that a lesion is benign.

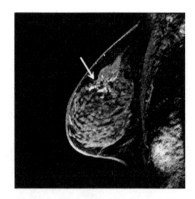

FIGURE 16.5 Sagittal postcontrast T1-weighted MRI with fat suppression reveals a focal area of nonmass-like heterogeneous enhancement (arrow). Pathology demonstrated DCIS.

Although MRI boasts extremely high sensitivity, the sensitivity of any imaging test is less than 100%; MRI is no exception. It is therefore critical to interpret each breast MRI while correlating with mammographic and ultrasound findings. The management of any lesion should be based on its worst feature, regardless of which imaging modality it appears on. Biopsy of a suspicious mammographic finding should not be deferred due to a negative breast MRI. Breast MRI should be used in conjunction with, and not as a replacement for, conventional imaging.

Figures 16.4 and 16.5 are included to serve as examples of typical MRI appearances of invasive cancer and DCIS. A complete discussion of the interpretation of breast MRI is beyond the scope of this chapter. While this section has presented principles and general rules of interpretation, it must be remembered that there are exceptions to the rules. This information is provided as a general overview, and other sources should be consulted if a more comprehensive review of interpretation is needed by the reader.

16.5 Indications for Breast MRI

16.5.1 High-Risk Screening

In 2007, the American Cancer Society issued guidelines (Saslow et al. 2007) concerning screening high-risk women with breast MRI. Annual screening MRI is now recommended for the following groups: patients with a known BRCA mutation, patients who are untested for the BRCA mutation but have a BRCA positive first degree relative, patients who have a history of radiation to the chest between the ages of 10 and 30 years old, patients with Li-Fraumeni syndrome and their first-degree relatives and patients with Cowden and Bannayan-Riley-Ruvalcaba syndrome and their first-degree relatives. Patients who have a lifetime risk of greater than 20%–25% for the development of breast cancer are also recommended to undergo annual screening MRI; this lifetime risk is generally calculated using risk models that depend largely on family history. It is important to note that screening MRI does not replace routine screening mammography, but rather should be performed in conjunction with mammography.

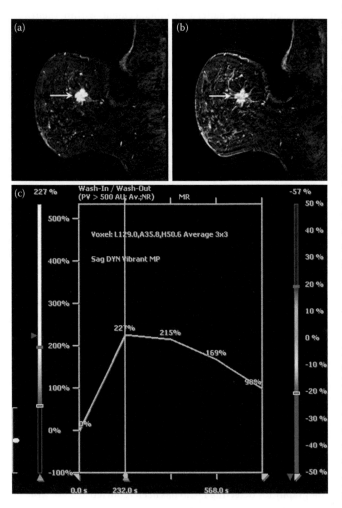

FIGURE 16.4 (See color insert.) Early (a) and late (b) sagittal subtraction images demonstrate a lobulated mass with spiculated margins (arrow). The enhancement intensity is greater in the early phase (a) than in the later phase (b), and wash-out of contrast is confirmed with the signal intensity curve (c). Pathology showed invasive ductal carcinoma with associated DCIS.

There is no consensus regarding the timing of mammographic and MRI screening; at our institution, mammography and MRI are typically alternated at 6-month intervals.

Patients who have a calculated lifetime risk of less than 15% for developing breast cancer are recommended not to undergo screening breast MRI. The guidelines for women calculated to have a lifetime risk of 15%–20% are less clearly defined. Patients in this group do not qualify for screening breast MRI, but may have clinical factors that increase their risk of breast cancer. Women with a history of atypical ductal hyperplasia, atypical lobular hyperplasia, and lobular carcinoma in situ are known to be at higher risk for developing breast cancer (Dupont and Page 1985, Page et al. 1991, Arpino et al. 2005). Women who have dense breasts are at a disadvantage not only because breast cancers can be undetectable on mammography due to obscuring dense tissue, but also because dense breast tissue has been reported as a risk factor in and of itself for the development of breast cancer (Harvey and Bovbjerg 2004).

Even women with a personal history of breast cancer may not meet the criteria to qualify for screening breast MRI according to the current guidelines. While it seems intuitive that these groups of patients at increased risk might benefit from additional screening in addition to routine mammography, the screening algorithm for these patients is a gray area, and at this time, there is insufficient evidence to recommend either for or against MRI screening in these groups of patients. There has been a suggestion that screening breast ultrasound may be of use in adjunctive screening along with mammography in these groups of patients (Berg et al. 2008).

16.5.2 Preoperative Evaluation of Newly Diagnosed Breast Cancer Patients

Although one of the most commonly cited indications for breast MRI, its use for local staging and preoperative evaluation of extent of disease in newly diagnosed breast cancer patients is also perhaps the most controversial (see also Chapter 8).

We have known for years that breast cancer patients can harbor more than a single malignant lesion. In 1985, Holland et al. (1985) published their results after examining mastectomy specimens of breast cancer patients. A minority of patients (37%) had a single malignant lesion; the remaining 63% were found to have additional sites of disease within the same breast. While 20% of additional foci were present within 2 cm of the index lesion, 43% were found more than 2 cm from the primary cancer.

The sensitivity of mammography for the detection of breast cancer ranges from 63% to 98% and is known to be as low as 30%–48% in dense breasts (Berg et al. 2004). Breast MRI has been shown to have a high sensitivity for detecting malignant lesions, including foci of cancer that are not evident on clinical examination or seen on conventional imaging. Liberman et al. (2003) retrospectively reviewed 70 patients with unilateral breast cancer who were considered candidates for breast conservation and who had preoperative MRI of the breast. Liberman et al. found that in 27% of patients, MRI detected additional unsuspected foci of

malignancy in the ipsilateral breast, separate from the primary cancer and occult on mammography and clinical examination. Berg et al. (2004) published the results of a prospective trial that compared the accuracy of clinical examination, mammography, ultrasound, and MRI in the preoperative evaluation of breast cancer extent. The limitation of mammography is highlighted in the fact that in the series reported by Berg et al. (2004), the sensitivity of mammography in heterogeneously dense breasts was 70%, and only 45% in extremely dense breasts. The combination of clinical examination, mammography, and MRI was the most sensitive for cancer detection, depicting 176 of 177 (99.4%) malignant foci. The performance of combined clinical examination, mammography, and MRI was significantly better than the combination of clinical examination, mammography, and ultrasound, which detected 93.2% of cancerous lesions. After clinical examination and mammography, MRI detected additional tumors that expanded appropriate surgery in 30% of breasts, and MRI performed after clinical examination, mammography, and ultrasound still showed additional tumors in 12% of breasts. As summarized in the meta-analysis published by Houssami et al. (2008), otherwise occult multifocal or multicentric foci of cancer have been detected in the ipsilateral breast by breast MRI only in 16% of women with breast cancer.

MRI detection of otherwise occult multifocal or multicentric foci of malignancy has been shown to affect surgical treatment. Bedrosian et al. (2003) reported that MRI findings prompted more extensive surgery in 26% of patients; 16.5% of patients were converted from planned breast conservation to mastectomy. Malignancy was confirmed by surgical pathology in 71% of the patients whose surgical management was changed by MRI; however, it is important to note that the remaining 29% were false-positive MRI findings and were proven benign after surgery. In the more recent meta-analysis by Houssami et al. (2008), the positive predictive value of MRI was 66%, meaning that a third of suspicious MRI findings were actually benign on final pathology. This illustrates the importance of performing percutaneous ultrasound-guided biopsy or MRI-guided biopsy to obtain proof of malignancy prior to extending surgery based on MRI findings.

In addition to evaluating the extent of disease in the ipsilateral breast, preoperative breast MRI performed in the newly diagnosed breast cancer patient offers opportunity to screen the contralateral breast for an occult cancer. In the meta-analysis reported by Brennan et al. (2009), a synchronous contralateral malignancy was diagnosed with MRI only in 4% of newly diagnosed breast cancer patients. The malignancies found in the contralateral breast were predominantly early stage tumors. It may be clinically significant to detect these occult contralateral cancers and allow treatment to occur for both breasts at the same time.

It is well validated that MRI detects more malignant lesions than conventional imaging such as mammography and ultrasound. The National Comprehensive Cancer Network guidelines suggest that MRI be considered for patients with a new diagnosis of breast cancer to evaluate the extent of disease and to screen the contralateral breast (Lehman et al. 2009). The benefits of preoperative breast MRI may include detection of mammographically occult disease which otherwise would be left intact within the breast and possibly put the patient at an increased rate of recurrence, better determination of extent of disease, and therefore improved surgical planning, reduced reoperation rates due to positive margins, and detection of unsuspected synchronous contralateral breast cancer.

16.5.3 Assessment for Residual Disease in Postlumpectomy Patients

Adequate surgical therapy is achieved only when pathologic review of the excised specimen reveals negative margins, or a rim of benign tissue, surrounding the malignancy. Although percutaneous biopsy is preferred for the initial diagnosis of breast cancer, some patients undergo surgical excisional biopsy for diagnosis. The literature regarding the use of MRI to evaluate location and extent of residual disease in this setting is limited. Lee et al. (2004) reviewed 82 breasts with proven malignancy; all patients had undergone prior excision with positive or close margins. Final surgical pathology revealed residual disease in 59.8% of breasts. MRI detected only 61.2% of the cases with residual disease, suggesting that the sensitivity of MRI in the postlumpectomy setting may be more limited. The probability that a suspicious lesion identified on MRI was indeed a cancer (positive predictive value) was only 33.3%. While MRI resulted in a change in the planned clinical management of these patients in 29% of cases and may be useful in some patients before reexcision, the overlap in the imaging appearance of benign postoperative change and residual malignant disease makes MRI interpretation in these patients challenging. The quoted low positive predictive value demonstrates the importance of performing percutaneous biopsy of any suspicious lesion identified on MRI to confirm its malignant nature prior to allowing its presence to alter surgical management. While the specificity of MRI may be increased by performing MRI after several weeks have passed since surgery, this is not feasible in clinical practice and the resultant delay in treatment is not acceptable to most patients or their surgeons.

16.5.4 Treatment Response Assessment

Preoperative or neoadjuvant chemotherapy offers advantages over traditional postoperative chemotherapy. Large tumors may respond well to preoperative chemotherapy and significantly decrease in size, allowing a woman to have breast conservation when otherwise her only surgical option would be mastectomy. Neoadjuvant treatment allows in vivo assessment of the effectiveness of the selected chemotherapy, as a patient can be serially imaged throughout their treatment to determine response or progression; their treatment may be altered accordingly if needed (Moreno-Aspitia 2011). A theoretical advantage of neoadjuvant chemotherapy is immediate systemic treatment of undetectable micrometastatic disease.

It has been shown that preoperative chemotherapy can be safely used instead of traditional adjuvant treatment. The National Surgical Adjuvant Breast and Bowel Project Protocol B18 trial compared the survival benefits of neoadjuvant versus adjuvant chemotherapy and found no significant survival benefit of one over the other (Wolmark et al. 2001, Rastogi et al. 2008). Importantly, a significant correlation between primary tumor response to neoadjuvant chemotherapy and long-term outcome was seen. Women who achieved a clinical response to chemotherapy had better disease-free and overall survival rates than those who had only partial response or no response to treatment. Overall survival at 9 years was 78% in those with a complete clinical response, 67% in those with a partial clinical response, and 65% in nonresponders (Wolmark et al. 2001). The outcome for women who had a complete pathologic response was better than those who had a complete clinical response but had residual invasive cancer found on pathology. Overall survival at 9 years in patients who had a complete pathologic response was 85%, compared with 73% in patients who had complete clinical response but residual invasive tumor at pathologic examination (Wolmark et al. 2001). Monitoring tumor response to chemotherapy therefore gives us information on clinical outcomes and prognosis; this is a significant advantage of neoadjuvant chemotherapy.

The majority of women who receive neoadjuvant chemotherapy will not achieve a pathologic complete response. In the B-18 trial, a pathologic complete response was documented in 13% of patients (Rastogi et al. 2008). The ability to identify women who are not responding to treatment gives clinicians an opportunity to change treatment plans, in the hopes of achieving response and optimizing the long-term outcomes for their patients.

Conventional assessment of response with mammography or ultrasound monitors changes in size and morphology; however, these changes may be seen only later in the course of treatment and may not be accurate. MRI has been shown to be more accurate than mammography or ultrasound for determining tumor size (Davis et al. 1996), but all imaging modalities may either underestimate or overestimate tumor size (Yang et al. 1997). Internal central necrosis may cause overestimation of the size of a responding tumor. Some infiltrative malignancies are difficult to accurately measure on imaging, some measurements are difficult to reproduce from one study to the next, and some imaging abnormalities may persist after treatment—it may be impossible to distinguish residual tumor from surrounding posttreatment fibrosis in the tumor bed on conventional imaging.

Functional imaging with PET (Chapter 17) or breast MRI may help us with these challenges. Changes in the physiology of breast cancers may occur when chemotherapy begins and become detectable before a reduction in tumor size is seen. Dynamic contrast-enhanced breast MRI allows not only assessment of size and morphology but also functional analysis of contrast uptake and wash-out within breast malignancies. As explained earlier, dynamic contrast-enhanced breast MRI is based on the neovascularity of breast malignancies. The level of perfusion, the density of blood vessels within a tumor, and the permeability of those vessels are all factors that affect the enhancement of a cancer. Newly formed blood vessels within a malignancy allow contrast to leak through and accumulate within the cancer. The high permeability of the vessels or arteriovenous shunting results in rapid wash-out of the contrast back into the bloodstream (Padhani and Khan 2010, O'Flynn and deSouza 2011). Based on these principles, MRI is able to yield insight into tumor perfusion and permeability. With the initiation of chemotherapy, changes in the vascular status of a malignancy may precede a reduction in its size; therefore, MRI evaluation of a tumor's vascular status may allow a much earlier assessment of response to treatment than conventional imaging. Successful chemotherapeutic treatment of a malignancy could result in a change toward normalization of blood flow, ultimately causing a decreased rate and magnitude of enhancement within the cancer.

In a study by Martincich et al. (2004), MRI was performed before initiation of chemotherapy, after two cycles of chemotherapy, and at the completion of treatment prior to surgery. The effects of chemotherapy on parameters including early contrast uptake were evaluated. The authors found that a >65% reduction in tumor volume and a decrease in the early enhancement ratio after two cycles of chemotherapy were associated with a major histopathologic response to chemotherapy and had a 93% accuracy for identifying those tumors that would undergo a complete pathologic response.

Kinetic parameters relating to tumor perfusion and vascular permeability have been described in other studies in the literature as well (Arlinghaus et al. 2010). For example, K^{trans} is a measure of transendothelial diffusion of contrast from the vascular space to the tumor interstitium and gives a measure of vascular permeability and blood flow within a breast cancer (Ah-See et al. 2008). Ah-See et al. (2008) studied 28 breast cancer patients who underwent MRI before chemotherapy and after two cycles. Ah-See et al. (2008) calculated quantitative kinetic parameters including K^{trans}, maximum gadolinium concentration, relative blood volume, and relative blood flow. Tumors that responded to chemotherapy showed reductions in these kinetic parameter values. While reductions in all these values after two cycles of treatment significantly correlated with clinical and pathologic response to chemotherapy, K^{trans} was the best predictor of response, and interestingly, reduction in tumor size failed to correlate with pathologic response. This study suggests that patients who will not ultimately respond to chemotherapy may be able to be identified with MRI after only two cycles of chemotherapy.

A study of 24 patients by Johansen et al. (2009) suggested that the relative signal intensity (RSI), another kinetic parameter, can be used to assess tumor response after only one course of chemotherapy. Patients who were responding to chemotherapy had decreased mean RSI values after initiation of treatment; no significant change in mean RSI was seen in nonresponders. Importantly, this study suggested that even

kinetic information gleaned from a baseline pretreatment MRI can have prognostic implications. Mean RSI values before chemotherapy were significantly lower in 5-year survivors compared with nonsurvivors, implying that analysis of enhancement characteristics may allow MRI prediction of survival in some patients even prior to the start of chemotherapy. Other studies have also concluded that MRI evidence of high perfusion and increased vascular permeability before the initiation of chemotherapy may portend a worse prognosis, with lower disease-free and overall survival rates (Pickles et al. 2009, Bone et al. 2003).

Most recently, Li et al. (2011) concluded that MRI-derived kinetic parameters could predict long-term outcomes after treatment with chemotherapy. MRI was performed in 62 women before initiation of chemotherapy and after two cycles, and the authors investigated kinetic parameters including K^{trans} and the initial area under the gadolinium concentration–time curve over 60 seconds ($IAUGC_{60}$). They found that higher posttreatment K^{trans} and larger $IAUGC_{60}$ were predictors of worse disease-free and overall survival; patients whose tumors remained highly vascular after two cycles of chemotherapy may be at higher risk of recurrence or mortality (Li et al. 2011).

Studies such as these suggest that there may be an increasing role for MRI in the evaluation of breast cancer patients prior to the initiation of chemotherapy and during treatment. The ability of MRI to assess physiological changes in addition to morphologic changes makes it an exciting tool for a future that may bring more molecular or functionally based treatments. The ability of MRI to identify nonresponding patients early in their treatment allows clinicians to optimize their therapy, and its potential ability to predict clinically important outcomes such as disease-free and overall survival could impart important prognostic information to clinicians and their patients.

16.5.5 Paget's Disease

Paget's disease presents as nipple changes, which can include pruritis, scaling, or erythema. This presentation accounts for 0.5%–4.3% (Morrogh et al. 2008) of breast cancers and is often associated with an underlying in situ or invasive breast cancer. Initial workup includes diagnostic mammography, which often does not reveal any abnormality. Treatment of Paget's disease may include either mastectomy or central segmentectomy with resection of the involved nipple–areolar complex and subsequent whole breast irradiation. Breast MRI may have a role in these patients, as it may identify mammographically occult disease and facilitate surgical and medical treatment planning (Figure 16.6).

16.5.6 Patients with Unknown Primary

Patients rarely present with their only clinical finding being a metastatic axillary lymph node; conventional imaging fails to demonstrate the primary malignancy. While this presentation accounts for less than 1% of breast cancers (DeMartini and

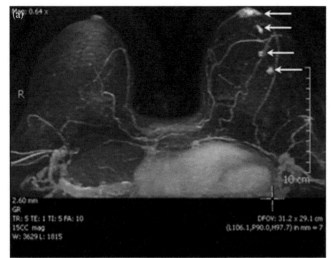

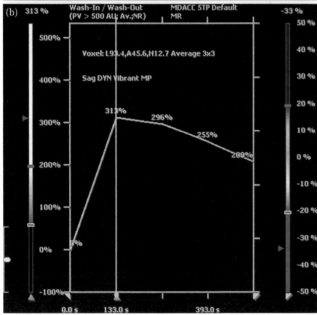

FIGURE 16.6 (See color insert.) (a) Maximum intensity projection image demonstrates enhancement of the left nipple (anterior arrow), asymmetric from the opposite breast. There are three additional irregular enhancing masses (arrows) more posteriorly within the breast. (b) The signal intensity curve demonstrates rapid wash-in and wash-out kinetics, typical of malignancy. Pathology demonstrated Paget's disease of the nipple with underlying DCIS.

Lehman 2008), it poses a management dilemma. If the primary malignancy is not identified, these patients will undergo mastectomy or whole breast irradiation. However, if the primary source can be identified, more accurate therapeutic and surgical planning can occur, and if the patient is an appropriate candidate, the patient may undergo breast conservation. Breast MRI has an important role in this situation, as it can successfully identify the primary malignancy in a large percentage of patients.

16.5.7 Problem Solving for Concerns on Clinical Examination, Mammography, or Ultrasound

Occasionally, breast MRI is used to address concerns generated by clinical examination or indeterminate findings on mammography or ultrasound. While the work-up of clinically suspicious nipple discharge must include mammography and galactography, in the absence of an explanation for the clinical findings, MRI may prove helpful for identifying (or excluding) a pathologic lesion and guiding biopsy, if needed (Van Goethem et al. 2009, Orel et al. 2000).

In most patients who present with palpable abnormalities, evaluation with diagnostic mammography and ultrasound is sufficient. However, if imaging shows no correlate for the palpable area, the level of clinical suspicion must be considered. If the palpable area remains suspicious clinically, then it should be addressed further and breast MRI plays a role in these patients. MRI may show a suspicious correlate and allow a means for biopsy (Figure 16.7). In light of the extremely high negative predictive value of breast MRI, a negative MRI in this context may allow us to dismiss the palpable area as benign. However, we must keep in mind that false-negative MRI examinations do exist. A palpable lesion that remains clinically suspicious must be biopsied even if MRI is negative.

Patients who present with nipple or skin retraction may also benefit from MRI if no cause can be determined on conventional imaging (Figure 16.8).

The use of MRI as a problem-solving tool when there is an equivocal finding on mammography or ultrasound has been suggested. There are limited data regarding this indication. Care must be taken to ensure that a negative MRI in this setting does not dissuade us from performing a biopsy of a suspicious

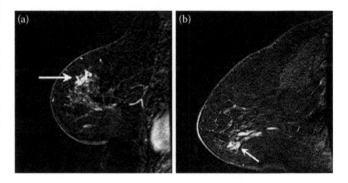

FIGURE 16.7 A 69-year-old woman presented with a palpable abnormality in the 12-o'clock region of the left breast. Mammography and ultrasound showed no correlates for the palpable abnormality. (a) Sagittal early subtraction MRI shows a focal area of nonmass-like enhancement with internal clumped enhancement (arrow) in the 12-o'clock position of the left breast, which correlated with the palpable abnormality. Pathology from the left breast showed DCIS. (b) Sagittal subtraction MRI of the right breast in the same patient demonstrates an area of nonmass-like enhancement with a linear distribution (arrow). Pathology from the right breast revealed DCIS.

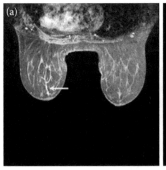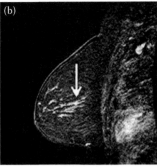

FIGURE 16.8 A 69-year-old woman presented with nipple inversion. No definite cause was identified on mammography or ultrasound. Delayed axial postcontrast T1-weighted MRI with fat suppression (a) and sagittal subtraction image (b) demonstrate nonmass-like enhancement with a segmental distribution (arrow). Pathology revealed DCIS.

mammographic or ultrasound finding (DeMartini and Lehman 2008).

16.5.8 Implant Imaging

MRI can be extremely helpful in evaluating the integrity of silicone implants, with a sensitivity of 74%–100% and a specificity of 63%–100% (Middleton 2000, Berg 2002, Venkataraman 2011). Silicone gel within an implant is homogeneously high in signal intensity on T2-weighted images and low in signal intensity on T1-weighted images, with a surrounding low signal envelope and fibrous capsule on all sequences (Berg 2002). Silicone-sensitive inversion recovery sequences or imaging with a modified 3-point Dixon technique may be used to detect high-signal silicone when evaluating for implant rupture. The approximate resonance frequency of silicone is 100 Hz lower than that of fat and 320 Hz lower than that of water (Venkataraman 2011).

Capsular contraction is the most common implant complication. Capsular contraction is more common in silicone implants compared to saline implants, secondary to foreign body reactions. In general, capsular contraction is a clinical diagnosis, and it is often not appreciated on imaging (Venkataraman 2011).

When a saline implant ruptures, the body resorbs the saline, leaving a collapsed envelope and a capsule. If the capsule is heavily calcified, it may not completely collapse. Regarding silicone implant rupture, there are two types of rupture: intracapsular and extracapsular (Middleton 2000). Intracapsular rupture is defined as rupture of the polyurethane envelope with silicone contained within the surrounding fibrous capsule. The most sensitive and specific test for intracapsular rupture is MRI. On MRI, the ruptured envelope is noted as multiple low-signal curvilinear lines within the high-signal silicone gel on T2-weighted and silicone-sensitive sequences, referred to as the "linguini sign." There may be subtle areas of focal separation of the envelope from the fibrous capsule, forming teardrop-shaped involutions of the envelope or subcapsular lines with intervening silicone, referred to as the teardrop sign or the keyhole sign.

In intracapsular rupture, silicone is not identified beyond the low-signal fibrous capsule (Venkataraman 2011).

Extracapsular rupture may be diagnosed with mammography, ultrasound, or MRI. In extracapsular rupture, free silicone is identified outside the implant margins. MRI is useful in evaluating the extent of mammographically occult extracapsular rupture, and MRI can delineate the extent of free silicone (Venkataraman 2011). Extracapsular rupture may be seen as globular masses or linear collections of high-signal intensity free silicone, separate from the implant in the breast tissue or in the axilla on T2-weighted and/or silicone-selective sequences.

16.6 Limitations, Challenges, and Controversies Surrounding Breast MRI

16.6.1 Controversy Surrounding Staging Breast MRI

There is controversy regarding the most appropriate use of MRI; perhaps the most contentious discussion surrounds its use for preoperative staging of newly diagnosed breast cancer patients (see also Chapter 8). While it seems intuitive that detecting additional cancer is beneficial, there have been challenges to that notion in the recent literature. There is no debate that MRI detects additional disease compared to that seen on other imaging studies or detected clinically. However, we have been aware of the existence of occult multifocal and multicentric breast cancer foci for many years, long before the advent of breast MRI. Opponents argue that the detection of additional malignant lesions provided by breast MRI does not necessarily result in improved patient outcomes.

The treatment of breast cancer has evolved significantly over the years. While mastectomy used to be considered the optimal surgical treatment, the majority of breast cancer patients now undergo breast conservation, followed by whole breast irradiation. Breast cancer patients may also receive systemic treatment (chemotherapy or endocrine therapy). Multiple trials have shown that appropriately selected patients treated with breast conserving surgery and subsequent radiation have survival rates equal to those of patients treated with mastectomy (White et al. 2008). Reported 10-year local recurrence rates after breast conservation surgery are in the range of 10% or less (Morrow 2010), which is much lower than the reported incidence of occult multifocal and multicentric disease. Breast conservation has been successful even with presumed residual occult foci of disease left in the breast following surgery. The argument against preoperative staging breast MRI is based on the fact that subclinical foci of cancer have been present all along and that these foci of disease have been successfully treated with radiation treatment.

Opponents of breast MRI in the staging setting argue that there is no convincing literature regarding a positive impact that preoperative MRI may have on the already low local recurrence rates. Although one published paper by Fischer et al. (2006) demonstrated that patients who underwent preoperative MRI had a statistically significant lower recurrence rate than patients who did not undergo MRI (1.2% recurred in the MRI group versus 6.8% in the non-MRI group), the study was flawed in that patients in the MRI group were more likely to have favorable tumors (smaller, node negative, and lower grade), and chemotherapy was given more often to patients in the MRI group. Other groups have published contradictory results and have demonstrated no improvements in recurrence rates in patients who undergo preoperative MRI. Solin et al. (2008) found a recurrence rate of 3% in 215 patients who had MRI prior to breast conservation surgery and a recurrence rate of 4% in 541 patients who did not have preoperative MRI; this was not statistically different. Hwang et al. (2009) also found no significant difference in recurrence rates between women with preoperative MRI and those without preoperative MRI.

MRI may be helpful in determining the extent of disease preoperatively and therefore aid a surgeon in obtaining clear margins without multiple surgeries. Unfortunately, there are few data to support this theory. The first multicenter, randomized, controlled study regarding this issue was the COMICE trial (Turnbull et al. 2010), designed to assess the clinical efficacy of MRI in women with breast cancer who were scheduled for breast conservation surgery. After 1623 women were "triple" assessed with clinical findings, mammographic and ultrasound findings, and pathology, patients scheduled for breast conservation were randomized to either receive MRI (816 patients) or to receive no additional imaging (807 patients). The authors found that preoperative MRI did not decrease the reoperation rate; 19% of patients in both groups required another operation. Published more recently, the results of another randomized controlled study, the MONET trial (Peters et al. 2011), also showed no decrease in the reoperation rate for patients who were evaluated preoperatively with MRI. In fact, patients in that study were more likely to have positive surgical margins and therefore require further surgery if they had undergone preoperative MRI. The data from these studies contradicts the theory that preoperative breast MRI in clinical practice may improve reexcision rates.

MRI often prompts a change in surgical management, usually toward more extensive resection or mastectomy (Houssami et al. 2008, Bedrosian et al. 2003, Brennan et al. 2009). Even though breast conservation is well-established to be effective and safe, mastectomy rates have been increasing in recent years; Mayo Clinic researchers reported an increase in the mastectomy rate beginning in 2004 (Katipamula et al. 2009). In the Mayo Clinic study, the mastectomy rate was 18% higher for patients who received MRI compared with patients who did not. In the study by Berg et al. (2004), 12% of mastectomies were considered potentially excessive surgical treatment. False-positive MRI findings may cause patients anxiety; patients may opt for mastectomy instead of awaiting ultrasound-guided biopsy or MRI-guided biopsy to prove that the lesion is indeed malignant. Since some studies have shown that the routine use of preoperative MRI is associated with delays in definitive treatment (Bleicher et al. 2009), every effort should be made by facilities performing breast MRI to perform additional workup studies and biopsies

in an efficient manner. Any delay that occurs while awaiting a second look ultrasound or MRI-guided biopsy only exacerbates patient anxiety. Without evidence of improved patient outcomes, the increase in more extensive surgery and mastectomies makes the use of preoperative staging breast MRI difficult to justify.

The role of MRI in the preoperative setting needs further refinement. Although the COMICE and MONET studies showed that MRI did not decrease reoperation rates, similar results should be validated with other large, prospective trials (Morris 2010). It is possible that certain subsets of patients may benefit from preoperative MRI, such as patients with invasive lobular cancer (Figure 16.9) (McGhan et al. 2010). Participants in the COMICE trial with invasive lobular cancer were more likely to undergo reoperation (Turnbull et al. 2010). Mann et al. (2010) retrospectively found that in patients with invasive lobular cancer, 9% of patients who underwent preoperative MRI required reexcision after breast conservation, while 27% of patients who did not undergo preoperative MRI required reexcision. Patients who did not have MRI were 3.3 times more likely to undergo reexcision. Mann et al. (2010) did not observe a higher mastectomy rate in patients who had MRI, and, in fact, mastectomy as final therapy was more common in patients who initially underwent breast conservation without preoperative MRI.

Critics of breast MRI have suggested that the rate of detection of occult lesions by MRI may be magnified in the literature due to patient selection bias. Patients were chosen to receive breast MRI because evaluation with mammography and ultrasound was thought to be inadequate; MRI may detect more lesions in these groups of patients than it would in the larger population of newly diagnosed breast cancer patients. A recent article from a center that routinely employs breast MRI in the preoperative assessment of all newly diagnosed breast cancer patients addressed this concern and reported that MRI detected otherwise occult malignant lesions distinct from the primary malignancy in 12% of women (Gutierrez et al. 2011).

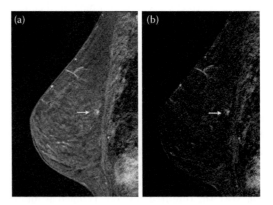

FIGURE 16.9 A 51-year-old woman presented with a new density in the upper left breast on mammography. (a) Contrast-enhanced sagittal T1-weighted MRI with fat suppression demonstrates an irregular enhancing mass (arrow). (b) Sagittal subtraction image shows the enhancing mass (arrow). Pathology revealed intermediate nuclear grade invasive lobular carcinoma.

Breast MRI is a relatively young technology, and the scientific data surrounding its use in clinical practice are limited. Most published studies have been retrospective, and we are in significant need of well-designed, prospective, multicenter trials to address many of the issues discussed. Neither the COMICE trial nor the MONET trial examined the impact of MRI on recurrence or overall survival, and these important questions must still be addressed with further studies.

16.6.2 Other Limitations and Challenges in Breast MRI

While the undisputable strength of breast MRI is its extremely high sensitivity, the specificity of MRI is lower. The reported specificities in the literature widely vary, but the specificity is thought to be in the range of 60%–70% (Weinstein and Rosen 2010). A spectrum of benign pathology can enhance after contrast administration, including fibroadenomas, papillomas, radial scars, fibrocystic change, pseudoangiomatous stromal hyperplasia (PASH), and lobular neoplasia. The reported positive predictive value of MRI suspicious findings ranges from 32% to 77% (Houssami et al. 2008, Brennan et al. 2009). False-positive findings do occur with some frequency, and it is therefore imperative that interpreting facilities have the capability to perform MRI-guided biopsy. Unnecessary modifications to surgical plans can be avoided by obtaining histology from any suspicious lesions and verifying that they do indeed represent malignancy. Biopsies must be performed in a timely manner, in order to avoid delays in definitive treatment (Bleicher et al. 2009).

The high sensitivity of MRI is often reassuring in clinical practice, but false-negative examinations do occur. A small percentage of cancers will show no enhancement after contrast administration, as documented in several studies (Shimauchi et al. 2010, Ghai et al. 2005, Schnall et al. 2006). Other malignancies will enhance, but will be unseen on MRI due to significant surrounding background enhancement (Uematsu et al. 2011). It is crucial that MRI be interpreted in conjunction with a current mammogram and ultrasound, if available, as conventional imaging will demonstrate a malignancy that is not seen on MRI in a minority of cases.

Accurate interpretation of MRI relies on the radiologist knowing the patient's clinical history and the indication for the current study, correlating with findings on conventional imaging, and correctly assessing morphologic and kinetic information; in addition, radiologists must be aware of common MRI artifacts that may affect the imaging findings (Ojeda-Fournier et al. 2007). Patient motion is problematic for breast MRI; it can cause misregistration artifacts and lead to incorrect interpretation of the subtracted images. It is important to review not only the subtraction images, but also the source data from which they are generated. It is sometimes technically challenging to obtain a good quality study, especially in patients with a larger body habitus. Some women may exceed the weight limit of the table or may not fit into the bore of the magnet and

are therefore unable to undergo MRI. Artifacts may be seen in patients with large breasts if the breast touches the coil or remains partially outside of the coil. Claustrophobia and anxiety are commonly cited reasons that patients refuse to undergo MRI (Berg et al. 2010). These concerns should be addressed and appropriately managed prior to beginning the study. Well-trained technologists are invaluable in minimizing these artifacts and obtaining high-quality studies. Technologists must be able to properly position the breast within the coil, while maximizing patient comfort and limiting motion. MRI technologists should have knowledge of good fat suppression techniques and must be able to identify an inadequate bolus of contrast or contrast extravasation after injection; these events lead to nondiagnostic studies.

While the gadolinium-chelate contrast agents have traditionally been thought of as being exceedingly safe, rare complications can occur. One in particular has become recognized recently. Nephrogenic systemic fibrosis (NSF) is a rare complication that can occur after gadolinium administration to patients with impaired renal function (Jalandhara et al. 2011). NSF causes fibrosis of the skin, which can lead to thickening and contractures. Although primarily a skin disease, visceral organs can be affected as well. If there is diaphragmatic or pulmonary involvement, respiratory failure may result. The U.S. Food and Drug Administration has issued warnings regarding the use of gadolinium-based contrast agents in patients with acute or severe kidney dysfunction, any level of renal dysfunction in patients due to hepatorenal syndrome, or in the perioperative liver transplantation period. In these groups of patients, contrast should be administered only if the information gleaned from the study is critical to patient management and cannot be obtained from any other study or without the use of contrast material. Increased awareness of this newly described disease has prompted radiology departments to screen patients for possible kidney disease; once renal disease is identified, contrast administration should be modified accordingly.

Discussion of breast MRI cannot be complete without at least a mention of its cost. In the study by Berg et al. (2010), 12% of women at high risk of breast cancer refused screening breast MRI due to financial concerns. The high cost of MRI is prohibitive for patients who do not have insurance coverage. Health insurance companies may refuse to pay for MRI studies, and radiologists must be aware of the implications that imaging recommendations have on patients. For instance, many insurers will refuse to pay for a follow-up breast MRI in 6 months. Although not always possible, every attempt should be made to define a lesion as benign or malignant and to avoid unnecessary 6-month follow-up examinations. Up to this point in time, the quality of breast MRI has varied depending on where it is performed. Different imaging parameters and protocols are used at institutions with differing levels of experience; these issues challenge radiologists who must render second opinions on breast MRI examinations performed at outside institutions. In some instances, the MRI study must be repeated due to an inadequate protocol, the need to perform additional kinetic analysis, or for biopsy planning. However, before routinely recommending a repeat MRI for all patients who present for second opinion, radiologists must keep in mind that insurers balk at the request to pay for another breast MRI within a short period. Hopefully, this issue should lessen with more standardized protocols across institutions participating in the American College of Radiology (ACR) breast MRI accreditation program.

16.7 Future Directions

16.7.1 Diffusion-Weighted MRI

The future of radiology will likely bring more physiologically based techniques, and examples are already described in the breast MRI literature. DWI may increase the specificity of breast MRI and may be another means of evaluating early tumor response to chemotherapy. DWI measures the motion of water molecules within tissues and is affected by the cellularity of a lesion and the integrity of its cell membranes. A breast cancer is typically more cellular than breast tissue or benign lesions and will therefore have restricted diffusion, manifested by increased signal on DWI and a decreased apparent diffusion coefficient (ADC).

The ADC is a quantitative measurement of diffusion that can be calculated if DWI is performed with at least two different b values. The ADC value is affected by both diffusion and perfusion factors as seen in the equation: ADC value = D + (f/b), where D is the diffusion coefficient and f is the perfusion factor (Woodhams et al. 2011). While perfusion is not a major contributing factor to the ADC value of normal breast tissue due to its low vascularity, perfusion factors may spuriously increase the ADC within a malignancy, which has undergone neoangiogenesis. Use of a higher b value minimizes the effect of perfusion. However, at the same time, as the b value increases, the signal intensity on DWI decreases. Thus, an optimal b value must be used to provide enough signal intensity to allow detection, while still minimizing the effect of perfusion as much as possible (Woodhams et al. 2011).

DWI may increase the specificity and positive predictive value of breast MRI (Ei Khouli et al. 2010, Partridge et al. 2009). Studies have reported differences in the ADC values of benign versus malignant breast lesions (Partridge et al. 2009, Guo et al. 2009, Marini et al. 2007, Woodhams et al. 2005); in general, malignancies have lower ADC values than benign lesions. Lower ADC values have been shown to be associated with more suspicious kinetics (Partridge et al. 2011), and the ADC value has been shown to be an independent predictor of benign versus malignant pathology (Partridge et al. 2011). In one study, the incorporation of DWI increased the positive predictive value (PPV) by 10% when compared to the PPV obtained using dynamic contrast-enhanced MRI alone (Partridge et al. 2009). Using an ADC cutoff defined in this particular study, biopsy could have been avoided in 33% of benign lesions without any cancers being missed (Partridge

et al. 2009). In another study, adding assessment of the ADC to the interpretation of the dynamic contrast-enhanced MRI decreased the false-positive rate from 36% to 24% (Ei Khouli et al. 2010).

DWI must be interpreted with caution. There is overlap in the ADC values of benign and malignant lesions. Cystic or necrotic malignancies and mucinous cancers may not demonstrate restricted diffusion. As many as 6%–37.5% (Partridge et al. 2009) of malignant breast lesions may not be visible on DWI. Careful correlation with the morphologic and kinetic information provided by a dynamic contrast-enhanced study should be made to avoid misinterpreting these lesions as benign. In the minority of patients who cannot receive IV gadolinium, it has been suggested that DWI could be used in lieu of dynamic contrast-enhanced breast MRI (Woodhams et al. 2011, Partridge et al. 2011).

The cellular lysis or necrosis that occurs after initiation of chemotherapy may also lead to increased diffusion (decreased signal on DWI and an increased ADC value), allowing an early determination of successful treatment (Padhani and Khan 2010, Li et al. 2011, Pickles et al. 2006). As discussed earlier, the ability to assess response to treatment early in the patient's course of therapy has important clinical implications.

16.7.2 Magnetic Resonance Spectroscopy

Magnetic resonance spectroscopy (MRS) is another emerging technique with the potential to increase the specificity of breast MRI. Like dynamic contrast-enhanced MRI and DWI, spectroscopy has shown potential for evaluating early tumor response to therapy (Tozaki et al. 2010). MRS displays information regarding the internal chemical characteristics of a lesion and, specifically, allows demonstration of a choline peak at 3.2 ppm (Weinstein and Rosen 2010). Choline levels have been shown to be significantly greater in malignant lesions than in benign lesions (Jacobs et al. 2004, Thakur et al. 2011, Dorrius et al. 2011). In a study of 57 breast lesions, all cancers demonstrated a choline peak as did three of 26 benign lesions (Bartella et al. 2006). In the study by Bartella et al. (2006), if biopsy was performed only on lesions with a choline peak, the PPV would have increased from 35% to 82%. No cancers would have missed, while sparing 58% of benign lesions a biopsy.

Fat and water resonances seen in MR spectra may also be helpful in distinguishing benign from malignant lesions. The water-to-fat (W/F) ratio in invasive ductal carcinomas has been shown to be higher than those in benign lesions or normal breast tissue (Thakur et al. 2011). Interestingly, the W/F ratio has been reported to be significantly higher in invasive ductal cancers than in invasive lobular cancers and, therefore, might be used to distinguish between these two cancer subtypes (Thakur et al. 2011). Analyzing both choline levels and W/F ratios may lead to even higher sensitivity and specificity (Thakur et al. 2011).

As of now, there is no defined protocol for clinical application of MRS. Researchers have used both single-voxel and multivoxel techniques to perform MR spectroscopy, and both qualitative and quantitative methods of analysis have been used. While some groups describe only the presence of a choline peak, others consider the height of the choline peak compared to the spectral noise level or SNR (Jacobs et al. 2004, Bartella et al. 2006). Since choline peaks have been reported in benign lesions (Bartella et al. 2006, Bolan et al. 2003) as well as in malignant lesions, there has been an attempt to define a choline concentration level that accurately distinguishes benign from malignant lesions; in one study, all benign lesions demonstrated a choline concentration of less than 1.5 mmol/L (Dorrius et al. 2011). It is important to note that these are relatively small studies, the results of which need to be validated in well-designed studies including large numbers of patients.

Although not yet routinely used in clinical practice, the potential ability of new techniques like DWI and MRS to increase the specificity and the PPV of MRI would benefit patients greatly. Although usually well-tolerated, biopsies can cause morbidity, financial burdens, and patient anxiety; any technique that decreases false-positive findings may lessen these problems. Further work regarding DWI and MRS must be done before incorporation into widespread clinical practice. Standardization of imaging parameters is needed for both techniques. There is no consensus regarding optimal *b* values to be used in the DWI, and there is no standardized technique for performing spectroscopy. There are no well-defined thresholds regarding cutoff ADC values or choline levels that should be considered suspicious for malignancy. Clinical practices will need guidance regarding how to implement these newer techniques into their already existing breast MRI protocols.

16.7.3 Possible Future Clinical Applications

Clinical applications of breast MRI will likely continue to become better defined, and new indications may emerge. Accelerated partial breast irradiation (APBI) is currently being investigated as an alternative to whole breast irradiation. Advantages of APBI include patient convenience with shorter overall treatment times and possibly decreased radiation doses to adjacent structures. As discussed earlier, occult multifocal and multicentric disease has been left intact within the breast since the onset of breast conservation; however, in most cases, multifocal and multicentric disease is successfully treated with whole breast radiation. APBI treats only the tumor bed; any occult foci away from the tumor bed may not be adequately treated and therefore leave the patient at an increased risk of recurrence. A recent study found that in patients otherwise eligible for APBI, MRI found unsuspected ipsilateral tumors in 6.2% of patients and contralateral tumors in 3.5% of patients. In 8.8% of the patients, preoperative MRI would have changed the radiation plan (Kurh et al. 2011). The high sensitivity of MRI may make it an important tool to identify otherwise occult tumor foci in patients who are being considered as candidates for APBI.

The status of the axillary lymph nodes in a newly diagnosed invasive breast cancer patient is important for staging purposes, ultimate prognosis, and for oncologic treatment decisions. In the absence of ultrasound visible metastatic lymph nodes, the patient must undergo sentinel node biopsy for evaluation of the axillary lymph nodes. As with any invasive procedure, sentinel node biopsy has postoperative complications, which can include infection, hematomas, and seromas; approximately 7% (Harnan et al. 2011) of patients may develop lymphedema. MRI has been investigated as a noninvasive technique to evaluate the axillary lymph nodes. Different MRI techniques for assessment of the axillary lymph nodes are described in the literature, including gadolinium enhanced MRI, ultrasmall superparamagnetic iron oxide (USPIO)-enhanced MRI, and MRS. A recent meta-analysis (Harnan et al. 2011) found the most promising results to come from studies investigating USPIO-enhanced MRI. After IV injection, normal lymph nodes will take up the USPIO nanoparticles and show decreased T2 signal; metastatic lymph nodes will not take up the nanoparticles and will show no change in signal (Vandermeer et al. 2007). USPIO-enhanced MRI had a sensitivity of 98% and specificity of 96%. Gadolinium enhanced MRI revealed a sensitivity of 88% and specificity of 73%, while MRS yielded the lowest sensitivity of 65% but a specificity of 100% (Harnan et al. 2011). The sensitivity of USPIO-enhanced MRI actually exceeded the sensitivity of sentinel node biopsy, although its specificity was lower than that of sentinel node biopsy. These data suggest that, with further study and development, USPIO-enhanced MRI may have clinical promise as a noninvasive method to accurately evaluate the axillary lymph nodes.

16.8 Summary

In summary, breast MRI has become an important clinical tool for the detection of breast cancer in specific patient populations. The clinical applications of breast MRI continue to be defined and in certain instances (such as preoperative staging) may be refined with further study. The majority of studies regarding MRI have been retrospective, and it is imperative that well-designed, prospective, multicenter trials be performed to examine controversial issues and emerging techniques. The effect that MRI has on important clinical outcomes such as recurrence and survival needs further study. Standardization of routine breast MRI protocols among institutions will likely be improved with the implementation of the ACR MRI accreditation process, but further research regarding the application and standardization of new imaging techniques such as DWI and MRS is needed. The functional capabilities of MRI allow not only for early evaluation of tumor response to treatment, but are also likely to become important tools for evaluating the efficacy of newly developed therapeutic agents.

Acknowledgments

We thank Dr. Malak Itani for assistance in preparing the figures and Joyce Bradley for assistance in manuscript preparation.

References

Ah-See ML, Makris A, Taylor NJ et al. 2008. Early changes in functional dynamic magnetic resonance imaging predict for pathologic response to neoadjuvant chemotherapy in primary breast cancer. Clin Cancer Res 14: 6580–6589.

American College of Radiology (ACR) 2003. ACR BI-RADS—Magnetic Resonance Imaging. In: ACR Breast Imaging Reporting and Data System, Breast Imaging Atlas. Reston, VA. American College of Radiology.

Arlinghaus LR, Li X, Levy M et al. 2010. Current and future trends in magnetic resonance imaging assessments of the response of breast tumors to neoadjuvant chemotherapy. J Oncol. doi: 10.1155/2010/919620.

Arpino G, Laucirica R, Elledge RM. 2005. Premalignant and in situ breast disease: biology and clinical implications. Ann Int Med 143: 446–457.

Bartella L, Liberman L, Morris EA, Dershaw DD. 2006. Nonpalpable mammographically occult invasive breast cancers detected by MRI. AJR Am J Roentgenol 186: 865–870.

Bartella L, Morris EA, Dershaw DD et al. 2006. Proton MR spectroscopy with choline peak as malignancy marker improves positive predictive value for breast cancer diagnosis: Preliminary study. Radiology 239: 686–692.

Bedrosian I, Mick R, Orel SG et al. 2003. Changes in the surgical management of patients with breast carcinoma based on preoperative magnetic resonance imaging. Cancer 98: 468–473.

Berg WA, Nguyen TK, Middleton MS, Soo MS, Pennello G, Brown SL. 2002. MR imaging of extracapsular silicone from breast implants: Diagnostic pitfalls. AJR Am J Roentgenol 178: 465–472.

Berg WA, Gutierrez L, NessAiver MS et al. 2004. Diagnostic accuracy of mammography, clinical examination, US, and MR imaging in preoperative assessment of breast cancer. Radiology 233: 830–849.

Berg WA, Blume JD, Cormack JB et al. 2008. Combined screening with ultrasound and mammography vs mammography alone in women at elevated risk of breast cancer. JAMA 299: 2151–2163.

Berg WA, Blume JD, Adams AM et al. 2010. Reasons women at elevated risk of breast cancer refuse breast MR imaging screening: ACRIN 6666. Radiology 254: 79–87.

Bleicher RJ, Ciocca RM, Egleston BL et al. 2009. Association of routine pretreatment magnetic resonance imaging with time to surgery, mastectomy rate, and margin status. J Am Coll Surg 209: 180–187.

Bolan PJ, Meisamy S, Baker EH et al. 2003. In vivo quantification of choline compounds in the breast with 1H MR spectroscopy. Magn Resonan Med 50: 1134–1143.

Boné B, Szabó BK, Perbeck LG, Veress B, Aspelin P. 2003. Can contrast-enhanced MR imaging predict survival in breast cancer? Acta Radiol 44: 373–378.

Brennan ME, Houssami N, Lord S et al. 2009. Magnetic resonance imaging screening of the contralateral breast in women with newly diagnosed breast cancer: Systematic review and meta-analysis of incremental cancer detection and impact on surgical management. J Clin Oncol 27: 5640–5649.

Cancer Statistics 2009 Presentation. American Cancer Society, accessed September 2, 2011, http://www.cancer.org/docroot/PRO/content/PRO_1_1_Cancer_Statistics_2009_Presentation.asp.

Davis PL, Staiger MJ, Harris KB et al. 1996. Breast cancer measurements with magnetic resonance imaging, ultrasonography, and mammography. Breast Cancer Res Treat 37: 1–9.

DeMartini W, Lehman C. 2008. A review of current evidence-based clinical applications for breast magnetic resonance imaging. Top Magn Reson Imag 19: 143–150.

Dorrius MD, Pijnappel RM, Jansen-van der Weide MC et al. 2011. Determination of choline concentration in breast lesions: quantitative multivoxel proton MR spectroscopy as a promising noninvasive assessment tool to exclude benign lesions. Radiology 259: 695–703.

Dupont WD, Page DL. 1985. Risk factors for breast cancer in women with proliferative breast disease. N Engl J Med 312: 146–151.

Ei Khouli RH, Jacobs MA, Mezban SD et al. 2010. Diffusion-weighted imaging improves the diagnostic accuracy of conventional 3.0-T breast MR imaging. Radiology 256: 64–73.

Ernster VL, Ballard-Barbash R, Barlow WE et al. 2002. Detection of ductal carcinoma in situ in women undergoing screening mammography. J Natl Cancer Inst 94: 1546–1554.

Fischer U, Baum F, Luftner-Nagel S. 2006. Preoperative MR imaging in patients with breast cancer: Preoperative staging, effects on recurrence rates, and outcome analysis. Magn Reson Imag Clin North Am 14: 351–362.

Ghai S, Muradali D, Bukhanov K, Kulkarni S. 2005. Nonenhancing breast malignancies on MRI: Sonographic and pathologic correlation. AJR Am J Roentgenol 185: 481–487.

Guo Y, Cai YQ, Cai ZL et al. 2009. Differentiation of clinically benign and malignant breast lesions using diffusion-weighted imaging. J Magn Reson Imag 16: 172–178.

Gutierrez RL, DeMartini WB, Silbergeld JJ et al. 2011. High cancer yield and positive predictive value: Outcomes at a center routinely using preoperative breast MRI for staging. AJR Am J Roentgenol 196: W93–W99.

Harnan SE, Cooper KL, Meng Y et al. 2011. Magnetic resonance for assessment of axillary lymph node status in early breast cancer: A systematic review and meta-analysis. Eur J Surg Oncol 37: 928–936.

Harvey JA, Bovbjerg VE. 2004. Quantitative assessment of mammographic breast density: Relationship with breast cancer risk. Radiology 230: 29–41.

Holland R, Veling SH, Mravunac M, Hendriks JH. 1985. Histologic multifocality of Tis, T1-2 breast carcinomas. Implications for clinical trials of breast-conserving surgery. Cancer 56: 979–990.

Houssami N, Ciatto S, Macaskill P et al. 2008. Accuracy and surgical impact of magnetic resonance imaging in breast cancer staging: Systematic review and meta-analysis in detection of multifocal and multicentric cancer. J Clin Oncol 26: 3248–3258.

Hwang N, Schiller DE, Crystal P, Maki E, McCready DR. 2009. Magnetic resonance imaging in the planning of initial lumpectomy for invasive breast carcinoma: Its effect on ipsilateral breast tumor recurrence after breast-conservation therapy. Ann Surg Oncol 16: 3000–3009.

Jacobs MA, Barker PB, Bottomley PA, Bhujwalla Z, Bluemke DA. 2004. Proton magnetic resonance spectroscopic imaging of human breast cancer: A preliminary study. J Magn Reson Imag 19: 68–75.

Jalandhara N, Arora R, Batuman V. 2011. Nephrogenic systemic fibrosis and gadolinium-containing radiological contrast agents: An update. Clin Pharmacol Therap 89: 920–923.

Johansen R, Jensen LR, Rydland J et al. 2009. Predicting survival and early clinical response to primary chemotherapy for patients with locally advanced breast cancer using DCE-MRI. J Magn Reson Imag 29: 1300–1307.

Katipamula R, Degnim AC, Hoskin T et al. 2009. Trends in mastectomy rates at the Mayo Clinic Rochester: Effect of surgical year and preoperative magnetic resonance imaging. J Clin Oncol 27: 4082–4088.

Kuhl CK. 2007. The current status of breast MR imaging. Part I. Choice of technique, image interpretation, diagnostic accuracy, and transfer to clinical practice. Radiology 244: 356–378.

Kuhl CK, Schild HH, Morakkabati N. 2005. Dynamic bilateral contrast-enhanced MR imaging of the breast: Trade-off between spatial and temporal resolution. Radiology 236: 789–800.

Kuhl CK. 2005. Dynamic Breast Magnetic Resonance Imaging. In Morris EA, Liberman L, eds. Breast MRI: Diagnosis and Intervention. pp. 79–139. New York: Springer Science + Business Media, Inc.

Kuhl CK, Schrading S, Bieling HB et al. 2007. MRI for diagnosis of pure ductal carcinoma in situ: A prospective observational study. Lancet 370: 485–492.

Kühr M, Wolfgarten M, Stölzle M et al. 2011. Potential impact of preoperative magnetic resonance imaging of the breast on patient selection for accelerated partial breast irradiation. Int J Radiat Oncol Biol Phys 81: e541–e546.

Lee JM, Orel SG, Czerniecki BJ, Solin LJ, Schnall MD. 2004. MRI before reexcision surgery in patients with breast cancer. AJR Am J Roentgenol 182: 473–480.

Lehman CD, DeMartini W, Anderson BO, Edge SB. 2009. Indications for breast MRI in the patient with newly diagnosed breast cancer. J Natl Comprehensive Cancer Netw 7: 193–201.

Le-Petross H, Lane D. 2011. Challenges and potential pitfalls in magnetic resonance imaging of more elusive breast carcinomas. Semin Ultrasound CT MR 32: 342–350.

Levman JE, Causer P, Warner E, Martel AL. 2009. Effect of the enhancement threshold on the computer-aided detection of breast cancer using MRI. Acad Radiol 16: 1064–1069.

Liberman L, Morris EA, Dershaw DD, Abramson AF, Tan LK. 2003. MR imaging of the ipsilateral breast in women with percutaneously proven breast cancer. AJR Am J Roentgenol 180: 901–91.

Li SP, Makris A, Beresford MJ et al. 2011. Use of dynamic contrast-enhanced MR imaging to predict survival in patients with primary breast cancer undergoing neoadjuvant chemotherapy. Radiology 260: 68–78.

Li XR, Cheng LQ, Liu M et al. 2011. DW-MRI ADC values can predict treatment response in patients with locally advanced breast cancer undergoing neoadjuvant chemotherapy. Med Oncol. [Epub ahead of print].

Mann RM, Loo CE, Wobbes T et al. 2010. The impact of preoperative breast MRI on the reexcision rate in invasive lobular carcinoma of the breast. Breast Cancer Res Treat 119: 415–422.

Marini C, Iacconi C, Giannelli M, Cilotti A, Moretti M, Bartolozzi C. 2007. Quantitative diffusion-weighted MR imaging in the differential diagnosis of breast lesion. Eur Radiol 17: 2646–2655.

Martincich L, Montemurro F, De Rosa G et al. 2004. Monitoring response to primary chemotherapy in breast cancer using dynamic contrast-enhanced magnetic resonance imaging. Breast Cancer Res Treat 83: 67–76.

McGhan LJ, Wasif N, Gray RJ et al. 2010. Use of preoperative magnetic resonance imaging for invasive lobular cancer: Good, better, but maybe not the best? Ann Surg Oncol 17: S255–S62.

Menell JH. 2005. Ductal Carcinoma In Situ. In Morris EA, Liberman L, eds. Breast MRI: Diagnosis and Intervention. pp. 164–172. New York: Springer Science + Business Media, Inc.

Middleton MS, McNamara MP Jr. 2000. Breast implant classification with MR imaging correlation. Radiographics 20: E1–E70.

Moreno-Aspitia A. 2011. Neoadjuvant therapy in early-stage breast cancer. Crit Rev Oncol/Hematol. May 24. [Epub ahead of print].

Morris EA. 2010. Should we dispense with preoperative breast MRI? Lancet 375: 528–530.

Morrogh M, Morris EA, Liberman L, VanZee K, Cody HS 3rd, King TA. 2008. MRI identifies otherwise occult disease in select patients with Paget disease of the nipple. J Am Coll Surg 206: 316–321.

Morrow M. 2010. Magnetic resonance imaging for screening, diagnosis, and eligibility for breast-conserving surgery: Promises and pitfalls. Surg Oncol Clin North Am 19: 475–492.

O'Flynn EAM, deSouza NM. 2011. Functional magnetic resonance: Biomarkers of response in breast cancer. Breast Cancer Res 13: 204.

Ojeda-Fournier H, Choe KA, Mahoney MC. 2007. Recognizing and interpreting artifacts and pitfalls in MR imaging of the breast. Radiographics S147–S164.

Orel SG, Dougherty CS, Reynolds C, Czerniecki BJ, Siegelman ES, Schnall MD. 2000. MR imaging in patients with nipple discharge: Initial experience. Radiology 216: 248–254.

Padhani AR, Khan AA. 2010. Diffusion-weighted (DW) and dynamic contrast-enhanced (DCE) magnetic resonance imaging (MRI) for monitoring anticancer therapy. Target Oncol 5: 39–52.

Page DL, Kidd TE Jr., Dupont WD, Simpson JF, Rogers LW. 1991. Lobular neoplasia of the breast: Higher risk for subsequent invasive cancer predicted by more extensive disease. Hum Pathol 22: 1232–1239.

Partridge SC, Rahbar H, Murthy R et al. 2011. Improved diagnostic accuracy of breast MRI through combined apparent diffusion coefficients and dynamic contrast-enhanced kinetics. Magn Reson Med 65: 1759–1767.

Partridge SC, DeMartini WB, Kurland BF, Eby PR, White SW, Lehman CD. 2009. Quantitative diffusion-weighted imaging as an adjunct to conventional breast MRI for improved positive predictive value. AJR Am J Roentgenol 193: 1716–1722.

Peters NH, van Esser S, van den Bosch MA et al. 2011. Preoperative MRI and surgical management in patients with nonpalpable breast cancer: the MONET—Randomised controlled trial. Eur J Cancer 47: 879–886.

Pickles MD, Gibbs P, Lowry M, Turnbull LW. 2006. Diffusion changes precede size reduction in neoadjuvant treatment of breast cancer. Magn Reson Imag 24: 843–847.

Pickles MD, Manton DJ, Lowry M, Turnbull LW. 2009. Prognostic value of pre-treatment DCE-MRI parameters in predicting disease free and overall survival for breast cancer patients undergoing neoadjuvant chemotherapy. Eur J Radiol 71: 498–505.

Rastogi P, Anderson SJ, Bear HD et al. 2008. Preoperative chemotherapy: Updates of National Surgical Adjuvant Breast and Bowel Project Protocols B-18 and B-27. J Clin Oncol 26: 778–785.

Saslow D, Boetes C, Burke W et al. 2007. American Cancer Society guidelines for breast screening with MRI as an adjunct to mammography. A Cancer J Clin 57: 75–89.

Schnall MD, Blume J, Bluemke DA et al. 2006. Diagnostic architectural and dynamic features at breast MR imaging: Multicenter study. Radiology 238: 42–53.

Shimauchi A, Jansen SA, Abe H, Jaskowiak N, Schmidt RA, Newstead GM. 2010. Breast cancers not detected at MRI: Review of false-negative lesions. AJR Am J Roentgenol 194: 1674–1679.

Solin LJ, Orel SG, Hwang WT, Harris EE, Schnall MD. 2008. Relationship of breast magnetic resonance imaging to outcome after breast-conservation treatment with radiation for women with early-stage invasive breast carcinoma or ductal carcinoma in situ. J Clin Oncol 26: 386–391.

Thakur SB, Brennan BS, Ishill NM et al. 2011. Diagnostic usefulness of water-to-fat ratio and choline concentration in malignant and benign breast lesions and normal breast parenchyma: An in vivo (1) H MRS study. J Magn Reson Imag 33: 855–863.

Tozaki M, Oyama Y, Fukuma E. 2010. Preliminary study of early response to neoadjuvant chemotherapy after the first cycle in breast cancer: Comparison of 1H magnetic resonance spectroscopy with diffusion magnetic resonance imaging. Jpn J Radiol 28: 101–109.

Turnbull L, Brown S, Harvey I et al. 2010. Comparative effectiveness of MRI in breast cancer (COMICE) trial: A randomized controlled trial. Lancet 375: 563–571.

Uematsu T, Kasami M, Watanabe J. 2011. Does the degree of background enhancement in breast MRI affect the detection and staging of breast cancer? Eur Radiol. doi: 10.1007/s0030-011-2175-6. [Epub ahead of print].

Vandermeer FQ, Bluemke DA. 2007. Breast MRI: State of the art. Cancer Investig 25: 384–392.

Van Goethem M, Verslegers I, Biltjes I, Hufkens G, Parizel PM. 2009. Role of MRI of the breast in the evaluation of the symptomatic patient. Curr Opin Obstet Gynocol 21: 74–79.

Venkataraman S, Hines N, Slanetz PJ. 2011. Challenges in mammography: Part 2, multimodality review of breast augmentation-imaging findings and complications. AJR Am J Roentgenol 197: W1031–W1044.

Warner E, Causer PA, Wong JW et al. 2011. Improvement in DCIS detection rates by MRI over time in a high-risk breast screening study. Breast J 17: 9–17.

Weinstein S, Rosen M. 2010. Breast MR imaging: Current indications and advanced imaging techniques. Radiol Clin North Am 48: 1013–1042.

White JR, Halberg FE, Rabinovitch R et al. 2008. American College of Radiology appropriateness criteria on conservative surgery and radiation: Stages I and II breast carcinoma. J Am Coll Radiol 5: 701–713.

Williams TC, DeMartini WB, Partridge SC, Peacock S, Lehman CD. 2007. Breast MR imaging: Computer-aided evaluation program for discriminating benign from malignant lesions. Radiology 244: 94–103.

Wolmark N, Wang J, Mamounas E, Bryant J, Fisher B. 2001. Preoperative chemotherapy in patients with operable breast cancer: Nine-year results from National Surgical Adjuvant Breast and Bowel Project B-18. J Natl Cancer Inst 96–102.

Woodhams R, Matsunaga K, Kan S et al. 2005. ADC mapping of benign and malignant breast tumors. Magn Reson Med Sci 4: 35–42.

Woodhams R, Ramadan S, Stanwell P et al. 2011. Diffusion-weighted imaging of the breast: Principles and clinical applications. Radiographics 31: 1059–1084.

Yang WT, Lam WW, Cheung H, Suen M, King WW, Metreweli C. 1997. Sonographic, magnetic resonance imaging, and mammographic assessments of preoperative size of breast cancer. J Ultrasound Med 16: 791–797.

Nuclear Medicine Imaging of the Breast: History, Clinical Application and Integration, and Future

Rachel F. Brem
The George Washington University

Caitrín Coffey
The George Washington University

Jessica Torrente
The George Washington University

Jocelyn Rapelyea
The George Washington University

17.1 Background

Breast cancer is the second leading cause of cancer death among women in the United States (American Cancer Society 2011) (Chapter 7). The mainstays of breast imaging, mammography (Chapter 1) and ultrasound (Chapter 15), rely on anatomically differentiating breast cancer from the heterogeneous pattern of normal tissue. However, overcoming the limitations of these modalities may require evaluating the function of the tissue rather than simply its appearance. Physiological breast imaging modalities hold the potential to improve both sensitivity as well as specificity of breast cancer detection. Nuclear medicine imaging of the breast, one modality that utilizes the physiological parameters of breast cancer, was developed in response to this challenge.

17.2 Nuclear Medicine Imaging of the Breast

Nuclear medicine imaging of the breast evaluates breast malignancies using a functional approach, based upon the premise that cancerous tissue can be distinguished from healthy breast parenchyma by biological changes within the cells. In functional imaging, the activity of breast tissue is assessed by administering a radiotracer isotope by injection, and imaging the radiotracer activity on a dedicated camera to reveal increased uptake in focal areas of abnormal proliferation.

Radiotracers are designed to accumulate in malignant tissues by specifically binding to mitochondria (Delmon-Moingeon et al. 1990). Increased mitochondrial density is a marker of hyperproliferation. The angioneogenesis surrounding a tumor also allows for increased delivery of a radiopharmaceutical to the lesion. An ideal radiotracer, then, preferentially accumulates at the site of a malignancy and shows minimal uptake within normal breast tissue and benign lesions. Currently, the most widely used radiotracer for breast imaging is Tc-99m sestamibi, a small cationic complex of technetium first used for cardiac imaging, which showed coincidental uptake in tumors (Schomacker and Schicha 2000). About 90% of its activity is concentrated in the mitochondria, but its uptake and retention in cancer cells also depends upon mitochondrial and plasma membrane potentials, tissue metabolism, and local angiogenesis and blood flow in addition to mitochondrial concentration (Maublant et al. 1993).

"Scintimammography" refers to the traditional use of a gamma camera and radiotracer to image the breast, before the development of high-resolution, breast-specific cameras. Scintimammography was performed with the patient lying prone with breasts pendulant and imaging performed in the lateral and anteroposterior positions. A review by Taillefer in over 2000 women from 20 studies found the sensitivity, specificity, negative predictive, and positive predictive value to be 85%, 89%, 84%, and 89%, respectively (Taillefer 1999). A meta-analysis of over 5000 patients imaged with a traditional gamma camera showed comparable results (Liberman et al. 2003), where the aggregate findings demonstrated a sensitivity of 85.2%, specificity of 86.6%, negative predictive value of 81.8%, positive predictive value of 88.2%, and accuracy of 85.9%. The majority (80%) of the studies reported sensitivity and specificity values over 80%, with nearly half of them yielding values over 90%. Similarly, a multi-institutional study in the United States led by Khalkali et al. (2000) reported overall sensitivity of 75.4% and specificity of 82.7%. The FDA approved Tc-99m sestamibi (Miraluma) for use in breast imaging in 1997 as a result of these findings. Miraluma is identical to Cardiolite, a radiopharmaceutical frequently used for cardiac imaging. Other radiopharmaceuticals used for breast imaging today include Tc-99m tetrofosmin and fluorodeoxyglucose.

However, scintimammography's main limitation was that it could not reliably detect subcentimeter breast cancers, due to its intrinsic inability to image lesions smaller than 1 cm. Resolution was diminished by distance between the breast and detector, and the far posterior and medial breast was largely inaccessible. Additionally, the breast could not be imaged in positions comparable to mammography, making the correlative, multimodality imaging necessary for both diagnosis and localization for sampling challenging. Therefore, these findings must be interpreted in the context of the size, stage, and type of the cancers detected. A multicenter study of 420 patients with 449 breast lesions reported sensitivity of 26%, 56%, 95%, and 97% for T1a, T1b, T1c, and T2 breast cancers, respectively (Scopinaro et al. 1997). Similarly in the report of his multi-institutional trial performed in the United States, Khalkhali showed significantly higher sensitivity for cancer >1 cm (74.2%) than for cancers <1 cm (48.2%), as well as significantly higher sensitivity for invasive cancers (82.0%) than for DCIS (45.9%) (2000). Scintimammography's ability to detect breast cancer was also better with palpable than nonpalpable breast cancers with sensitivities of palpable and nonpalpable cancers reported as 87% and 61%, respectively; this difference is likely due to the larger size of palpable breast cancers. The utility of nuclear medicine imaging of the breast as a clinical tool depends on its ability to detect early, small, and nonpalpable breast cancer. Studies of scintimammography showed unreliable detection of subcentimeter and nonpalpable cancers, yet the results encouraged the

further development of nuclear medicine imaging of the breast as an adjunct imaging modality. Crucially, scintimammography was not impacted by breast density as is mammography; the technique was equally sensitive in women with dense and fatty breasts (Khalkhali et al. 2002). Still, until gamma cameras were developed such that they could reliably image subcentimeter and early-stage cancers as well as produce images in positions comparable to mammography, nuclear medicine imaging of the breast could not meaningfully be integrated into clinical practice.

The known limitations of traditional scintimammography led to the development of a high-resolution, small-field-of-view, breast-specific gamma camera (HRBSG) (Scopinaro et al. 1999) (Figure 17.1). This type of camera allows improved flexibility and comfort in patient positioning, imaging in views comparable to mammography, and eliminates dead space by employing the space of the entire detector. Significant improved contrast of small lesions was also attained with the HRBSG camera along with marked improvement in spatial resolution. Currently, the technique of imaging the breast with a gamma-emitting radiopharmaceutical and a high-resolution, breast-specific gamma camera is referred to as either breast-specific gamma imaging (BSGI) or molecular breast imaging (MBI). Available HRBSG cameras now include free standing and mobile units (Figure 17.1), and a camera mounted on a modified mammography system.

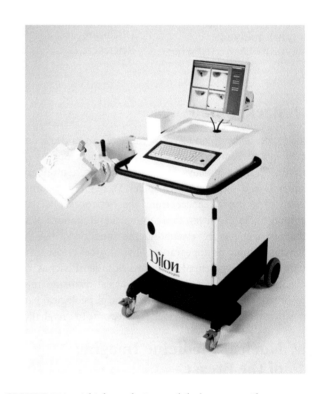

FIGURE 17.1 A high-resolution, mobile, breast-specific gamma camera unit. (Images courtesy of Dilon Technologies, Inc.)

17.3 The BSGI/MBI Procedure

The patient is injected with 15–25 mCi of technetium-99m sestamibi immediately prior to the start of imaging. Historically, the radiotracer was administered in the pedal vein to minimize risk of axillary lymph node uptake due to extravasation during the injection; however, patient discomfort as well as increased ability to differentiate extravasation from metastatic lymph nodes (Werner et al. 2009) has changed this practice. The radiotracer is generally administered in the antecubital vein of either arm for a surveillance examination, but in the case of a known breast abnormality should be administered into the contralateral arm to avoid axillary lymph node uptake due to extravasation during the injection.

The BSGI/MBI procedure is well-tolerated by patients. The examination requires no fasting or preparation beforehand, and is performed with the patient comfortably seated. Positioning for the examination includes the craniocaudal (CC) (Figure 17.2a) and mediolateral oblique (MLO) (Figure 17.2b) views of each breast, obtained in positions comparable to mammography, although the breast is only lightly compressed by the gamma camera's plate. The initial CC and MLO views are obtained and reviewed by the radiologist, who may request additional images be obtained as needed, comparable to the approach of additional images obtained with diagnostic mammography. As the detector is on an articulating arm or mounted on a mammography unit, a true lateral, exaggerated CC, or any other view possible with mammography can be acquired. The examination takes approximately 45 minutes but can vary in length depending upon breast size and additional views requested. The breast is gently compressed by a plate while images are obtained to 100,000 counts per image (about 6 minutes each). BSGI produces between four and 10 images in total. With additional improvements in camera technology, it is expected that in the future, even shorter scan time will be able to produce images of equivalent quality.

BSGI/MBI can be performed following percutaneous biopsy—FNA, core biopsy, or vacuum-assisted biopsy (Chapter 8)—and will not be affected by postprocedural changes or inflammation (personal experience/unpublished data). Nuclear medicine imaging can be performed even days following surgery and will show a thin rim of radiotracer uptake around the postoperative seroma (Figure 17.3). However, nodular or irregular uptake at the seroma site is notable and suggestive of residual disease. As with MRI (Chapter 16), for premenopausal women who are undergoing BSGI/MBI for high-risk surveillance or following a diagnosis of breast cancer, the procedure should optimally be scheduled between days 7 and 14 of the menstrual cycle to minimize background uptake due to hormonal fluctuations. In the case of a clinical finding, imaging can be performed during any time of the menstrual cycle.

When a focal area of radiotracer uptake is identified on BSGI/MBI, particularly in the case of a normal mammogram, targeted ultrasound can be used to evaluate the area in question and determine whether biopsy is warranted (Figure 17.4). The percentage of lesions apparent on BSGI/MBI that are able to be targeted with ultrasound has not yet been determined, but likely parallels the 56%–57% reported for MRI (Meissnitzer et al. 2009, Abe et al. 2010). Additionally, the mammogram can be reevaluated to determine whether any subtle mammographic finding is present that can be targeted for stereotactic biopsy (Figure 17.5). If a lesion corresponding to the area of focal increased radiotracer uptake is not identified with targeted ultrasound or targeted mammographic reevaluation, an MRI can be performed with MRI guided biopsy used to sample the lesion. However, some BSGI/MBI cameras now have FDA-approved direct gamma guided biopsy devices that, similar to MRI-guided biopsy devices, allow for direct targeting of lesions identified with BSGI.

A major concern regarding BSGI/MBI is radiation exposure. Initially, administration of 25–30 mCi of technetium 99-m sestamibi was recommended for the examination, based on the

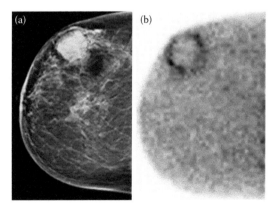

FIGURE 17.3 A 51-year-old woman with stereotactic biopsy-proven DCIS and a 3.5-cm postbiopsy hematoma in the right breast. (a) A 3.5-cm hematoma is present on the left CC view of postbiopsy mammogram. (b) BSGI performed to assess extent of disease reveals ring-like enhancement in the left CC view indicating hematoma but no other foci of breast cancer.

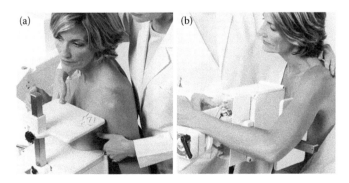

FIGURE 17.2 Patient undergoing a BSGI examination. (a) Patient positioned for image acquisition in the craniocaudal (CC) view. (b) Patient positioned for image acquisition in the mediolateral oblique (MLO) view. (Images courtesy of Dilon Technologies, Inc.)

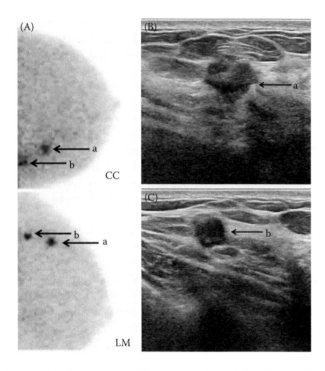

FIGURE 17.4 A 45-year-old woman underwent BSGI for evaluation of a palpable mass in the 11:00-o'clock region of the left breast. (A) BSGI images in the CC and lateral medial (LM) views demonstrate two foci of focal increased radiotracer uptake in the upper outer quadrant of the left breast for which directed ultrasound was recommended. (B) Second-look ultrasound of lesion a, 3-cm from the nipple, demonstrates a 1.2-cm hypoechoic mass with irregular borders highly suspicious for malignancy. Ultrasound-guided core biopsy revealed invasive ductal carcinoma, grade II/III. (C) Second-look ultrasound of the second lesion (b) identified with BSGI, 4 cm from the nipple, demonstrates another suspicious 1-cm hypoechoic mass corresponding to the patient's palpable lump. This lesion was also proven by pathology to be invasive ductal carcinoma, grade II/III.

dose use for cardiac imaging. However, in our practice, we currently use 15–20 mCi and experience no consequence to imaging. Even lower doses would be better, and ongoing studies are evaluating doses as low as 4–10 mCi (O'Connor et al. 2011), although no studies describing imaging with doses this low have been published to date, and there is no clinical data to support the use of extremely low doses, i.e., those less than 15 mCi. It is highly likely that in the future radiation exposure will be significantly reduced to comply with the principle of as low as reasonably achievable (ALARA).

A study by Hendrick et al. (2010) reviewed the radiation dose of a number of screening breast imaging procedures. Hendrick himself noted that the effective dose from BSGI studies equals approximately 2–3 years of natural background radiation exposure. However, it is critical to note that BSGI is not a screening tool, except in the very-high-risk population, and therefore cannot be compared to screening mammography. The dose used for BSGI is reasonable for the ability to detect early, curable breast cancer, and it is substantially lower than for other nuclear medicine imaging techniques, such as positron emission mammography (PEM). Even so, significant ongoing research aims to determine what the safest, lowest dose of radiotracer is that will not impact the detection of the smallest breast cancers. These findings will likely be integrated into clinical practice in the near future.

Equipment cost, radiopharmaceutical cost, space requirements, and reimbursement are important for the utility of new equipment in a clinical facility. The different equipment used in nuclear medicine imaging of the breast can range from $300,000 to nearly $1,000,000, with BSGI/MBI equipment falling at the lower end of this range. The cost of a dose of sestamibi usually is about $75 while the cost of a dose of FDG for PEM is usually about $600 and can be as high as $950. BSGI uses a mobile unit about the size of an ultrasound unit and does not require a

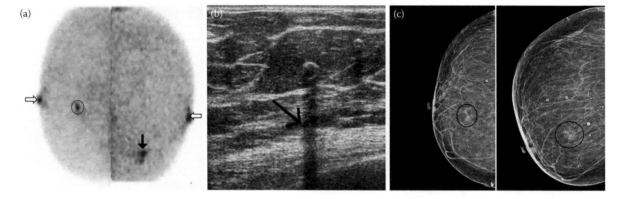

FIGURE 17.5 A 41-year-old patient with biopsy-proven LEFT ductal carcinoma in situ. (a) Right and left BSGI images in the CC view demonstrate normal physiological bilateral nipple uptake (hollow arrows). The BSGI demonstrates focal increased radiotracer uptake (solid arrow) in the left breast corresponding to the biopsy-proven DCIS. On the right breast, an occult focus of DCIS (circle) was identified for which directed ultrasound was recommended. (b) Ultrasound directed to the inferomedial right breast was unremarkable, with a focal area of shadowing (arrow) due to a benign coarse calcification. (c) CC and MLO mammographic views of the RIGHT breast demonstrate a very vague asymmetry (circle) initially interpreted as normal. Stereotactic biopsy of this area demonstrated atypical ductal hyperplasia, and at surgical excision DCIS was identified.

dedicated room, although injection of any radiotracer does. The equipment can be easily situated in a breast imaging center and generally requires no additional shielding, although shielding requirements vary by state. In the case of some institutions, the radiotracer injection can be administered in nuclear medicine and the imaging completed in breast imaging. Finally, BSGI/MBI uses recognized CPT codes and is generally reimbursed, although reimbursement is dependent upon carrier and local differences.

17.4 The Evolution of High-Resolution, Small-Field-of-View Gamma Imaging Camera

A prototype HRBSG camera for the detection of small, non-palpable breast lesions was initially evaluated in comparison with a general-purpose gamma camera in 50 women with various clinical indications for scintimammography (Brem et al. 2002). Examination with the HRBSG camera was performed after traditional scintimammography and 28 cancers total were included, 71% of which were not palpable, and which ranged in size from 3 to 60 mm (mean 11 mm). The study demonstrated a 14.3% improvement in sensitivity for all cancers and 16.7% improvement in sensitivity for the detection of nonpalpable breast cancers with the HRBSG camera. This study was the first to demonstrate the feasibility of detecting subcentimeter cancers with BSGI and imaging in positions comparable to mammography, and the results suggested that functional integration of BSGI into clinical practice could augment the tools available for breast imaging. A commercial HRBSG camera was developed with the support of this research and was approved by the FDA in 1999.

The discussion of more widely implementing BSGI/MBI as an adjunct breast imaging modality must include the technology's sensitivity and specificity, which has varied in the literature. In a study evaluating 146 consecutive women who underwent BSGI and subsequent biopsy for a variety of clinical indications, 167 suspicious lesions were identified, 83 of which were malignant (Brem et al. 2008). BSGI correctly identified 80 lesions as malignant for a sensitivity of 96.4% for all cancers; 65 of 67 (mean size = 20 mm) invasive cancers for a sensitivity of 97.0% and 15 of 16 DCIS (mean size = 18 mm) for a sensitivity of 93.8%. The smallest invasive cancer and DCIS measured 1 mm each. The sensitivity of BSGI for subcentimeter cancers was 88.9%; moreover, for the 5 lesions ≤5 mm, the sensitivity was 100%.

Similarly, a study evaluating a dual-headed MBI camera in 88 women with 128 cancers reported a sensitivity of 90% for all cancers and 82% in subcentimeter cancers. The sensitivity of the single-headed camera was 80% for all cancers as compared to 68% for subcentimeter cancers (Druska et al. 2008). These results seem to favor some single-headed cameras whose sensitivity for small breast cancers is higher than the dual-headed camera (Brem et al. 2008).

Most recently, the sensitivity of BSGI with respect to size of invasive tumors was evaluated in a retrospective review of 139 women with 149 invasive cancers, mean cancer size 1.8 cm (Tadwalkar et al. 2011). BSGI detected 98% of all cancers and 100% of cancers >7.1 mm. The 3 of 149 (2%) of cancers not detected by BSGI were all <7 mm and all pathologic grade 1. All cancers of grade 2 and 3 and as small as 2 mm were detected. Further research is needed to correlate sensitivity of BSGI/MBI to tumor characteristics, particularly as there is increasing interest in potentially using functional breast imaging as a marker for tumor biology.

17.5 Uses for Nuclear Medicine Imaging of the Breast

17.5.1 BSGI/MBI as a Screening Tool for High-Risk Women

Functional breast imaging has been evaluated as a screening tool for women at high risk for breast cancer, especially in those who have dense breasts (Brem et al. 2008, Coover et al. 2004, Rhodes et al. 2011). In one study, 94 high-risk women with normal mammograms and clinical breast examinations underwent BSGI to investigate its ability to detect mammographically occult breast cancer (Brem et al. 2008). Of the 94 women, 16 had a positive BSGI and targeted ultrasound to the areas of suspicious focal radiotracer uptake demonstrated sonographic abnormalities in 11. Eleven ultrasound-guided biopsies yielded benign findings in nine women and invasive ductal carcinoma measuring 9 and 10 mm, respectively, in two women. Both of these patients had previously been diagnosed with breast cancer, one with a contralateral cancer and the other a recurrence in the site of lumpectomy. Benign pathology in the other nine biopsies demonstrated fibrocystic breast tissue in seven, fat necrosis in one, and fibroadenoma in one. BSGI demonstrated a sensitivity of 100%, specificity of 84%, and the NPV of 100% in this study. In the five of 16 patients with abnormal BSGI examinations but no sonographic correlate, 6-month and 1-year follow-up visits were normal, supporting the use of BSGI/MBI in conjunction with anatomic imaging modalities for comparison. Coover's smaller study of 37 women with dense breasts, family histories of breast cancer, and no suspicious mammographic or clinical findings revealed five positive examinations with an HRBSG camera, three of which corresponded to mammographically occult cancers (8.1%) (Coover et al. 2004). Recently, a prospective study by Rhodes et al. (2011) using MBI with a dual-headed, high-resolution CZT-based gamma camera for 936 asymptomatic women at high risk for breast cancer and with heterogeneously or extremely dense breast tissue reported a supplemental yield of 7.5 cancers per 1000 women screened. Sensitivity of gamma imaging alone was reported to be 85%; the sensitivity was 91% when combined with mammography.

Research focused on defining the sensitivity and specificity of BSGI/MBI for detecting mammographically occult cancer, particularly in populations of women at increased risk for breast cancer where incidence of breast cancer is expected to be higher than in a normal screening population, is ongoing. Future prospective, multi-institutional studies are needed to most accurately evaluate BSGI/MBI.

In our practice, one use of BSGI is to evaluate high-risk women with normal mammograms and physical examinations. A retrospective evaluation of this population in our practice showed that use of BSGI yielded a 2.1% incremental increase in cancer detection, mainly in women with >50% breast density (unpublished data). BSGI provides an adjunct imaging modality for the more than 20% of women who cannot undergo MRI due to implantable devices, renal insufficiency, body habitus, or severe claustrophobia (Berg et al. 2010). Furthermore, it has been reported that the specificity of BSGI is superior to that of MRI (Brem et al. 2007b, Kim 2011) and may be advantageous in selected patients. Of note is that MRI and BSGI can in fact be complementary examinations, such as in the clinical situation of a premenopausal woman with newly diagnosed breast cancer and multiple enhancing lesions identified on preoperative MRI. In such situations BSGI can be used to identify which, if any, lesions require further investigation. Furthermore, in women who have marked diffuse parenchymal enhancement on MRI, which results in higher rates of short-interval follow-up (Hambly et al. 2011) than for women with mild or moderate background enhancement and may impact diagnostic accuracy (Uematsu et al. 2011), BSGI can be helpful to identify a concerning lesion (Figure 17.6).

BSGI/MBI offers other advantages over MRI including far less physician time for interpretation, due to the generation of 4–10 images with BSGI as compared to up to 2,000 images with MRI, with the potential to improve workflow at breast imaging centers. Of course, MRI has the anatomic information that BSGI/MBI lacks due to the intrinsic differences in the imaging modalities.

When considering BSGI/MBI as a screening tool to image high-risk women, the negative predictive value can be very significant, particularly in women with challenging mammograms. If a high-risk woman is seen with a normal mammogram, but the interpretation is limited due to breast density, multiple findings, or distortion from prior breast surgeries, a normal BSGI/MBI examination can be extremely reassuring that the absence of disease is 94%–98% (Brem et al. 2008, personal communication, J. Weigert, MD). Physiological imaging can also corroborate a

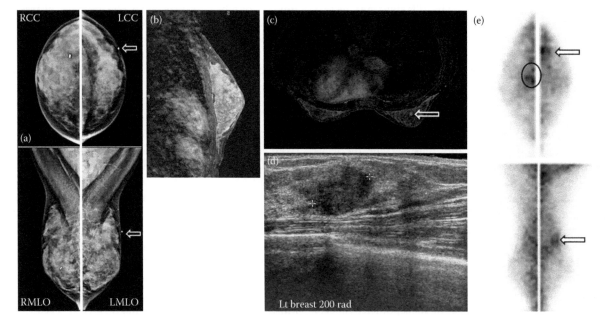

FIGURE 17.6 A 38-year-old woman with a strong family history presents with a palpable abnormality in the left breast. (a) Bilateral mammogram with bee-bee marker (arrow) in area of palpable abnormality appears stable and demonstrates extremely dense tissue and marker from prior benign biopsy in the right breast. (b) Recent MRI performed for high-risk screening demonstrates diffuse parenchymal enhancement and no enhancing lesion in the left breast. (c) Recent MRI demonstrated one enhancing lesion in the right 6:00-o'clock position (arrow), which was proven by MRI-guided biopsy to be a fibroadenoma. (d) Targeted ultrasound of the left 2:00-o'clock position evaluating area of patient's palpable abnormality reveals a large hypoechoic mass at the left 2:00-o'clock position that is suspicious for malignancy. (e) BSGI demonstrates focally increased radiotracer uptake in the 6:00-o'clock position (circle), at site of previously biopsied fibroadenoma, as well as in the upper outer quadrant of the left breast (arrow) corresponding to patient's palpable mass and the mass seen on ultrasound at the left 2:00-o'clock position. Ultrasound-guided core biopsy of this area revealed a grade III invasive ductal carcinoma not detected with MRI.

benign second-look ultrasound. Of course, the NPV of BSGI/ MBI, ranging from 94% to 98%, does not rule out the potential presence of breast cancer with complete certainty.

17.5.2 Detection of Occult Cancerous Foci in Women Newly Diagnosed with Breast Cancer

Presurgical BSGI/MBI can successfully identify additional occult foci of breast cancer in newly diagnosed women (Figure 17.7). In one study (Brem et al. 2008), 7.2% of patients with breast cancer had an additional, occult focus of breast cancer identified with BSGI. A retrospective review of 159 women who underwent BSGI for suspicious mammography or physical examination findings reported detection of additional occult foci of breast cancer in 9% of women with breast cancer; 6% in the same breast, and 3% in the contralateral breast (Brem et al. 2010). Similarly, Zhou et al. (2009) reported their findings in 138 women with breast cancer where an additional 11 (7%) foci of breast cancer were detected. Recently, O'Connor et al. (2011) demonstrated that MBI offered to newly diagnosed patients altered surgical management in 12.2%. Findings for use of contrast-enhanced breast MRI (CE-MRI) for detecting additional, occult foci of

breast cancer were similar, where MRI revealed contralateral occult disease in 3% of newly diagnosed women (Lehman et al. 2007, Hollingsworth et al. 2009) and multicentric disease in 7.7% (Hollingsworth et al. 2009). Both BSGI/MBI and MRI are used clinically for the detection of additional foci in newly diagnosed breast cancer patients, but additional, multi-institutional studies are needed to directly compare these modalities.

Nuclear medicine imaging of the breast holds potential to alter surgical options for women diagnosed with breast cancer. Women with small breast tumors that do not exceed one quadrant of the breast can be offered lumpectomy and radiation therapy, which are as effective as mastectomy for long-term survival (Damle et al. 2011). However, the recommended surgical treatment for cancers that have spread beyond one quadrant of the breast is mastectomy. At the time of initial diagnosis, it is important to eliminate the possibility of additional, occult foci of disease for the purposes of both optimal surgical management and partial breast radiation therapy. If bilateral disease is identified, as it is in 3% of women (Chaudary et al. 1984), optimal surgical and reconstructive surgery can be planned and undertaken. In the particular clinical situation where additional foci of contralateral disease are discovered following a TRAM reconstruction, bilateral TRAM is no longer an option as fibrosis prohibits subsequent surgical dissection of the rectus. In one study, 82 patients underwent BSGI for newly diagnosed breast cancer, and surgical management was altered in 22% as a result of findings from the BSGI (Killelea et al. 2009). BSGI/MBI can be helpful in optimally identifying all foci of breast cancer in women with newly diagnosed breast cancer prior to definitive surgery.

17.5.3 Detection of Invasive Lobular Carcinoma

Invasive lobular carcinoma is the second most commonly diagnosed invasive breast cancer and often has an insidious presentation both clinically and with imaging. Lobular cancers originate from the lobular epithelium of the breast and malignant cells infiltrate between normal tissue and do not incite a desmoplastic reaction. Mammographically, ILC can present as a vague asymmetry, often lacks suspicious calcifications, and is difficult to detect with both mammography and sonography (Sickles 1991, Krecke and Gisvold 1993). Hilleren et al. (1991) reported that 16% of invasive lobular carcinomas were mammographically occult. As a result, these tumors may not be detected until they are large and node-positive (Selinko et al. 2004). MRI has been shown to detect ILC with a lower sensitivity than for other invasive cancers (Mann et al. 2004). Twenty-eight women from four institutions with pure ILC were evaluated, all with mammography and BSGI and some with ultrasound and MRI. The sensitivity of mammography, ultrasound, and MRI for ILC was 79%, 68%, and 83%, respectively. The sensitivity of BSGI for the detection of ILC was 93% in this study and has been reported at 91% (Hruska et al. 2008) for MBI and 100% (Tadwalkar et al. 2011) for BSGI in other recent studies (Figure 17.8).

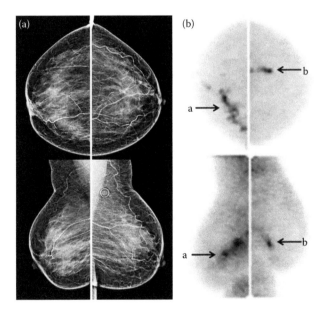

FIGURE 17.7 A 68-year-old woman presents with right nipple discharge. (a) Bilateral mammogram appears stable and demonstrates benign-appearing calcifications. (b) Bilateral BSGI shows focal radiotracer uptake in a ductal distribution in the upper inner quadrant of the right breast (a) as well as increased radiotracer uptake in a linear distribution in the lower outer quadrant of the left breast (b). Both lesions were localized and sampled using ultrasound and biopsy showed grade II DCIS bilaterally. The patient thus opted for bilateral mastectomies and pathology showed a 2.4-cm DCIS with a 0.7-cm focus of invasive ductal carcinoma on the right side and a 0.8-cm DCIS on the left.

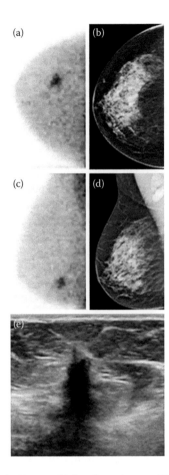

FIGURE 17.8 A 60-year-old female underwent BSGI for high-risk screening (a, c). Recent mammogram demonstrates dense tissue with stable-appearing calcifications in the inferomedial right breast (b, d). BSGI demonstrates focally increased radiotracer uptake in the lower outer quadrant of the right breast. Second-look ultrasound of this area showed a hypoechoic mass at the 7:00-o'clock position of the right breast (e) and ultrasound-guided core biopsy revealed invasive lobular carcinoma that was mammographically occult.

17.5.4 Detection of Ductal Carcinoma In Situ

In situ cancers account for approximately 20% of all breast cancers (American Cancer Society 2011). The sensitivity of BSGI/MBI for DCIS ranges from 91% to 100% (Brem et al. 2007a, 2008, O'Connor et al. 2009, Hruska et al. 2008). Recently, a small prospective trial by Keto et al. (2011) reported no difference between the sensitivities of MRI and BSGI in patients newly diagnosed with DCIS; however, larger multi-institutional trials are needed to substantiate these findings.

17.6 Alternate Physiological Imaging Modalities

17.6.1 Positron Emission Mammography

Nuclear medicine imaging of the breast also includes positron emission tomography (mammography) (PEM), a procedure that uses a dedicated breast PET imager and follows administration of 2-deoxy-2-[18F]-fluoro-D-glucose (FDG). Analogous to conventional gamma cameras, PET scans using FDG use a metabolic approach but are not designed for breast-specific imaging and lacked the resolution for detection of small lesions. PEM has been developed to resolve the limitation in resolution and utilizes PET imaging with improved resolution of breast lesions as an adjunct imaging modality. PEM also allows for imaging in positions comparable to mammography.

The PEM device consists of two detector heads that obtain tomographic images with a 23 × 17 cm field of view. Since PEM is a PET examination, as with other PET studies, patients' blood glucose levels must not be elevated, and the technique requires that patients fast for a minimum of 4 hours before study initiation. Patients must rest quietly for approximately 1–3 hours and must void before beginning the examination. Ten to 15 mCi of FDG are injected in the arm contralateral to the side of known abnormality, as in MBI/BSGI studies; however, in contrast, there is no preparation required for BSGI/MBI studies and imaging may begin immediately following radiotracer injection.

Berg et al. (2006) included 77 women with either known breast cancer or suspicious breast lesions imaged with PEM. The sensitivity of PEM was 89% for invasive cancers and 91% for DCIS. Interestingly, PEM detected 50% of T1a cancers, and 67% of T1b lesions in the study. Tafra et al. (2005) evaluated 44 patients imaged with PEM with suspected or proven breast cancer (2005). In this series, the sensitivity of PEM for detecting the index lesion was 89%. Four lesions were not detected measuring 3, 6, and 10 mm, as well as 1 mm for a breast lymphoma. However, the smallest reported breast cancers detected with BSGI/MBI (<2 mm) (Tadwalkar et al. 2011) are smaller than those identified by PEM. Investigators of PEM have suggested integrating mammography and ultrasound studies with PEM results to improve the specificity of the technique (Berg et al. 2006). Recently, a prospective study reported that integrating PEM and MR imaging for women with one diagnosis of breast cancer increased sensitivity for detecting additional cancerous lesions in the ipsilateral breast from 60% with MRI alone to 74% (Berg et al. 2011). However, the use of MRI and PEM in all women is not likely to be widely integrated into clinical practice due to high cost of these examinations.

Equipment costs are a concern, especially in times of economic austerity. The cost of a BSGI unit is approximately $300,000; MBI, approximately $600,000; and PEM, approximately $800,000. PEM devices require shielding, while BSGI/MBI units do not. In the two largest clinical trials of PEM, the images of 18% and 53% of the patients enrolled were not included for analysis, due to inability of readers to interpret the studies due to the quality of the images (Berg et al. 2006, Tafra et al. 2005). Additional research and clinical use will allow for more consistent interpretation of PEM examinations. Furthermore, reimbursement for any PET examination requires a diagnosis of cancer and, therefore, PEM cannot yet be used for screening of high-risk populations or for further evaluation of equivocal mammographic or physical examination findings as BSGI and MBI studies can.

17.6.2 Breast MRI

Like BSGI/MBI, breast MRI is a physiological imaging modality used clinically as a screening tool for high-risk women, to further evaluate indeterminate mammographic, sonographic or physical examination findings, and to assess extent of disease in newly diagnosed breast cancer patients (Chapter 16). Generally studies have shown that MRI and BSGI have comparable sensitivities, and comparison of the effectiveness of the two modalities remains important. In a study of 201 women with 122 lesions biopsy-proven to be malignant or atypical, the sensitivities of MRI and BSGI for cancerous lesions were 90.6% and 91.6%, respectively. There were six "indeterminate" lesions with MRI and three lesions with BSGI that were not able to be categorized as positive or negative, and MRI alone yielded more false-positives than BSGI alone (17 vs. 10) (Lanzkowsky et al. 2009). A smaller study of 23 women with 33 equivocal breast findings that required physiological imaging showed sensitivities of 89% and 100% and specificities of 71% and 25% for BSGI and MRI, respectively (Brem et al. 2007). Similarly, Bertrand reported equivalent sensitivity but higher specificity of BSGI, compared with MRI (Bertrand 2008). Notably, 4 (5.3%) patients in this study were not able to undergo MRI while all complied with BSGI examinations.

The American Cancer Society now recommends screening MRI for women at 20%–25% or greater increased lifetime risk for breast cancer (Saslow et al. 2007). The increased specificity of BSGI could be a significant improvement particularly in this population who can greatly benefit from increased screening. For any screening examination to be practical, both false-positives and false-negatives must be minimized; inconclusive results can result in challenging patient management as well as patient anxiety. BSGI has been reported to both confirm the need for biopsy and reduce the number of false-positive biopsies from MRI, and in some cases may obviate the need for a costly short-term follow-up MRI (Lanzkowsky et al. 2008). Although MRI and BSGI detect breast cancer with comparable sensitivities, BSGI has a number of benefits over MRI. In addition to reported higher specificity, there are no contraindications to BSGI/MBI, and all patients can be imaged comfortably provided that they have venous access. In contrast, MRI is contraindicated in the following clinical situations: implanted cardiac or CNS devices such as pacemakers or implanted defibrillators, aneurysm clips, large body habitus, which prohibits imaging the patient in the MRI bore, claustrophobia, and renal insufficiency or decreased glomerular filtration rate (GFR), due to the risk of nephrogenic systemic fibrosis (NSF).

Furthermore, a single headed high-resolution breast-specific gamma camera is less costly than MRI equipment, as is the BSGI examination itself. In a 2008 study, the national Medicare reimbursement averages were reported for BSGI to be $219.43 and MRI, $994.43. For 75 patients imaged with both MRI and BSGI, the use of BSGI alone would have saved $58,107 with no reduction in cancer detection (Bertrand 2008). From a workflow standpoint, the portability of a BSGI unit makes this equipment easier to integrate into a breast imaging practice, the unit requires little space and a dedicated room is not required save for the radiotracer injection. Additionally, a BSGI/MBI study generates only 4–10 images, as compared with upwards of 1000 images generated by MRI. In our experience BSGI studies require far shorter time for radiologist interpretation and are, thus, beneficial to the surgeon, who can quickly and easily appreciate the extent of disease and incorporate this information into surgical management. Still, the advantage of the more detailed MRI study lies in its depiction of the breast anatomy, which may be more useful to presurgically determine features of disease, such as involvement of the pectoralis muscle.

Well-designed, prospective, and multi-institutional studies are needed to compare BSGI/MBI and MRI to usefully inform imaging recommendations. Currently, our clinical experience indicates that BSGI detects breast cancer with equal sensitivity to and better specificity than MRI, especially in premenopausal women with dense breasts (Figure 17.6), but that BSGI/MBI is not intended to replace CE-MRI. These techniques along with PEM and traditional anatomic imaging form a complement of tools aimed at optimizing breast cancer detection, and can be used in conjunction for the highest accuracy. Clinically, BSGI and MRI may be used together; for example, in the case of numerous suspicious foci on MRI examination, a BSGI can mitigate the need to biopsy all lesions. This principle is important in women with newly diagnosed breast cancer as well as those at increased risk of breast cancer, such as *BrCA*1/2 genetic mutation carriers, who can drastically benefit from physiological surveillance but may not want to endure the possibility of false-positive results or unnecessary biopsies. Additionally, if a focal area of increased radiotracer uptake is identified with BSGI that cannot be targeted with ultrasound, evaluation with MRI and MRI-guided biopsy is an option. With the advent of FDA approval of a gamma-guided biopsy system (GammaLoc, Dilon Technologies, Newport News, VA) that can be used to target BSGI-detected lesions, use of MRI for this purpose will likely not be necessary.

17.7 Biopsy of Lesions Prompted by BSGI/MBI

Should an abnormal area of focally increased radiotracer uptake present on BSGI/MBI examination, the area must be localized for biopsy. Until recently, radiologists relied upon multimodality imaging to identify the abnormality using comparison with mammography, directed ultrasound, or MRI. Ultrasound localization allows the lesion to be biopsied using ultrasound-guided core needle biopsy (Figure 17.8e). Reevaluation of the patient's mammogram can occasionally reveal a subtle abnormality unremarkable upon initial review of the films, which can be sampled using stereotactic, vacuum-assisted guidance (Figure 17.5). If the abnormality on BSGI does not correspond to any area on ultrasound or mammography an MRI and subsequent MRI-guided biopsy may be performed. All of these modalities for sampling lesions detected on BSGI are used at our institution. Recently, FDA approval of a gamma-guided, stereotactic localization device (GammaLoc) (Figure 17.9) has expanded the

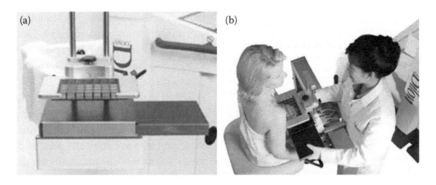

FIGURE 17.9   GammaLoc device. (a) GammaLoc grid system. (b) Patient being positioned for a GammaLoc-guided biopsy. (Courtesy of Dilon Technologies, Inc.)

options available for biopsy and now allows for direct biopsy of lesions using the BSGI camera.

17.8 Clinical Indications for BSGI/MBI

Metabolic/physiological imaging is an important tool that can substantially improve breast cancer detection, particularly in certain populations of women. Nuclear medicine imaging of the breast is equally sensitive in women with dense and fatty breasts (Khalkhali et al. 2002), and is thus helpful for evaluating women with dense tissue who are up to six times more likely to develop breast cancer in their lifetime and whose mammograms are more difficult to read with any level of certainty. Women with breast implants (Figures 17.10 and 17.11) or silicone injections (Figure 17.12), for whom mammography is also insufficient in ruling out disease, may benefit. BSGI is also used for high-risk women with normal mammograms, for added surveillance, or

in any woman with an indeterminate mammographic or clinical finding in order to rule out a malignant process. In our practice women who are *BrCA1*- or *BrCA2*-positive are recommended to receive mammography and BSGI examinations annually, as well as MRI annually at the 6-month interval, although additional research is needed to substantiate recommendations for this population. BSGI and MRI can be used interchangeably for the majority of patients, but high-risk women who begin supplemental surveillance with either technology are recommended to continue annually with the same examination for comparison purposes. Women who have had numerous enhancing areas on prior MRIs are recommended to undergo BSGI for future examinations, due to its higher specificity (Bertrand et al. 2008, Brill et al. 2008).

In our institution, every woman with a newly diagnosed breast cancer undergoes either MRI or BSGI to determine extent of disease for treatment purposes, to identify additional occult

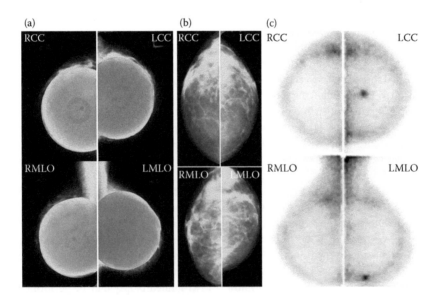

FIGURE 17.10   A 44-year-old patient with bilateral implants and a palpable mass in the left breast. (a) CC and MLO mammographic views with implant intact are unremarkable. (b) CC and MLO implant-displaced views demonstrate normal-appearing dense breast tissue. (c) BSGI demonstrates focally increased radiotracer uptake in the left breast corresponding to a mammographically occult focus of invasive ductal carcinoma.

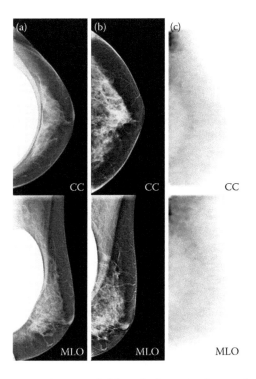

foci of cancer, if present, and to aid with surgical planning. We have also begun to use BSGI to assess progress in women undergoing neoadjuvant chemotherapy, as correlation between tumor size on ultrasound at the time of diagnosis, at surgical excision after therapy, and on BSGI at both stages are quite accurate upon preliminary investigation. In a study of 11 patients, BSGI measurements before chemotherapy were within 5 mm of ultrasound measurements in six out of 10 patients with available ultrasound measurements, and within 1 cm in all but one patient (Damle et al. 2010), suggesting that BSGI may be used to monitor response to neoadjuvant therapy and can serve as an indicator of progress during therapy (Figure 17.13). In a similar study of MBI, postchemotherapy imaging was concordant with pathology in eight (67%) of 12 cases (Boughey et al. 2010). Further studies of BSGI used for patients undergoing neoadjuvant chemotherapy are ongoing.

The Society of Nuclear Medicine has outlined practice guidelines that include the above-mentioned indications (Goldsmith et al. 2010). BSGI/MBI is an FDA-approved imaging modality for which CPT codes exist. Reimbursement rates vary based on insurance type and location, but in our experience, BSGI/MBI is generally reimbursed. Recently, the American College of Radiology has implemented a certification program for BSGI, further demonstrating the increasingly mainstream role BSGI has in the clinical imaging of breast cancer.

FIGURE 17.11 Patient with bilateral silicone implants and a strong family history of breast cancer. (a) CC and MLO views of the left breast demonstrate subpectoral silicone implants and dense breast tissue. (b) Implant-displaced views of the left breast demonstrate normal, dense breast tissue. (c) BSGI of the left breast shows mild radiotracer uptake along the edge of the implant with no suspicious focus of radiotracer uptake. BSGI can reassure that disease is not present in women with implants whose mammograms are more challenging to interpret.

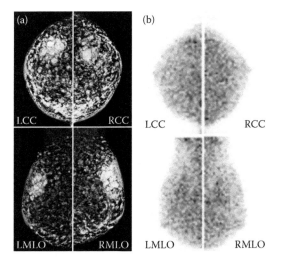

FIGURE 17.12 Patient with direct silicone injection and BSGI. (a) Bilateral mammogram demonstrates extensive foreign body granulomas making interpretation very challenging. (b) BSGI demonstrates no abnormalities. BSGI is not impacted by direct silicone injection based on its physiological approach to imaging.

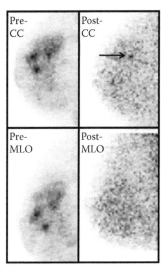

FIGURE 17.13 Patient being monitored with BSGI during neoadjuvant chemotherapy. Patient with a palpable mass in the right breast underwent ultrasound-guided core biopsy, which revealed grade III invasive ductal carcinoma. Neoadjuvant chemotherapy was recommended before surgery. Prechemotherapy BSGI (pre-CC and pre-MLO) showed focally increased radiotracer uptake indicative of multicentric disease. Postchemotherapy BSGI (post-CC and post-MLO) showed resolution of radiotracer activity with the exception of one small focus in the CC view (arrow, post-CC). Right simple mastectomy revealed a 0.4-cm residual focus of IDC.

17.9 Summary

Physiological/metabolic imaging of the breast is an exciting adjunct imaging technology and continues to constitute an important role in the armamentarium of tools available for breast cancer detection and diagnosis. BSGI/MBI is highly specific and sensitive and has shown to be effective in screening high-risk women, particularly those with dense breast tissue, breast implants, or who have had direct silicone injections; evaluating extent of disease in newly diagnosed breast cancer patients; identifying additional, occult foci of cancer; and elucidating indeterminate findings from traditional imaging or clinical breast examinations. Additionally, PEM has shown to be a highly sensitive tool for detecting breast cancer although this modality is currently only indicated for women with a diagnosis of breast cancer, limiting its use in screening other populations. Physiological imaging techniques are optimally used in a multimodal fashion with currently available anatomic and physiological technologies, and their clinical integration will allow for improved ability to diagnose and treat early, small, and node-negative breast cancers. Nuclear medicine imaging of the breast has been frequently compared to breast MRI and research suggests that BSGI/MBI is equally sensitive and has the advantages of higher specificity, lower equipment cost, ease of interpretation, and the ability to image all women regardless of body habitus or claustrophobia, implanted devices, or impaired renal function.

Nuclear medicine imaging of the breast is an FDA-approved imaging modality currently being clinically integrated to enhance the methods available to detect breast cancer. Our practice, as well as others that frequently use nuclear medicine imaging of the breast in standard clinical protocols, has found the tool crucial for providing optimal breast cancer screening and diagnosis to women. Research continues to evaluate the role of nuclear medicine in breast imaging, to further define indications for its use, and to compare the examinations to other available modalities for breast cancer detection. However, the physiological/metabolic approach of nuclear medicine breast imaging studies yields many advantages, and its utility in clinical situations where other imaging modalities are limited makes nuclear medicine imaging an important part of the set of tools available for optimal breast cancer diagnosis and outcomes.

Disclosure

Dr. Brem is a consultant for and is on the Board of Managers of Dilon Technologies, Inc.

References

Abe H, Schmidt RA, Shah RN et al. 2010. MR-directed ("second-look") ultrasound examination for breast lesions detected initially on MRI: MR and sonographic findings. Radiology 194(2): 370–377.

American Cancer Society. 2011–2012. Breast Cancer Facts and Figures.

Berg WA, Blume JD, Adams AM, Jong JA, Barr RG, Lehrer DE et al. 2010. Reasons women at elevated risk of breast cancer refuse breast MR imaging screening: ACRIN 6666. Radiology 254(1): 79–87.

Berg WA, Blume JD, Cormack JB et al. 2008. Combined screening and mammography vs ultrasound alone in women at elevated risk of breast cancer. JAMA 299(18): 2151–2163.

Berg WA, Madsen KS, Schilling K et al. 2011. Breast cancer: Comparative effectiveness of positron emission mammography and MR imaging in presurgical planning for the ipsilateral breast. Radiology 258(1): 59–72.

Berg WA, Weinberg IN, Naraanan D et al. 2006. High-resolution fluorodeoxyglucose ositron emission tomography with compression ("positron emission mammography") is highly accurate in depicting primary breast cancer. Breast J 12(4): 209–323.

Berry DA, Cronin KA, Plevritis SK, et al. 2005. Effect of screening and adjuvant therapy on mortality from breast cancer. N Engl J Med 353(17): 1784–1792.

Bertrand M. 2008. First year's experience using breast-specific gamma imaging: A comparative anaylsis with mammography, ultrasound and MRI in the detection of breast cancer. In Miami Breast Cancer Conference. February 20–23.

Bertrand M et al. 2008. Breast-specific gamma imaging compared to breast MRI in the presurgical planning of patients with known cancer diagnosis. In Radiological Society of North America Annual Meeting.

Boughey J, Hruska C, Rhodes D, Tortorelli C, O'Connor M, Wahner-Roedler D. 2010. Molecular breast imaging for assessing response to neoadjuvant chemotherapy. In (Abstract) The American Society of Breast Surgeons 11th Annual Meeting Proceedings. Vol. 36. <http://www.breast surgeons.org/upload/ASBS_Official_Proceedings_2010 .pdf> Accessed 11/7/2011.

Brem RF, Schoonjans JM, Kieper DA et al. 2002. Evaluation of a high-resolution, breast-specific gamma camera: A pilot study. J Nucl Med 43(7): 909–915.

Brem RF, Rapelyea JA, Zisman G et al. 2005. Occult breast cancer: Scintimammography with high-resolution breast-specific gamma camera in women at high risk for breast cancer. Radiology 237(1): 274–280.

Brem RF, Fishman M, Rapelyea JA. 2007a. Detection of ductal carcinoma in situ with mammography, breast specific gamma imaging, and magnetic resonance imaging: A comparative study. Acad Radiol 14: 945–950.

Brem RF, Petrovich I, Rapelyea JA et al. 2007b. Breast specific gamma imaging with 99m technitium sestamibi and magnetic resonance imaging in the diagnosis of breast cancer: A comparative study. Breast J 13: 465–469.

Brem RF, Floerke AC, Rapelyea JA et al. 2008. Breast specific gamma imaging as an adjunct imaging modality for the diagnosis of breast cancer. Radiology 247: 651–657.

Brem RF, Shahan C, Rapelyea JA, Donnelly CA, Rechtman LR, Kidwell AB et al. 2010. Detection of occult foci of breast cancer using breast-specific gamma imaging in women with one mammographically or clinically suspicious breast lesion. Acad Radiol 17(6): 735–743.

Brill K et al. 2008. Breast-specific gamma imaging compared to breast MRI in patients requiring diagnostic imaging after screening mammography. In ASCO Annual Meeting. September 2008.

Burstein HJ, Polyak K, Wong JS et al. 2004. Ductal carcinoma in situ of the breast. N Engl J Med 350: 1430–1441.

Chaudary MA, Millis RR, Hoskins EOL, Halder M, Bulbrook RD, Cuzick J et al. 1984. Bilateral primary breast cancer: A prospective study of disease incidence. Br J Surg 71(9): 711–714.

Coover LR, Caravaglia DO, Kuhn P. 2004. Scintimammography with dedicated breast camera detects and localizes occult carcinoma. J Nucl Med 45(4): 553–558.

Damle S, McSwain AP, Brem RF, Rapelyea JA, Torrente J, Tatarian T, Teal CB. 2010. Clinical utility of breast-specific gamma imaging in patients receiving neoadjuvant chemotherapy: An institutional review. In (Abstract) The American Society of Breast Surgeons 11th Annual Meeting Proceedings. Vol. 60. <http://www.breastsurgeons.org/upload/ASBS_Official_Proceedings_2010.pdf>. Accessed 11/7/2011.

Damle S, Teal CB, Lenert JJ et al. 2011. Mastectomy and contralateral prophylactic mastectomy rates: An institutional review. Ann Surg Oncol 18: 1356–1363.

Delmon-Moingeon LI, Piwnica-Worms D, Van den Abbeele AD, Holman BL, Davison A, Jones AG. 1990. Uptake of the cation hexakis(2-methoxyisobutylisonitrile)-technetium-99m by human carcinoma cell lines in vitro. Cancer Res 50(7): 2198–202.

Druska CB, Phillips SW, Whaley DH et al. 2008. Molecular breast imaging: Use of a dual-head dedicated gamma camera to detect small breast tumors. Am J Roentgenol 191(6): 1805–1815.

Duffy SW, Tabar L, Chen H et al. 2002. The impact of organized mammography service screening on breast carcinoma mortality in seven Swedish countries. Cancer 95(3): 458–469.

Goldsmith SJ, Parsons W, Guiberteau MJ et al. 2010. SNM Practice Guideline for breast scintigraphy with breast-specific gamma cameras 1.0. J Nucl Med Technol 38(4): 219–224.

Hambly NM, Liberman L, Dershaw DD et al. 2011. Background parenchymal enhancement on baseline screening breast MRI: Impact on biopsy rate and short-interval follow-up. AJR 196(1): 218–224.

Hendrick RE. 2010. Radiation doses and cancer risks from breast imaging studies. Radiology 257: 246–253.

Hilleren DJ, Andersson IT, Lindholm K, Linnell FS. 1991. Invasive lobular carcinoma: Mammographic findings in a 10-year experience. Radiology 178: 149–154.

Hollingsworth AB, Stough RG, O'Dell CA, Brekke CE. 2009. Breast magnetic resonance imaging for preoperative locoregional staging. Am J Surg 197(5): 691–693.

Hruska CB, Phillips SW, Whaley DH et al. 2008. Molecular breast imaging: Use of a dual-head dedicated gamma camera to detect small breast tumors. AJR 191(6): 1805–1815.

Keto JL, Kirstein L, Sanchez DP et al. 2011. MRI versus breast-specific gamma imaging (BSGI) in newly diagnosed ductal cell carcinoma-in-situ: a prospective head-to-head trial. Ann Surg Oncol. DOI 10.1245/s10434-011-1848-3.

Khalkhali I, Baum JK, Villanueva-Meyer J et al. 2002. (99m)-Tc sestamibi breast imaging for the examination of patients with dense and fatty breasts: Multicenter study. Radiology 222(1): 149–155.

Khalkhali I, Villanueva-Meyer J, Edell SL et al. 2000. Diagnostic Accuracy of 99m Tc-sestamibi brast imaging: Multicenter trial results. J Nucl Med 41(12): 1972–1979.

Killelea BK, Gillego A, Kirstein LJ, Asad J, Shpilko M, Shah A et al. 2009. George Peters Award: How does breast-specific gamma imaging affect the management of patients with newly diagnosed breast cancer? Am J Surg 298: 470–474.

Kim BS. 2011. Usefulness of breast-specific gamma imaging as an adjunct modality in breast cancer patients with dense breast: a comparative study with MRI. Ann Nucl Med. DOI: 10.1007/s12149-011-0544-5. Published online 10/18/2011.

Kolb TM, Lichy J, Newhouse JH. 2002. Comparison of the performance of screening mammography, physical examination, and breast US and evaluation of factors that influence them: An analysis of 27,825 patient evaluations. Radiology 225(1): 165–175.

Kopans D, Moore R, McCarthy K et al. 1996. Positive predictive value of breast biopsy performed as a result of mammography: There is no abrupt change at age 50 years. Radiology 200: 357–360.

Krecke KN, Gisvold JJ. 1993. Invasive lobular carcinoma of the breast: Mammograhic findings and extent of disease at diagnosis in 184 patients. Am J Roentgenol 161: 957–960.

Lanzkowsky L et al. 2009. The use of gamma imaging compared to MRI in breast patients needing additional diagnostic imaging. In National Consortium of Breast Centers Annual Conference. March 15–18.

Lanzkowsky L, Rubin D, Fu K et al. 2008. Breast specific gamma imaging in the management of indeterminate lesions detected on breast MRI. In National Consortium of Breast Centers Annual Conference. March 2–5.

Lehman CD, Gatsonis C, Kuhl CK et al. 2007. MRI evaluation of the contralateral breast in women with recently diagnosed breast cancer. N Engl J Med 356(13): 1295–303.

Liberman M, Sampalis F, Mulder DS et al. 2003. Breast cancer diagnosis by scintimammography: A meta-analysis and review of the literature. Breast Cancer Res Treat 80(1): 115–126.

Mandelblatt JS, Cronin KA, Berry DA et al. 2011. Modeling the impact of population screening on breast cancer mortality in the United States. Breast 20(3): S75–S81.

Mann RM, Hoogeveen YL, Blickman JG et al. 2004. The role of MRI in invasive lobular carcinoma. Breast Cancer Res Treat 86: 31–37.

Maublant J, Zhang Z, Rapp M et al. 1993. In vitro uptake of technetium-99m-teboroxime in carcinoma cell lines and normal cells: Comparison with technetium-99m-sestamibi and thallium-201. J Nucl Med 21: 1949–1952.

Meissnitzer M, Dershaw DD, Lee CH, Morris EA. 2009. Targeted ultrasound of the breast in women with abnormal MRI findings for which biopsy has been recommended. Radiology 193(4): 1025–1029.

Miglioretti DL, Rutter CM, Geller BM. 2004. Effect of breast augmentation on the accuracy of mammography and cancer characterisitcs. J Am Med Assoc 291(4): 442–450.

O'Connor MK, Hruska CB, Rhodes DJ et al. 2011. Molecular breast imaging (MBI) in the preoperative evaluation of women with biopsy-proven breast cancer. Abstract: Breast Cancer Imaging: State of the Art 2011. J Nucl Med 52(4): 670.

O'Connor MK, Hruska CB, Rhodes DJ, Weinmann AL, Tortorelli CL, Maxwell R et al. 2011. Low dose molecular breast imaging. Abstract: J Nucl Med 52(4): 670.

O'Connor M, Rhodes D, Hruska C. 2009. Molecular breast imaging. Exp Anticancer Ther 9(8): 1073–1080.

Pisano ED, Gatsonis C, Endrick E et al. 2005. Diagnostic performance of digital versus film mammography for breast cancer screening. N Engl J Med 353: 1773–1783.

Raza S, Vallejo M, Chikarmane SA et al. 2008. Pure ductal carcinoma in situ: A range of MRI features. Am J Roentgenol 191: 689–699.

Reddy M. Given-Wilson R. 2006. Screening for breast cancer. Women's Health Med 3(1): 22–27.

Rhodes DJ, Hruska CB, Phillips SW, Whaley DH, O'Connor MK. 2011. Dedicated dual-head gamma imaging for breast cancer screening in women with mammographically dense breasts. Radiology 258(1): 106–118.

Rosenberg RD, Hunt WC, Williamson MR et al. 1998. Effects of age, breast density, ethnicity, and estrogen replacement therapy on screening mammographic sensitivity and cancer stage at diagnosis: a review of 183,134 screening mammograms in Albuquerque, New Mexico. Radiology 209: 511–518.

Saslow D, Boetes C, Burke W et al. 2007. American Cancer Society guidelines for breast screening with MRI as an adjunct to mammography. CA: Cancer J Clin 57(2): 75–89.

Schomacker K, Schicha H. 2000. Use of myocardial imaging agents for tumour diagnosis—A success story? Eur J Nucl Med Molecul Imag 34: 1845–1863.

Scopinaro F, Pani R, De Vicentis G, Soluri A, Pellegrini R, Porfiri LM. 1999. High-resolution scintimammography improves the accuracy of technetium-99m methoxyisobutylisonitrile scintimammography: Use of a new dedicated gamma camera. Eur J Nucl Med 26: 1279–1288.

Scopinaro F, Schillaci O, Ussof W et al. 1997. A three center study on the diagnostic accuracy of Tc-99m MIBI scintimammography. Anticancer Res 17(3b): 1631–1634.

Selinko VL, Middleton LP, Dempsey PJ. 2004. Role of sonography in diagnosing and staging invasive lobular carcinoma. J Clin Ultrasound 32: 323–333.

Sickles EA. 1991. The subtle and atypical mammographic features of invasive lobular carcinoma. Radiology 153: 25–26.

Tadwalkar RV, Rapelyea JA, Torrente J, Rechtman LR, Teal CB, McSwain AP et al. 2011. Breast-specific gamma imaging as an adjunct modality for the diagnosis of invasive breast cancer with correlation to tumour size and grade. Brit J Radiol. doi: 34392802v1.

Tafra L, Cheng Z, Uddo J et al. 2005. Pilot clinical trial of 18F-fluorodeoxyglucose positron-emission mammography in the surgical management of breast cancer. Am J Surg 190: 628–632.

Taillefer R. 1999. The role of Tc-99m and other conventional radiopharmaceuticals in breast cancer diagnosis. Semin Nucl Med 29(1): 16–40.

Uematsu T, Kasami M, Watanabe J. 2011. Does the degree of background enhancement in breast MRI affect the detection and staging of breast cancer? Eur Soc Radiol 21: 2261–2267.

Weinmann AL, Hruska CB, O'Connor MK. 2011. Design of optimal collimation for dedicated molecular breast imaging systems. Med Phys 36(3): 845–856.

Werner J, Rapelyea JA, Yost KG. 2009. Quantification of radiotracer uptake in axillary lymph nodes using breast specific gamma imaging (BSGI): Benign radio-tracer extravasation versus uptake secondary to breast cancer. The Breast Journal 15(6): 579–582.

Williams MB, Williams B, Goode AR et al. 2000. Performance of a PSPMT based detector for scintimammography. Phys Med Biol 45(3): 781–800.

Zhou JN, Gruner S et al. 2009. Clinical utility of breast specific gamma imaging for evaluating disease extent in the newly diagnosed breast cancer patient. Am J Surg 197(2): 159–163.

V

Future Directions

18

Breast Elasticity Imaging

Salavat Aglyamov
The University of Texas at Austin

Richard Bouchard
The University of Texas MD Anderson Cancer Center

Iulia Graf
The University of Texas at Austin

Stanislav Emelianov
The University of Texas at Austin

18.1 Introduction

Elasticity imaging (or elastography) is a general name for a group of diagnostic methods capable of remote evaluation of tissue mechanical properties. Biochemical, molecular, cellular, and functional (i.e., microscopic) changes in biological tissues leading to pathologies (e.g., breast cancer) often result in macroscopic changes in tissue properties, such as tissue elasticity, viscosity, and other mechanical attributes of tissue. The fact that many pathologies manifest themselves as changes in stiffness serves as the fundamental basis for diagnostic elasticity imaging. From ancient times to the present, physicians have used palpation as a diagnostic tool to detect breast cancer because tumors are generally stiffer than normal tissue. Thus, manual palpation can be considered the oldest form of elastography. In the case of manual palpation, many of the components necessary for modern elasticity imaging are present. First of all, a physician compresses the breast to introduce deformation and stress in the tissue. The physician's fingers are then used as a sensitive tool to assess the mechanical response of the underlying tissue to this external excitation. Finally, based on his or her experience, the physician makes a conclusion about the presence of pathologies and their properties. The sensitivity of the physician's fingers, however, is limited and small lesions or lesions that lie deep within the breast are difficult to palpate. In addition, palpation is not objective and provides only qualitative—not quantitative—information. To this end, elasticity imaging technologies offer an improvement over palpation in detecting and differentiating tumors. They promise to provide objective and quantitative information that is operator independent.

All elasticity imaging techniques, in general, follow a similar approach, as shown in Figure 18.1 The first step is the application of a mechanical excitation, which can be externally applied, as is the case with compression by an ultrasound transducer, can be internally generated, as is the case with acoustic radiation force, or can be naturally occurring, as is the case with mechanical waves generated by cardiac motion. Different measurement methods (e.g., ultrasound imaging, Chapter 15; magnetic resonance imaging, or MRI, Chapter 16; optical imaging, Chapter 20; the assessment of surface pressure) can be used to assess tissue response. Interpretation of this mechanical response is the final step of elastography. It is often the most critical step as it generally relies on a set of assumptions that dictate how the measured mechanical response relates to elasticity parameters. Different tissue excitation sources can be combined with a multitude of tissue motion estimation methods to offer a broad variety of breast elasticity imaging techniques.

Although elasticity imaging is not yet established as an independent clinical tool and many of the new approaches are still in the early stages of development, some of these techniques are already available on the market as a complementary modality for existing clinical systems. Elasticity imaging has a number of potential clinical uses (Greenleaf et al. 2003, Insana 2006, Ophir et al. 2011, Parker 2011, Parker et al. 2010, Sarvazyan et al. 2011); yet, since the introduction of elastography in 1991 (Ophir et al. 1991), elasticity imaging of the breast has been the most widely pursued application. There is little doubt that with time breast elastography will become an important tool in clinical practice.

18.2 Biophysical Principles of Breast Elasticity Imaging

Before breast elasticity imaging techniques are described in detail, necessary terminology and simplifications used in elastography will first be defined. In general, to fully describe the mechanics of breast tissue, very sophisticated mathematical

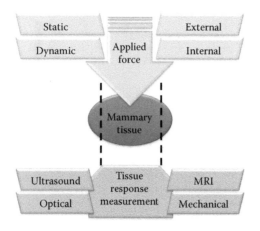

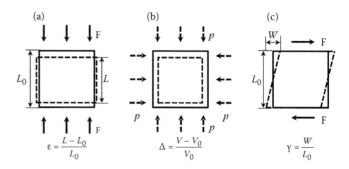

FIGURE 18.2 (a) Uniaxial compression, (b) hydrostatic compression, and (c) shear deformation.

FIGURE 18.1 Diagram generalizing various elasticity imaging techniques.

models are required; these should include not only elastic but also anisotropic, viscous, nonlinear, and poroelastic properties for organs as complex as the breast (Fung 1993). In most practical cases, however, a comprehensive mechanical characterization of the tissue is not required, and breast tissue is often considered as an isotropic, linear, elastic medium. For this idealized medium, its mechanical properties are independent of direction, there is a linear dependence between load and tissue response, and there is no dissipation of the energy during deformation (Christensen 1971, Timoshenko and Goodier 1951). Such an approach is a first approximation and is sufficient to address many of the biomechanical problems arising in elasticity imaging. These simplifications are applied not only to the characteristics of the tissue, but also to the description of the mechanical load. It is assumed that deformations are small and vary slowly with respect to the measurement duration. During the last decade, a number of new theoretical approaches based on more comprehensive mechanical models have been developed (Doyley 2012, Hall et al. 2011, Nightingale 2011, Palmeri and Nightingale 2011, Parker et al. 2010, Sarvazyan et al. 2011), but a linear, isotropic, elastic model will be used here to explain the basis of elasticity imaging. More advanced models of elasticity imaging will be shortly discussed at the end of the chapter.

Elasticity is the property of a body that causes it to be restored to its original shape after deformation. When a body is deformed (i.e., changes shape and/or volume) an external force, *F* internal elastic forces arise to maintain the body's original shape. These internal forces are higher for "stiff" materials; consequently, deformation of stiff materials requires more significant effort. Therefore, the elasticity of a material can be described as its resistance to deformation.

Figure 18.2a presents the uniaxial deformation of a cubic sample, where normal forces are applied on the top and bottom surfaces of the sample. The force divided by the cross-sectional area is called "stress," which commonly has units of pascals (Pa) or newtons per a meter squared (N/m²). The change in shape or volume that a stress induces can be quantified by "strain." Normal or axial strain is defined as a relative change in length

under deformation. Strain is dimensionless and is assumed to be negative under compression and positive under tension. Hooke's law for elastic media defines the relationship between applied stress, σ, and strain, ε, as (Timoshenko and Goodier 1951)

$$\sigma = E\varepsilon, \tag{18.1}$$

where *E* is the ratio of uniaxial stress to the corresponding uniaxial strain and is called the Young's modulus of elasticity. Young's modulus is a measure of the stiffness of an elastic material and is the most common elastic modulus used in elasticity imaging to describe elastic properties of breast tissue. To deform stiffer tissue, increased stress must be applied; therefore, these tissues have a greater Young's modulus. Other elastic moduli commonly used in the theory of elasticity are bulk modulus, *K*, and shear modulus, μ (Timoshenko and Goodier 1951). Because strain is dimensionless, elastic moduli have the units of stress (e.g., Pa).

The bulk elastic modulus, *K*, accounts for the resistance of the material to changing volume, (i.e., it is a measure of a tissue's compressibility), as shown in Figure 18.2b. For hydrostatic compression, when normal stresses or pressure are applied to all surfaces of the body, the only way for the body to deform is through a change in volume. Contrary to this, a shear modulus quantifies a deformation without a change in volume, where only the shape of the body changes. A pure shear deformation is shown in Figure 18.2c, where a shear stress, τ, is applied sideways on the sample and is defined as *F/A*, where *A* is the area over which the parallel force vector, *F*, is applied. The shear strain, γ, is defined as the tangent of a change of angle between two initially orthogonal planes. Shear elasticity is an essential property of solids as the shear modulus of liquids is zero, which reflects the fact that liquids are unable to hold their shape. In general, two independent parameters are required to describe the elasticity of a linear, isotropic medium under a small deformation condition. One parameter quantifies compressibility (bulk modulus), while the other describes a material's resistance to shear deformation (shear modulus):

$$p = K\Delta,$$
$$\tau = \mu\gamma, \tag{18.2}$$

where *p* is hydrostatic pressure on the sample surface and Δ is a volume change during hydrostatic compression.

The elastic constants K and μ of soft tissue reflect the scales of structural tissue composition. A comprehensive review of the role of different levels of tissue organization in elasticity constants is available (Sarvazyan 2001). While bulk modulus is determined by molecular composition and molecular interaction forces, shear modulus is determined by cellular and higher levels of structural organization of soft tissues (Sarvazyan 2001). The molecular composition of soft tissue varies much less than its structure; therefore, bulk modulus does not change significantly for most types of soft tissues. The speed of sound in a medium, c, is generally defined by the bulk modulus and density, ρ, of the tissue (i.e., $K \sim \rho c^2$). The speed of sound is approximately constant for most soft tissues, including cancerous and normal breast tissues, indicating that bulk modulus of tissues does not change significantly with pathological changes in tissue (Goss et al. 1978, 1980). On the contrary, the range of variability of structural features of tissues, such as geometrical parameters of cells and their degree of heterogeneity, is significantly greater (Sarvazyan 2001). Shear modulus is therefore very sensitive to structural changes in the breast and may change by hundreds of a percent during, for example, the development of a malignant tumor.

Because water is the major constituent of all soft biological tissues, tissues are characterized by a low degree of compressibility. This means that the bulk modulus is orders of magnitude larger than the shear modulus (i.e., $K \gg \mu$). Thus, breast tissues are generally assumed to be nearly incompressible materials, i.e., volume change is minimal for any type of load. The ratio of the bulk and Young's moduli defines the Poisson's ratio, ν, as

$$\nu = 0.5 - E/(6K). \qquad (18.3)$$

Poisson's ratio characterizes a material's compressibility and is approximately 0.5 for soft tissues and liquids. A material's shear modulus is coupled with its Young's modulus through Poisson's ratio as

$$\mu = E/2(1 + \nu). \qquad (18.4)$$

If a soft tissue is assumed to be nearly incompressible (i.e., $\nu \approx 0.5$), a consequence of this relationship is that its Young's modulus, E, is three times its shear modulus, μ (i.e., $E = 3\mu$). This is a useful relationship as it allows the elastic behavior of breast tissue to be characterized by only one elastic constant—the Young's or shear modulus; additionally, this constant can be used to quantify tissue elasticity regardless of the type of deformation. Given that Young's modulus is most closely associated with what is sensed during a manual palpation examination, it can be said, with some simplification, that elasticity imaging evaluates the spatial distribution of Young's modulus in breast tissue, much the same way x-ray imaging quantifies a spatial distribution of density.

Several studies have been performed to evaluate the Young's modulus of breast tissue using direct mechanical tests (Krouskop et al. 1998, O'Hagan and Samani 2009, Samani et al. 2003,

Samani and Plewes 2004, Samani and Plewes 2007, Samani et al. 2007, Skovoroda et al. 1995, Wellman et al. 1999) (see also review papers by Aglyamov and Skovoroda 2000, Sarvazyan 2001). Unfortunately, direct mechanical measurement of Young's modulus in vivo is difficult; therefore, these studies were made ex vivo. Despite the obvious limitations of these measurements (e.g., dehydration of the tissue samples, change in temperature), the results suggest that the Young's modulus for breast tissue varies by two orders of magnitude, from 1 to 100 kPa. In these studies, a large number of breast tissue samples, including fat and fibroglandular tissue as well as a range of benign and malignant breast cancer types, were characterized with ex vivo mechanical testing (Krouskop et al. 1998, O'Hagan and Samani 2009, Samani et al. 2007, Skovoroda et al. 1995). It must be noted that the variation of reported Young's modulus values for these studies is significant (for example, the Young's modulus of normal glandular tissue was estimated to be 1 ± 0.5 kPa (Skovoroda et al. 1995), 3.24 ± 0.61 kPa (Samani et al. 2007), and 28 ± 14 kPa (Krouskop et al. 1998) in three different studies). This variation could be explained by different experimental conditions and/or different levels of strain used in each test. For all studies, however, there is a significant difference in Young's modulus values between cancerous and benign breast tissues. For example, compared with normal glandular tissue, infiltrating ductal carcinoma is about 2 times (Skovoroda et al. 1995), 2.5 times (Wellman et al. 1999), or 4 times (Krouskop et al. 1998) stiffer than healthy breast tissue. In another study (Samani et al. 2007), different grades of this type of cancer were found to be from three to 13 times stiffer than healthy tissue. The Young's modulus of fat has been found to be less than or similar to the Young's modulus of fibroglandular tissues in all studies (Krouskop et al. 1998, Samani et al. 2007, Wellman et al. 1999). Despite the limited data from direct mechanical tests and the high interstudy variability found in the literature, the results of ex vivo studies indicate the potential of breast elasticity imaging given the consistent increase observed in the Young's modulus of cancerous tissues.

Currently, no imaging modality can provide direct measurements of a tissue's Young's modulus in vivo. Therefore, all elastography approaches are indirect in that a model is assumed to define the relationship among the mechanical excitation, the tissue response, and the Young's modulus. Often, it is not necessary to quantitatively measure Young's modulus values, but rather providing an image that qualitatively relates to the elasticity distribution in tissue is sufficient.

Based on the type of mechanical excitation, the methods of elasticity imaging are divided into static (or quasi-static), in which case breast tissue is deformed slowly, and dynamic, in which case the load significantly depends on time. Palpation, for example, is typically categorized as a static method. The type of the mechanical excitation used in an elasticity imaging approach establishes the expected relationship between Young's modulus and the measured tissue response.

Static methods usually rely on breast deformation (also often referred to as compression) from a clinician's hand or some compression device (Ophir et al. 2011). If axial strain is measured inside tissue, Hooke's law (Equation 18.1) can be used to evaluate Young's modulus, which is inversely proportional to strain (given the assumptions of a static equilibrium and a constant stress distribution in the tissue). In dynamic methods, the breast is perturbed by harmonic vibrations or impulsive loading using mechanical devices or acoustic radiation force. After measuring tissue motion, the Young's modulus (or shear modulus) can be evaluated based on the estimated speed of a plane shear wave, c_t, traversing through an infinite, homogenous medium as

$$c_t = \sqrt{\frac{\mu}{\rho}} = \sqrt{\frac{E}{3\rho}}. \tag{18.5}$$

A notable difference between Equations 18.1 and 18.5 is that stress measurements are required for static methods to estimate Young's (or shear) modulus (Equation 18.1), while stress need not be known to estimate such a modulus for methods that rely on shear wave velocity measurements (Equation 18.5). Consequently, static (or quasi-static) methods are often used to estimate the *relative* spatial distribution of elasticity in tissue. Both equations, however, are based on significant oversimplifications. Boundary conditions and tissue heterogeneity can lead to a multitude of artifacts or even complete loss of contrast in an image (Barbone and Bamber 2002, Skovoroda et al. 1995). In general, a Young's modulus distribution must be reconstructed based on a solution to an inverse elasticity problem (Aglyamov et al. 2007, Barbone and Oberai 2010, Doyley 2012, Sarvazyan et al. 2011, Skovoroda 2006). This solution, however, is often limited by noise, relies on an incomplete set of experimental data, or requires significant computational resources.

The mechanical response of tissue to an excitation (i.e., induced displacement, velocity, strain, or stress) can be detected using various approaches (Figure 18.1). Therefore, elastography methods are grouped based on imaging techniques used to assess tissue response. The use of optical methods, however, is limited to the penetration depth for quasi-ballistic photons, which is not sufficient for a full interrogation of the breast, and they are generally not considered for breast applications. Historically, ultrasound was first applied for breast elasticity imaging purposes (Céspedes et al. 1993, Garra et al. 1997), and it is still the most common modality used for breast elastography.

18.3 Ultrasound Elasticity Imaging

Ultrasound is especially attractive for elasticity imaging because it is a nonionizing, noninvasive, inexpensive, portable, and real-time method. In contrast to manual palpation, where stress on the breast surface is sensed by the clinician's fingertips, ultrasound imaging permits the measurement of another type of tissue mechanical response—the internal motion of tissue under stress. Ultrasound systems are used to measure displacements, strains, and the speed of mechanical shear waves propagating in the breast.

18.3.1 Detection of Tissue Motion

Motion tracking techniques were originally introduced for the examination of blood velocity profiles and were later adopted and improved for elasticity imaging applications (Hein and O'Brien 1993, Hoskins 1999, Hoskins 2002). Tissue motion results in changes to the received echo pattern from consecutive ultrasound frames. In general, the motion may be estimated in two ways: either the velocity is evaluated using the Doppler effect, whereby motion produces a shift in frequency between the frequency of the transmit pulse and the frequency of the received echo, or the change in the location of a small kernel in the echo pattern is tracked using speckle tracking techniques (Hoskins 1999). While the Doppler effect is primarily used for the measurement of blood flow and tissue velocity in dynamic elastography, speckle tracking techniques are applied in both static and dynamic approaches to measure induced displacements.

The major advantage of using Doppler-based methods for elasticity imaging is that Doppler color flow imaging is already available on most medical ultrasound systems and can be used "as is" for elastography. For example, one of the earliest real-time clinical demonstrations of breast elasticity imaging was the utilization of the internal vibrations induced by a patient's voice and their effect on power Doppler images (Sohn and Baudendistal 1995, Sohn et al. 1994, Svensson and Amiras 2006). Recently, such an approach, called a power Doppler vocal fremitus test, was used to differentiate benign from malignant lesions based on vibration patterns of the lesions (Yildirim et al. 2011).

Doppler-based methods, however, have low spatial resolution and are not directly applicable in quasi-static elastography. As a result of these limitations, most of the current tracking methods in elastography are based on speckle tracking techniques, which tend to provide higher resolution and lower noise images. Despite these advantages, however, most speckle tracking methods are computationally costly, and thus their clinical implementation for ultrasound elasticity imaging was previously limited by unacceptably long off-line data processing times.

To measure local displacements, speckle tracking estimation procedures are applied to two consecutive ultrasound frames: one acquired before and one after compression of the tissue (see Figure 18.3). A small segment (kernel) of echo reflected from a group of scatters with size, T, is identified in the first image. This kernel is then tracked within a search region in the second image, and the best match to the kernel within the search defines the time shift, Δt_n, at a point, n, in the predeformed frame. This time shift can then be used to calculate the axial displacement (i.e., along the axis of the ultrasound beam), d_n, as

$$d_n = c\Delta t_n/2 \tag{18.6}$$

where c is speed of sound. When this procedure is repeated for other kernels (i.e., kernels at different axial locations along a beam and at other beams entirely) throughout a region of interest (ROI), a displacement map is produced. A strain image can then be generated by displaying the spatial derivatives from this

(a)

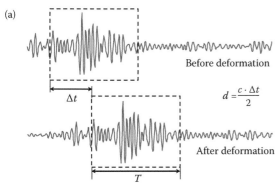

$$d = \frac{c \cdot \Delta t}{2}$$

(b)

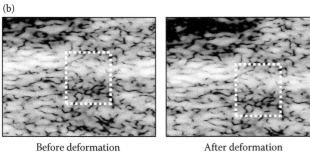

Before deformation   After deformation

FIGURE 18.3 Illustration of speckle tracking algorithm applied for (a) 1-D ultrasound signal and (b) 2-D ultrasound image. In 1-D case, the change in the position of kernel length, T, before and after compression permits the evaluation of time lag, Δt, and axial displacement, d. In 2-D case, both axial and lateral displacements are evaluated. Kernel region is indicated schematically by dotted box.

displacement field. Axial strain, ε_n, can be calculated numerically using, for example, the central difference formula as

$$\varepsilon_n = (d_{n+k} - d_{n-k})/(t_{n+k} - t_{n-k}), \tag{18.7}$$

where k is the step size. Pattern matching can be performed in one, two, or three dimensions, depending on the dimensions of the ultrasound data available (Chen et al. 2005). Most of the currently available medical systems provide 2-D data, and both the axial (along the beam direction) and lateral (perpendicular to the beam direction in the imaging plane) components of the

displacement vector can be estimated. However, the estimation of lateral displacement is much less accurate and usually increases the computational cost of the tracking method. To provide fine spatial resolution, kernel regions on the one hand should be as small as possible (typically, several wavelengths). On the other hand, small kernel size leads to high noise levels, and therefore kernel size choice is an unavoidable tradeoff. The quality of the match between two kernels can be defined using various estimation functions: sum absolute differences, sum squared differences, or different types of correlation functions (Cespedes et al. 1995, Chen et al. 2005, Chen et al. 2004, deJong et al. 1991, Hall et al. 2003, Hein and O'Brien 1993, Lubinski et al. 1999, Pesavento et al. 2000, Pesavento et al. 1999, Zhu and Hall 2002). Radiofrequency signals have a finite sampling interval, but the impact of the finite sampling interval can be reduced by using different types of interpolation, which allows for sub-sample tracking accuracy.

The inherent tradeoff between the speed of tracking calculations and quality of the resulting displacement and strain images has been the focus of researchers for the past few decades. To this end, there has been significant effort dedicated toward the development of real-time speckle tracking algorithms that are able to provide quality elasticity images (Hall et al. 2003, Jiang and Hall 2007, Shiina et al. 2002). Progress in developing real-time speckle tracking algorithms has led to commercialization of elasticity imaging, which has allowed this promising technology to move from the laboratory to the clinic. Currently, there are several commercially available ultrasound systems, including systems manufactured by Hitachi Medical Systems, Siemens Medical Solutions, Philips Healthcare, Ultrasonix Medical Corporation, General Electric Healthcare, and SuperSonic Imagine, in which a real-time elastography tool has been made available to clinicians (Piccoli and Forsberg 2011). In most of these systems, the elasticity imaging mode is realized as a static compressive elastography technique, a general implementation of which is described in the next section.

18.3.2 Static Methods of Elasticity Imaging

The general concept behind static elastography is presented in Figure 18.4 (Ophir et al. 1991). The method is based on applying

(a)

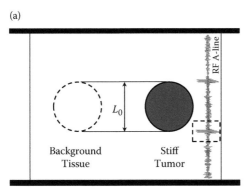

(b)

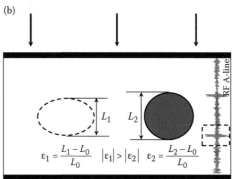

FIGURE 18.4 Schematic representation of static ultrasound elastography: compliant background tissue with a stiff lesion (a) before and (b) after compression. Axial strain in the stiff lesion is lower than strain in the more compliant background tissue.

a small compressive force to the surface of the breast that is continuously imaged with ultrasound. Speckle tracking techniques are then performed on ultrasound images acquired before and after compression to estimate the internal axial strains within the tissue. Based on assumptions discussed earlier, the induced axial strain at a point is inversely proportional to the Young's modulus at that point and, therefore, an axial strain image (or "elastogram") reflects the distribution of the elasticity within the interrogated tissue. In Figure 18.5, ultrasound and axial strain images of a tissue-mimicking phantom (see Chapter 11) with a stiff circular inclusion embedded in a homogeneous background are shown. Please note that the inclusion is too deep to be palpated from the phantom's top surface. In this example, a linear array transducer was pressed into the top surface of the phantom, and motion within was tracked to estimate the displacement field, which was then used to calculate a strain image. Because it is not possible to measure the stress distribution inside a medium, such an approach only provides a relative indication of the Young's modulus distribution.

Figures 18.6 and 18.7 show the results of the first implementation of elastography for breast lesion imaging (Céspedes et al. 1993, Garra et al. 1997). The data acquisition system employed a modified mammography compression paddle; all signal processing to achieve strain estimations was performed off-line.

(a) (b)

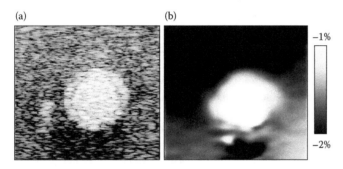

−1%

−2%

FIGURE 18.5 **(See color insert.)** (a) B-mode ultrasound and (b) axial strain image or "elastogram" of a tissue-mimicking phantom with a stiff inclusion.

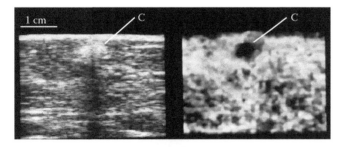

FIGURE 18.6 B-mode ultrasound image (left) and axial strain image (right) of the right breast of a 62-year-old patient. The strain image shows a well defined stiff (black) area, approximately 8 mm in diameter, within the more compliant (white) fat. In the ultrasound image, the cancer nodule appears as a hyperechoic area, with a relatively strong shadow. (From Céspedes I, Ophir J, Ponnekanti H, Maklad N. (1993). *Ultrason Imaging* 15: 73. With permission.)

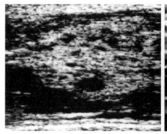

FIGURE 18.7 B-mode ultrasound image (left) and strain image (right) in a patient with a deep fibroadenoma, which is well defined on both images and is firmer than the surrounding (i.e., black in the strain image). (From Garra BS, Cespedes EI, Ophir J, Spratt SR, Zuurbier RA, Magnant CM, Pennanen MF. (1997). *Radiology* 202: 79–86. With permission.)

This initial clinical test confirmed the applicability of static elastography for breast imaging (Céspedes et al. 1993). Figure 18.6 presents B-mode ultrasound and strain images from a breast with an infiltrating ductal carcinoma.

In the first significant clinical test, 46 lesions from 45 patients were examined with an analysis that combined strain imaging data with normal B-mode imaging data (Garra et al. 1997). The study suggested that cancers are stiffer (based on strain images) than benign fibroadenomas and many other benign lesions. It was also found that the size of cancerous tumors was larger on elastograms than their size as estimated by ultrasound imaging or by measurements obtained once the lesions were surgically excised. In comparison, the size of fibroadenomas in the strain image was the same or smaller than the size of corresponding hypoechoic lesions noted in the B-mode image.

Figure 18.8 presents the result of a free-hand elastography test performed on a ductal carcinoma (Hiltawsky et al. 2001). In this particular study, 84 patients with 53 breast lesions were examined, and normalized axial strains were used to produce elasticity images. There was a significant difference in normalized strain between solid lesions and their surrounding tissue (p < 0.01); however, no significant normalized strain difference between benign and malignant lesions was found.

Since static ultrasound elastography has been offered by multiple manufacturers, the number of clinical studies utilizing elasticity imaging in the breast has increased exponentially. With current static (or quasi-static) elastography modes, examination of the breast is a freehand, real-time, and interactive process that affords the clinician the opportunity to adjust his or her compression technique during the exam.

Currently, elastography is generally limited to the characterization of lesions that have already been identified through an independent screening method. The goal of the elastography examination is to differentiate benign from malignant tumors and potentially eliminate unnecessary biopsies. For this purpose, researchers are still exploring useful criteria in differentiating lesions. Criteria used to date include the strain contrast between a lesion and its surrounding tissue, the relative size of the lesion

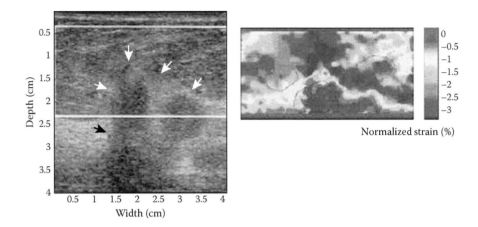

Normalized strain (%)

FIGURE 18.8 (**See color insert.**) B-mode ultrasound image (left) and normalized strain image (right) of a ductal carcinoma in a 70-year-old patient. Average normalized strain values inside and outside the lesion were calculated to be 0.25% and 1.28%, respectively. (From Hiltawsky KM, Kruger M, Starke C, Heuser L, Ermert H, Jensen A. (2001). *Ultrasound Med Biol* 27: 1461–1469. With permission.)

in B-mode versus strain images, and characteristic patterns in the strain image (Barr 2010, Barr and Lackey 2011, Burnside et al. 2007, Fleury et al. 2009, Hall et al. 2003, Hiltawsky et al. 2001, Itoh et al. 2006, Leong et al. 2010, Regner et al. 2006, Sohn et al. 2009, Tan et al. 2008, Yoon et al. 2011, Zhi et al. 2010). An elasticity score based on a contrast and distribution of strain has also been proposed (Itoh et al. 2006) and used in clinical tests (Kumm and Szabunio 2010, Raza et al. 2010, Schaefer et al. 2011, Sohn et al. 2009, Tan et al. 2008, Zhi et al. 2010). Elastography in combination with B-mode ultrasound and Doppler color imaging could improve the diagnostic accuracy of distinguishing benign from malignant breast masses (Cho et al. 2012, Leong et al. 2010, Sohn et al. 2009).

Cancers infiltrate into surrounding tissue and thus appear less mobile than benign tumors. Therefore, shear strains at the tumor boundary are higher for malignant lesions, a feature which could be used in breast cancer diagnosis (Konofagou et al. 2000, Thitaikumar et al. 2008, Thitaikumar and Ophir 2007, Xu et al. 2010). Clinical tests for eight patients with malignant tumors and 33 patients with fibroadenomas indicated that shear strain area is significantly larger for malignant tumors when compared to benign masses such as fibroadenomas (Xu et al. 2010).

Reported data for the sensitivity and specificity of ultrasound elastography vary from study to study. In a recent multicenter study of 779 cases, an overall specificity of 89.5%, and a positive predictive value of 86.8% were demonstrated for elastography compared to using B-mode ultrasound alone (76.1% and 77.2%, respectively) for detecting malignant lesions (Wojcinski et al. 2010). In one study of 186 patients with 200 lesions (Raza et al. 2010), the sensitivity and specificity of elastography were found to be 92.7% and 85.8%, respectively, with four false-negative and 16 false-positive results. In another study of 415 women with 550 breast lesions, a sensitivity of 78.0%, specificity of 98.5%, and an overall accuracy of 93.8% were reported (Tan et al. 2008) for elastography.

Ultrasound elasticity imaging can also be used to improve the characterization of nonsimple cysts (Barr 2010, Barr and Lackey 2011, Booi et al. 2008). One such approach to achieve this

is based on using a "bull's-eye" strain artifact, in which case a black lesion featuring a gray spot in the center and a bright posterior spot is apparent in the strain image. Of the 62 lesions that generated a bull's-eye artifact in the strain image, all were confirmed to be benign cysts by pathology. By using this screening method in such cases, one could expect a decrease from 63.4% to 47.3% in the biopsy rate (Barr and Lackey 2011).

Despite the demonstrated diagnostic value of the method, there are several limitations of static elastography. Perhaps the most significant is that boundary conditions—in addition to a tissue's elasticity distribution—influence what is observed in the resulting strain image. As seen in Figure 18.5, the strain distribution is not uniform in the homogeneous region of the phantom because of the difference in the boundary conditions for the top and bottom surfaces of the phantom. Such a strain distribution suggests that stiffness varies through this homogenous region, an effect that is clearly an artifact related to mechanical boundary conditions. Another limitation is the relative nature of the generated strain images. A more direct elasticity metric, such as Young's modulus, cannot be provided based on only strain distribution; knowledge of the applied stress field must also be known. In addition, conventional ultrasound transducers do not account for out-of-plane motion, which can produce significant decorrelation between ultrasound frames. This problem, however, can likely be addressed with the development of 2-D ultrasound transducers and techniques that allow for the acquisition of volumetric strain images (Fisher et al. 2010).

Due to the aforementioned limitations, a lesion is generally first identified through an independent screening method (i.e., either sonographically or with x-ray tomography) before a clinician attempts to characterize it with elastography (Garra 2007). Because there can be many low-strain areas in the breast that do not correlate with the presence of an actual lesion, elastography is currently not used to screen for cancer as too many false positives would likely result. Yet, as the sensitivity and specificity of elasticity imaging techniques improve, the use of elastography as a screening mechanism may increase in the future.

18.3.3 Dynamic Methods

Dynamic methods for imaging breast elasticity stimulate tissues with harmonic or impulse forces, while tissue motion is measured using the Doppler effect or speckle tracking algorithms (Greenleaf et al. 2003, Palmeri and Nightingale 2011, Parker 2011, Parker et al. 2010, Sarvazyan et al. 2011). A Young's modulus distribution is evaluated based on induced motion amplitude, similar to static approaches (Equation 18.1), or based on shear wave phase or group velocity (Equation 18.5). For dynamic methods, viscous effects play a significant role in tissue motion, and the assumption of a pure elastic body should be made cautiously.

In early implementations, a low-frequency harmonic vibration (20–1000 Hz) was externally applied to excite internal vibrations within tissue, and Doppler detection algorithms were utilized to make real-time vibration images (Krouskop et al. 1987, Lerner and Parker 1987). Since acoustic radiation force (ARF) was introduced (Sarvazyan 1998) to locally disturb tissue, ARF-based dynamic methods have steadily progressed in their development (Palmeri and Nightingale 2011), using advantages afforded by high-frame-rate ultrasound imaging.

18.3.3.1 Acoustic Radiation Force-Based Techniques

Instead of applying a force to the external surface of the breast, it is also possible to generate a force within the breast using ARF. ARF is produced whenever an acoustic wave encounters absorbing or scattering elements in its propagation path (Torr 1984). This force is generated due to a transfer of momentum from the propagating wave to the interacting (i.e., through absorption or scattering) medium. In the case of pure absorption, which is a condition often assumed for soft tissue, an ARF is created in the direction of acoustic wave propagation (Lyons and Parker 1988, Westervelt 1951). If a pure-absorption and plane-wave assumption is made, force application at a point in space is related to the absorption coefficient and temporal average intensity at that point (Nyborg 1965) as

$$F_{ARF} = \frac{2\alpha I}{c}, \tag{18.8}$$

where α is acoustic absorption, I is temporal average intensity of the acoustic beam, and c is speed of sound.

It is possible to achieve elasticity imaging through the generation of an impulsive or harmonic ARF in soft tissue. To this end, the dynamic response of tissue is quantified by tracking the ARF-induced displacement either within the volume of the beam ("on-axis") or at laterally flanking locations ("off-axis"). Measureable ARF does not require the use of dramatically increased transmit intensities; clinically relevant B-mode imaging intensities can be utilized to achieve real-time ARF-based imaging (Bouchard et al. 2009). Instead of increasing transmit intensity, the transmit pulse length is generally increased—to tens or hundreds of microseconds—to achieve a sufficient transfer of momentum to induce dynamics that can be easily tracked. In this section, three different ARF-based elasticity imaging

techniques that have been implemented in breast imaging studies are presented and discussed.

18.3.3.2 ARFI Imaging

If an ARF is only generated for a short period of time (on the order of several milliseconds or less), it is commonly referred to as an ARF impulse or an "ARFI." For the purposes of elasticity imaging, it is possible to generate an ARFI in soft tissue to induce micrometer displacement in the region of the acoustic wave propagation and in the direction away from the face of the transmitting transducer. The resulting dynamic response induced by an ARFI can then be tracked with correlation-based methods for the purpose of elasticity imaging. The peak displacement achieved within the acoustic beam profile, or "on-axis," and resulting from an ARFI has been shown to be inversely proportional to tissue stiffness (Palmeri et al. 2005). A significant amount of research has been published on such an impulsive, on-axis approach, commonly referred to as ARFI imaging, for the purpose of qualitatively characterizing viscoelastic elements of breast, liver, and vascular disease (Behler et al. 2009, Melodelima et al. 2006, Nightingale et al. 2001, Palmeri et al. 2008). ARFI imaging is able to provide a displacement map of an ROI that indicates relative stiffness within a tissue. Although it is unable to provide Young modulus estimates, it can inform a clinician that one region of an ROI is stiffer (or more compliant) than another region.

Nightingale et al. (1995) first utilized ARFI imaging to differentiate cysts from solid lesions in breast tissue in vivo. In this preliminary investigation, in six of seven participants it was possible to induce streaming with ARF in cystic fluid; this response was then tracked with Doppler-based methods for characterization purposes. Similar techniques were then utilized to perform elasticity imaging on the breast of a volunteer with a palpable lesion (Nightingale et al. 2002). Results from this study are illustrated in Figure 18.9. In a more comprehensive study, ARFI imaging methods were then used to characterize 27 in vivo masses prior to needle biopsy (Sharma et al. 2004). Based off these results, it was determined that different ARFI imaging metrics—including peak induced displacement, tissue recovery time, and ARFI imaging lesion size versus B-mode imaging size—could be used to assess lesion malignancy. In another preliminary clinical study, 86 women with 92 lesions were examined with a commercially available ARFI imaging tool offered by Siemens Medical Solutions and called Virtual Touch™ Tissue Imaging (VTI) (Meng et al. 2011). Investigators analyzed the ratio of the B-mode lesion area to the area yielded by VTI; this metric was termed the area ratio (AR). In the 92 lesions imaged, a significant difference in AR values was found between malignant versus benign subsets, with benign lesions having a mean AR of 1.08 and malignant lesions having a mean AR of 1.99.

Despite the demonstrated clinical utility of ARFI imaging, the elasticity imaging method is hindered by a few significant limitations. First, ARFI imaging results tend to give the operator a qualitative—not quantitative—indication of local tissue stiffness. Such a restriction prohibits the clinician from comparing the results of an imaging study to known stiffness metrics for a

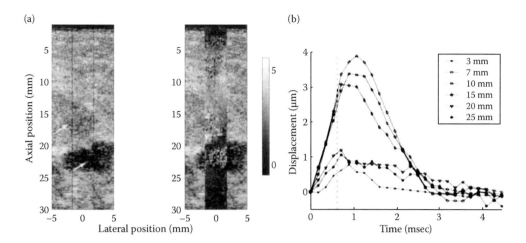

FIGURE 18.9 (a) ARFI image of tissue displacement at 0.8 ms (right) and matched B-mode image (left) in an in vivo female breast. The transducer is located at the top of the images, and the color bar scale is in micrometers. There is a lesion located on the right side of the images between 20 and 25 mm that is evident as a darker region of tissue in the B-mode image (lower arrow). This lesion was palpable and, upon aspiration, was determined to be an infected lymph node. In the ARFI image, the lesion boundary appears stiffer than its interior and the tissue above it (i.e., it exhibits smaller displacements). In addition, the half oval structure in the B-mode image immediately above and to the left of the lesion (upper arrow) appears to be outlined as a softer region of tissue than its surroundings in the ARFI image. (b) Displacement through time at different depths in the center of the ARFI image (0 mm laterally). The tissue at depths of 7, 10, and 15 mm exhibits faster excitation and recovery velocities than that at 20 and 25 mm. (From Nightingale K, Soo MS, Nightingale R, Trahey G. (2002). *Ultrasound Med Biol* 28: 227–235. With permission.)

given pathology (e.g., this region has a specific Young's modulus, which is consistent with an infiltrating ductal carcinoma). Additionally, the repeated transmission of relatively long ultrasound pulses can result in a significant amount of transducer face heating, frequently in excess of a few degrees (Bouchard et al. 2009). Real-time application of ARF-based techniques might therefore require more conservative frame rates or optimized pulse sequencing to reduce the number of acoustic radiation force impulses transmitted during an imaging study.

18.3.3.3 Shear Wave Elasticity Imaging

It is also possible to characterize off-axis, ARF-induced tissue dynamics outside of the acoustic beam profile for the purpose of quantitative elasticity imaging. If one interrogates laterally offset locations slightly later in time (i.e., typically milliseconds following the ARFI transmission), displacement due to the production of shear waves, which travel in the direction transverse to longitudinal wave propagation, can be measured (Sarvazyan et al. 1998). In a linear, isotropic, homogeneous, elastic medium, the speed of these shear waves as a function of tissue elasticity can be expressed using Equation 18.5.

In soft tissue, shear wave velocities generally range from 1 to 5 m/s (Bishop et al. 1998). By tracking these ARFI-induced shear waves, it is possible to perform a quantitative elasticity imaging technique commonly referred to as shear wave elasticity imaging (SWEI), which was first introduced by (Sarvazyan et al. 1998). This imaging technique is based on the relationship presented in Equation 18.5 and allows for the quantification of tissue stiffness (i.e., by providing a Young's modulus estimate) if an elasticity model is assumed (Bercoff et al. 2004, Nightingale et al. 2003, Sarvazyan et al. 1998).

In their first implementation of SWEI in in vivo breast tissue, Bercoff et al. (2003) generated shear waves by mounting an ultrasound transducer to a mechanical vibration unit that caused the face of the transducer to oscillate at 60 Hz. This surface vibration then coupled into the underlying breast tissue to induce shear waves throughout the ROI. The speed of these shear waves was then estimated by tracking axial tissue displacement with conventional correlation-based methods. In an initial pilot study of 15 patients, this low-frequency shear wave technique was able to detect six of the 15 tumors present. The authors attributed this relatively poor detection rate to low frame rates of the system, which did not permit optimal user feedback for the clinician, and due to insufficient coupling of the low-frequency vibration to the ROI.

To improve the robustness of the technique, the mechanical actuation of the transducer face was replaced with a succession of ARF pulses to induce shear waves in the underlying tissue (Bercoff et al. 2004). This technique, termed supersonic shear imaging (SSI), utilizes a series of ARF pulses focused successfully deeper in tissue to allow for the constructive interference of the resulting shear waves along a Mach cone to produce intense quasi-planar shear waves. This imaging technique was then performed in a pilot study with 15 female patients who were undergoing breast biopsy due to the presence of a suspicious lesion that had been identified with mammography. A representative example of an SSI velocity map of an infiltrating ductal carcinoma grade III is shown in Figure 18.10 (Tanter et al. 2008). Recall that the presented shear wave velocities directly correlate to the Young's modulus of the tissue, as dictated by Equation 18.5. Therefore, the red region in the image, with shear wave velocities in excess of 8 m/s, denotes the presence of a stiffer lesion.

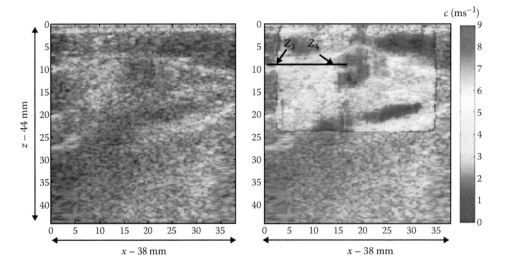

FIGURE 18.10   (See color insert.) Comparison between B-mode ultrasound and quantitative elasticity map performed by the SSI mode. Case 1: Infiltrating ductal carcinoma grade III. Hypoechoic lesion with indistinct margins, slightly posterior shadowing classified as BI-RADS category 5. According to the criteria of elastography interpretation, this is a highly suspicious lesion with a high Young's modulus. (From Tanter M, Bercoff J, Athanasiou A, Deffieux T, Gennisson JL, Montaldo G, Muller M, Tardivon A, Fink M. (2008). *Ultrasound Med Biol* 34: 1373–1386. With permission.)

In a more recent clinical study, suspicious lesions in 46 women were examined with SSI (Athanasiou et al. 2010). Of the 48 lesions imaged, all were successfully characterized as either benign or malignant; malignant lesions presented with a mean Young's modulus value of 146.6 kPa, while benign lesions presented with a much lower mean elasticity of 45.3 kPa.

SWEI can suffer from the same transducer heating issues that can affect all ARF-based imaging methods (Bouchard et al. 2009). Additionally, boundary interfaces within the ROI that cause reflections of the propagating shear wave can confound interpretation of an estimated stiffness modulus. Lastly, SWEI acquisition is limited by the speed of the slow-moving waves that it generates and tracks (Bouchard et al. 2011). Given the relative static nature of breast imaging, however, sampling times are likely more than sufficient for such an application. To overcome this limitation, SuperSonic Imagine offers a SWEI tool that is capable of transmitting a plane-wave tracking pulse to sample the entire field of view (FOV) at once. Such a high-speed tracking method permits one to track the entire FOV at kilohertz frame rates. Siemens Medical Solutions also offers a SWEI tool, called Virtual Touch™ Tissue Quantification, on their commercial ultrasound platform.

18.3.3.4 Vibro-Acoustography

If the transmitted acoustic pulse length is significantly longer than in the aforementioned impulsive cases, it is possible to investigate the vibrational or harmonic response of a tissue with ARF (Urban et al. 2011). The response of breast tissue to low-frequency (i.e., kHz range) vibrations induced by ARF was investigated with a technique known as vibro-acoustography (Fatemi and Greenleaf 1999). A low-frequency ARF field was generated by transmitting from two confocal transducers driven at slightly

different frequencies. At the transmit focus, the beams of these two transducers overlapped and resulted in a harmonic ARF at the difference frequency, which ranged from 22 to 25 kHz, of the two driving functions. The amplitude of the induced vibration at the ARF application point was measured by an off-axis hydrophone. This response could then be used to interrogate mechanical resonances within a tissue. Using vibro-acoustography, Fatemi et al. (2002) were able to detect stiff microcalcifications as small as 110 micrometers in an excised human breast specimen. Vibro-acoustography has also been shown to be able to distinguish ducts, vessels, and Cooper's ligaments in addition to being sensitive to the presence of lesions, with an example of a visualized infiltrating ductal carcinoma offered in a recent study (Alizad et al. 2008). Despite the unique mechanical contrast it provides, however, vibro-acoustography is hindered by a few limitations, perhaps the most significant of which is scanning time (Alizad et al. 2006). Due to the point-by-point raster scanning nature of the method, image acquisition time is on the order of minutes.

18.4 Magnetic Resonance Elastography

In some of the aforementioned elasticity imaging techniques, a harmonic force is applied externally to the breast, and the induced motion is measured with ultrasonic-based methods. This motion in the breast tissue can also be tracked using magnetic resonance imaging (MRI) with a technique called magnetic resonance elastography (MRE) (Litwiller et al. 2012, Mariappan et al. 2010, Sinkus et al. 2012). One of the early implementations of MRI to measure tissue motion was for the assessment of cardiac function and pathologies using MR tagging techniques (Axel and Dougherty 1989, Zerhouni et al. 1988). During the

1990s, several investigators showed the feasibility of using MR methods to measure small displacements, thus demonstrating the potential of MRI for elasticity imaging (Chenevert et al. 1998, Fowlkes et al. 1995, Muthupillai et al. 1995, 1996, Plewes et al. 1995, Steele et al. 2000). Currently, this motion tracking technology is becoming available as an upgrade on conventional MRI scanners (Mariappan et al. 2010).

MRI has several advantages over ultrasound with respect to elasticity imaging. Although ultrasound accurately measures motion along the beam axis, lateral motion is measured with a resolution that is limited by a transducer's depth-dependent beam width. Additionally, out-of-plane motion is generally not considered. These restrictions compromise the quality of the displacement data and constrain the type of model that can be used to produce an elasticity image. And although ultrasound boasts the advantages of low-cost and real-time imaging, MRI gives one the ability to measure 3-D displacements within an object and does this at a higher overall resolution than clinical ultrasound. These advantages not only increase the quality of images when compared to ultrasound, but they often allow for a more accurate Young's modulus distribution, which provides the clinician truly quantitative information about breast elasticity. A solution to an inverse problem is very sensitive to incomplete or noisy experimental data; therefore, MRE techniques permit a more successful application of reconstructive methods as the full 3-D set of elasticity equations can be solved (Chenevert et al. 1998).

One of the first implementations of static MRE for measuring internal tissue strains was used for saturation tagging (Fowlkes et al. 1995). This method, however, suffered from a spatial resolution that was limited by the tagged grid size, which was restricted to 2-D motion. A quasi-static method using bipolar gradient phase encoding (Plewes et al. 1995, 2000) and stimulated echo MRI (Chenevert et al. 1998, Steele et al. 2000) has also been used to measure 2-D and 3-D displacement fields. Preliminary strain images of a normal volunteer were obtained using a quasi-static implementation of this method (Plewes et al. 2000).

In 1995, Muthupillai et al. (1995) proposed a dynamic approach to MRE based on a phase contrast imaging method, which they tested on tissue-mimicking phantoms. Currently, dynamic MRE is more commonly used than static approaches for the purpose of breast elasticity imaging. A low-frequency, harmonic excitation is externally coupled to the subject and is used to generate shear waves at the ROI. Vibrations of a single frequency (typically, within the audible frequency range) are generated by the mechanical driver. The electrical signal for these devices is created by a signal generator triggered by and synchronized to the MR pulse sequence. Several driving mechanisms that externally couple to the patient have been developed (Braun et al. 2003, Muthupillai et al. 1995, Tse et al. 2009), while ARF-based methods, which create shear waves directly at the ROI, have also been used (Wu et al. 2000).

With MRE, estimation of an elasticity distribution is possible based on the phase and amplitude of the harmonic motion. The absolute value of the shear modulus can be computed from local

estimates of phase speed (Equation 18.5) based on the speed of a shear wave in an infinite, homogeneous, elastic medium; a similar assumption is made for SWEI and sonoelasticity (Kruse et al. 2000, Lorenzen et al. 2002, Manduca et al. 2001, 2003, McKnight et al. 2002, Muthupillai et al. 1995). But as has been discussed previously, the influence of boundary conditions significantly complicates such an approach for in vivo conditions.

Another reconstructive approach is based on the solution of an inverse problem. In this case, elastic coefficients are considered unknown and are reconstructed using an experimentally measured distribution of the displacement vector (Sinkus et al. 2000, 2005a, 2005b, 2007, 2012, van Houten et al. 2000, Weaver et al. 2001, Xydeas et al. 2005). An example of such a reconstructive approach for a 60-Hz harmonic excitation is presented in Figure 18.11 (Sinkus et al. 2000). The lesion is very well visualized, with a maximum Young's modulus value of about 3.5 kPa.

MRE has shown promise in the detection and differential diagnosis of cancer in the breast (Lorenzen et al. 2002, McKnight et al. 2002, Sinkus et al. 2005a, 2005b, 2007, Xydeas et al. 2005). The potential of MRE as a means to characterize breast lesions in vivo was first explored in a study of 20 patients in which a 65-Hz mechanical driver was utilized (Xydeas et al. 2005). Results indicated that MRE was able to accurately differentiate benign from malignant lesions based on elasticity differences. In another study that included 15 female patients with malignant breast tumors, five women with benign breast tumors, and 15 healthy volunteers, malignant invasive breast tumors presented with the highest estimated Young's modulus values, with a median of 15.9 kPa and a range between 8 and 28 kPa reported (Lorenzen et al. 2002). In contrast, benign breast lesions presented with significantly lower values of elasticity (median value: 7.0 kPa) than their malignant counterparts. It was reported that a combination of Breast Imaging Reporting and Data System (BI-RADS) categorization with MRE-derived viscoelastic information leads to a substantial rise in specificity (Sinkus et al. 2007). In this study, the use of MRE in the analysis of 39 malignant and 29

(a)   (b)

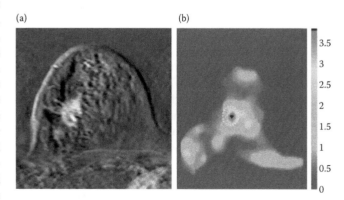

FIGURE 18.11 (See color insert.) (a) Gadolinium-DTPA-enhanced substraction image from an MR mammography. (b) In vivo Young's modulus values from a patient suffering from a breast carcinoma. (Adapted from Sinkus R, Lorenzen J, Schrader D, Lorenzen M, Dargatz M, Holz D. (2000). *Phys Med Biol* 45: 1649–1664.)

benign lesions offered a significant diagnostic gain, with specificity and sensitivity increases of approximately 20% and 100%, respectively, reported.

18.5 Mechanical Imaging

Mechanical imaging is a branch of elasticity imaging that visualizes internal structures of the breast by measuring stress patterns on the surface of tissue (Egorov et al. 2009, Egorov and Sarvazyan 2008, Kaufman et al. 2006, Sarvazyan 1998, Sarvazyan and Egorov 2012, Wellman et al. 1999). This is achieved by compressing the tissue with a probe that contains a pressure sensor array mounted on its contact surface. Mechanical imaging, which is also called "stress imaging" or "tactile imaging," most closely mimics manual palpation. The measurement probe acts similarly to human fingers during a clinical examination by compressing the breast tissue and measuring the resulting pressure distribution on the surface. Unlike other elasticity imaging modalities, where internal tissue strain is measured, mechanical imaging depends on the characterization of surface stress. Under deformation, a lesion produces increased stress on the breast surface; this increase can then be detected by pressure sensors. In many cases, when tumors are located within a few centimeters of the tissue surface, the efficiency of this approach is comparable to other methods. This method, however, cannot be used for imaging lesions located well below the limit of manual palpability.

There are several advantages of stress-based approaches over other methods of elasticity imaging. Measurements of stress provide potential possibilities to evaluate absolute—and not just relative—values of elasticity. Additionally, mechanical imaging does not require highly intensive and computationally costly calculations to implement motion tracking algorithms, which are often required for the aforementioned ultrasound- and MRI-based approaches. Consequently, due the simplicity of the hardware and software required for their implementation, stress-based methods are relatively low-cost. Using cost analysis of currently used breast screening and diagnostic modalities, it was found that mechanical imaging may be over ten times more cost-effective than mammography (Sarvazyan et al. 2008). Disadvantages of stress-based approaches include low spatial resolution and an inability to detect deep lying lesions. Currently, a mechanical imaging clinical device for breast examination is available on the market through Medical Tactile Inc. Using this system, a sensitivity of 91.4% and a specificity of 86.8% in the differentiation of benign from malignant breast lesions in a study of 147 benign and 32 malignant lesions was reported (Egorov et al. 2009).

18.6 Other Mechanical Characteristics of Breast Tissue

In most of the elasticity imaging approaches presented, an overly simplified tissue model of a linear, isotropic, elastic medium is assumed to allow for more straightforward and tractable interpretation of elasticity results. In reality, however, breast tissue under stress demonstrates a much more complicated dynamic behavior than would an idealized elastic body. On the one hand, these often ignored material parameters—such as viscosity or anisotropy—can confound elasticity results and lead to difficulties in interpreting elasticity imaging studies. On the other hand, these additional tissue parameters used to describe tissue behavior open new possibilities for diagnosis. Despite its name, elasticity imaging has not been exclusively considered as a means to measure elasticity (or Young's modulus), but rather it has also been utilized as a way to estimate a set of different mechanical parameters that may be used—in conjunction with linear elasticity—in the diagnosis of breast cancer. Here we shall briefly consider several such parameters.

Hooke's law (Equation 18.1) suggests a linear strain–stress dependence. This assumption, however, is only true for relatively small strains. When strain levels increase, this relationship becomes nonlinear, and a tissue's Young's modulus tends to increase with applied strain (Erkamp et al. 2004, Skovoroda et al. 1999). Breast tissues demonstrate highly nonlinear stress–strain behavior. For example, results of ex vivo tests show that Young's modulus values can almost double when strain increases from 5% to 20% (Krouskop et al. 1998). Tumors tend to be more nonlinear than healthy tissue (Hall et al. 2011). Consequently, the elastic contrast between normal and cancerous tissues significantly increases with an increase in strain. For example, for a 1% deformation, a ductal carcinoma in situ is four times stiffer than normal glandular tissue; for a 15% deformation, this difference doubles to an 8:1 stiffness contrast (Wellman et al. 1999). Recent in vivo studies exhibited that elasticity contrast between a fibroadenoma and the surrounding tissue decreases from 10 for low strain levels to seven for higher (i.e., 11%) levels of strain (Oberai et al. 2009). In the case of invasive ductal carcinomas, however, the contrast does not change. The results of nonlinear reconstructions demonstrate that nonlinear elastic parameters are three times higher for carcinomas than for fibroadenomas. Therefore, tumor differentiation based on imaging the nonlinear elastic properties of tissue has significant diagnostic potential.

Viscosity is another important mechanical characteristic of biological tissues (Fung 1993). Viscoelastic behavior of tissue manifests itself in the temporal behavior of a tissue's dynamic response to mechanical excitation. A generalization of Hooke's law for viscoelastic mediums suggests a complex relationship among stress, strain, and their temporal derivatives (Christensen 1971). Evaluation of breast tissue viscosity has the potential to improve benign versus malignant differentiation of breast lesions (Qiu et al. 2008).

For large time scales, a creep test can be used to measure a tissue's viscous response. In this test, a constant stress is suddenly applied to the breast, and the increase in deformation with time is measured until saturation. Results from an initial clinical test performed on 21 patients show that strain in cancers saturate faster (creep time: 1.4 ± 0.2 s) compared to benign lesions (creep time: 3.9 ± 0.7 s) and normal glandular tissue (creep time: 3.2 ± 0.8 s) (Qiu et al. 2008, Sridhar and Insana 2007).

Measurement of the shear wave propagation in breast tissue also has the potential to estimate additional mechanical parameters (beyond just the Young's or shear modulus), including viscous and anisotropic parameters (Sinkus et al. 2005). Figure 18.12 presents the result of a reconstructed MRE image based on the model of a viscoelastic anisotropic medium. Note that a fibroadenoma detected by MRI (Figure 18.12a) exhibits an increase in elastic (Figure 18.12b), anisotropic (Figure 18.12c, e, f) and

viscous (Figure 18.12d) properties in comparison with the surrounding tissue.

18.7 Conclusions

While breast biopsy remains the gold standard for cancer diagnosis, a large portion of biopsy specimens reveal benign tumors. Breast elasticity imaging is a promising approach in differentiating benign from malignant lesions, and thus potentially reducing the number of unnecessary biopsies performed. In the last several years, breast elasticity imaging has been aggressively developed and thoroughly investigated in clinical trials. Given the encouraging initial clinical results and ease of integration with existing clinical imaging systems, elasticity imaging may soon become a routine clinical tool for breast cancer imaging, diagnosis, and differentiation.

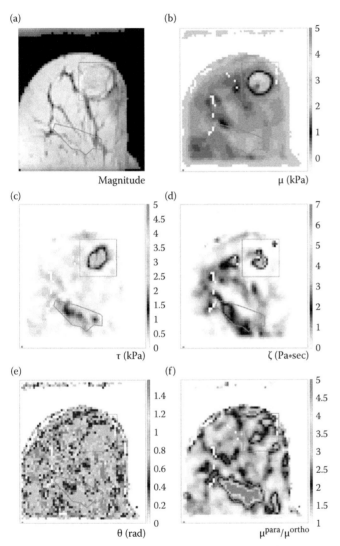

FIGURE 18.12 (See color insert.) In vivo results from a patient with a fibroadenoma. (a) The lesion is easily visible in the MR magnitude image (red rectangle) and the corresponding image of the shear modulus (b) shows the tumor well delimited from the surrounding fatty tissue. The MR magnitude image indicates that the lesion is traversed by septae, which probably lead to the enhance values of the anisotropy (c). The core of the lesion exhibits increased viscous properties (d) and the angle of rotation θ indicates a preferred direction of the fibers inside the lesion (e). (f) The ratio μ^{para}/μ^{ortho} shows that parts of the lesion exhibit strong anisotropic properties. (Adapted from Sinkus R, Tanter M, Catheline S, Lorenzen J, Kuhl C, Sondermann E, Fink M. (2005). *Magn Reson Med* 53: 372–387.)

References

Aglyamov SR, Skovoroda AR. 2000. Mechanical properties of soft biological tissues. Biophycics 45: 1103–1111. Pergamon (English Transl. from Russian).

Aglyamov SR, Skovoroda AR, Xie H, Kim K, Rubin JM, O'Donnell M, Wakefield TW, Myers D, Emelianov SY. 2007. Model-based reconstructive elasticity imaging using ultrasound. Int J Biomed Imag 2007: 35830-1–35830-11.

Alizad A, Whaley DH, Greenleaf JF, Fatemi M. 2006. Critical issues in breast imaging by vibro-acoustography. Ultrasonics 44: e217–e220.

Alizad A, Whaley DH, Greenleaf JF, Fatemia M. 2008. Image features in medical vibro-acoustography: In vitro and in vivo results. Ultrasonics 48(6–7): 559–562.

Athanasiou A, Tardivon A, Tanter M, Sigal-Zafrani B, Bercoff J, Deffieux T, Gennisson JL, Fink M, Neuenschwander S. 2010. Breast lesions: Quantitative elastography with supersonic shear imaging—Preliminary results. Radiology 256: 297–303.

Axel L, Dougherty L. 1989. MR imaging of motion with spatial modulation of magnetization. Radiology 171: 841–845.

Barbone PE, Bamber JC. 2002. Quantitative elasticity imaging: What can and cannot be inferred from strain images. Phys Med Biol 47: 2147–2164.

Barbone PE, Oberai AA. 2010. A review of the mathematical and computational foundations of biomechanical imaging, In De S, Guilak F, Mofrad MRK, eds. Computational Modeling in Biomechanics. Dordrecht: Springer Science + Business Media.

Barr RG. 2010. Real-time ultrasound elasticity of the breast: Initial clinical results. Ultrasound Q 26(2): 61–66.

Barr RG, Lackey AE. 2011. The utility of the "bull's-eye" artifact on breast elasticity imaging in reducing breast lesion biopsy rate. Ultrasound Q 27: 151–155.

Behler RH, Nichols TC, Zhu H, Merricks EP, Gallippi CM. 2009. ARFI imaging for noninvasive material characterization of atherosclerosis part II: Toward in vivo characterization. Ultrasound Med Biol 35: 278–295.

Bercoff J, Chaffai S, Tanter M, Sandrin L, Catheline S, Fink M, Gennisson JL, Meunier M. 2003. In vivo breast tumor detection using transient elastography. Ultrasound Med Biol 29: 1387–1396.

Bercoff J, Tanter M, Fink M. 2004. Supersonic shear imaging: A new technique for soft tissue elasticity mapping. IEEE Trans Ultrason Ferroelectr Freq Control 51: 396–409.

Bishop J, Poole G, Leitch M, Plewes DB. 1998. Magnetic resonance imaging of shear wave propagation in excised tissue. J Magn Reson Imag 8(6): 1257–1265.

Booi RC, Carson PL, O'Donnell M, Roubidoux MA, Hall AL, Rubin JM. 2008. Characterization of cysts using differential correlation coefficient values from 2D breast elastography: Preliminary study. Ultrasound Med Biol 34(1): 12–21.

Bouchard R, Dahl J, Hsu S, Palmeri M, Trahey G. 2009. Image quality, tissue heating, and frame rate trade-offs in acoustic radiation force impulse imaging. IEEE Trans Ultrason Ferroelect Freq Contr 56: 63–76.

Bouchard RR, Hsu SJ, Palmeri ML, Rouze NC, Nightingale KR, Trahey GE. 2011. Acoustic radiation force-driven assessment of myocardial elasticity using the displacement ratio rate (DRR) method. Ultrasound Med Biol 37(7): 1087–1100.

Braun J, Braun K, Sack I. 2003. Electromagnetic actuator for generating variably oriented shear waves in MR elastography. Magn Reson Med 50: 220–222.

Burnside E, Hall T, Sommer A, Hesley G, Sisney G, Svensson W, Fine J, Jiang J, Hangiandreou N. 2007. Differentiating benign from malignant solid breast masses with US strain imaging. Radiology 245(2): 401–410.

Céspedes I, Huang Y, Ophir J, Spratt S. 1995. Methods for estimation of subsample time delays of digitized echo signals. Ultrason Imag 17: 142–171.

Céspedes I, Ophir J, Ponnekanti H, Maklad N. 1993. Elastography: Elasticity imaging using ultrasound with application to muscle and breast in vivo. Ultrason Imag 15: 73–88.

Chen X, Xie H, Erkamp R, Kim K, Jia C, Rubin JM, O'Donnell M. 2005. 3-D correlation-based speckle tracking. Ultrason Imag 27: 21–36.

Chen X, Zohdy MJ, Emelianov SY, O'Donell M. 2004. Lateral speckle tracking using synthetic lateral phase. IEEE Trans Ultrason Ferroelectr Freq Control 51: 540–550.

Chenevert TL, Skovoroda AR, O'Donnell M, Emelianov SY. 1998. Elasticity reconstructive imaging by means of stimulated echo MRI. Magn Reson Med 39: 482–490.

Cho N, Jang M, Lyou CY, Park JS, Choi HY, Moon WK. 2012. Distinguishing benign from malignant masses at breast US: combined US elastography and color Doppler US—Influence on radiologist accuracy. Radiology 262: 80–90.

Christensen RM. 1971. Theory of Viscoelasticity. An Introduction. New York and London: Academic Press.

de Jong PG, Arts T, Hoeks AP, Reneman RS. 1991. Experimental evaluation of the correlation interpolation technique to measure regional tissue velocity. Ultrason Imag 13: 145–161.

Doyley MM. 2012. Model-based elastography: A survey of approaches to the inverse elasticity problem. Phys Med Biol 57: R35–R73.

Egorov V, Kearney T, Pollak SB, Rohatgi C, Sarvazyan N, Airapetian S, Browning S, Sarvazyan A. 2009. Differentiation of benign and malignant breast lesions by mechanical imaging. Breast Cancer Res Treat. doi: 10.1007/s10549-009-0369-2.

Egorov V, Sarvazyan AP. 2008. Mechanical imaging of the breast. IEEE Trans Med Imag 27: 1275–1287.

Erkamp RQ, Emelianov SY, Skovoroda AR, O'Donnell M. 2004. Nonlinear elasticity imaging: Theory and phantom study. IEEE Trans Ultrason Ferroelectr Freq Control 51: 532–539.

Fatemi M, Greenleaf JF. 1999. Vibro-acoustography: An imaging modality based on ultrasound-stimulated acoustic emission. Proc Natl Acad Sci U S A 96: 6603–6608.

Fatemi M, Wold LE, Alizad A, Greenleaf JF. 2002. Vibro-acoustic tissue mammography. IEEE Trans Med Imag 21: 1–8.

Fisher TG, Hall TJ, Panda S, Richards MS, Barbone PE, Jiang J, Resnick J, Barnes S. 2010. Volumetric elasticity imaging with a 2-D CMUT array. Ultrasound Med Biol 36(6): 978–990.

Fleury EF, Rinaldi JF, Piato S, Fleury JC, Roveda JD. 2009. Appearance of breast masses on sonoelastography with special focus on the diagnosis of fibroadenomas. Eur Radiol 19(6): 1337–1346.

Fowlkes JB, Emelianov SY, Pipe JG, Skovoroda AR, Carson PL, Adler RS, Sarvazyan AP. 1995. Magnetic-resonance imaging techniques for detection of elasticity variation. Med Phys 22: 1771–1778.

Fung YC. 1993. Biomechanics. In Mechanical Properties of Living Tissues. New-York: Springer.

Garra BS. 1993. Imaging and estimation of tissue elasticity by ultrasound. Ultrasound Q 23: 255–268.

Garra BS, Cespedes EI, Ophir J, Spratt SR, Zuurbier RA, Magnant CM, Pennanen MF. 1997. Elastography of breast lesions: Initial clinical results. Radiology 202: 79–86.

Goss SA, Johston RL, Dunn F. 1978. Comprehensive compilation of empirical ultrasonic properties of mammalian tissues. J Acoust Soc Am 64: 423–457.

Goss SA, Johston RL, Dunn F. 1980. Compilation of empirical ultrasonic properties of mammalian tissues II. J Acoust Soc Am 68: 93–108.

Greenleaf JF, Fatemi M, Insana M. 2003. Selected methods for imaging elastic properties of biological tissues. Annu Rev Biomed Eng 5: 57–78.

Hall TJ, Barbone PE, Oberai AA, Jiang J, Dord J-F, Goenezen S, Fisher TG. 2011. Recent results in nonlinear strain and modulus imaging. Curr Med Imag Rev 7(4): 313–327.

Hall TJ, Zhu Y, Spalding CS. 2011. In vivo real-time freehand palpation imaging. Ultrasound Med Biol 29: 427–435.

Hein IA, O'Brien WJ. 1993. Current time-domain methods for assessing tissue motion by analysis from reflected ultrasound echoes: A review. IEEE Trans Ultrason Ferroelect Freq Contr 40: 84–102.

Hiltawsky KM, Kruger M, Starke C, Heuser L, Ermert H, Jensen A. 2001. Freehand ultrasound elastography of breast lesions: Clinical results. Ultrasound Med Biol 27: 1461–1469.

Hoskins PR. 1999. A review of the measurement of blood velocity and related quantities using Doppler ultrasound. Proc Inst Mech Eng Part H: J Eng Med 213: 391–400.

Hoskins PR. 2002. Ultrasound techniques for measurement of blood flow and tissue motion. Biorheology 39: 451–459.

Insana MF. 2006. Elasticity imaging, In Akay M, ed. Wiley Encyclopedia of Biomedical Engineering. Hoboken: John Wiley & Sons.

Itoh A, Ueno E, Tohno E, Kamma H, Takahashi H, Shiina T, Yamakawa M, Matsumura T. 2006. Breast disease: Clinical application of U.S elastography for diagnosis. Radiology 239: 341–350.

Jiang J, Hall TJ. 2007. A parallelizable real-time motion tracking algorithm with applications to ultrasonic strain imaging. Phys Med Biol 52: 3773–3790.

Kaufman CS, Jacobson L, Bachman B, Kaufman L. 2006. Digital documentation of the physical examination: Moving the clinical breast exam to the electronic medical record. Amer J Surg 192: 444–449.

Konofagou EE, Harrigan T, Ophir J. 2000. Shear strain estimation and lesion mobility assessment in elastography. Ultrasonics 38: 400–404.

Krouskop TA, Dougherty DR, Levinson SF. 1987. A pulsed Doppler ultrasonic system for making noninvasive measurements of the mechanical properties of soft tissues. J Rehab Res Biol 24: 1–8.

Krouskop TA, Wheeler TM, Kallel F, Garra BS, Hall T. 1998. Elastic moduli of breast and prostate tissues under compression. Ultrason Imag 20: 260–274.

Kruse SA, Smith JA, Lawrence AJ, Dresner MA, Manduca AJFG, Ehman RL. 2000. Tissue characterization using magnetic resonance elastography: Preliminary results. Phys Med Biol 45: 1579–1590.

Kumm TR, Szabunio MM. 2010. Elastography for the characterization of breast lesions: initial clinical experience. Cancer Control 17(3): 156–161.

Leong LC, Sim LS, Lee YS, Ng FC, Wan CM, Fook-Chong SM, Jara-Lazaro AR, Tan PH. 2010. A prospective study to compare the diagnostic performance of breast elastography versus conventional breast ultrasound. Clin Radiol 65(11): 887–894.

Lerner RM, Parker KJ. 1987. Sonoelasticity images derived from ultrasound signals in mechanically vibrated targets. In Proc. 7th European Communities Workshop.

Litwiller DV, Mariappan YK, Ehman RL. 2012. Magnetic resonance elastography. Curr Med Imag Rev 8(1): 46–55.

Lorenzen J, Sinkus R, Lorenzen M, Dargatz M, Leussler C, Roschmann P, Adam G. 2002. MR elastography of the breast: Preliminary clinical results. Rofo 174: 830–834.

Lubinski MA, Emelianov SY, O'Donnell M. 1999. Speckle tracking methods for ultrasonic elasticity imaging using short-time correlation. IEEE Trans Ultrason Ferroelectr Freq Control 46: 82–96.

Lyons ME, Parker KJ. 1988. Absorption and attenuation in soft tissues. II. Experimental results. IEEE Trans Ultrason Ferroelect Freq Contr 35: 511–521.

Manduca A, Lake DS, Kruse SA, Ehman RL. 2003. Spatio-temporal directional filtering for improved inversion of MR elastography images. Med Image Anal 7: 465–473.

Manduca A, Oliphant TE, Dresner MA, Mahowald JL, Kruse SA, Amromin E, Felmlee JP, Greenleaf JF, Ehman RL. 2001. Magnetic resonance elastography: Non-invasive mapping of tissue elasticity. Med Image Anal 5: 237–254.

Mariappan YK, Glaser KJ, Ehman RL. 2010. Magnetic resonance elastography: A review. Clin Anat 23: 497–511.

McKnight AL, Kugel JL, Rossman PJ, Manduca A, Hartmann LC, Ehman RL. 2002. MR Elastography of breast cancer: Preliminary results. AJR 178: 1411–1417.

Melodelima D, Bamber JC, Duck FA, Shipley JA, Xu L. 2006. Elastography for breast cancer diagnosis using radiation force: System development and performance evaluation. Ultrasound Med Biol 32: 387–396.

Meng W, Zhang G, Wu C, Wu G, Song Y, Lu Z. 2011. Preliminary results of acoustic radiation force impulse (ARFI) ultrasound imaging of breast lesions. Ultrasound Med Biol 37: 1436–1443.

Muthupillai R, Lomas DJ, Rossman PJ, Greenleaf JF, Manduca A, Ehman RL. 1995. Magnetic resonance elastography by direct visualization of propagating acoustic strain waves. Science 269: 1854–1857.

Muthupillai R, Rossman PJ, Lomas DJ, Greenleaf JF, Riederer SJ, Ehman RL. 1996. Magnetic resonance imaging of transverse acoustic strain waves. Magn Reson Med 36: 266–274.

Nightingale K. 2011. Acoustic radiation force impulse (ARFI) imaging: A review. Curr Med Imag Rev 7(4): 328–39.

Nightingale K, McAleavey S, Trahey G. 2003. Shear-wave generation using acoustic radiation force: In vivo and ex vivo results. Ultrasound Med Biol 29: 1715–1723.

Nightingale K, Soo MS, Nightingale R, Trahey G. 2002. Acoustic radiation force impulse imaging: in vivo demonstration of clinical feasibility. Ultrasound Med Biol 28: 227–235.

Nightingale KR, Kornguth PJ, Walker WF, McDermott BA, Trahey GE. 1995. A novel ultrasonic technique for differentiating cysts from solid lesions: Preliminary results in the breast. Ultrasound Med Biol 21: 745–751.

Nightingale KR, Palmeri ML, Nightingale RW, Trahey GE. 2001. On the feasibility of remote palpation using acoustic radiation force. J Acoust Soc Am 110: 625–634.

Nyborg WL. 1965. Acoustic streaming. Phys Acoustics 2: 265–331.

Oberai AA, Gokhale NH, Goenezen S, Barbone PE, Hall TJ, Sommer AM, Jiang J. 2009. Linear and nonlinear elasticity imaging of soft tissue in vivo: demonstration of feasibility. Phys Med Biol 54: 1191–1207.

O'Hagan JJ, Samani A. 2009. Measurement of the hyperelastic properties of 44 pathological ex vivo breast tissue samples Phys Med Biol 54: 2557–2569.

Ophir J, Cespedes I, Ponnekanti H, Yazdi Y, Li X. 1991. Elastography: A quantitative method for imaging the elasticity of biological tissues. Ultrason Imag 13: 111–134.

Ophir J, Srinivasan S, Righetti R, Thittai A. 2011. Elastography: A decade of progress (2000–2010). Curr Med Imag Rev 7(4): 292–312.

Palmeri ML, Nightingale KR. 2011. Acoustic radiation force-based elasticity imaging methods. Interface Focus 1: 553–564.

Palmeri ML, Sharma AC, Bouchard RR, Nightingale RW, Nightingale KR. 2005. A finite-element method model of soft tissue response to impulsive acoustic radiation force. IEEE Trans Ultrason Ferroelect Freq Contr 52: 1699–1712.

Palmeri ML, Wang MH, Dahl JJ, Frinkley KD, Nightingale KR. 2008. Quantifying hepatic shear modulus in vivo using acoustic radiation force. Ultrasound Med Biol 34(4): 546–558.

Parker KJ. 2011. The evolution of vibration sonoelastography. Curr Med Imag Rev 7(4): 283–291.

Parker KJ, Doyley MM, Rubens DJ. 2010. Imaging the elastic properties of tissue: The 20 year perspective. Phys Med Biol 56: R1–R29.

Pesavento A, Lorenz A, Siebers S, Ermert H. 2000. New real-time strain imaging concepts using diagnostic ultrasound. Phys Med Biol 45: 1423–1435.

Pesavento A, Perrey C, Krueger M, Ermert H. 1999. A time-efficient and accurate strain estimation concept for ultrasonic elastography using iterative phase zero estimation. IEEE Trans Ultrason Ferroelectr Freq Control 46: 1057–1067.

Piccoli CW, Forsberg F. 2011. Advanced ultrasound techniques for breast Imaging. Semin Roentgenol 46(1): 60–67.

Plewes DB, Betty I, Urchuk SN, Soutar I. 1995. Visualizing tissue compliance with MR imaging. J Magn Reson Imag 5: 733–738.

Plewes DB, Bishop J, Samani A, Sciarretta J. 2000. Visualization and quantification of breast cancer biomechanical properties with magnetic resonance elastography. Phys Med Biol 45: 1591–1610.

Qiu Y, Sridhar M, Tsou JK, Lindfors KK, Insana MF. 2008. Ultrasonic viscoelasticity imaging of nonpalpable breast tumors: Preliminary results. Acad Radiol 15(12): 1526–1533.

Raza S, Odulate A, Ong EM, Chikarmane S, Harston CW. 2010. Using real-time tissue elastography for breast lesion evaluation: Our initial experience. J Ultrasound Med 29(4): 551–563.

Regner DM, Hesley GK, Hangiandreou NJ, Morton MJ, Nordland MR, Meixner DD, Hall TJ, Farrell MA, Mandrekar JN, Harmsen WS, Charboneau JW. 2006. Breast lesions: Evaluation with US strain imaging—Clinical experience of multiple observers. Radiology 238(2): 425–437.

Samani A, Bishop J, Luginbuhl C, Plewes DB. 2003. Measuring the elastic modulus of ex vivo small tissue samples. Phys Med Biol 48: 2183–2198.

Samani A, Plewes D. 2004. A method to measure the hyperelastic parameters of ex vivo breast tissue samples. Phys Med Biol 49: 4395–4405.

Samani A, Plewes D. 2007. An inverse problem solution for measuring the elastic modulus of intact ex vivo breast tissue tumours. Phys Med Biol 52(5): 1247–1260

Samani A, Zubovits J, Plewes D. 2007. Elastic moduli of normal and pathological human breast tissues: An inversion-technique-based investigation of 169 samples. Phys Med Biol 52: 1565–1576.

Sarvazyan A. 1998. Mechanical imaging: A new technology for medical diagnostics. Int J Med Inf 49: 195–216.

Sarvazyan A. 2001. Elastic properties of soft tissue, In Levy M, Bass H, Stern R, eds. Handbook of Elastic Properties of Solids, Liquids and Gases. London: Academic Press. pp. 107–127.

Sarvazyan A, Egorov V. 2012. Mechanical Imaging—A technology for 3-D Visualization and characterization of soft tissue abnormalities: A review. Curr Med Imag Rev 8(1): 64–73

Sarvazyan A, Egorov V, Son JS, Kaufman CS. 2008. Cost-effective screening for breast cancer worldwide: Current state and future directions. Breast Cancer: Basic Clin Res 1: 91–99.

Sarvazyan A, Hall TJ, Urban MW, Fatemi M, Aglyamov SR, Garra BS. 2011. An overview of elastography-an emerging branch of medical imaging. Curr Med Imag Rev 7(4): 255–282.

Sarvazyan AP, Rudenko OV, Swanson SD, Fowlkes JB, Emelianov SY. 1998. Shear wave elasticity imaging: A new ultrasonic technology of medical diagnostics. Ultrasound Med Biol 24: 1419–1435.

Schaefer FK, Heer I, Schaefer PJ, Mundhenke C, Osterholz S, Order BM, Hofheinz N, Hedderich J, Heller M, Jonat W, Schreer I. 2011. Breast ultrasound elastography—Results of 193 breast lesions in a prospective study with histopathologic correlation. Eur J Radiol 77(3): 450–456.

Sharma AC, Soo MS, Trahey GE, Nightingale KR. 2004. Acoustic radiation force impulse imaging of in vivo breast masses. Proc IEEE Ultrason Symp 1: 728–731.

Shiina T, Nitta N, Ueno E, Bamber JC. 2002. Real time elasticity imaging using the combined autocorrelation method. J Med Ultrason 29: 119–128.

Sinkus R, Daire J-L, Vilgrain V, van Beers BE. 2012. Elasticity imaging via MRI: basics, overcoming the waveguide limit, and clinical liver results. Curr Med Imag Rev 8(1): 56–63.

Sinkus R, Lorenzen J, Schrader D, Lorenzen M, Dargatz M, Holz D. 2000. High-resolution tensor MR elastography for breast tumour detection. Phys Med Biol 45: 1649–1664.

Sinkus R, Siegmann K, Xydeas T, Tanter M, Claussen C, Fink M. 2007. MR elastography of breast lesions: Understanding the solid/liquid duality can improve the specificity of contrast-enhanced MR mammography. Magn Reson Med 58: 1135–1144.

Sinkus R, Tanter M, Catheline S, Lorenzen J, Kuhl C, Sondermann E, Fink M. 2005a. Imaging anisotropic and viscous properties of breast tissue by magnetic resonance-elastography. Magn Reson Med 53: 372–387.

Sinkus R, Tanter M, Xydeas T, Catheline S, Bercoff J, Fink M. 2005b. Viscoelastic shear properties of in vivo breast lesions measured by MR elastography. Magn Reson Imag 23: 159–165.

Skovoroda AR. 2006. The Problems of Theory of Elasticity for Diagnosing Pathology of Soft Biological Tissues. Moscow: Fizmatlit.

Skovoroda AR, Emelianov SY, O'Donnell M. 1995. Reconstruction of tissue elasticity based on ultrasound displacement and strain images. IEEE Trans Ultrason Ferroelectr Freq Control 42: 747–765.

Skovoroda AR, Klishko AN, Gusakyan DA, Mayevskii YI, Yermilova VD, Oranskaya GA, Sarvazyan AP. 1995. Quantitative analysis of the mechanical characteristics of pathologically changed soft biological tissues. Biophysics, Pergamon 40: 1359–1364.

Skovoroda AR, Lubinski MA, Emelianov SY, O'Donnell M. 1999. Reconstructive elasticity imaging for large deformations. IEEE Trans Ultrason, Ferroelectr, Freq Control 46: 523–535.

Sohn C, Baudendistal A. 1995. Differential diagnosis of mammary tumours with vocal fremitus in sonography: Preliminary report. Ultrasound Obstet Gynaecol 6: 205–207.

Sohn C, Baudendistal A, Kaufmann M, Bastert G. 1994. The positive vocal fremitus in malignant breast tumours in colour MEM ultrasound imaging—An exciting artefact in confirming the diagnoses? Geburtsh Frauenneilkd 54: 427–431.

Sohn YM, Kim MJ, Kim EK, Kwak JY, Moon HJ, Kim SJ. 2009. Sonographic elastography combined with conventional sonography: How much is it helpful for diagnostic performance? J Ultrasound Med 28(4): 413–420.

Sridhar M, Insana MF. 2007. Ultrasonic measurements of breast viscoelasticity. Med Phys 34: 4757–4767.

Steele DD, Chenevert TL, Skovoroda AR, Emelianov SY. 2000. Three-dimensional static displacement, stimulated echo NMR elasticity imaging. Phys Med Biol 45: 1633–1648.

Svensson WE, Amiras D. 2006. Ultrasound elasticity imaging. Breast Cancer 1–7. Online 2006. doi: 10.1017/S1470903106002835.

Tan SM, Teh HS, Mancer JF, Poh WT. 2008. Improving B mode ultrasound evaluation of breast lesions with real-time ultrasound elastography—A clinical approach. Breast 17(3): 252–257.

Tanter M, Bercoff J, Athanasiou A, Deffieux T, Gennisson JL, Montaldo G, Muller M, Tardivon A, Fink M. 2008. Quantitative assessment of breast lesion viscoelasticity: Initial clinical results using supersonic shear imaging. Ultrasound Med Biol 34: 1373–1386.

Thitaikumar A, Mobbs L, Kraemer-Chant C, Garra B, Ophir J. 2008. Breast tumor classification using axial shear strain elastography: A feasibility study. Phys Med Biol 53(17): 4809–4823.

Thitaikumar A, Ophir J. 2007. Effect of lesion boundary conditions on axial strain elastograms: A parametric study. Ultrasound Med Biol 33: 1463–1467.

Timoshenko S, Goodier JN. 1951. Theory of Elasticity. New York: McGraw-Hill Book Company.

Torr G. 1984. The acoustic radiation force. Amer J Phys 52: 402.

Tse ZT, Janssen H, Hamed A, Ristic M, Young I, Lamperth M. 2009. Magnetic resonance elastography hardware design: A survey. Proc Inst Mech Eng H 223: 497–514.

Urban MW, Alizad A, Aquino W, Greenleaf JF, Fatemi M. 2011. A review of vibro-acoustography and its applications in medicine. Curr Med Imag Rev 7(4): 350–359.

van Houten E, Miga MI, Weaver JB, Kennedy FE, Paulsen KD. 2000. Three-dimensional subzone-based reconstruction algorithm for MR elastography. Magn Reson Med 217: 827–837.

Weaver JB, van Houten EE, Miga MI, Kennedy FE, Paulsen KD. 2001. Magnetic resonance elastography using 3D gradient echo measurements of steady-state motion. Med Phys 28: 1620–1628.

Wellman PS, Howe RD, Dalton E, Kern KA. 1999. Breast tissue stiffness in compression is correlated to histological diagnosis. Technical Report. Cambridge: Harvard BioRobotics Laboratory, Harvard University.

Westervelt PJ. 1951. The theory of steady forces caused by sound waves. J Acoust Soc Am 23: 312.

Wojcinski S, Farrokh A, Weber S, Thomas A, Fischer T, Slowinski T, Schmidtand W, Degenhardt F. 2010. Multicenter study of ultrasound real-time tissue elastography in 779 cases for the assessment of breast lesions: Improved diagnostic performance by combining the BI-RADSR-US classification system with sonoelastography. Ultraschall Med 31: 484–491.

Wu T, Felmlee JP, Greenleaf JF, Riederer SJ, Ehman RL. 2000. MR imaging of shear waves generated by focused ultrasound. Magn Reson Med 43: 111–115.

Xu H, Rao M, Varghese T, Sommer A, Baker S, Hall TJ, Sisney GA, Burnside ES. 2010. Axial shear strain imaging for differentiating benign and malignant breast masses. Ultrasound Med Biol 36: 1813–1824.

Xydeas T, Siegmann K, Sinkus R, Krainick-Strobel U, Miller S, Claussen CD. 2005. Magnetic resonance elastography of the breast: Correlation of signal intensity data with viscoelastic properties. Invest Radiol 40(7): 412–420.

Yildirim D, Gurses B, Ekci B, Kaur A. 2011. Power Doppler vocal fremitus breast sonography: Differential diagnosis with a new classification scheme. Journal of Cancer Therapy 2: 243–252.

Yoon JH, Kim MH, Kim E-K, Moon HJ, Kwak JY, Kim MJ. 2011. Interobserver variability of ultrasound elastography: How it affects the diagnosis of breast lesions. Am J Roentgenol 196: 730–736.

Zerhouni EA, Parish DM, Rogers WJ, Yang A, Shapiro EP. 1988. Human heart: Tagging with MR imaging—A method for noninvasive assessment of myocardial motion. Radiology 169: 59–63.

Zhi H, Xiao XY, Yang HY, Ou B, Wen YL, Luo BM. 2010. Ultrasonic elastography in breast cancer diagnosis: Strain ration versus 5-point scale. Acad Radiol 17: 1227–1233.

Zhu Y, Hall TJ. 2002. A modified block matching method for real-time freehand strain imaging. Ultrason Imag 24: 161–176.

Electrical Impedance Imaging and Spectroscopy of the Breast

Bin Zheng

University of Pittsburgh

Lihua Li

College of Life Information Science and Instrument Engineering

19.1 Introduction

A number of imaging and nonimaging-based breast abnormality detection or cancer screening technologies have been investigated as the alternative and/or adjunct approaches to the mammography (Chapter 1) in order to improve the performance and efficacy of breast cancer screening and early detection (Chapter 7). Among them, electrical impedance (EI) property of breast tissue has been recognized and proved as potentially useful biomarker that directly relates to the biological processes of developing breast abnormalities and/or cancer at the molecular or cellular levels. EI technology is also a safe (nonradiation), possibly noninvasive, low cost, fast (real time), and easily implemented cancer screening method. Hence, EI technology has been investigated and evaluated by a number of research groups around the world for the purpose of developing better or more sensitive systems and screening approaches to detect breast abnormalities or classify suspicious breast lesions. This chapter reviews the basic concept of applying EI technology to distinguish between normal and malignant breast tissue as well as the development and evaluation of two noninvasive EI approaches namely, electronic impedance tomography (EIT) and electronic impedance spectroscopy (EIS), to detect breast abnormalities or cancer by generating EI tomography or mapped images as well as multifrequency EI scanned sweep signals. The potential advantages and limitations of current EI technology, as well as its future perspectives, in detecting breast cancer are also discussed.

19.2 Electrical Impedance Measurement of Breast Tissue

To discuss the basic concept of electrical impedance (EI) measurement of human organ (i.e., bone, brain, breast, heart, lung, and thyroid) or the corresponding tissues, a RC series, or parallel circuit is often used to represent a simplified electrical impedance measurement loop involving the tested human organ or tissue specimen. In analyzing the basic electrical properties of measurement loops, four terms, namely resistance (R), conductance (σ), capacitance (C), and permittivity (ε), are commonly used. Specifically, resistance opposes current flow and conductance is the inverse of resistance. Capacitance is a property that opposes a change in voltage or electric potential across an object and acts to store energy. A capacitor usually consists of two conductors that are oppositely charged and separated by the dielectric material. Thus, the capacitance of a common two-plate capacitor is expressed as $C = \varepsilon A/d$, where A is the area of each plate and d is the distance between two plates. Permittivity is a property of the dielectric material reflecting the ability of charges in the material to move in response to an electrical field. In a simplified RC series circuit, the impedance (Z) is a complex value expressed as

$$Z = R + jX_C, \qquad (19.1)$$

where $X_C = \dfrac{1}{2\pi fC}$. The impedance can also be expressed in the form of polar coordinates, $Z = Z/\theta$, where $Z = \sqrt{R^2 + X_C^2}$ and $\theta = \arctan\left(\dfrac{X_C}{R}\right)$. In an RC parallel circuit, the admittance (Y) is also a complex value expressed as

$$Y = G + j\frac{1}{X_C}, \qquad (19.2)$$

where $Y = \dfrac{1}{Z}$ and $G = \dfrac{1}{R}$ (conductance). Hence, in both RC series and parallel circuits, the electrical impedance (or admittance) is frequency (*f*) dependent. Thus, selecting the optimal frequency range of the applied electric current plays an important role in determining the performance of applying EI technology to detect diseases or abnormal tissues depicted in different human organs.

Human and animal tissues consist of cells and extracellular medium with unique electrical characteristics. For example, under normal conditions, blood typically has good conductivity with a lower impedance level around 150 Ω cm (resistivity); muscle has medium impedance level around 530 Ω cm, and fat has substantially high impedance level in the range of 2060 to 2720 Ω cm. Thus, the change of electrical impedance level in a tissue might be a useful indicator for detecting abnormalities or diseases (Hope et al. 2004). Specifically, the cell consists of the cellular membrane and the intracellular medium. The cellular membrane is basically composed of a lipid bilayer and proteins, which primarily show the electrical property of capacitance with the selective permeability that is a unique physiological property of living cells to define membrane conductivity. The intracellular material consists primarily of ionic solutions with microscopic structures and proteins that can be charged and moved in response to applied external electrical fields. The extracellular medium consists of ionic liquid with higher conductance. Due to different biological processes in living tissues (or cells), their frequency-dependent behavior of the electrical impedance is more complicated than a simple man-made RC series or parallel circuit with fixed capacitance or conductivity. The conductivity of biological tissues is not constant. The tissue conductivity can often be expressed as

$$\sigma' = \sigma + j\,\pi\varepsilon_0\varepsilon \tag{19.3}$$

where σ is the real portion of the conductivity, ε_0 is the permittivity of free space, and ε is the relative permittivity of the measured tissue. Thus, at different frequencies the contribution to the impedance by the different tissue components (cellular membrane, intracellular, and extracellular medium) also varies.

The tissue impedance level varies with the varying frequencies of the applied electrical field and the change of the electrical properties of the breast tissue can be divided into three categories as represented in α-, β-, and γ-dispersion. The α-dispersion is mainly affected by the ionic environment surrounding the cells, which occurs at a relatively low frequency range (10 Hz to 10 kHz) in which the ionic extracellular medium with primarily conductance values dominates the measurement. As the electrical field frequency increases from 10 kHz to 10 MHz (entering β-dispersion with a cell structure relaxation), the permittivity and capacitance of the cell membrane gradually become the dominative forces that form a large portion of the electrical impedance being measured. Typically, from above 30 kHz to 30 MHz, the capacitive charging of the cell membrane and dipolar relaxation of proteins in the measured tissue determine the overall permittivity. At even higher frequency ranges, the γ-dispersion is mainly related to the movement of water molecules inside the tissue. In medical application, the frequencies of the applied electrical field are typically limited to the α- and β-dispersion regions. In addition, the measured impedance values of human tissue also vary with many other factors including

temperature and time. The measurement results are often anisotropic (Foster et al. 1989). Hence, measurements under multiple frequencies are often required and compared in investigating the electrical impedance properties of the human tissues.

Applying electrical impedance measurement in breast cancer research started as early as the 1920s (Fricke et al. 1926), in which the differences of electrical impedance properties (capacitance) between malignant and benign breast tissues were detected and identified. The first operational electrical impedance scanning system for analyzing breast tissues was developed in the early 1980s (Chaundhary et al. 1984). Using this system, the researchers detected and found that malignant breast tissues might have 20- to 40-fold higher conductivity and capacitance than the negative (healthy) breast tissues. Since then, a large number of in vitro studies have been conducted and reported to measure various electrical impedance properties from the excised normal and abnormal (including malignant) breast tissue specimens (Zou et al. 2003). Table 19.1 summarizes examples of a few reported studies and the experimental results of measuring electrical impedance of breast tissues in vitro.

These studies found that in a wide frequency range the differences in conductivity (or electrical impedance) between benign and malignant breast tissue were primarily caused by a

TABLE 19.1 Examples of the Studies in Measuring Electrical Impedance of Breast Tissues In Vitro

Study	Frequency	Experimental Results
Fricke et al. 1926	20 kHz	The measured capacitances ranged from 107 to 583 pF for negative tissues and from 545 to 2860 pF for malignant tissues acquired from 58 patients.
Chaundhary 1984	3 MHz to 3 GHz	The measured conductivities under 100 MHz ranged from 1.5 to 3 mS/cm for the negative tissues and from 7.5 to 12 mS/cm for the malignant tissues.
Jossinet et al. 1985	1 kHz	Average impedance was 200 Ω cm for the malignant tissues and 400 Ω cm for the surrounding nonpathologic fatty tissues.
Surowiec et al. 1988	20 kHz to 100 MHz	The measured conductivities ranged from 0.3 to 0.4 mS/cm for the negative peripheral tissues and from 2 to 8 mS/cm for tissues extracted from the malignant tumor center. Also permittivity values ranged from 8 to 800 for the negative tissues and from 80 to 10000 for the malignant tissues.
Campbell et al. 1992	3.2 GHz	The study found no significant difference in measured conductivity and permittivity levels between the tissues extracted from benign and malignant tumors.
Jossinet 1996	0.5 kHz to 1 MHz	At this frequency range, the measured impedance ranged from 243 ± 77 to 389 ± 97 Ω cm for malignant tissues, from 1747 ± 283 to 2188 ± 338 Ω cm for adipose fatty tissues, from 859 ± 306 to 1109 ± 371 Ω cm for connective tissues, and from 200 ± 52 to 245 ± 70 Ω cm for fibroadenoma tissues. At frequency >100 kHz, the malignant tissues had the most capacitive response.

number of factors including changes in (1) water and electrolyte content (which are related to hormonal changes and angiogenesis), (2) membrane permeability and polarization, as well as (3) orientation and packing density of cells. With regard to comparison with surrounding normal tissues, most of these studies concluded that the tissues located inside malignant tumors generally had higher conductivity (or permittivity) and lower impedance. These studies also indicated that at the higher frequencies (in the β-dispersion region), the electrical impedance measurement showed higher diagnostic value.

To locate breast lesions and detect tissue electrical impedance in vivo, researchers have also developed and tested various invasive EI measurement devices. Figure 19.1 illustrates a simplified diagram that shows the working concept of an invasive EI device. In the EI device, two electrodes are embedded within a coaxial needle probe that is inserted into the suspicious breast lesion site to conduct the test. The electrical signal (input AC current) at the selected frequency is applied to the probe and the response (output voltage) signal is detected by the detection electrode. Another reference electrode probe is placed on a woman's abdomen. The electrical impedance values are computed and the corresponding measurement signals are displayed by the electrical signal processing unit in the EI device. Using similar EI devices, a number of studies have been conducted and reported to measure electrical impedance of breast lesions. For example, in one study (Morimoto et al. 1993), EI examinations over the input current frequency range of 0 to 200 kHz were conducted on 54 patients just before biopsy under general or local anesthesia. Using a model circuit that consists of the extracellular resistance in parallel with a series combination of the intracellular resistance and the capacitance, the researchers computed the extracellular resistance (R_e), the intracellular resistance (R_i), and membrane capacitance (C_m) based on the measured complex impedance signals. Despite the overlap of the computed values for the malignant and benign tissues, the study reported a statistically significant difference between malignant and benign tissues ($p < 0.01$) in which the malignant tissues had significantly higher resistance (R_e and R_i) and lower capacitance (C_m) than the benign tissues. The researchers concluded that such an electrical

impedance measurement might be able to distinguish between the malignant and benign breast lesions in vivo. This or similar invasive EI devices can be relatively low cost and have fast response. By combining with needle biopsy, such EI measurement may aid in the better localization of the biopsied tumor and increase the accuracy of the diagnosis (Morimoto et al. 1993).

However, considerable variability has also been reported in tissue impedance measurements conducted under varying conditions using frequency ranging from 20 kHz to 100 MHz (Jossinet 1996). Despite the reported variability, these previous studies involving both in vitro and in vivo tests established an important foundation for better understanding the concept and feasibility of applying electrical impedance technology to detect and diagnose breast cancer. A consensus has been established and widely accepted. It indicates that the cancer cells exhibit altered dielectric properties and, as a result, the tissues extracted from malignant tumors demonstrate significantly higher capacitance and conductivity, with associated decreased electrical impedance.

19.3 Electrical Impedance Tomography (EIT) for Breast Cancer Screening and Detection

To overcome the difficulty and limitation of the invasive measurement approach, in the last decade noninvasive EI methods have also been investigated and developed for detecting breast abnormalities and/or cancer by directly applying electrodes to the breast skin and measuring EI signal distributions inside the breast. Electrical impedance tomography (EIT) is a popular approach for this purpose. EIT is a technology applied to reconstruct or map the impedance distribution in an object of interest from the electrical signals measured from the surface of the object. EIT enables observers to visualize the spatial distribution of electrical impedance inside the human body including the breast. Thus, an EIT device basically uses voltage measurements on the object's surface when an electric current passes through the volume, as initial data for the image reconstruction. A suitable operation of the EIT device requires both fast and effective reconstruction algorithms and high accuracy of electrical measurements. Currently, due to the advance of EIT technology (including both hardware and software), EIT images can be reconstructed quickly, which enables observers to monitor many processes (such as heart pulsation) in real time.

Since the 1990s, a number of research groups have developed several EIT experimental systems in the laboratories and investigated the feasibility of applying EIT to detect breast cancer (Zou et al. 2003). Among them, a few examples are discussed in this chapter. First, a research group at Dartmouth College (http://www-nml.dartmouth.edu/biomedprg/EIS/index.html, Hanover, NH, U.S.A.) has built a broadband high-frequency EIT system for breast imaging (Halter et al. 2008). The system includes three electronic subsystems arranged hierarchically, which are an EIT measurement channel module, a signal control module, and a signal processing computing module. The EIT

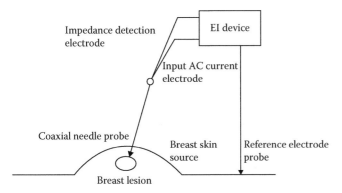

FIGURE 19.1 Illustration of applying an invasive coaxial needle probe to measure electrical impedance of a suspicious breast lesion.

measurement module includes 64 Ag/AgCl electrodes incorporated into a mechanical framework uniquely designed for breast imaging. The framework uses multiple rings of the electrodes that were designed to compress the breast to a specific geometry by considering the information on both the locations of the electrodes and the shape of the breast. This EIT system simultaneously applies electrical signals to all electrodes and can be operated under frequencies ranging from DC to 10 MHz. Using a preestablished imaging protocol, each breast was examined at 20 logarithmically spaced frequencies ranging from 10 kHz to 10 MKz. The measured output electrical signals from the breast are separately grouped into 16 circuit boards (four electrodes or channels each) to be processed by a 32-bit digital signal processor (DSP) and field programming gate arrays (FPGA). A digital matched filter is implemented to determine the in-phase and quadrature components of the measured signals. In order to sample high frequency signals, a multiperiod undersampling technique is incorporated to record the data at approximately the full bandwidth of the ADC (analog-to-digital converter), which results in a maximum frame rate of 182.7 frames per second when collecting 15 spatial patterns. These methods increase the data processing efficacy and enable the system to examine one breast and produce EIT images in approximately 2 to 3 minutes using all 64 electrodes.

The researchers reported that in a clinical trial this EIT system had been used to image 96 women with diverse breast density (including 11 fatty, 56 scattered fibroglandular dense, 24 heterogeneously dense, and five extremely dense breasts). The volumetric conductivity distributions within the breast were reconstructed for each breast at each of the 20 examined frequencies for comparison. The results showed that the mean absolute conductivity values for all examined breasts ranged from 0.0237 S/m at 10 kHz to 0.2174 S/m at 10 MHz indicating an almost linear increase of conductivity as the function of the logarithm increase of the applied electrical field frequency. As breast density increased, the average impedance decreased as expected. The results also showed that at lower frequencies, less current was able to penetrate into the deeper breast tissue due to the higher resistance (or impedance) of the breast skin and the peripheral fatty tissue, which reduces the signal discriminatory ability. At higher frequencies, more current was able to be delivered into the deeper breast tissue region due to reduction of contact impedance and, thus, the internal breast tissue composition became visible and the modeling errors were also reduced. The study concluded that significant differences between tissue type conductivities at higher frequencies could be made due to the better model–data match. Despite the significant improvement in system development (including both hardware and data processing software), the acquired information displayed on EI images could only correspond relatively well with some of the given clinical information. No specific performance level of cancer detection or classification between malignant and benign tumors was reported, but the conclusion was that the EIT examinations showed potential to yield higher sensitivity and quite low specificity due to the difficulty in distinguishing between benign and malignant lesions.

Second, a joint group involving many researchers from several engineering departments at Rensselaer Polytechnic Institute (Troy, NY, U.S.A., http://www.ecse.rpi.edu/homepages/saulnier/eit/) has developed and extensively modified a different EIT system. The system (named as ACT 4) is a multifrequency (up to 1 MHz), multichannel (up to 72 channel) instrument. Currently, the system is configured to support 60 electrodes in two 5 × 6 radiolucent arrays (Liu et al. 2005). Each electrode is driven by a 16-bit precision voltage source and has a connected sensor to measure the resulting (output) current and voltage signals. The electrical sensors and data processing device are digitally controlled to produce and measure signals at 5 kHz to 1 MHz with the magnitude and phase signals independently controlled. This group has also published a number of studies that focused on developing new image reconstruction algorithms to build better and easily interpretable 2-D and 3-D EIT images to show the internal electrical impedance structure and distribution of a breast with improved resolution. Specifically, to help radiologists more effectively read and interpret EIT images, the researchers have developed and tested a new EIT image reconstruction algorithm based on a data linearization approach and a forward model to build the EIT images in mammography geometry (Choi et al. 2007). When applying this reconstruction algorithm to a phantom tank test, a 5-mm-cube metal target and a 6-mm-cube agar target were visible and detectable at the depth of 15 mm.

Recently, by integrating the EI electrodes into the mammography machine, the researchers have investigated and tested the feasibility of conducting EIT examination simultaneously with breast tomosynthesis imaging examination (Chapter 4) to improve performance in classifying between benign and malignant breast lesions (Kim et al. 2007). During an examination, the breast was compressed by the two regular mammography plates that have radiolucent electrode arrays attached. The breast was first scanned six times by the ACT4 EI system at 5, 10, 30, 100, 300, and 1000 kHz. Voltage up to 0.5 V (at peak level) were applied during each scan. The current applied to the electrodes ranged from 0.4 mA peak at 5 kHz to 4.9 mA at 1 MHz. The ACT4 system has voltage sources that apply orthogonal sets of voltage patterns to the breast under test. The data receiver unit of the ACT4 system measured both the applied voltages and the resulting currents through the electrodes. Thus, each scan at a specific frequency obtained the EI signal spectrum information with 59 orthonormal excitation patterns that aim to maximize the discriminatory power of the EIT examination. After the EIT examination, a digital mammography or a breast tomosynthesis scan was then applied. Using the acquired EI measurement signal data, a set of Cole–Cole plots were generated. Each Cole–Cole plot is a graph with a complex graph axis format of an imaginary part versus the real part of measured electrical admittance. This combined test was conducted on two breasts including one positive (depicting the ductal carcinoma) and one negative. The results showed that the reactive component of the positive breast was larger than that of the negative breast. In addition, the shape of the plot curves was different with a semicircular shape for the negative breast and a nearly straight line type shape for the positive breast.

Third, an international team (involving the Institute of Radio-Engineering and Electronics of the Russian Academy of Sciences and the TCI Medical Inc., Albuquerque, NM, U.S.A.) also developed a 3-D EIT system, referred to as the breast cancer detection device (BCDD) (Cherepenin et al. 2001). This EIT system was installed with total 256 electrodes arranged in a square matrix of 12×12 cm^2. The protruding electrodes are placed on a rigid sensor plane to improve the contact with the breast skin when the sensor plane is pressed against the breast. A microprocessor in the control system of the EIT device is programmed to automatically determine which electrode current is driven and which electrode is selected to measure the response (output) electrical signal from the breast. The output signals are detected and processed through an analogue synchronous detector and a switchable integrator converts input AC signal to DC signal before analogue-to-digital conversion. A synchronous detector measures real part of impedance and a 16-bit A/D converter measures voltage difference signals. A full EIT scan and impedance measurement takes less than 20 seconds. Then, the dedicated image reconstruction software (based on the back-projection concept) is applied to reconstruct the 3-D conductivity distribution maps (the tomographic cross sections or image slices that are parallel to the electrode sensor plane). The resulting EIT image slice shows breast tissue conductivity in a grayscale distribution. In each image slice the dark areas indicate low conductivity, while the light areas represent the high conductivity measurement. Specifically, the regions depicting normal (negative) breast tissues typically show the "mosaic" type EI images due to the complicated anatomical tissue structure, while the abnormal regions depicting breast tumors often show the clear (focal) bright areas due to the increase of local conductivity of the breast tumors.

The BCDD EIT system was first calibrated by conducting tests on a saline-filled tank and the breast of a volunteer. The system was then applied to 21 women with a breast tumor detected in one breast from the mammograms. By analyzing the reconstructed EIT images, the researchers reported that the clear focal enhancement EI signals that represent breast tumors could be seen in 14 cases and the unfocal abnormal EI signal patterns that are different from the rest of the negative regions could also be detectable in the other four cases, which generated an approximately 86% (18/21 cases) detection sensitivity.

Although the great research and development efforts in applying EIT technology to breast cancer detection have been made by a number of groups around the world, the T-Scan 2000, originally developed by Mirabel Medical Systems Ltd, Israel, is the only commercialized EIT system that has been approved by the U.S. Food and Drug Administration (FDA) (1999) as an adjunct breast cancer screening tool to mammography to date. The system measures surface currents on the breast and uses the information to distinguish tissue inside the breast. To be used in the clinical test, the system is able to map noninvasively the local distribution of breast tissue electrical impedance at various frequencies in real time (Assenheimer et al. 2001). The T-Scan 2000 system includes two electrode probes. A voltage source is connected to a handheld electrode probe touched to the woman's arm and the measuring probe with multiple electrodes arranged on a rectangular grid is pressed on the breast surface. Thus, an alternating electric field is established between the woman's arm and the breast under test. The current travels from the woman's arm to the highly conducting pectoralis muscle that can be defined as an "isopotential" plane. Hence, a roughly parallel electric configuration is established between the pectoralis muscle and the plate (measuring probe) pressed on the breast surface. The measuring probe has an eight by eight pad array with a total footprint of 32×32 mm. At each pad the current is measured using a transimpedance measurement method, in which the evoked current under a particular pad is measured, while all other pads are kept at the ground potential. The voltage applied between the handheld probe and the measuring probe placed on the breast is in the range of 1 to 2.5 V with the frequencies spanning from 100 Hz to 100 kHz. The measured signals, coupled with the system's transfer function, are utilized to compute the electrical impedance at each pad sensor. The conductance and capacitance-related maps are also calculated from the measured data and used as the EIT images displayed on the associated computer screen. Due to the system's capability to produce real-time EIT maps or images, physicians can use the displayed EIT images as guidance to manipulate the placement of the measuring probe on the breast surface during the EIT test. The T-Scan 2000 EIT system has been calibrated to be able to detect a 5% capacitance change in the range of 20 pF to 1 nF or a conductance value change in the range of 1 to 20 μS, making it suitable for detecting and/or measuring subtle changes of the electrical properties of breast tissues. The system generally has higher signal-to-noise ratio at lower frequencies (e.g., 70 dB at 100 Hz) than that at higher frequencies (e.g., 40 dB at 100 kHz). Similar to the use of the other EIT systems, the diagnosis made using the T-Scan 2000 is also based on the spectral behavior of the amplitude and phase of the measured current distribution at various frequencies.

To test and approve the feasibility and/or clinical utility of using the T-Scan 2000 system as an adjunct tool to mammography in breast cancer screening and detection, a clinical trial examining 504 biopsied breasts that consisted of 179 malignant and 325 benign findings was conducted and reported. By conducting and comparing three examination models, namely the T-Scan 2000 alone, mammography alone, and the combination of T-Scan 2000 and mammography, the study results showed that using T-Scan 2000 as adjunct tool to mammography increased both the detection sensitivity of using mammography alone from 82% to 88% and the specificity from 39% to 51% ($p < 0.01$) as shown in the U.S. FDA report of TransScan T-Scan 2000–P970033 (http://www.fda.gov/cdrh/pdf/p970033.html). Since the T-Scan 2000 became commercially available, a number of research groups have independently evaluated the performance and the other clinical issues when applying T-Scan 2000 systems to different test databases. Table 19.2 summarizes a few example studies in this field.

Despite the fact that several studies have shown promising results in using the T-Scan 2000 to conduct EIT examination as

TABLE 19.2 Examples of Studies Evaluating the Performance of the T-Scan 2000 for Detecting Breast Cancer and/or Classifying between Malignant and Benign Breast Lesions

Study	Database	Performance	Comment
Melloul et al. 1999	18 cancer and 103 negative cases	72% sensitivity and 67% specificity	No improvement in cancer detection rate was made when adding to mammography
Malich et al. 2001	210 women depicted 103 malignant and 137 benign lesions	88% sensitivity and 66% specificity	Adding EIT to mammography increased the sensitivity but reduced the specificity
Wersebe et al. 2002	117 patients depicted 71 malignant and 58 benign lesions	62% sensitivity and 69% specificity	Quantitative analysis of EI signals did not help to differentiate TP and FP lesions
Martin et al. 2002	74 patients with suspicious findings in mammography	59% malignant lesions showed high EI signals	Mammography and EI test had the comparable false-positive rate (~17%)
Diebold et al. 2005	256 patients involving 118 malignant lesions	75% sensitivity and 42% specificity	EIT sensitivity was higher for small lesions (<10 mm) than for large lesions (>10 mm)
Stojadinovic et al. 2005	29 cancer cases and 1074 negative cases 52% women were younger than 40 years old	17% sensitivity and 90% specificity	Sensitivity in younger women (<40 years old) was much higher than in older women, 50% (3/6) versus 9% (2/23)

an adjunct to mammography, a number of limitations in using this system were also identified, which include (1) the higher rate of false-positive detections caused by image artifacts (i.e., air bubbles, interfering bones, muscles) that also result in high conductivity measurement or enhancement spots, (2) the limited maximum depth of tissue electrical impedance measurement (e.g., <3–3.5 cm), and (3) the lack of ability to locate lesions for biopsy (Zou et al. 2003).

19.4 Electrical Impedance Spectroscopy (EIS) for Breast Cancer Screening and Detection

Although great efforts have been made to develop and improve breast EIT technologies, how to interpret the EIT images is difficult due to two reasons. First, the specificity of current EIT images is quite low. Since EIT images often contain many false-positive "enhancement" spots, it is difficult to correctly detect the true-positive signals. Second, unlike mammograms,

radiologists, and other healthcare professionals have not been trained to read and interpret the information displayed on EIT images. As a result, generating and displaying EIT images may not be helpful in clinical practice. To solve this practical issue in applying EIT technology, a simplified EI information display method has been used to replace the old approach of showing multiple EIT images. For example, the modified version of the commercialized T-Scan 2000 system (named as T-Scan 2000ED) does not produce and display any image or data that need to be interpreted by the clinicians to make a diagnostic decision. Instead, it only shows the clinicians either a green or a red indictor bar highlighted on the system output "image" display screen. The green bar indicates a negative breast and the red bar denotes a "suspicious" breast that may depict tumors leading to the cancer detection (Stojadinovic et al. 2008).

Instead of generating or mapping the internal EI signal distribution images of a breast, developing simple noninvasive electrical impedance signal or spectrum measurement methods by connecting external electrodes and/or an impedance bridge to the breast skin to directly detect breast lesions and classify between malignant and negative breasts has also attracted research interest as early as the 1970s (Singh et al. 1979). For example, one study performed noninvasive EI measurements on 24 women by applying a four-electrode sensor device to the breast skin surface and reported that distinguishing between the normal and abnormal breasts was possible by measuring the average mammary EI signal levels (Ohmine et al. 2000).

Without generating and using EIT images, a research group in the Department of Radiology at the University of Pittsburgh recently developed and tested a different approach to measure electrical impedance signals directly detected from the breast and then analyze the electrical impedance signals to detect breast abnormalities that are likely to be recommended for biopsy and may eventually lead to the cancer detection and verification (Zheng et al. 2008, 2011). Since breasts typically have quite small capacitance, and directly measuring breast capacitance is difficult and often unreliable, this new approach focuses on detecting and analyzing output electrical impedance signals (including both amplitude and phase signals) at or near the resonance frequency of the breast tissue being measured. For this purpose, an inductor is added. The EIS is then developed based on a simple model involving a series of RLC circuit sections that simulate an input electronic device (probe) and the measured breast (Figure 19.2). Although a capacitor ($C1$) is included in the probe, it has substantially greater capacitance value than the breast capacitance, (C). Since in serial combination of capacitances the impedance is dominated by the smallest capacitance, the breast capacitance (C) dominates the impedance measurement. At lower frequencies, breast capacitance (C) is a dominant factor. As the frequency increases, the inductive reactance (L) gradually increases its weight in determining the overall impedance. At a special frequency, the positive reactance of the inductor cancels the negative reactance of the capacitor and the phase signal changes from the negative (current leading voltage) to neural (phase is zero). This frequency is defined as the

Input electronic probe

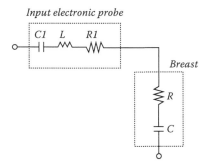

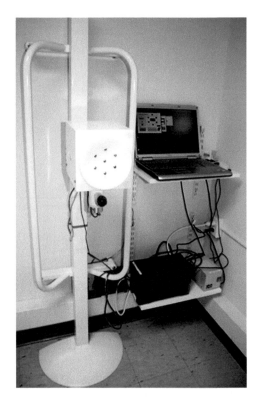

FIGURE 19.2 A simplified diagram illustrating the EIS concept in measuring resonance frequency of breasts.

resonance frequency, $f = \dfrac{1}{2\pi\sqrt{LC}}$. At the resonance frequency, the inductor and capacitor combination becomes invisible and the remaining resistor (R) becomes the total impedance of the system. Above the resonance frequency, the inductor dominates the system characteristics and phase turns to positive (voltage leading current). Hence, by selecting the proper inductor in the electronic probe device, the resonance frequency can be set up at an optimal level to increase sensitivity or signal-to-noise ratio in detecting the small change of breast capacitance that may relate to a developing breast abnormality or cancer.

After building a prototype EIS system with a single pair of detection probes and achieving promising preliminarily results in testing a group of 150 women (Zheng et al. 2008), a new multi-probe-based EIS system has been built and installed in the breast imaging clinical facility at the University of Pittsburgh Medical Center (Figure 19.3). The major components of the system consist of a specially designed mechanical support, an electronic device with two sensor module cups, power control device, and laptop computer installed with system control and management software. The two sensor module cups have different surface curvatures; namely, one fits breasts with small brassiere cup sizes and one fits breasts with larger brassiere cup sizes. Each sensor cup includes seven mounted metallic probes. The center probe is intended to enable easy contact with the nipple during the EIS examination and the other six probes are uniformly distributed along an "outer" circle and are intended to enable contact with six points on the breast skin surface at different locations with a fixed distance from the center probe. The sensor supporting cups are separately mounted on two (front and back) sides of the electronic box and the latter can be easily rotated 180° to allow the operator to select an optimal sensor cup for the individual woman being examined. The sensor box can also be freely moved up or down in a weight balance manner along a vertical rail system to adjust for the height of the breast of each woman being examined. The maximum electric voltage and current applied to the sensor probes (breast skin) are less than 1.5 V and 30 mA, respectively. Unlike some other electronic devices (e.g., electrocardiography) that are operated at relatively low frequency, the EIS system is operated at a much higher frequency range (e.g., from 200 to 800 kHz) by using an appropriately adjusted matching inductor

FIGURE 19.3 A picture of a multiprobe-based EIS system installed in the breast imaging clinical facility.

in the sensor probe. Since at high frequency range the capacitance of the breast tissue determines the measured impedance and minimizes the impact of resistive noise of the skin, a conductive gel is not used in EIS measurement, which makes the EIS examination more efficient and less operator dependent.

After the EIS system detects that all seven probes are in adequate contact with the breast skin, the EIS scanning starts automatically. The EIS examination takes 12 seconds to scan one breast and the same scanning process is typically conducted on both left and right breasts in one complete examination. The multiple REIS signal sweeps generated between six different pairs of sensor probes applied to each breast are automatically recorded and saved in a specific data file named with the test ID number in a database. Each EIS signal sweep records 121 output signals including signal amplitude (a), signal phase (p), and signal magnitude ($I = \sqrt{a^2 + p^2}$) ranging from 200 to 800 kHz at a 5-kHz increment. The resonance frequency of each EIS signal sweep recorded in one sensor pair is detected where both the phase of the impedance signal sweep crosses the horizontal line and the amplitude of the impedance signal sweep reaches the minimum value. Figure 19.4 shows one example of the six sets of EIS output signal sweeps (including both signal amplitude and phase) acquired from six pairs of sensor probes applied to one breast. These curves show the range of the detected resonance frequency shift among six pairs of sensor probes. In this system configuration, the detected EIS-generated resonance frequencies are typically measured in the range from 300 to 600 kHz.

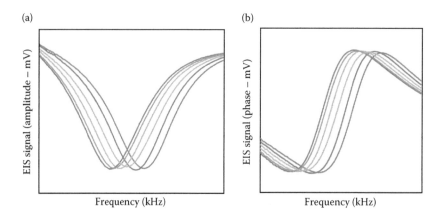

FIGURE 19.4   An example of a complete set of six pairs of EIS output signal sweeps including (a) EIS amplitude signals and (b) EIS phase signals acquired from scanning one breast.

Since the measured EIS signals including EIS-generated resonance frequencies are primarily determined by breast capacitance levels that are affected by both genotype and phenotype characteristics of a breast, the measured EIS signal reflects different biological characteristics and breast tissue structure among different women, differences across the life cycle of a given woman, as well as different measurement paths of a breast. As a result, using the absolute measurement of EIS signals to detect and identify breast abnormalities or cancer is quite difficult and probably unreliable due to the substantial resonance–frequency shifts among the different women being tested under varying conditions.

To improve the detection performance (including both sensitivity and specificity) of EIS technology for suspicious breast abnormalities or cancers, a different approach based on measuring and analyzing the asymmetrical EIS signals acquired from two bilateral (left and right) breasts was developed and tested. The concept of this approach is supported by several underlying scientifically known evidences (Scutt et al. 2006) including that (1) humans naturally show bilateral symmetry in paired morphological traits, including breasts (e.g., interbreast size difference has been shown

to be associated with breast cancer developing risk) and (2) breast tissue pattern asymmetry is an important radiographic image phenotype related to the biological processes (e.g., changes due to hormones, such as estrogen), as well as the observations of clinical practice including that (1) radiologists routinely examine the region-based bilateral mammographic tissue asymmetry and the asymmetric tissue pattern changes over time when making cancer detection related clinical decisions (Blanks et al. 1999, Sumkin et al. 2003) and (2) since there is considerable interobserver and intraobserver variability in subjectively assessing mammographic tissue density and patterns (Berg et al. 2000), computerized schemes have been proven to be able to achieve more objective, reliable, and consistent results in assessing mammographic tissue density by avoiding observer variability (Wei et al. 2011). Hence, the basic hypothesis of this EIS signal processing approach for detection of breast abnormalities is that breasts containing lesions are likely to have higher bilateral EIS signal (in particular, at the resonance frequency) difference between the left and the right breasts.

Figure 19.5 shows an example of comparing two EIS examinations performed on one 50-year-old woman who had negative

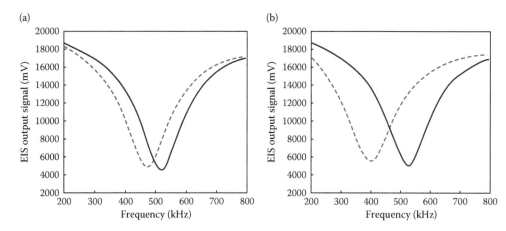

FIGURE 19.5   Comparison of two EIS output signal sweeps that have the maximum resonance frequency difference among the six pairs of sensor probes applied to the right (solid curve) and left (dash curve) breast in an EIS examination. The figure includes EIS signal sweeps acquired from a 50-year-old woman who had negative mammograms during the screening mammography examination followed by the EIS examination (a) and EIS signal sweeps acquired from a 47-year-old woman recommended for biopsy and verified as having breast cancer (b).

screening mammograms following the EIS test and one 47-year-old woman who was recommended for biopsy due to a suspicious finding from the mammograms and verified as having cancer. The mammographic tissue density of both women was rated by radiologists during interpreting their screening mammograms as "heterogeneously dense" (BIRADS 3). In the first negative examination, the resonance frequencies measured from two breasts ranged from 416 to 546 kHz and the resonance frequency differences between the six mirror-matched pairs of EIS output signal sweeps measured from the left and right breasts varied from 6 to 45 kHz. In the second positive case, the measured resonance frequencies ranged from 320 to 525 kHz, with the paired resonance frequency differences changed from 2 to 120 kHz. Two pairs of the EIS output signal sweeps with the maximum resonance frequency difference are plotted and shown in Figure 19.5 for these two cases.

In a reported preliminary study, a data set involving 140 sets of EIS examination data was used (Zheng et al. 2011). All EIS examinations were performed on consenting women aged between 30 and 50 years old. The cases were classified based on actual diagnostic results into three groups. The "positive" group includes 56 women who had been recommended for biopsy following an imaging-based diagnostic workup (BIRADS 4 or 5) and the EIS examination was performed prior to the biopsy. Among these 56 biopsy cases, 37 cases depicted suspicious masses, asymmetric density, or architectural distortions, eight depicted microcalcification clusters alone, and 11 depicted both masses and microcalcification clusters. In addition, these "positive" cases have diverse characteristics including 16 palpable masses and varying lesion locations namely that nine biopsied lesions were marked near the nipple (e.g., periareolar or retroareaolar), three were close to the chest wall, and the remaining 44 were diversely distributed and marked by the radiologists in different breast regions (from 1- to 12-o'clock regions) in the original diagnostic reports. The biopsy results divided the group into three types of cases, namely, nine verified cancer cases, nine high-risk (precancer) cases (e.g., lobular carcinoma in situ) with recommendation for surgical excision of the lesions, and 38 biopsy-proved benign cases. Among the nine cancer cases, four depicted masses only, two were associated with microcalcification clusters only, and three depicted both types of abnormalities. Among the nine high-risk cases, six depicted masses only, two depicted microcalcification clusters only, and one depicted both types of abnormalities. The second group includes 63 women who had negative screening mammography examinations that followed the EIS test. The third group includes 21 women who had been recalled for a diagnostic follow-up (BIRADS 0) due to the suspicious findings from the screening mammograms but were later (after a diagnostic workup) determined not to require a biopsy (BIRADS 1 or 2).

To detect and classify whether the case was recommended for biopsy based on the EIS test results, a set of 33 features was initially extracted around the resonance frequencies of six EIS output signal sweeps generated from each breast. Except that one feature was defined as the maximum absolute difference between the ranges of all six resonance frequencies for each of the breasts as computed for the left (*L*) and right (*R*) breasts ($F_1 = Max |\Delta f_L - \Delta f_R|$), the remaining 32 features were divided into two distinct groups. In the first group, 16 features were computed from the differences of the averaged EIS signal values of the six EIS output signal sweeps (including both signal magnitude and phase) for each breast, while in the second group, 16 similar features were computed from the difference of one pair of mirror-matched EIS signal sweeps, which shows the maximum resonance frequency difference among all six matched pairs of bilateral EIS output signal sweeps.

A genetic algorithm (GA)-based feature selection protocol was then applied to select an optimal feature set from the initial feature set and build an optimal artificial neural network (ANN) to classify between the "positive" (biopsied) and "negative" (non-biopsied) cases. The ANN includes three layers. The first (input) layer included *N* neurons that connect to *N* selected features, the second layer included *M* hidden neurons, and the third (decision) layer included one neuron that generates a likelihood score of a test case being "positive." A binary coding method and a GA fitness criterion based on the area under a receiver operating characteristic (ROC) curve (AUC) was implemented in GA to select an optimal feature set and determine the ANN internal structure. During the GA optimization process, the initial 100 chromosomes in the first generation were randomly generated. After a GA chromosome was selected, a leave-one-case-out method was used to assess the performance of the ANN-based classifier in which 139 cases were used to train the ANN and one was applied to test the trained ANN and generate a classification score. After iteratively repeating this process 140 times, each case in the data set was used once to test the ANN and the 140 classification scores were processed by a ROC fitting and analysis program to compute AUC value. Thus, after each training cycle the GA generated a performance score (AUC) for each GA chromosome. The GA chromosomes that produce higher AUC values have higher probabilities of being selected in generating new chromosomes using the method of crossover and mutation. The GA optimization process was terminated when it reached a global maximum performance level or a predetermined number of growth generations or iterations.

The study reported that GA selected 14 features from the initial feature pool. The distribution of these 14 selected features shows that two features represented resonance frequency differences, six represented EIS signal magnitude differences around the resonance frequencies, and six represented EIS signal phase differences. To classify between 56 biopsy ("positive") cases and 84 nonbiopsy ("negative") cases, the ANN yielded a classification performance of AUC = 0.830 with a 95% confidence interval for AUC [755, 0.890].

From the experimental results (Figure 19.6), a number of interesting observations regarding the potentially unique characteristics of applying this resonance frequency-based EIS technology to detect breast abnormalities or cancer can be made. First, EIS yielded higher performance in detecting cancer or precancer high-risk lesion cases than biopsy-proved benign cases despite the fact that the ANN classifier was trained using a

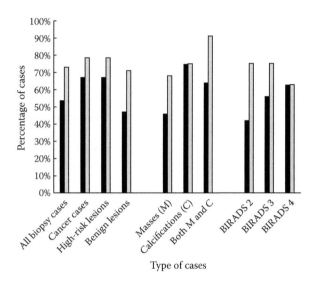

FIGURE 19.6 Summary of sensitivity levels of an ANN-based EIS signal classifier by biopsy outcome, type of abnormalities, and subjectively rated breast density (BIRADS) at 90% (black bars) and 80% specificity (gray bars) levels.

suboptimal optimization criterion for classifying between biopsied and nonbiopsied (rather than cancer and noncancer) cases. For example, at 90% specificity level the ANN correctly detected 30 biopsied cases as "EIS-positive" that included 67% (6/9) of the cancer cases, 67% (6/9) of the high-risk cases, and 47% (18/38) of biopsy-proven benign cases. The sensitivity for cancer or high-risk cases was higher than that of the biopsy-proven benign cases. Second, the classifier was able to detect different types of breast abnormalities. The detected 30 biopsy cases at 90% specificity included 17 that depicted masses only, six that depicted calcifications only, and seven that depicted both types of abnormalities representing 46%, 75%, and 64% of cases in these three types of abnormalities, respectively. Third, the performance of the EIS signal classifier was not breast density dependent. Among the 30 detected biopsy cases at 90% specificity, five were rated as BIRADS 2, 20 as BIRADS 3, and five as BIRADS 4 representing 42%, 56%, and 63% of cases in the three breast density groups in the data set, respectively. Fourth, the detection sensitivity of the classifier also did not depend on the lesion size and location. At 90% specificity, the mean and standard deviation of the detected mass size was 1.32 ± 1.41 cm (ranging from 0.4 to 7.0 cm). The detected lesions were also diversely distributed in different breast regions. Aside from the four reported being located near the nipple (retroareaolar) and two close to the chest wall, the 9- and 10-o'clock regions depicted four lesions each, the 1-, 2-, and 11-o'clock regions contained three lesions each, the 3-, 5-, and 12-o'clock regions had two lesions each, and the 4-o'clock region depicted one lesion.

Besides an ANN-based classifier, researchers have also investigated and tested several other classifiers based on different machine learning concepts or statistical models to analyze and classify the resonance frequency-based EIS signals, which include support vector machine (SVM) and Gaussian mixture

model (GMM) (Lederman et al. 2011a) applying to the extended EIS data set (Lederman et al. 2011b). The study results indicated that by using an optimal fusion method to combine detection results generated from multiple classifiers, the classification performance using the bilateral asymmetrical features computed based on the resonance frequency of EIS output signal sweeps could be further improved.

19.5 Limitations and Future Perspectives

Developing electrical impedance technologies and systems as an adjunct to mammography to assist and improve breast cancer screening and detection has been attracting wide research interest over the past several decades. Despite the significant progress that has been made in many areas of system development and clinical tests as well as the availability of a commercialized system (T-Scan 2000), EIT or EIS has not been routinely accepted and applied in the clinical practice due to its relatively low detection sensitivity and/or specificity. In addition, a number of technical issues have not been fully investigated and remained unsolved. For example, first, the resolution of the EIT system falls significantly with the increasing distance from the measuring electrodes, which makes conventional EIT measurement schemes unable to match mammography to date. Although increasing the number of electrodes could in general help increase EIT image resolution and size of imaging field, it could lead to enormous increase of the complexity and cost of the system hardware as well as the computation time for image mapping and/or reconstruction. Second, measurement results of electrical impedance values of the breasts are often inconsistent because the measured breast tissue impedance levels depend on many external factors (e.g., breast skin surface temperature, breast tissue density, hormonal influence level, and the phase of menstrual cycle). These variations or uncertainties limit the sensitivity and specificity of applying EI technology as a stand-alone and reliable breast cancer screening and detection tool to date. Therefore, how to optimally compensate the impact of these factors remains an open research topic in this field.

Recently, there has been increasing interest in developing personalized mammography or individualized screening programs with varying screening intervals (Schousboe et al. 2011). Towards this end, the development of new breast cancer risk stratification tools with significantly improved discriminatory power or positive predictive values (PPV) is important. Hence, one new research direction of developing and applying EI technology is to better assess the risk of individual women developing breast cancer in the near-term, in particular for younger women and/or women with dense breasts (Stojadinovic et al. 2008, Zheng et al. 2008). For example, in a large two-cohort and prospectively controlled clinical trial involving 25 institutions in the United States and Israel (Stojadinovic et al. 2008), 1751 women (aged between 30 and 39 years old) in the specificity cohort and 390 women (aged between 30 and 45 years old) in the sensitivity cohort were enrolled in the study. In the specificity cohort each recruited woman originally visited her obstetrician/gynecologist for an annual physical examination and did not exhibit any breast cancer related signs or symptoms, while in the sensitivity cohort, each woman had a palpable and/or mammography-detected suspicious breast finding and was scheduled to undergo a breast biopsy. Each woman was examined using T-Scan 2000ED system that reports only either a green bar indicating a "negative" case or a red bar indicating a "positive" (or "high-risk") case. The study reported 94.7% specificity (95% confidence interval [CI], 93.7%–95.7%) in the specificity cohort and 26.4% sensitivity (with 95% CI, 17.4%–35.4%) in the sensitivity cohort (ranging from 0% to 53% in different test sites or participated institutions). A trend of increasing sensitivity with decreasing tumor size (e.g., 35.6% sensitivity for tumors with size ≤2.0 cm versus 22.2% sensitivity for tumors with size >2.0 cm) is an interesting finding that may indicate one advantage of applying EI technology over the other existing methods in detecting early (small) cancers. When using the T-Scan 2000ED scanning result as a risk indicator, the reported odds ratio for cancer detection was 5.0 in this large cohort study, which also indicated that EI signals measured from the breast could be a promising risk factor that has not been investigated or integrated in any existing breast cancer risk models to date (Amir et al. 2010).

Unlike mammography or other imaging technologies, the T-Scan 2000ED or other EIS-based risk assessment tools do not directly detect breast lesions nor does the system identify the location of the abnormality in question. These systems only detect changes (e.g. distortions) in electrical fields that may stem from biological changes (e.g., ductal epithelial changes) inside a breast. Hence, such EIS-based examinations are not intended in any way to compete with imaging-based screening mammography and it is not intended for screening women with known high-risk factors. However, since the majority of the breast cancer cases (e.g., >60% [Madigan et al. 1995]) are likely to be detected in the women without carrying any known risk factors, EIS detection results may benefit to this large fraction of breast cancer patients. As an adjunct (or supporting) tool to the imaging-based screening, the use of EIS examination results at the current performance level should only be considered as a "rule in" approach, namely given a "positive" EIS result the physicians should consider recommending an imaging-based follow-up examination in particular for the younger women who do not participate in regular mammography or other imaging-based screening programs. Currently, the EIS cannot be used as a "rule out" approach (i.e., women should not use a negative EIS result as an excuse to avoid imaging-based screening).

In summary, despite the limitations of EIS at current performance levels for breast cancer screening or diagnosis, continued research and improvement of EIS technology may produce a new effective risk indicator, especially if the developed EIS systems are low-cost, widely accessible, and easy to use. Therefore, in combination with the other known breast cancer risk factors, EIS may eventually play an important role to help the establishment of the optimal personalized breast cancer screening

programs. For this purpose, more research and development work is needed in this field.

References

Amir E, Freedman OC, Seruga B, Evans DG. 2010. Assessing women at high risk of breast cancer: A review of risk assessment models. J Natl Cancer Inst 102: 680–691.

Assenheimer M, Laver-Moskovitz O, Malonek D et al. 2001. The T-Scan technology: Electrical impedance as a diagnostic tool for breast cancer detection. Physiol Meas 22: 1–8.

Berg WA, Campassi C, Langenberg P, Sexton MJ. 2000. Breast imaging reporting and data system: Inter- and intra-observer variability in feature analysis and final assessment. Am J Roentgenol 174: 1769–1777.

Blanks RG, Wallis MG, Given-Wilson RM. 1999. Observer variability in cancer detection during routine repeat (incident) mammographic screening in a study of two versus one view mammography. J Med Screen 6: 152–158.

Campbell AM, Land DV. 1992. Dielectric properties of female breast tissue measured in vitro at 3.2 GHz. Phys Med Biol 37: 193–210.

Chaundary SS, Mishra RK, Swarup A, Thomas JM. 1984. Dielectric properties of normal and malignant human breast tissue at radiowave and microwave frequencies. Ind J Biochem Biophys 21: 76–79.

Cherepenin V, Karpov A, Korjenevsky A et al. 2001. A 3D electrical impedance tomography (EIT) system for breast cancer detection. Physiol Meas 22: 9–18.

Choi MH, Kao T, Isaacson D et al. 2007. A reconstruction algorithm for breast cancer imaging with electrical impedance tomography in mammography geometry. IEEE Tran Biomed Eng 54: 700–710.

Diebold T, Jacobi V, Scholz B et al. 2005. Value of electrical impedance scanning (EIS) in the evaluation of BI-RADS III/IV/V-lesions. Technol Cancer Res Treat 4: 93–97.

Foster KR, Schwan HP. 1989. Dielectric properties of tissue and biological materials: A critical review. Crit Rev Biomed Eng 17: 25–104.

Fricke H and Morse S. 1926. The electrical properties of tumors of the breast. J Cancer Res 16: 310–376.

Halter RJ, Hartov A, Paulsen KD. 2008. A broadband high-frequency electrical impedance tomography system for breast imaging. IEEE Trans Biomed Eng 55: 650–659.

Hope TA, Lles S. 2004. Technology review: The use of electrical impedance scanning in the detection of breast cancer. Breast Cancer Res 6: 69–74.

Jossinet J, Lobel A, Michoudet C, Schmitt M. 1985. Quantitative technique for bio-electrical spectroscopy. J Biomed Eng 7: 289–294.

Jossinet J. 1996. Variability of impedivity in normal and pathological breast tissue. Med Biol Eng Comput 34: 346–350.

Kim BS, Isaacson D, Xia H et al. 2007. A method for analyzing electrical impedance spectroscopy data from breast cancer patients. Physiol Meas 28: S237–S246.

Lederman D, Zheng B, Wang X, Wang XH, Gur D. 2011a. Improving breast cancer risk stratification using resonance-frequency electrical impedance spectroscopy through fusion of multiple classifiers. Ann Biomed Eng 39: 931–945.

Lederman D, Zheng B, Wang X, Sumkin JH, Gur D. 2011b. A GMM-based breast cancer risk stratification using a resonance-frequency electrical impedance spectroscopy. Med Phys 38: 1649–1659.

Liu N, Saulnier GJ, Newell JC, Isaacson D, Kao TJ. 2005. ACT4: A high-precision, multi-frequency electrical impedance tomography. In Proc 6th Conf. on Biomedical Appl. Electrical Impedance Tomography. University College London, London, UK. 22–24 June.

Madigan MP, Ziegler RG, Benichou J et al. 1995. Proportion of breast cancer cases in the United States explained by well-established risk factors. J Natl Cancer Inst 87: 1681–1685.

Malich A, Boehm T, Facius M et al. 2001. Differentiation of mammographically suspicious lesions: Evaluation of breast ultrasound, MRI mammography and electrical impedance scanning as adjunctive technologies in breast cancer detection. Clin Radiol 56: 278–283.

Martin G, Martin R, Brieva MJ et al. 2002. Electrical impedance scanning in breast cancer imaging: Correlation with mammographic and histologic diagnosis. Eur Radiol 12: 1471–1478.

Melloul M, Paz A, Ohana G et al. 1999. Double phase TC-sectamibi scintimammography and Tran-Scan in diagnosing breast cancer. J Nucl Med 40: 376–380.

Morimoto T, Kimura S, Konishi Y et al. 1993. A study of the electrical bioimpedance of tumors. J Invest Surg 6: 25–32.

Ohmine Y, Morimoto T, Kinouchi Y et al. 2000. Noninvasive measurement of electrical bioimpedance of breast tumors. Anticancer Res 20: 1941–1946.

Schousboe JT, Kerlikowske K, Lob A, Cummings SR. 2011. Personalizing mammography by breast density and other risk factors for breast cancer: Analysis of health benefits and cost-effectiveness. Ann Inter Med 155: 10–21.

Scutt D, Lancaster GA, Manning JT. 2006. Breast asymmetry and predisposition to breast cancer. Breast Cancer Res 8: R14.

Singh B, Smith CW, Hughes R. 1979. In vivo dialectic spectrometer. Med Biol Eng Comput 17: 45–60.

Stojadinovic A, Nissan A, Gallimidi Z et al. 2005. Electrical impedance scanning for early detection of breast cacner in young women: Preliminary results of a multicenter prospective clinical trial. J Clin Oncol 23: 2703–2715.

Stojadinovic A, Moskovitz O, Gallimidi Z et al. 2006. Prospective study of electrical impedance scanning for identifying young women at risk for breast cancer. Breast Cancer Res Treat 97: 179–189.

Stojadinovic A, Nissan A, Shriver CD et al. 2008. Electrical impedance scanning as a new breast cancer risk stratification tool for young women. J Surg Oncol 97: 112–120.

Sumkin JH, Holbert BL, Hermann JS et al. 2003. Optimal reference mammography: A comparison of mammograms obtained 1 and 2 years before the present examination. Am J Roentgenol 180: 343–346.

Surowiec AJ, Stuchly SS, Barr JB, Swarup A. 1988. Dielectric properties of breast carcinoma and the surrounding tissues. IEEE Trans Biomed Eng 35: 257–263.

Wei J, Chang HP, Wu Y et al. 2011. Association of computerized mammographic parenchymal pattern measure with breast cancer risk: A pilot case-control study. Radiology 260: 42–49.

Wersebe A, Siegmann K, Krainick U et al. 2002. Diagnostic potential of targeted electrical impedance scanning in classifying suspicious breast lesions. Invest Radiol 37: 65–72.

Zheng B, Zuley M, Sumkin JH et al. 2008. Detection of breast abnormalities using a prototype electrical impedance spectroscopy system: A preliminary study. Med Phys 35: 3041–3048.

Zheng B, Lederman D, Sumkin JH et al. 2011. A preliminary evaluation of multi-probe resonance-frequency electrical impedance based measures of breast. Acad Radiol 18: 220–229.

Zou Y, Guo Z. 2003. A review of electrical impedance techniques for breast cancer detection. Med Eng Phys 25: 79–90.

20

Optical Imaging of the Breast

Kelly E. Michaelsen
Dartmouth College

Michael A. Mastanduno
Dartmouth College

Ashley M. Laughney
Darthmouth College

Brian W. Pogue
Dartmouth College

Keith D. Paulsen
Dartmouth College

20.1 Introduction to Optical and Near-Infrared Imaging

Optical imaging is often used generically to imply the use of either visible wavelengths (400–650 nm) or near-infrared (NIR) wavelengths (650–1000 nm) of light to image or measure transmittance or reflectance in order to characterize the absorption and scattering properties of the breast. The main difference between these two wavelength bands is the depth of tissue through which light can propagate. Visible light does not penetrate very far due to high absorption and scatter, whereas NIR light is highly scattered but only weakly absorbed (see Figure 20.1). Thus, NIR light can be used for imaging through thicker tissue such as the breast. As a less energetic form of electromagnetic radiation, NIR light has the benefit of being nonionizing, and thus, poses less of a health risk to patients and medical personnel relative to traditional x-rays (see Chapter 1). NIR light undergoes multiple scattering events as it travels in tissue, whereas x-rays and visible light rarely scatter more than once. Index of refraction differences in tissue components such as mitochondria and collagen fibrils contribute to Mie and Rayleigh scattering of NIR light. The transport mean free path describes the average distance between photon interactions and is usually on the order of 1 mm. The scattering is quantified by the scattering coefficient, μ_s, which describes the number of scattering events per unit length of tissue. The reduced scattering coefficient, μ_s', is more frequently used when considering multiple scattering where directionality is lost. Individual scattering events are highly anisotropic with average cosine of the scattering angle, g, being close to 0.9 in most tissues, whereas multiple scattering can appear isotropic in the far field and is parameterized through the transport or reduced scattering coefficient defined as $\mu_s', = (1 - g)\mu_s$. This pervasive multiple scattering limits the spatial resolution of NIR imaging techniques.

Absorption of light can also attenuate the signal, both in the visible wavelengths as well as in the NIR when light is transmitted through thick tissues. In the NIR regime, the absorption coefficient is about two orders of magnitude lower than its scattering counterpart whereas in the visible regime the two are more comparable (see Figure 20.1). Hence, light can travel through several centimeters of tissue and still have enough intensity to be detected; thus, its spectrum (which is altered by absorption and scatter) can be used for breast imaging. The absorption coefficient, μ_a, represents the inverse of the mean free distance for exponential attenuation, estimated in the absence of scatter. Several tissue molecular components contribute significantly to light absorption in the NIR region including oxy- and deoxy-hemoglobin, water, and lipids. At each wavelength these individual chromophores contribute linearly to the total

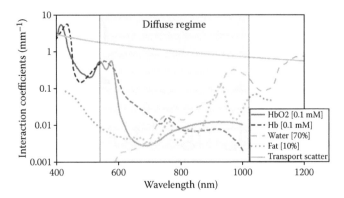

FIGURE 20.1 Absorption and scattering as a function of wavelength in the visible and NIR regime for the major tissue chromophores in breast tissue. (From Pogue BW, Leblond F, Krishnaswamy V, Paulsen KD. 2010. *American Journal of Roentgenology*, 195, 321–332. With permission.)

absorption coefficient, weighted by their respective concentrations and molar extinction coefficients.

20.1.1 Optical Spectroscopy

Tissue absorption and scattering coefficients vary as a function of wavelength. Spectroscopy is a technique that acquires measurements at multiple wavelengths to gain additional information about the properties of tissue. Each of the major absorbers has a characteristic molar extinction spectrum in the NIR, which is its absorption coefficient normalized by the concentration at each wavelength. Thus, if measurements of the absorption coefficient are recorded at different wavelengths, concentrations of absorbers such as hemoglobin, water, beta-carotene, bilirubin, and lipids can be determined based on the known extinction coefficients for each contributor at the different wavelengths.

Spectroscopic techniques provide insight into the metabolic state of tissue. Knowledge of local oxy- and deoxy-hemoglobin, water, beta-carotene, bilirubin, and lipid concentrations along with scattering parameters is useful in distinguishing healthy from diseased tissue. Normal breast parenchyma is largely composed of two constituents: fibroglandular and adipose tissue. Higher hemoglobin, water, and scattering are typically found in its fibroglandular component due to higher metabolic demands relative to fat (Brooksby et al. 2006). Similarly, malignant tissue also possesses higher hemoglobin, water, and scattering levels, roughly twice those found in normal breast tissue. It has also a lower concentration of lipids when compared to healthy breast tissue (Cerussi et al. 2006). These differences are likely due to the angiogenesis and increased vascularity that are characteristic of neoplasms.

20.1.2 Exogenous Fluorescence Imaging

Contrast between normal and diseased tissue can be enhanced through the injection of fluorescent agents with high retention

in malignant lesions. Two contrast agents, indocyanine green (ICG), and omocyanine, are currently being used for their absorption and fluorescence properties in the NIR window. They bind with high affinity to serum proteins and aggregate in areas of hypervascularity, which are common in cancerous tissues. Also, tumor vasculature is typically more permeable than normal blood vessels, and thus, contrast agents extravasate into malignancies over time. Initial studies have shown lesion localization is possible with fluorescence imaging using ICG and that uptake and release of the contrast agent differ among benign, malignant, and normal breast conditions (Intes et al. 2003). More recently, an examination of fluorescence at two different time points in 13 malignant and eight benign lesions has shown increased contrast levels. Contrast enhancement was accomplished by correcting for absorption differences and homogenizing the background for easier visualization of abnormal structures. Significant differences in visibility was evident in malignant versus benign lesions leading to increased specificity to 75% from 25% when compared to mammography alone (Poellinger et al. 2011). Fluorescence imaging of the breast, while in the early stages of development, shows promise as a technique for diagnostic imaging provided contrast dyes with high specificity can be approved for medical use.

20.2 Optical Imaging: Spectral Diagnosis and Imaging during Intervention

Optical spectroscopy has been studied extensively for surface imaging of breast during interventional procedures. Surgical applications of spectroscopy have largely evolved around fiber optic probes and multispectral imaging techniques; the former samples local tissue volumes (~1 mm³) with high spatial (μm) and spectral (nm) resolution while the latter offers a large field of view (FOV) at discrete wavelengths with lower spatial resolution (mm). Localized optical spectroscopy via fiber optic probes is well suited for elucidating the underlying characteristics of disease and providing sensing feedback during tissue sampling—the most immediate application is stereotactic guidance for biopsy sampling of mammographically detected lesions (see Chapter 8). Multispectral imaging provides rapid, wide-field assessment of tissue optical properties and therefore may more accurately assess disease extent, heterogeneity and multifocality. While some probe-based techniques have been extended into imaging modalities via raster scanning, truly wide-field imaging systems have only recently been considered for surgical resection guidance (Bogaards et al. 2007, Gioux et al. 2010, Themelis et al. 2009) and mapping of sentinel lymph nodes (Sevick-Muraca et al. 2008, Hirche et al. 2010, Murawa et al. 2009, Troyan et al. 2009). Progress in this application has been enabled by the rapid development of highly sensitive detectors and specific near-infrared (NIR) probes (Ntziachristos et al. 2003). Optical images of breast cancer acquired during surgery at microscopic, mesoscopic, and macroscopic scales are shown in Figure 20.2.

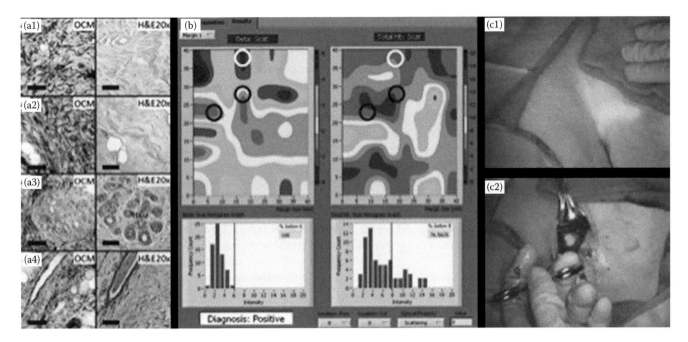

FIGURE 20.2 **(See color insert.)** Representative images of intraoperative optical imaging of breast cancer at microscopic (a), mesoscopic (b), and macroscopic (c) sampling volumes. (a1–4) Microscopic optical coherence microscopy acquired 50 μm below the tissue surface and corresponding histology. Scale bar is 100 μm. (From Zhou C, Cohen DW, Wang Y et al. 2010. *Cancer Res*, 70, 10071–10079. With permission.) (b) Intraoperative breast tumor margin assessment using quantitative diffuse reflectance imaging; image of a pathologically confirmed margin. (From Brown JQ, Bydlon TM, Richards LM et al. 2010. *IEEE Journal of Selected Topics in Quantum Electronics*, 16, 530–544. With permission.) Subcutaneous (c1) and intraoperative (c2) identification of the sentinel lymph node following ICG injection using fluorescence imaging. (From Troyan SL, Kianzad V, Gibbs-Strauss SL et al. 2009. *Annals of Surgical Oncology*, 16, 2943–2952. With permission.)

Both fiber-mediated spectroscopy and multispectral imaging focus on enabling sensitivity at greater depths, increasing the contrast enhancement of exogenous agents, and demonstrating a correlation between spectral parameters and the diagnostic gold standard, pathology. Signatures of cancer available for optical detection include increased metabolism, (2-deoxyglucose uptake, decreased tumor oxygenation due to elevated metabolic demands), sustained angiogenesis (increased hemoglobin concentrations, increased vascular permeability), and increased proteolytic activity (decreased green fluorescence due to collagen degradation).

Here, we briefly review applications of diffuse reflectance spectroscopy (DRS), fluorescence spectroscopy (FS), Raman spectroscopy (RS), and optical coherence tomography (OCT) delivered by fiber optic probes as potentially powerful alternatives to standard histological processing for monitoring treatment response and biopsy guidance. We also discuss the imaging extension of probe-based techniques and multispectral imaging of NIR fluorescence as they apply to surgical guidance and mapping of sentinel lymph nodes.

20.2.1 Localized Spectroscopy for Biopsy Guidance

Suspicious lesions detected by mammography are biopsied for microscopic assessment of disease type, grade, and receptor status (see Chapter 8). Biopsy is performed by surgical excision of the suspicious lesion or by core needle sampling, where 5–12 cores of tissue, typically 1 mm in diameter and several centimeters in length are acquired from the area of concern (Liberman et al. 1994). Core sampling is preferred because it is minimally invasive, but it is subject to undersampling that has been reported to result in a frequency of missed cancers ranging from 0.3% to 8.2% and repeat biopsies in 9% to 18% of patients (Zhu et al. 2009). Additionally, 70%–90% of mammographically detected lesions are found to be benign upon biopsy, suggesting an unnecessary sampling of nonmalignant tissues (Haka et al. 2005). Incorporation of fiber-based diffuse reflectance and/or fluorescence probes into a biopsy needle provides nondestructive, optical evaluation of suspicious tissue at multiple sites to guide the sampling of tissues most likely to be cancerous and to reduce the number of acquisitions needed for a confirmed diagnosis.

Bigio et al. (2000) were the first to measure UV–visible diffuse reflectance via a biopsy needle probe at a short source-to-detector separation (350 μm). The detected scattering features are sensitive to refractive index variations associated with malignant transformations such as hyperproliferation of epithelium, nuclear crowding and enlargement, and subcellular compositional changes in the stromal matrix (Wilson et al. 2005, Perelman et al. 1998, Mourant et al. 2001, Wax et al. 2002, Drezek et al. 2003). Absorption bands in the remitted spectrum contain physiological information about local tissue blood concentration and oxygenation, and the presence of lipids and water in the probed tissue volumes. Model-based approaches to photon transport have been developed to describe light transport in tissue for these small source-to-detector separations, where the diffusion approximation is invalid (Palmer and Ramanujam 2006, Zhu et al. 2005, Zonios and Dimou 2006, Pfefer et al. 2003, Amelink et al. 2003, van Veen et al. 2005). These models typically employ an empirical approximation to Lorenz-Mie scattering (Wang et al. 2005) combined with a Beer-Lambert attenuation factor to account for absorption by chromophores in the probed tissue volume (Amelink and Sterenborg 2004, Krishnaswamy et al. 2009). Multivariate statistical analysis can then be used to recognize patterns in the spectra and render a diagnostic decision for discriminating between malignant and benign tissues.

The fundamental limitation in this empirical approach to spectral analysis is that accurate knowledge of the photon pathlength is required in order to estimate chromophore concentrations within the optically sampled volume, and robust estimation of this pathlength is not possible in highly scattering media such as tissue without additional information. Van Veen overcame this limitation by measuring diffuse reflectance at multiple source–detector separations with a fiber optic probe that allowed the pathlength dependence on optical absorption and scattering in vivo to be assessed (van Veen et al. 2005). Alternatively, Brown employed a scalable Monte Carlo solution to account for the nonlinear wavelength-dependent effect of pathlength that confounds quantification of diffuse reflectance sampled via a fiber optic probe delivered through a biopsy needle (Brown et al. 2009). He proposed oxygen monitoring of the vascular compartment of breast tissue at the time of diagnostic biopsy to predict clinical or therapeutic response.

UV-visible endogenous fluorescence has also been explored as a method of enhancing specific contrast between benign and malignant tissues using fiber-optic probes. The wavelength-dependent influence of heterogeneous tissue absorption and scattering has a nonlinear effect on the recovered fluorescence, limiting excitation–emission quantification. Volynskaya et al. (2008) achieved more stratified pathologic classification (discriminating among normal breast tissue, fibrocystic change, fibroadenomas, and infiltrating ductal carcinoma) through the absorption and scattering parameters recovered from diffuse reflectance spectroscopy to extract measures of intrinsic, or undistorted fluorescence. While quantification of intrinsic fluorescence enhances diagnostic sensitivity, its signal is weak and spectral features are broadly shaped, which makes its diagnostic potential difficult to harness.

In contrast to fluorescence, many Raman active molecules are present in breast tissue with sharp and well-delineated spectral features (Mahadevan-Jansen et al. 1998). The Raman effect detects chemical changes specific to breast cancer by sensing nonlinear energy transfer to and from molecular vibration modes. Its spectral response is characterized by high information content relative to its diffuse reflectance and fluorescence

counterparts, but its signal strength is orders of magnitude weaker, resulting in long acquisition times (~30 s per pixel). Haka et al. (2005) was the first to demonstrate in vivo collection of Raman spectra from breast tissue, and while the authors suggested feasibility for intraoperative margin assessment, Raman's long integration times, sparse sampling, and limited depth penetration are barriers to clinical adoption in this application.

Optical coherence tomography (OCT) is another emerging high-resolution imaging technique that may serve as a powerful alternative and/or complement to histology during interventional procedures. While depth of imaging is limited to a few millimeters, OCT has demonstrated transverse resolutions less than 1 μm (Boppart et al. 2004) sufficient for capturing the architectural morphology visualized in fixed hematoxylin and eosin (H&E)-stained sections. OCT employs a technique known as low-coherence interferometry, performing optical ranging with NIR wavelengths in a manner analogous to ultrasound (see Chapter 15). The method has been used to perform optical biopsy during breast tumor resection (Nguyen et al. 2009) and axillary lymph node dissection (Nguyen et al. 2010). Axial data are acquired rapidly and at full margin depth (up to 2 mm); however, transverse sampling speed is currently too slow for wide-field assessment of the surgical margin.

20.2.2 The Imaging Extension of Spectroscopy for Surgical Margin Assessment

Point-based spectroscopy provides a nondestructive, real-time alternative to microscopic evaluation of tissue; however, localized sampling is too inefficient for effective assessment of breast resection margins (several cubic centimeters) from the surgeon's perspective. The imaging extension of probe-based spectroscopic techniques has been realized through multiplexing or raster scanning to achieve wider coverage of surgical margins. The mesoscopic sampling volumes assume the malignant phenotype provides disease-specific contrast in volume-averaged measures; consequently, microscopic residual disease may not be detected. Sensing depth varies from several millimeters in the UV-visible (absorption dominant) to several centimeters in the NIR. Within each wavelength band, altering the illumination-detection geometry may also vary the sensing depth.

Ramanujam and colleagues pioneered the development of an optical spectral imaging platform to rapidly and nondestructively create molecular composition maps of the surfaces of the resected surgical specimens (Brown et al. 2009, Bydlon et al. 2010, Wilke et al. 2009). Excised breast tissue is imaged in a Plexiglas box with ports for an 8-channel probe that samples diffuse reflectance over an area of 2 × 4 cm with 5-mm resolution. Remitted light is measured as a function of wavelength and the shape and magnitude of this spectrum is rapidly parameterized via a scalable Monte Carlo model (Palmer and Ramanujam 2007, Palmer et al. 2006, Zhu et al. 2006). Results from a 121-patient study indicate detection of cancer at the margin of the excised specimen with 80% sensitivity, residual disease within 1 mm of the margin with 73% sensitivity and an overall specificity of 65%

(Bydlon et al. 2009). In other efforts to more comprehensively sample lymph nodes and surgical margins, stepper motors have been employed to raster scan the sampling beam across the tissue surface (Austwick et al. 2010, Keshtgar et al. 2010, Laughney 2010). These methods remain too slow for clinical adoption, and still sample only a fraction of the typical lumpectomy specimen ($\sim$1 cm^2). An inherent tradeoff between spatial resolution, sampling volume and acquisition time exists for all imaging extensions of spectroscopy, and the optimization of these parameters with respect to the clinical utility of an imaging approach warrants further investigation.

20.2.3 Multispectral Image-Guided Surgery and Sentinel Lymph Node Detection

Multispectral imaging systems that detect diffuse reflectance, and particularly NIR fluorescence, appear to have the greatest potential to impact breast conserving surgery and sentinel lymph node (SLN) detection because they extract functional and molecularly specific information at sufficient resolution over a wide field. Breast conserving therapy (BCT), which includes local excision and radiation treatment of the breast, has been the standard of care for treatment of early invasive breast cancers (stages I and II) since Fisher and Veronisi demonstrated that survival after BCT is equivalent to mastectomy for most patients when surgical margins are clear of residual disease (Fitzal et al. 2009); however, the consequences of local recurrence are not insignificant. BCT maximizes removal of malignant tissues while minimizing damage to healthy, viable breast. However, if the excised tissue is positive for residual disease (determined in pathology subsequent to surgery), the probability of local recurrence increases, and the standard of care is a reexcision procedure (Schnitt et al. 1994, Scopa et al. 2006, Smitt et al. 1995, Spivack et al. 1994). The majority of retrospective studies report 20% to 40% reexcision rates due to positive or close margins (Pleijhuis et al. 2009), but the impact of multiple excisions on local recurrence is not clear because the definition of the surgical margin is not consistent between institutions and breast imaging and radiation treatment have progressed significantly over the last two decades. It is clear that surgical margin assessment is important for disease control particularly because risk of local recurrence is significantly higher for those patients in which extent of disease is underestimated (Mirza et al. 2000). However, the two studies investigating the effect of multiple excisions for local control on recurrence rates are contradictory. In a study of 459 patients by Menes et al. (2005), the risk of local recurrence after breast conservation for breast cancer increased progressively with the number of reexcisions needed to achieve clear margins; however this study included only 28 local recurrences and 22 patients who underwent two or more reexcisions. In a study of 2700 patients who underwent breast conserving surgery to maintain local control, the 5- and 10-year local recurrence rates for patients with two or more reexcisions (137 patients), patients with one reexcision (1514 patients), and patients with no reexcision (1119 patients) were 5.5%, 1.9%, 2.5%, and 10.5%,

5.7%, and 5.6%, respectively. Patients were excluded from this study if they underwent neoadjuvent chemotherapy, had positive margins, or required mastectomy to achieve negative margins. When controlling for these and other factors in the multivariate analysis, the need for multiple reexcisions to obtain negative margins in women undergoing BCS did not impact the incidence of local recurrence (Morrow et al. 2007). Optical diagnosis, particularly multispectral fluorescence, could improve completeness of resection during BCT because it enhances surgical vision, identifies molecularly specific contrast, and has high detection sensitivity.

A typical fluorescence imaging system involves a spectrally resolved source to excite exogenous or endogenous tissue fluorophores, the capturing of fluorescence emissions with a sensitive camera, and the filtering of excitation light. NIR imaging systems currently available for preclinical and clinical applications are reviewed by Gioux et al. (2010); of particular note is the fluorescence-assisted resection and exploration (FLARE) system pioneered by Frangioni (Troyan et al. 2009). FLARE measures reflectance at multiple NIR wavelengths, enabling correction for photon–tissue interactions and spectral unmixing of the fluorescence signal. Targeted molecular imaging is achieved through indocyanine green (ICG) moieties because ICG fluoresces sufficiently through several millimeters of tissue and has been approved for other indications by the FDA. The development of other NIR fluorescent probes has undoubtedly expedited the exploration of intraoperative applications of fluorescence imaging. Excellent reviews of the three major types of labels used in optical imaging—fluorescent proteins, bioluminescence, and fluorescent dyes—can be found in the references (te Velde et al. 2010, Kobayashi et al. 2010). Spectral unmixing dramatically improves the sensitivity and spatial accuracy of NIR cameras (Xu and Rice 2009, Mayes et al. 2008). Traditionally, fluorescence is quantified postacquisition by ratiometric methods; however, Themelis et al. (2009) developed a real-time spectral unmixing technique to improve fluorescence quantification via concurrent collection of fluorescence, light attenuation at the excitation wavelength, and color images through one lens and over the same FOV (three cameras operate in parallel). While early hardware realizations may seem cumbersome and cost-prohibitive (development of the FLARE system cost approximately $120,000 USD; Gioux et al. 2010), Liu et al. (2011) recently developed wireless goggles for NIR fluorescence-guided surgery costing $1200 USD. A CCD-based consumer-grade night-vision viewer was integrated with a modified compact head lamp as the light source. Fluorescence images directly project onto the goggle headset and video data can be wirelessly transferred from the headset for remote analysis.

NIR fluorescence in optical imaging of sentinel lymph node (SLN) mapping is also rapidly expanding in clinical oncology. Lymphatic spread of breast cancer is a strong indicator of both distant metastases and patient survival (Austwick et al. 2010). Standard-of-care previously had been removal of all axillary lymph nodes to control local disease and screen for metastases, but this invasive procedure has now been replaced with SLN biopsy. If the first node to receive lymphatic draining from a tumor site does not contain tumor cells, then lymphatic metastasis is assumed not to have occurred. Currently, the SLN is identified by injecting the subareola region with both a radioactive colloid (99m-technetium) and blue dye (lymphazurin); gamma cameras are employed to trace the lymphatic vessels and the blue dye aids in visualization of the node intraoperatively (only superficially visible due to increased tissue attenuation at shorter wavelengths). However, fluorescence imaging with ICG enables both transcutaneous and real-time lymphography with intraoperative lymph node visualization (Troyan et al. 2009, Hirche et al. 2010, Murawa et al. 2009).

20.3 NIR Systems for Whole Breast Tomography

A number of NIR imaging systems have been developed with unique configurations of hardware technologies, source detector setups, and breast positioning geometries. Technical approaches have been used to determine absorption and scattering coefficients of breast tissue based on continuous wave, time domain, and frequency domain methods (see Figure 20.3).

In continuous wave imaging, light is emitted into tissue at a constant amplitude (or with very low frequency amplitude modulation) and detected at another position. It is the simplest, fastest, most compact, and least expensive technique. However, unlike the other approaches, it is unable to reliably distinguish between absorption and scattering in the tissue.

Time domain imaging involves injection of a short light pulse with subsequent detection of the temporal distribution of photons that pass through the tissue. The detected distribution has a greater temporal spread than the initial light pulse due to differences in the path lengths taken by the collected photons as they scatter between the source and detector. From the shape of this distribution, absorption and scattering properties can be determined. Time domain systems are most sensitive to low light levels but are also the most expensive technology.

Intensity modulation of a laser source and measurement of the baseline amplitude and shift in phase are used in frequency domain imaging techniques. Measurements at many source and detector positions or at multiple frequencies are used in determining absorption and scattering. With frequency domain imaging, additional wavelengths can be added at relatively low cost and the instrumentation is typically more stable than time domain components but with lower signal sensitivity (Nissila et al. 2006).

Two types of source detector arrangements are most often used in optical imaging. In a reflectance geometry, the sources and detectors are on the same side of the tissue, typically attached to the same probe. Transmission mode occurs when the light source and detector are on opposing sides of the breast. This configuration achieves greater penetration depth; however, it also requires more sensitive detectors.

Patient positioning schemes have been variable in different optical imaging systems. In some cases, the patient lies prone with the breast pendant either in an optically matched liquid or in an array of fibers that directly contact the breast (Intes 2005, Pogue et al. 2004).

Alternatively, the patient can be imaged from a seated or standing position. The breast is sometimes compressed during the scan (Fang et al. 2009). Most systems scan a single breast at a time, although one approach has focused on bilateral breast acquisitions to compare the temporal changes between the two sides (Schmitz et al. 2005, Flexman et al. 2008).

20.3.1 NIR for Diagnostic and Therapeutic Breast Monitoring

Despite differences in system design, optical imaging methods seek to assist in clinical decision-making by providing unique information about the functional status of breast tissue. From risk assessment to screening, diagnosis, therapy monitoring, and follow-up imaging, diffuse optics has potential clinical applicability.

Women with more dense breasts have a higher incidence and mortality from breast cancer (Chiu et al. 2010). Scattering estimates from NIR spectroscopy have shown correlation with breast density (Srinivasan et al. 2003). Thus, optical imaging of younger

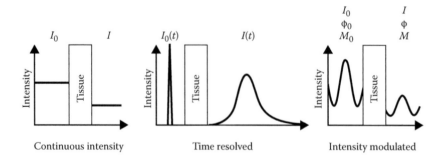

FIGURE 20.3 Different methods for obtaining absorption and scattering information used in diffuse optical imaging: continuous wave, time domain, and frequency domain. (From Delpy DT, Cope M. 1997. *Philosophical Transactions of the Royal Society of London Series B-Biological Sciences*, 352, 649–659. With permission.)

women may be able to assess their risk of breast cancer without ionizing radiation. One screening study compared optical and normal mammograms of 292 women to determine if the optical scans could characterize parenchymal density. Comparing the results for individuals with more than 75% dense tissue to those with less, the area under the curve (AUC) of a receiver operating characteristic (ROC) analysis of the NIR results was 0.92. Optical imaging appears to distinguish women with dense breasts, and thus, a higher risk of cancer development. Despite the potential to determine breast density, how a population wide assessment with NIR imaging would alter breast screening programs remains unclear.

Since optical imaging does not use ionizing radiation, it could be used in screening younger women as well as those at higher risk of developing cancer without the potential negative effects of repeated mammograms. However, because of its poor spatial resolution and variable contrast between regions of adipose and glandular tissue, NIR imaging does not appear to identify lesions well enough to be adopted clinically as a stand-alone approach.

NIR imaging can assist in the diagnosis of breast suspicions after normal mammographic screening. Much of the research on optical breast imaging has focused on this potential use as an inexpensive, noninvasive, and safe adjunct to regular screening as will be discussed in more detail later in this chapter.

Monitoring therapy using NIR light is another area that has been extensively studied. Although many chemotherapy regimens are prescribed by physicians for the treatment of breast cancer, no consensus exists on a noninvasive technique for determining the treatment response. The availability of such a technique would allow patients who may not be responding to a particular drug combination to be identified early so they could switch to another therapeutic cocktail that may be more effective. In a study conducted at University of California, Irvine (UCI), a patient was followed weekly over the course of 18 weeks of chemotherapy and her optical imaging results showed a 50% decrease in tissue optical index (a metric involving hemoglobin, water, and lipid content) during this time (Cerussi et al. 2010). The UCI investigators also reported a longitudinal study of 11 patients undergoing neoadjuvant chemotherapy, which showed significant decreases in total hemogloblin levels for patients with complete pathologic response to treatment relative to patients with only a partial response (Pakalniskis et al. 2011).

A multicenter trial sponsored by the American College of Radiology Imaging Network (ACRIN) is currently underway to examine the ability of diffuse optical imaging to assess patient response to presurgical neoadjuvant chemotherapy. This study will enroll 60 patients at five academic medical centers examined with identical breast scanning devices (Tromberg et al. 2008). The patients will be imaged four times over the course of the study: once before, twice during, and once after chemotherapy. Optical results will be compared with pathology obtained during surgery to determine the effectiveness of NIR spectroscopy in predicting tumor response to chemotherapy.

Another possible application of optical imaging is surveillance for recurrence after cancer treatment. Patient follow-up currently consists of physical examinations and semiannual or annual mammography. Including optical spectroscopy as an adjunct to traditional mammography could be particularly effective in this group of patients because of the elevated risk of developing recurrent cancer in the ipsilateral and contralateral breast.

Optical imaging is a noninvasive, inexpensive, and safe technique that provides functional information about breast tissue, which is unique amongst current clinically available modalities. More extensive studies like the ACRIN trial and standardization of hardware are required before it will gain widespread adoption despite the benefits that have already been demonstrated.

20.4 Image Reconstruction Algorithms for NIR Tomography of the Breast

The main difference between imaging breast tissue with optical versus x-ray photons is the amount of scattering that occurs. Since x-rays are only weakly scattered, they pass nearly straight through tissue and collimators can ensure that the measured attenuation is almost independent of scatter. Scattering is the dominant effect with optical photons and occurs 100 times more frequently than absorption. Individual photons follow a torturous path through tissue, and collectively they appear to be diffusive, much like milk being poured into coffee. Modeling is essential to account for scatter and correctly quantify the nonlinear relationship between the measured light signals and the optical properties of the tissue traversed. Topics important to understanding light transport through tissue and its use in breast imaging include diffusion theory, reconstruction algorithms, and packaged software for modeling.

20.4.1 NIR Transport Modeling

Optical photon transport through tissue can be modeled in several ways. Analytical solutions exist for simple shapes such as circles and rectangles but are very difficult to adapt to geometries of clinical relevance (Arridge 1995). Numerical techniques such as Monte Carlo and finite element methods (FEM) are generally preferred. Monte Carlo is the most accurate and adaptable approach for modeling light transport, but it can be prohibitively slow computationally (Jacques and Pogue 2008, Wang and Jacques 1993, Wang et al. 1995). A modeling domain is established and the paths of individual photons (ballistic particles that can be redirected in this case) are calculated in very small steps based on the probability of absorption and scattering events. After a sufficient number of individual photons are simulated, usually millions, results can be reliably interpreted. Because of the computational costs associated with Monte Carlo calculations, especially in image reconstruction, it is used mostly to validate other methods.

Diffusion theory based on FEM is widely applied to clinical light modeling because it is fast, adaptable to any geometry, and accurate to within a few percent as long as the diffusion approximation is valid. In this method, light propagation is represented as a concentration of optical energy that moves down a gradient. The optical radiance in tissue can be predicted by a specific form of the Boltzman equation, the radiative transport equation (RTE) (Duderstadt and Hamilton 1976, Jacques and Pogue 2008). Diffusion theory depends on three underlying assumptions: (1) the radiance is only linearly anisotropic, (2) the rate of change of the flux is much lower than the collision frequency, and (3) the radiance can be represented by an isotropic fluence rate plus a small directional flux (Jacques and Pogue 2008). With these assumptions, the RTE can be reduced to the diffusion equation,

$$\frac{1}{c}\frac{\partial\Phi(r,t)}{\partial t} - \nabla\cdot D\nabla\Phi(r,t) = -\mu_a\Phi(r,t) + S(r,t)$$

which predicts the light fluence rate at a point $\mathbf{r}$ and time t in a highly scattering medium. Pogue et al. (2001) give a complete derivation for the time and frequency domains. The terms in the equation represent the change in flux, diffusion, loss due to absorption, and the input light. The approximations are valid as long as the imaging domain (and distance from the light source) is significantly larger than the mean reduced scattering length ($1/\mu_s'$) and scatter dominates absorption as is always the case in breast tissue. Important relations between basic quantities and equations are shown in Table 20.1.

Clinical implementation of FEM requires discretization of the imaging domain into a grid, or mesh, as illustrated in Figure 20.4. The diffusion equation is solved for fluence represented as a piecewise continuous basis function expansion over the mesh. Detailed discussion of finite element methods is beyond the scope of this chapter but can be found in the literature (Arridge and Hebden 1997, Lynch 2005). Since the diffusion problem is solved discretely on a mesh, the accuracy of the solution is governed by the resolution of the mesh (i.e., spacing between nodes). A finer mesh (more nodes) will have higher accuracy since the solution is obtained at more points but will take longer to compute.

Unlike x-ray fluence, which follows a linear Beer's Law attenuation profile, optical fluence is nonlinearly related to the optical properties of the tissue. Estimates of optical properties obtained from measured fluence data are generally reconstructed

FIGURE 20.4 Three-dimensional tetrahedral finite element mesh of a compressed breast. This mesh is an example of a patient-specific mesh for image-guided NIRS.

iteratively over the mesh. Several numerical techniques are available to converge on a solution such as Newton or conjugate gradient methods. A general procedure for obtaining an optical property image is summarized in Figure 20.5.

20.4.2 Image Reconstruction

When imaging the breast using optical tomography, a model domain that matches the imaging geometry when the data were collected is critical. During the reconstruction process, measured data are compared with modeled data, and the two domains must be coincident in order to recover optical properties accurately. A circle or a cylinder of known radius is sufficient to represent a breast positioned in a circular fiber array (Jiang et al. 2009). Then simulated optical source and detector locations are placed on the surface of the mesh to match the clinical configuration.

An accurate reconstruction requires that the measured data be as close to the modeled data as possible. The imaging geometry is represented through the meshing process and the raw data are adjusted through calibration. The differences in source/detector combinations due to fiber–patient coupling, detector sensitivity, and source strength must be removed. A homogeneous tissue phantom is measured before each patient examination and

TABLE 20.1 Important Quantities, Their Symbols, Units, and Relationships in Diffusion-Based NIR Transport Theory

Quantity	Symbol	Units	Quantity	Symbol	Units
Absorption coefficient	μ_o	cm^{-1}	Reduced scattering coefficient	$\mu_s' = \mu_s(1-g)$	cm^{-1}
Scattering coefficient	μ_s	cm^{-1}	Transport mean free path	$MFP' = 1/(\mu_o - \mu_l')$	cm
Anisotropy of scattering	G	Dimensionless	Diffusion length	$D = MFP'/3$	cm
Refractive index	N	Dimensionless	Optical penetration depth	$\delta = \sqrt{D/\mu_o}$	cm

Source: Jacques SL, Pogue BW. 2008. *Journal of Biomedical Optics*, 13. With permission.
Note: More thorough discussions can be found in the substantial literature on the subject.

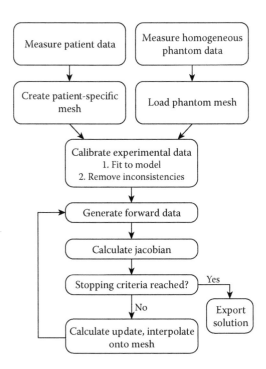

FIGURE 20.5 Workflow for complete optical data acquisition and reconstruction. After measuring data from a patient and homogeneous phantom, a mesh is created for each data set. The phantom measurements are used to calibrate the patient data and then an iterative reconstruction is performed.

patient data are calibrated by adjusting the measured fluence to vary only with path length using the homogeneous tissue phantom data as a reference. Source strength in the model is constant, and the experimental source strength is adjusted to match the model, removing any DC bias that would introduce error into the image reconstruction. As with other imaging modalities, data calibration is important for accurate results, and many groups have invested considerable time to develop optimal techniques (McBride et al. 2001).

Once an imaging domain has been defined, a forward problem can be solved. Here, the tissue optical properties and the simulated light source locations are assumed to be known and the boundary data at the detector positions are unknown. Forward data are calculated at the locations of simulated detectors based on the diffusion equation. Forward solutions are unique, well-posed and do not require iteration (unless an iterative solver is used, often the case for large 3-D meshes). The inverse problem is framed as having boundary data at known source and detector locations with the optical properties of each node in the mesh as unknowns. Since the number of nodes is typically an order of magnitude (or more) greater than the number of measurements and the functional relationship between the known boundary information and the unknown optical properties is nonlinear, inverse problems are nonunique, ill-posed, and challenging to solve. Thus, they must be stabilized and solved iteratively by calculating a forward model, comparing the simulated data with

the measured data, updating the optical properties, and repeating the process until a stopping criterion is reached. Emerging computational tools, such as GPU processing, may allow these types of methods for optical spectroscopic reconstruction to evolve more quickly and gain clinical adoption.

20.4.3 Advanced Algorithms: Image-Guided Recovery

Image-guided algorithms are becoming more popular for many multimodal instruments (Carpenter et al. 2008, Fang et al. 2009). They are attractive for two reasons: computational efficiency and higher accuracy. These techniques require an image of tissue structure be known a priori and coregistered with the optical scan. The spatial information is fed into the optical reconstruction in the form of a segmented map where glandular, adipose, and tumor tissue types are identified. Optical properties are allowed to vary between regions, but often not within. Making these assumptions yields a smaller computational problem because the bulk optical properties of the segmented regions are found rather than the optical properties at each node of the mesh. The other advantage of these methods is that the optical properties are estimated more accurately on the spatial scale defined by the segmentation of the complementary imaging modality. These algorithms enable detection of smaller lesions with lower contrasts on a spatial scale approaching 3 mm (Pogue 2006). Optical imaging systems have been combined with MRI (see Chapter 16) and x-ray tomosynthesis (see Chapter 4) for breast applications (Carpenter et al. 2007, Fang et al. 2009) and may represent the most promising clinical approaches.

20.4.4 Software

Software packages have been developed to assist researchers and clinicians with the computations involved in optical tomography and spectroscopy. Examples include Virtual Photonics Technology (tissue spectroscopy), TOAST (time and frequency domain), and NIRFAST (frequency domain and fluorescence) (Dehghani and Pogue 2011, Schweiger and Arridge 2011). These software packages are designed to be flexible and are intended to accelerate the transition of optical tomography into more widespread clinical use. They are frequently open source and guide a user although the entire reconstruction process with a GUI-driven approach.

NIRFAST is an open-source modeling and reconstruction package capable of single and multiwavelength optical or functional imaging from measured data (Dehghani et al. 2008). Data analysis and calibration tools are included, as well as mesh creation for simple and more advanced shapes. The package presents a GUI that allows user customization and provides documentation on data formats and reconstruction procedures. In addition to basic functionality, NIRFAST offers image-guided spectroscopy. Users are able to constrain absorption and fluorescence reconstructions based on tissue anatomy obtained from multimodal imaging systems. As more advanced imaging techniques rely on multiple modalities, packaged reconstruction tools with these capabilities will become increasingly important.

20.5 Optical and X-Ray Mammography

Combining x-ray and optical imaging of the breast provides complementary information about the tissue (see Chapter 1 for more on mammography). The high spatial resolution of mammography leads to lesion localization while optical tomography offers insight into the functional status of tissue. Several systems have been developed to take advantage of the synergistic relationship between x-ray and optical imaging.

They can be divided into two categories: those in which the optical scan is performed as an adjunct to x-ray mammography and those that represent fully integrated optical

and mammographic systems (see Figure 20.6). One example of an adjunct approach was developed at the University of Pennsylvania (Chance et al. 2005). This system uses a three wavelength LED source surrounded by eight detectors in a circle around the source. Placement of the probe is based on tumor location as determined from the x-ray mammogram. A second example of an adjunct system, in use at Dartmouth College is shown in Figure 20.7.

A fully integrated system has been developed at Massachusetts General Hospital (MGH) (Fang et al. 2009). It is comprised of both continuous wave and frequency domain instrumentation operating at three and two wavelengths, respectively, and over 40 source locations with light collection at 32 detectors. The breast is compressed with the sources and detectors in transmission mode on either side of the tissue. After the optical

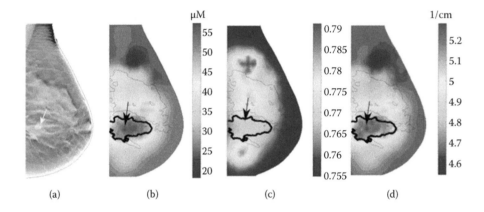

FIGURE 20.6 **(See color insert.)** Results of a reconstruction from an integrated x-ray mammography and optical spectroscopy system at the MGH from a breast with a 2.5-cm invasive ductal carcinoma indicated by the arrow. (a) Tomosynthesis image. (b) Total hemoglobin (micromoles per liter). (c) Oxygen saturation. (d) Scattering coefficient (cm⁻¹). (From Fang Q, Selb J, Carp SA et al. 2011. *Radiology*, 258, 89–97. With permission.)

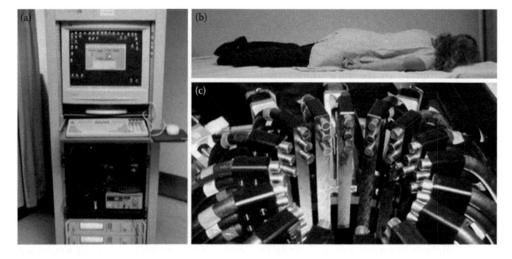

FIGURE 20.7 Dartmouth DOT stand-alone breast imaging-system. (a) Frequency domain NIR instrumentation rack for light generation and detection. (b) Patient examination table. (c) Breast interface.

scan, the source and detector probes are removed before x-ray imaging is performed. No movement of the target breast occurs between scans. An integrated system benefits from complete coregistration of optical and x-ray data sets (see Figure 20.6). However, spatial and other constraints create challenges when adding optical capabilities to traditional x-ray systems. Another fully integrated system is currently being built at Dartmouth College.

The promising results of combined x-ray and optical technology has led to the development of commercial systems as well. Advanced Research Technologies (ART) and Imaging Diagnostic Systems have separately developed the instrumentation. The ART platform uses single-photon counting TD diffuse tomography based on measurements at four wavelengths from 760 to 850 nm. This system has been approved for sale in Europe and Canada and is currently undergoing clinical trials in the United States in preparation for FDA approval (Intes 2005).

20.5.1 Optical Properties of Normal Breast Tissue

X-ray-guided optical tomography has also been used to characterize the properties of normal breasts. In addition to studies on breast density, differences in age, body mass index (BMI), menopausal status, and breast thickness have been quantified with respect to optical absorption and scattering. Studies of tissue absorption in terms of hemoglobin concentrations have provided the most robust results. Inverse relationships between hemoglobin levels and age, BMI, and breast thickness have been reported (Intes 2005, Fang et al. 2011). These results likely reflect the higher concentration of fatty tissue (which has lower metabolic requirements) compared with fibroglandular content in the breasts of older women, those with higher BMI or larger breasts. Scattering is also inversely correlated with BMI and breast thickness (Intes 2005, Srinivasan et al. 2003), likely on the account of fewer scattering centers in adipose tissue. Lastly, fluctuations of around 10% in hemoglobin levels have been observed in premenopausal women during the menstrual cycle, with higher hemoglobin occurring in the secretory phase consistent with histologically observed cellular proliferation during this period (Pogue et al. 2004, Shah et al. 2001).

20.5.2 Optical Properties of Diseased Breast Tissue

Measureable differences in the optical properties of malignant, benign, and normal tissues are essential for widespread adaptation of optical imaging in the clinic. Overall, increases in hemoglobin concentrations and scattering have shown the most promise for distinguishing malignancy from normal tissue while oxygen saturation and water concentrations have not been as consistent. Direct comparisons of absorption coefficients have shown that they are 2.5 times higher in malignant

tissue compared to surrounding breast, whereas scattering is 20% higher when examining a group of 87 histologically confirmed carcinomas (Grosenick et al. 2005). Hemoglobin levels in tumors larger than 6 mm were measured to be on average 1.5 times that of the same area of normal tissue in the contralateral breast (Poplack et al. 2007). A twofold increase in tissue optical index between tumor and normal tissue was also found in 58 patients (Cerussi et al. 2006). Determination of an average contrast between malignant and normal breast tissue is difficult due to tumor heterogeneity, differences between the imaging systems and chromophore quantification schemes used to date, but breast cancers appear to contain about twice the hemoglobin of normal tissue, which is not surprising because malignancies typically exhibit hypervascularization due to angiogenesis.

Distinguishing between malignant and benign conditions has proved more challenging and has been limited to differences in hemoglobin contrasts. Rather than directly comparing optical absorption or hemoglobin concentration in the region of interest (ROI) across patients, evaluations of the ratio of the ROI to background tissue values have been used in order to account for intersubject variability. In 26 malignant, 17 solid benign, and eight cystic cases, a group at MGH showed significantly higher contrast in the malignancies compared to both the benign lesions and cysts (Fang et al. 2011). A comparison of 11 malignant cases with 12 benign conditions revealed a significantly higher deoxygenated hemoglobin contrast in the malignant group (Intes 2005). These data indicate that benign lesions can be distinguished based on ratios of hemoglobin contrast between lesion and background tissue.

20.5.3 Comparing X-Ray and Optical Mammography

Estimates of sensitivity and specificity are important in determining the clinical utility of optical mammography. An analysis of sensitivity and specificity of an NIR spectroscopy system has been performed with 116 patients including 44 verified cancers at two different centers (Chance et al. 2005). The approach utilized two metrics based on region of interest to contralateral contrast of oxy and deoxygenated hemoglobin measurements. Another study of 60 patients, divided roughly equally into malignant, benign, and normal cases, has also been reported (Kukreti et al. 2010). A summary of the results from these two clinical efforts is shown in Table 20.2. These results are quite promising, indicating that optical tomography can effectively determine the presence or absence of malignancy.

TABLE 20.2 Accuracy of Malignancy Detection Using Diffuse Optical Imaging

Study	Sensitivity	Specificity	PPV	NPV	AUC
Chance et al. 2005	96	93	89	97	95
Kukreti et al. 2010	91	95	95	89	—

To understand the potential benefits of optics when combined with x-ray mammography, comparison of the performance of the combination relative to current clinical standards is necessary. A study of 110 patients, 58 with abnormal mammograms, showed a considerable improvement in the positive predictive value of mammography when combined with optical imaging (Poplack et al. 2007). Another study compared the ROC curve of a mammography system alone or when combined with a commercially available optical system based on a study of 82 patients, 37 with benign, and 42 with malignant histology (Poellinger et al. 2008). The ROC curve for mammography alone had an AUC of 0.72 while the combined technique had an AUC of 0.80, a significant improvement over mammography alone. Based on these results, optical tomography as an adjunct to mammography can improve the diagnostic accuracy of breast lesion classification.

Optical imaging provides valuable additional information on the metabolic status of tissue when included as an adjunct to mammography. It improves the positive predictive value and AUC of mammography alone. More large-scale studies to further quantify the benefit of optical imaging are needed. Also, a cost/benefit analysis of adding optical imaging to clinical examinations is important before widespread adaptation of optical imaging of the breast is possible.

20.6 Optical Breast Imaging and MRI

Currently, dynamic contrast enhanced magnetic resonance (DCE-MR) is used to image high-risk patients for breast cancer because of its high sensitivity, relative to x-ray mammography (see Chapter 16 for more on breast MRI). While the sensitivity has been reported as high as 100% (Kuhl 2007), its ability to distinguish between benign and malignant breast tissue is more modest. Breast MRI's low specificity, reported to be 60%–80% (Kuhl 2007), reveals a large number of false positive results that require unnecessary biopsies. Modern radiology is trending toward multimodal imaging because information from combined systems provides more complete tissue assessments for clinical decisions. One of the first widely accepted combinations to provide structure and function of tissue was PET/CT, which integrated the structural imaging from x-ray computed tomography (CT) with the function of glucose uptake, as measured by positron emission tomography (PET). PET/CT is not specifically used for breast imaging, yet has become an essential tool for detecting cancer metastases. With so few functional imaging tools available clinically, the potential for introducing novel optical imaging systems to monitor specific molecules, cellular activity, drug uptake, and tissue physiology is high.

Since optical spectroscopy is a highly specific functional imaging modality, it can be combined with more spatially resolved imaging modalities. Ntziachristos et al. (2000, 2002) and Brooksby et al. (2004, 2006) have developed combined MRI-NIRS systems for concurrent MRI and optical imaging.

These two modalities are well suited for integration because MRI has high soft tissue resolution but more limited functional use whereas optical imaging has the opposite characteristics. Since both are nonionizing, risk of their repeated use on patients is minimal. Their combination has been successful in increasing the information available from clinical MRI examinations, showing distinctions between malignant and benign lesions in several case studies (Carpenter et al. 2007, Ntziachristos et al. 2002). As a functional add-on to MRI, these bimodality examinations are high in cost per scan, making them less appealing for breast screening than their stand-alone optical counterparts or combinations with x-ray mammography. However, they are very well suited to quantifying tissue properties in small suspicious areas prior to biopsy as a follow-up to mammographic screening, and studies to investigate their benefits are underway.

20.6.1 Challenges in MRI Coupling

Since MRI scanners must remain in magnetically shielded rooms and ferrous metals cannot be introduced, significant challenges arise in coupling them with optical systems. Fiber optic technology allows light to be delivered and collected with minimal attenuation through long fiber optic cables (10 m or more), which keeps the rest of the optical instrumentation outside the MR bay. Space limitations imposed by clinical breast coils and MR bores require that optical fibers be housed below the prone patient in a small space around the pendant breast. Both circular and parallel plate fiber arrays have been incorporated into clinical breast coils, although the parallel plate configuration has been shown to improve data quality and it shares the same geometry with MR-guided biopsy (Carpenter et al. 2008). Optode (optical sensor) positioning can be difficult for lesions that are near the chest wall or for women with smaller breasts because of the physical structure of MR breast coil platform, which prevents optodes from reaching the chest wall. Fiber positioning for nongrid arrays also relies on general knowledge of the tumor location from previous imaging studies or scout acquisitions because examination length often prohibits repositioning.

20.6.2 Algorithm Development: Prior Information

After successful demonstration by Ntziachristos et al. several important contributions were made by Brooksby et al. and Carpenter et al. working at Dartmouth. Optical information clearly enhances the data available from MR imaging, but the synergy between the modalities is twofold. Merging MRI information into the tomographic reconstruction can enhance the inherently low spatial resolution of optical imaging. The concept of using a priori information was introduced by Brooksby et al. (2003) to improve standard optical imaging using the method outlined in Figure 20.8. Two types of prior information are now

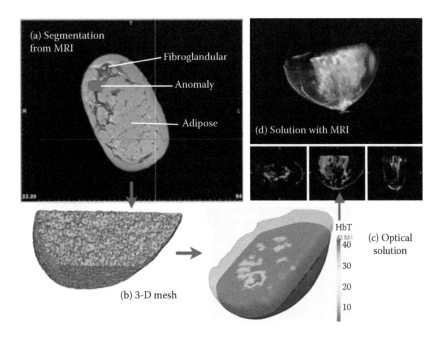

FIGURE 20.8 (**See color insert.**) Mesh creation and inclusion of prior information in MRI-NIR imaging. Tissue types are segmented based on MRI images (a) and a 3-D mesh is created (b) with these assignments preserved. The optical image reconstruction is completed in which the optical parameters in each region are estimated (c). The optical solution is merged with the MRI image (d).

commonly used in MRI-guided diffuse optical imaging: spatial priors and spectral priors.

20.6.2.1 Spatial Priors

Spatial priors are best suited for multimodal systems, but they are also required for stand-alone systems where the outer boundary of the breast must be known. In a circular imaging geometry, the breast is approximated as a cylinder where the radius is known as prior information. With more sophisticated combined systems, the tissue boundary is recovered as a complex shape and adipose, glandular, and tumor tissues are separated. Here, hard or soft priors reconstructions can be used to improve the resolution of the optical imaging to nearly 3 mm (Pogue 2006).

20.6.2.2 Spectral Priors

Spectral priors can be used simultaneously with spatial priors to yield molecular specific information from spectroscopic data. Including information about the absorption spectrum in the reconstruction yields more accurate tissue quantification than incorporating only structural information (Brooksby et al. 2005, Srinivasan et al. 2004), but the most accurate tissue quantification is obtained by combining both types of prior information.

Prior information in optical imaging is ideally obtained from the acquisition of optical data from multiple wavelengths in a highly spatially resolved modality such as MRI and is utilized in the tomographic reconstruction process to constrain the tissue optical property estimates both spectrally and spatially. With the incorporation of priors, spectral systems can report tissue parameters such as oxygenated and deoxygenated hemoglobin, water and lipid concentrations, and scatter amplitude and power rather than μ_a and μ_s at multiple wavelengths. These concepts are especially beneficial to image-guided systems, as they can quantify key physiological parameters with high spatial resolution as shown in Figure 20.9.

20.6.3 Future Directions

Several optical imaging techniques been developed to increase the information content available during MR imaging of the breast with the ultimate goal of increasing the specificity of contrast MR examinations before biopsy. Future directions in optical breast imaging will focus on techniques currently employed in small animal imaging or stand-alone optical imaging such as fluorescence imaging. Fluorescence molecular tomography is an especially promising modality that has already been demonstrated in humans by Corlu et al. (2007) and can be combined with MRI to monitor exogenous dyes or endogenous tissue fluorescence such as protoporphyrin IX emission (Gibbs-Strauss et al. 2009). These properties are not recoverable with current imaging modalities commonly used to examine the breast for cancer, and combined analysis of MRI and vascular and molecular parameters will provide a complete metabolic profile of tumors.

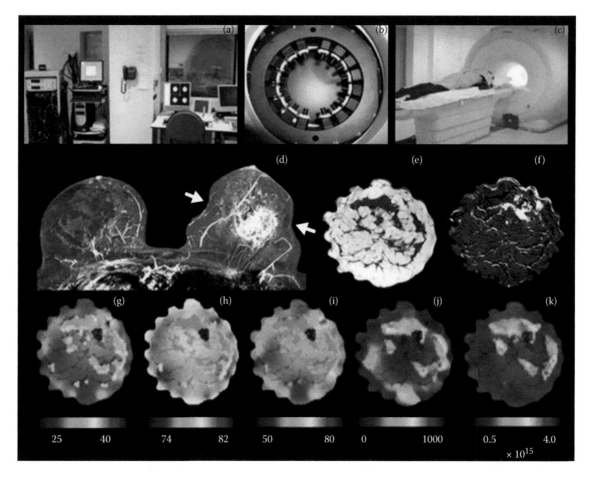

FIGURE 20.9   (See color insert.) Clinical example of MRI-NIR imaging. MRI and NIR imaging are performed simultaneously using a clinical scanner (a, c) and circular NIR breast array (b). MRI images (d–f) are used to locate the tumor and assign tissue properties for optical reconstructions of total hemoglobin (g), oxygen saturation (h), percent water (i), scatter amplitude (j), and scatter power (k). (From Carpenter CM, Pogue BW, Jiang SJ et al. 2007. *Optics Letters*, 32, 933–935.)

20.7 Conclusions

The field of optical imaging in breast cancer is diverse and is under development for a range of potential disease management stages. Use of optics in breast management has largely occurred over two wavelength ranges, visible (400–650 nm), and near-infrared (650–1000 nm), as they offer very different interactions with tissue. Since visible light is highly attenuated, it is largely utilized for surface imaging and spectroscopy with the greatest potential for systems to be developed for application in surgery or biopsy. Near-infrared light is highly scattered but only weakly absorbed, so it has substantial potential for use in tomography applications. The integration of spectral tomography methods with conventional imaging systems such as x-ray and MRI is emerging as a major area of interest where the synergy of the combined information may indeed prove valuable for the diagnosis of complex diseases such as breast cancer.

References

Amelink A, Bard MPL, Burgers SA, Sterenborg HJCM. 2003. Single-scattering spectroscopy for the endoscopic analysis of particle size in superficial layers of turbid media. Appl Opt 42: 4095–4101.

Amelink A, Sterenborg H. 2004. Measurement of the local optical properties of turbid media by differential path-length spectroscopy. Appl Opt 43: 3048–3054.

Arridge SR. 1995. Photon-measurement density-functions: I. Analytical forms. Appl Opt 34: 7395–7409.

Arridge SR, Hebden JC. 1997. Optical imaging in medicine: II. Modelling and reconstruction. Phys Med Biol 42: 841–853.

Austwick MR, Clark B, Mosse CA et al. 2010. Scanning elastic scattering spectroscopy detects metastatic breast cancer in sentinel lymph nodes. J Biomed Opt 15: 047001.

Bigio I, Bown S, Briggs G et al. 2000. Diagnosis of breast cancer using elastic-scattering spectroscopy: Preliminary clinical results. J Biomed Opt 5: 221–228.

Bogaards A, Sterenborg HJCM, Trachtenberg J, Wilson BC, Lilge L. 2007. In vivo quantification of fluorescent molecular markers in real-time by ratio imaging for diagnostic screening and image-guided surgery. Lasers Surg Med 39: 605–613.

Boppart SA, Luo W, Marks DL, Singletary KW. 2004. Optical coherence tomography: Feasibility for basic research and image-guided surgery of breast cancer. Breast Cancer Res Treat 84: 85–97.

Brooksby B, Jiang S, Dehghani H et al. 2004. Magnetic resonance-guided near-infrared tomography of the breast. Rev Sci Instrum 75: 5262–5270.

Brooksby B, Pogue BW, Jiang S et al. 2006. Imaging breast adipose and fibroglandular tissue molecular signatures using hybrid MRI-guided near-infrared spectral tomography. Proc Natl Acad Sci USA 103: 8828–8833.

Brooksby B, Srinivasan S, Jiang S et al. 2005. Spectral-prior information improves Near-Infrared diffuse tomography more than spatial-prior. Opt Lett 30: 1968–1970.

Brooksby BA, Dehghani H, Pogue BW, Paulsen KD. 2003. Near-infrared tomography breast image reconstruction with a priori structural information from MRI: Algorithm development for reconstructing heterogeneities. IEEE JSTQE 9: 199–209.

Brown J, Wilke L, Geradts J, Kennedy S, Palmer G, Ramanujam N. 2009. Quantitative optical spectroscopy: A robust tool for direct measurement of breast cancer vascular oxygenation and total hemoglobin content in vivo. Cancer Res 69: 2919–2926.

Brown JQ, Bydlon TM, Richards LM et al. 2010. Optical assessment of tumor resection margins in the breast. IEEE J Sel Top Quantum Electron 16: 530–544.

Bydlon T, Kennedy S, Richards L et al. 2010. Performance metrics of an optical spectral imaging system for intra-operative assessment of breast tumor margins. Opt Exp 18: 8058–8076.

Bydlon TM, Brown JQ, Barry WT et al. 2009. Rapid optical imaging of breast tumor margins: Final results from a 100-patient clinical study. Cancer Res 69: 770s–771s.

Carpenter C, Srinivasan S, Pogue B, Paulsen K. 2008. Methodology development for three-dimensional MR-guided near infrared spectroscopy of breast tumors. Opt Exp 16: 17903–17914.

Carpenter CM, Pogue BW, Jiang SJ et al. 2007. Image-guided spectroscopy provides molecular specific information in vivo: MRI-guided spectroscopy of breast cancer hemoglobin, water, and scatterer size. Opt Lett 32: 933–935.

Cerussi A, Shah N, Hsiang D, Durkin A, Butler J, Tromberg BJ. 2006. In vivo absorption, scattering, and physiologic properties of 58 malignant breast tumors determined by broadband diffuse optical spectroscopy. J Biomed Opt 11: 044005.

Cerussi AE, Tanamai VW, Mehta RS, Hsiang D, Butler J, Tromberg BJ. 2010. Frequent optical imaging during breast cancer neoadjuvant chemotherapy reveals dynamic tumor physiology in an individual patient. Acad Radiol 17: 1031–1039.

Chance B, Nioka S, Zhang J et al. 2005. Breast cancer detection based on incremental biochemical and physiological properties of breast cancers: A six-year, two-site study. Acad Radiol 12: 925–933.

Chiu SY, Duffy S, Yen AM, Tabar L, Smith RA, Chen HH. 2010. Effect of baseline breast density on breast cancer incidence, stage, mortality, and screening parameters: 25-year follow-up of a Swedish mammographic screening. Cancer Epidemiol Biomarkers Prev 19: 1219–1228.

Corlu A, Choe R, Durduran T et al. 2007. Three Dimensional in vivo fluorescence diffuse optical tomography of breast cancer in humans. Opt Exp 5: 6696–6716.

Dehghani H, Eames ME, Yalavarthy PK et al. 2008. Near infrared optical tomography using NIRFAST: Algorithm for numerical model and image reconstruction. Commun Numer Methods Eng 25: 711–732.

Dehghani H, Pogue BW. 2011. NIRFAST [Online]. Dartmouth College. Available: http://www.dartmouth.edu/~nir/nirfast/.

Delpy DT, Cope M. 1997. Quantification in tissue near-infrared spectroscopy. Philos Trans Roy Soc Lond Ser B-Biol Sci 352: 649–659.

Drezek R, Guillaud M, Collier T et al. 2003. Light scattering from cervical cells throughout neoplastic progression: Influence of nuclear morphology, DNA content, and chromatin texture. J Biomed Opt 8: 7–16.

Duderstadt JJ, Hamilton LJ. 1976. Nuclear Reactor Analysis. New York, Wiley.

Fang Q, Carp SA, Selb J et al. 2009. Combined optical imaging and mammography of the healthy breast: Optical contrast derived from breast structure and compression. IEEE Trans Med Imaging 28: 30–42.

Fang Q, Selb J, Carp SA et al. 2011. Combined optical and X-ray tomosynthesis breast imaging. Radiology 258: 89–97.

Fitzal F, Riedl O, Jakesz R. 2009. Recent developments in breast-conserving surgery for breast cancer patients. Langenbeck's Arch Surg 394: 591–609.

Flexman ML, Li Y, Bur AM et al. 2008. The design and characterization of a digital optical breast cancer imaging system. In Conf Proc IEEE Eng Med Biol Soc. pp. 3735–3738.

Gibbs-Strauss SL, O'Hara JA, Hoopes PJ, Hasan T, Pogue BW. 2009. Noninvasive measurement of aminolevulinic acid-induced protoporphyrin IX fluorescence allowing detection of murine glioma in vivo. J Biomed Opt 14: 014007–014008.

Gioux S, Choi HS, Frangioni JV. 2010. Image-guided surgery using invisible near-infrared light: Fundamentals of clinical translation. Mol Imag 9: 237–255.

Grosenick D, Wabnitz H, Moesta KT, Mucke J, Schlag PM, Rinneberg H. 2005. Time-domain scanning optical mammography: II. Optical properties and tissue parameters of 87 carcinomas. Phys Med Biol 50: 2451–2468.

Haka AS, Shafer-Peltier KE, Fitzmaurice M, Crowe J, Dasari RR, Feld MS. 2005. Diagnosing breast cancer by using Raman spectroscopy. Proc Natl Acad Sci USA 102: 12371–12376.

Hirche C, Murawa D, Mohr Z, Kneif S, Huenerbein M. 2010. ICG fluorescence-guided sentinel node biopsy for axillary nodal staging in breast cancer. Breast Cancer Res Treat 121: 373–378.

Intes X. 2005. Time-domain optical mammography SoftScan: Initial results. Acad Radiol 12: 934–947.

Intes X, Ripoll J, Chen Y, Nioka S, Yodh AG, Chance B. 2003. In vivo continuous-wave optical breast imaging enhanced with Indocyanine Green. Med Phys 30: 1039–1047.

Jacques SL, Pogue BW. 2008. Tutorial on diffuse light transport. J Biomed Opt 13: 041302.

Jiang S, Pogue BW, Carpenter CM et al. 2009. Evaluation of breast tumor response to neoadjuvant chemotherapy with tomographic diffuse optical spectroscopy: Case studies of tumor region-of-interest changes. Radiology 252: 551–560.

Keshtgar MRS, Chicken DW, Austwick MR et al. 2010. Optical scanning for rapid intraoperative diagnosis of sentinel node metastases in breast cancer. Br J Surg 97: 1232–1239.

Kobayashi H, Ogawa M, Alford R, Choyke PL, Urano Y. 2010. New strategies for fluorescent probe design in medical diagnostic imaging. Chem Rev 110: 2620–2640.

Krishnaswamy V, Hoopes PJ, Samkoe KS, O'Hara JA, Hasan T, Pogue BW. 2009. Quantitative imaging of scattering changes associated with epithelial proliferation, necrosis, and fibrosis in tumors using microsampling reflectance spectroscopy. J Biomed Opt 14: 014004.

Kuhl CK. 2007. Current status of breast MR imaging—Part 2. Clinical applications. Radiology 244: 672–691.

Kukreti S, Cerussi AE, Tanamai W, Hsiang D, Tromberg BJ, Gratton E. 2010. Characterization of metabolic differences between benign and malignant tumors: High-spectral-resolution diffuse optical spectroscopy. Radiology 254: 277–284.

Laughney AM, Krishnaswamy V, Garcia-Allende PB et al. 2010. Automated classification of breast pathology using local measures of broadband reflectance. J Biomed Opt. (in press).

Liberman L, Dershaw DD, Rosen PP, Abramson AF, Deutch BM, Hann LE. 1994. Stereotaxic 14-gauge breast biopsy—How many core biopsy specimens are needed. Radiology 192: 793–795.

Liu Y, Bauer AQ, Akers WJ et al. 2011. Hands-free, wireless goggles for near-infrared fluorescence and real-time image-guided surgery. Surgery 149: 689–698.

Lynch D. 2005. Numerical Partial Differential Equations for Environmental Scientists and Engineers: A First Practical Course. New York: Springer.

Mahadevan-Jansen A, Mitchell MF, Ramanujam N et al. 1998. Near-infrared Raman spectroscopy for in vitro detection of cervical precancers. Photochem Photobiol 68: 123–132.

Mayes PA, Dicker DT, Liu YY, El-Deiry WS. 2008. Noninvasive vascular imaging in fluorescent tumors using multispectral unmixing. Biotechniques 45: 459.

Mcbride TO, Pogue BW, Jiang S, Osterberg UL, Paulsen KD. 2001. Development and calibration of a parallel modulated near-infrared tomography system for hemoglobin imaging in vivo. Rev Sci Instrum 72: 1817–1824.

Menes TS, Tartter PI, Bleiweiss I, Godbold JH, Estabrook A, Smith SR. 2005. The consequence of multiple re-excisions to obtain clear lumpectomy margins in breast cancer patients. Ann Surg Oncol 12: 881–885.

Mirza NQ, Vlastos G, Meric F et al. 2000. Ductal carcinoma-in-situ: Long-term results of breast-conserving therapy. Ann Surg Oncol 7: 656–664.

Morrow M, O'Sullivan MJ, Li T, Freedman G. 2007. The effect of multiple reexcisions on the risk of local recurrence after breast conserving surgery. Ann Surg Oncol 14: 3133–3140.

Mourant JR, Johnson TM, Freyer JP. 2001. Characterizing mammalian cells and cell phantoms by polarized backscattering fiber-optic measurements. Appl Opt 40: 5114–5123.

Murawa D, Hirche C, Dresel S, Huenerbein M. 2009. Sentinel lymph node biopsy in breast cancer guided by indocyanine green fluorescence. Br J Surg 96: 1289–1294.

Nguyen F, Zysk A, Chaney EJ et al. 2009. Intraoperative evaluation of breast tumor margins with optical coherence tomography. Cancer Res 69: 8790–8796.

Nguyen FT, Zysk AM, Chaney EJ et al. 2010. Optical coherence tomography the intraoperative assessment of lymph nodes in breast cancer. IEEE Eng Med Biol Mag 29: 63–70.

Nissila I, Hebden JC, Jennions D et al. 2006. Comparison between a time-domain and a frequency-domain system for optical tomography. J Biomed Opt 11: 064015.

Ntziachristos V, Bremer C, Weissleder R. 2003. Fluorescence imaging with near-infrared light: New technological advances that enable in vivo molecular imaging. Eur Radiol 13: 195–208.

Ntziachristos V, Yodh AG, Schnall M, Chance B. 2000. Concurrent MRI and diffuse optical tomography of breast after indocyanine green enhancement. Proc Natl Acad Sci USA 97: 2767–2772.

Ntziachristos V, Yodh AG, Schnall MD, Chance B. 2002. MRI-guided diffuse optical spectroscopy of malignant and benign breast lesions. Neoplasia 4: 347–354.

Pakalniskis MG, Wells WA, Schwab MC et al. 2011. Tumor angiogenesis change estimated by using diffuse optical spectroscopic tomography: Demonstrated correlation in women undergoing neoadjuvant chemotherapy for invasive breast cancer. Radiology 259: 365–374.

Palmer GM, Ramanujam N. 2006. Monte Carlo-based inverse model for calculating tissue optical properties. Part I: Theory and validation on synthetic phantoms. Appl Opt 45: 1062–1071.

Palmer GM, Ramanujam N. 2007. Monte Carlo-based inverse model for calculating tissue optical properties. Part I: Theory and validation on synthetic phantoms. Appl Opt 46: 6847–6847.

Palmer GM, Zhu C, Breslin TM, Xu F, Gilchrist KW, Ramanujam N. 2006. Monte Carlo-based inverse model for calculating tissue optical properties. Part II: Application to breast cancer diagnosis. Appl Opt 45: 1072–1078.

Perelman LT, Backman V, Wallace M et al. 1998. Observation of periodic fine structure in reflectance from biological tissue: A new technique for measuring nuclear size distribution. Phys Rev Lett 80: 627–630.

Pfefer TJ, Matchette LS, Bennett CL et al. 2003. Reflectance-based determination of optical properties in highly attenuating tissue. J Biomed Opt 8: 206–215.

Pleijhuis R, Graafland M, De Vries J, Bart J, De Jong J, Van Dam G. 2009. Obtaining adequate surgical margins in breast-conserving therapy for patients with early-stage breast cancer: Current modalities and future directions. Ann Surg Oncol 16: 2717–2730.

Poellinger A, Burock S, Grosenick D et al. 2011. Breast cancer: Early- and late-fluorescence near-infrared imaging with indocyanine green—A preliminary study. Radiology 258: 409–416.

Poellinger A, Martin JC, Ponder SL et al. 2008. Near-infrared laser computed tomography of the breast first clinical experience. Acad Radiol 15: 1545–1553.

Pogue BW. 2006. Image analysis methods for diffuse optical tomography. J Biomed Opt 11: 33001.

Pogue BW, Geimer S, Mcbride TO, Jiang S, Osterberg UL, Paulsen KD. 2001. Three-dimensional simulation of near-infrared diffusion in tissue: Boundary condition and geometry analysis for finite-element image reconstruction. Appl Opt 40: 588–600.

Pogue BW, Jiang S, Dehghani H et al. 2004. Characterization of hemoglobin, water, and NIR scattering in breast tissue: Analysis of intersubject variability and menstrual cycle changes. J Biomed Opt 9: 541–552.

Pogue BW, Leblond F, Krishnaswamy V, Paulsen KD. 2010. Radiologic and near-infrared/optical spectroscopic imaging: Where is the synergy? AJR Am J Roentgenol 195: 321–332.

Poplack SP, Tosteson TD, Wells WA et al. 2007. Electromagnetic breast imaging: Results of a pilot study in women with abnormal mammograms. Radiology 243: 350–359.

Schmitz CH, Klemer DP, Hardin R et al. 2005. Design and implementation of dynamic near-infrared optical tomographic imaging instrumentation for simultaneous dual-breast measurements. Appl Opt 44: 2140–2153.

Schnitt SJ, Abner A, Gelman R et al. 1994. The relationship between microscopic margins of resection and the risk of local recurrence in patients with breast cancer treated with breast-conserving surgery and radiation therapy. Cancer 74: 1746–1751.

Schweiger M, Arridge SR. 2011. TOAST: Image Reconstruction in Optical Tomography. [Online]. University College London. Available: http://web4.cs.ucl.ac.uk/research/vis/toast/index.html [Accessed].

Scopa CD, Aroukatos P, Tsamandas AC, Aletra C. 2006. Evaluation of margin status in lumpectomy specimens and residual breast carcinoma. Breast J 12: 150–153.

Sevick-Muraca EM, Sharma R, Rasmussen JC et al. 2008. Imaging of lymph flow in breast cancer patients after microdose administration of a near-infrared flurophore: Feasibility study. Radiology 246: 734–741.

Shah N, Cerussi A, Eker C et al. 2001. Noninvasive functional optical spectroscopy of human breast tissue. Proc Natl Acad Sci USA 98: 4420–4425.

Smitt MC, Nowels KW, Zdeblick MJ et al. 1995. The importance of the lumpectomy surgical margin status in long-term results of breast-conservation. Cancer 76: 259–267.

Spivack B, Khanna MM, Tafra L, Juillard G, Giuliano AE. 1994. Margin status and local recurrence after breast-conserving surgery. Arch Surg 129: 952–956.

Srinivasan S, Pogue BW, Jiang S et al. 2003. Interpreting hemoglobin and water concentration, oxygen saturation, and scattering measured in vivo by near-infrared breast tomography. Proc Natl Acad Sci USA 100: 12349–12354.

Srinivasan S, Pogue BW, Jiang S, Dehghani H, Paulsen KD. 2004. Spectrally constrained chromophore and scattering NIR tomography improves quantification and robustness of reconstruction. Appl Opt 44: 1858–1869.

Te Velde EA, Veerman T, Subramaniam V, Ruers T. 2010. The use of fluorescent dyes and probes in surgical oncology. EJSO 36: 6–15.

Themelis G, Yoo JS, Soh K-S, Schulz R, Ntziachristos V. 2009. Real-time intraoperative fluorescence imaging system using light-absorption correction. J Biomed Opt 14: 064012.

Tromberg B. 2011. Virtual Photonics Technology Initiative [Online]. Beckman Laser Institute. Available: http://www.virtualphotonics.org/Pages/Page/Item?slug=welcome-to-the-virtual-photonics-technology-initiative [Accessed].

Tromberg BJ, Pogue BW, Paulsen KD, Yodh AG, Boas DA, Cerussi AE. 2008. Assessing the future of diffuse optical imaging technologies for breast cancer management. Med Phys 35: 2443–2451.

Troyan SL, Kianzad V, Gibbs-Strauss SL et al. 2009. The FLARE((TM)) intraoperative near-infrared fluorescence imaging system: A first-in-human clinical trial in breast cancer sentinel lymph node mapping. Ann Surg Oncol 16: 2943–2952.

Van Veen RLP, Amelink A, Menke-Pluymers M, Van Der Pol C, Sterenborg H. 2005. Optical biopsy of breast tissue using differential path-length spectroscopy. Phys Med Biol 50: 2573–2581.

Volynskaya Z, Haka AS, Bechtel KL, Fitzmaurice M et al. 2008. Diagnosing breast cancer using diffuse reflectance spectroscopy and intrinsic fluorescence spectroscopy. J Biomed Opt 13: 024012.

Wang LH, Jacques SL. 1993. Hybrid model of monte-carlo simulation and diffusion-theory for light reflectance by turbid media. J Opt Soc Am A-Opt Image Sci Vis 10: 1746–1752.

Wang LH, Jacques SL, Zheng LQ. 1995. MCML—Monte-Carlo modeling of light transport in multilayered tissues. Comput Methods Progr Biomed 47: 131–146.

Wang X, Pogue BW, Jiang SD et al. 2005. Approximation of Mie scattering parameters in near-infrared tomography of normal breast tissue in vivo. J Biomed Opt 10: 051704.

Wax A, Yang CH, Backman V, Kalashnikov M, Dasari RR, Feld MS. 2002. Determination of particle size by using the angular distribution of backscattered light as measured with low-coherence interferometry. J Opt Soc Am A-Opt Image Sci Vis 19: 737–744.

Wilke LG, Brown JQ, Bydlon TM, Kennedy SA et al. 2009. Rapid noninvasive optical imaging of tissue composition in breast tumor margins. Am J Surg 198: 566–574.

Wilson JD, Bigelow CE, Calkins DJ, Foster TH. 2005. Light scattering from intact cells reports oxidative-stress-induced mitochondrial swelling. Biophys J 88: 2929–2938.

Xu H, Rice BW. 2009. In-vivo fluorescence imaging with a multivariate curve resolution spectral unmixing technique. J Biomed Opt 14: 064011.

Zhou C, Cohen DW, Wang Y et al. 2010. Integrated optical coherence tomography and microscopy for ex vivo multiscale evaluation of human breast tissues. Cancer Res 70: 10071–10079.

Zhu C, Burnside ES, Sisney GA et al. 2009. Fluorescence spectroscopy: An adjunct diagnostic tool to image-guided core needle biopsy of the breast. IEEE Trans Biomed Eng 56: 2518–2528.

Zhu C, Palmer G, Breslin T, Harter J, Ramanujam N. 2006. Diagnosis of breast cancer using diffuse reflectance spectroscopy: Comparison of a Monte Carlo versus partial least squares analysis based feature extraction technique. Lasers Surg Med 38: 714–724.

Zhu CF, Palmer GM, Breslin TM, Xu FS, Ramanujam N. 2005. Use of a multiseparation fiber optic probe for the optical diagnosis of breast cancer. J Biomed Opt 10: 024032.

Zonios G, Dimou A. 2006. Modeling diffuse reflectance from semi-infinite turbid media: Application to the study of skin optical properties. Opt Exp 14: 8661–8674.

21

Outlook for Computer-Based Decision Support in Breast Cancer Care

Gautam S. Muralidhar
The University of Texas at Austin

Alan C. Bovik
The University of Texas at Austin

Mia K. Markey
The University of Texas at Austin

Computer-based decision support systems (CDSS) for breast cancer screening have been in clinical practice for the last 20 years, helping radiologists detect and diagnose signs of early breast cancer. CDSS are routinely used by radiologists to detect breast cancer on screening mammograms (Chapter 1) and are commonly referred to as computer-aided detection (CADe) systems (Sampat et al. 2005, Giger et al. 2008) (Chapter 14). CDSS are also used for quantitatively analyzing breast magnetic resonance (MRI) (Chapter 16) and ultrasound (Chapter 15) images (Sampat et al. 2005, Giger et al. 2008). CDSS for breast MRI and ultrasound are often loosely referred to as computer-aided diagnoses systems (CADx) even though they have been regulated by the U.S. Food and Drug Administration (FDA) only for the purposes of quantitative image analysis and visualization, and not for making diagnostic decisions. As discussed in Chapter 14, a number of studies have shown varying degrees of effectiveness of CDSS in helping radiologists interpret breast imaging data. In this chapter, we will provide a perspective on the future of CDSS in breast imaging.

21.1 CADe for Mammography

Mammographic CADe has been around for more than a decade now. Numerous laboratory and clinical studies have shown mammographic CADe systems to be generally effective in improving both the sensitivity and the specificity of breast cancer detection (for example, see Freer and Ulissey 2001, Gromet 2008, Karssemeijer and te Brake 1996, Qian et al. 1995, Qian et al. 1999, Sampat et al. 2008, Skaane et al. 2007). However, in a recently published article in the *Journal of National Cancer Institute*, Fenton et al. (2011) raised questions on the effectiveness of CADe systems in helping radiologists detect signs of early cancer on mammography. The results from this study suggested that when CADe was used in the interpretation of screening mammograms, there was a statistically significant drop in specificity and positive predictive value, while there was a non-statistically significant increase in sensitivity (Fenton et al. 2011). The increase in sensitivity was attributed mainly to ductal carcinoma in situ, which often manifests as microcalcifications on mammography, while the use CADe did not improve the sensitivity for invasive cancer (Fenton et al. 2011). This study raises certain uncomfortable questions such as "does the use of CADe for interpreting screening mammograms really benefit women?" and "what about all the laboratory and clinical studies conducted till date that have shown CADe to be promising for improving the sensitivity and specificity of breast cancer detection?"

The low incidence rate of cancer makes it an extremely challenging task for a radiologist when confronted with a large number of mammograms on a daily basis to correctly detect an existing abnormality (Birdwell 2009). Oversight errors due to the rare occurrence of cancer are often attributed to be one of the main reasons for missed cancer detection (Birdwell 2009). Though practical, mammography is not a perfect modality. The process of projecting the 3-D breast onto a 2-D image plane introduces anatomical noise due to overlapping out-of-plane tissue structures. Anatomical noise could not only mask existing abnormalities, but could also create false visual relationships between different healthy tissue structures suggesting the presence of cancer, when in reality none exists. Hence, reading mammograms to detect breast cancer, whose incidence rate is about four in 1000 mammograms is a very stressful job for radiologists. To make matters worse, a study published by Bassett et al. in July 2011 has shown that since 2000, the number of medical residents who would consider breast imaging as a subspecialty has decreased. This trend is concerning since the number of mammograms generated on a daily basis is large and fewer qualified breast imaging radiologists are available to read them. The study by Wing and Langlier (2009) suggested that the ratio of number of breast imaging radiologists to the number of women undergoing routine mammographic screening is expected to drop by 10% in the next few years. This makes the role of CADe systems as second readers for interpreting screening mammograms all that more important. The numerous laboratory and clinical studies that have shown the benefits of CADe cannot be ignored. Then, what is wrong with present-day CADe systems for their effectiveness to be called into contention by Fenton et al. (2011)? In addition to being an opportunity to explore better ways to evaluate CAD systems (Berry 2011), the study of Fenton et al. has to be viewed as an opportunity to assess the shortcomings of present-day CADe systems. In fact, the study by Fenton et al. implicitly presents a case for certain improvements to be made in the development and evaluation of present-day CADe systems. Chief among those improvements are the need for understanding the breast anatomy better, better algorithms for the detection of invasive cancer, benchmark data sets, and online adaptation of CADe systems.

21.1.1 The Present and Future of Mammographic CADe

21.1.1.1 Better Understanding of Breast Anatomy

An interesting avenue for future research is to quantitatively gain a better understanding of breast anatomy. For instance, quantifying the orientation or the anisotropy of the breast structures could yield insights into the formation of early stage cancers. Preliminary work in this direction by Reiser et al. (2011) has shown that, on average, the breast structure has a preferred orientation of pointing towards the nipple. The power spectral density of mammographic tissue has been shown to follow a power law $P(f) = A/f^\beta$ (Burgess et al. 2001). Reiser et al. (2011) presented an approach to estimate the power law coefficients A and β by taking into account the inherent anisotropy of the breast tissue in local regions of interest. This contrasts with earlier work on power law coefficient estimation wherein the anisotropy was not accounted for. The work by Reiser et al. (2011) is significant as it could pave the way for the development of better algorithms for the detection of cancers that present as architectural distortions and spiculated masses as one could strongly conjecture that the presence of these lesions could change the power law coefficients A and β due to a disruption of the normal anisotropy of the breast tissue.

21.1.1.2 Better Detection of Invasive Cancers

Much of the current work in mammographic CADe is focused on the development of algorithms for the better detection of

invasive cancers (Sampat et al. 2008, Rangayyan and Ayres 2006, Banik et al. 2011). Common manifestations of invasive cancers on mammography include noncalcified lesions such as masses (spiculated and nonspiculated) and architectural distortions. Spiculated masses are characterized by radiating lines or spicules from a central mass region. Spiculated masses account for about 14% of all biopsied lesions, and about 81% of these are malignant (Liberman et al. 1998). Hence, spiculated masses carry a much higher risk of malignancy than calcifications or other types of masses. Spiculated lesions can also form without a central mass region being visible on mammography. Such lesions, referred to as architectural distortions, are also strongly suggestive of malignancy; approximately 48%–60% of architectural distortions that are biopsied are found to be cancer (Baker et al. 2003, Orel et al. 1999), and about 80% of those cancers are invasive (Baker et al. 2003). Thus, the detection of these lesions is critical. Recent clinical studies to evaluate the performance of commercial CADe systems for microcalcification detection have reported sensitivity close to 100% with false positive per image (FPI) rate being less than 1.0 (Kim et al. 2010, Sadaf et al. 2009), where as for mass detection the reported sensitivities are only in the range of 67% to 89% with the FPI ranging from 0.40 to 0.74 (Freer and Ulissey 2001, Kim et al. 2010, Sadaf et al. 2009, Yang et al. 2007, Burhenne et al. 2000). Thus, the rate of detection of invasive lesions is not adequate on mammography by present-day CADe systems.

Figure 21.1 illustrates the typical components that make up a present-day mammographic CADe system. Most CADe systems employ a two-stage design. The first stage comprises of a high sensitivity–low specificity stage and employs image processing methods to detect candidate lesion pixels in the image (Sampat et al. 2005). The second stage is a false positive reduction stage, where in the objective is to retain only those pixels detected in the first stage that have a high likelihood of belonging to the lesion, while eliminating the remaining pixels from consideration (Sampat et al. 2005).

The sensitivity of a CADe system is essentially determined by the performance of the first stage. A number of algorithms have been developed for the detection of invasive noncalcified lesions in the first stage of processing. For a description of some of these algorithms, the interested reader is referred to the review article by Sampat et al. (2005). Despite a lot of research, the performance

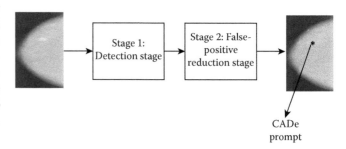

FIGURE 21.1 Schematic of a present-day mammographic CADe system.

of these algorithms is not optimal. One of the main reasons for this is that invasive noncalcified masses and architectural distortions have an extremely varied appearance on mammograms, and a detection algorithm has to be able to detect these lesions irrespective of how they appear on the mammogram. Figure 21.2 illustrates four examples of spiculated masses as seen on different mammograms.

As can be seen from Figure 21.2, it is evident that the spiculated masses have highly variable appearances on different mammograms. By contrast to mass lesions, calcifications usually appear as clusters of bright white spots and are not very varied in their appearance. Again, the interested reader is referred to the article by Sampat et al. (2005) for a description of the algorithms that have been developed for detecting calcifications.

It is interesting to note that radiologists use the BI-RADS® categorization (D'Orsi et al. 2003) for describing the appearance of both calcified and noncalcified lesions as seen on mammography. For example, a noncalcified lesion such as the spiculated mass usually has one of the following BI-RADS® mass shape assignments: round, oval, and irregular, while its mass margin assignment would most often be "spiculated." In other words, spiculated masses have a good higher order representation in that these are either round, oval, or irregular masses with lines called spicules that radiate outward from the mass boundary. Indeed, prior studies by Sampat et al. (2006) and Muralidhar et al. (2010) have shown that radiologists have good agreement in annotating spiculated masses on mammograms. This raises an interesting question: given that there exists a well-defined higher-order representation in the form of BI-RADS® categories for invasive noncalcified lesions, shouldn't a good

detection algorithm be able to reliably detect different instances of the same category of noncalcified lesion such as a spiculated mass on different mammograms? The problem is not due to a lack of higher order representation, but more due to a lack of understanding of the distribution of attributes that make up this higher order representation. Indeed, most detection algorithms that are designed to detect invasive noncalcified lesions such as spiculated masses implicitly assume that there exists a higher order representation for these masses (such as round or oval blobs with radiating lines) (Karssemeijer and te Brake 1996, Sampat et al. 2008, Rangayyan and Ayres 2006, Li et al. 1997, Liu et al. 2001). However, most detection algorithms are not based on probabilistic models of the attributes or properties of the higher order representation. Examples of such attributes include the diameter of the masses, length and width of the spicules, and the joint statistics of the diameter and the length of the spicules. Learning the distribution of such attributes is one approach to understanding the varied shape appearance of noncalcified lesions such as spiculated masses and architectural distortions, and the attribute distributions could in turn help in the design of image processing algorithms for the detection of these lesions. Sampat et al. (2008) and Muralidhar et al. (2010) have explored such an attribute distribution-based approach for developing detection and annotation algorithms for spiculated masses on mammography. A model-driven detection approach is an avenue for future research and more work is needed in this direction to understand the highly stochastic and varied appearance of invasive lesions.

Historically, CADe systems for mammography have been designed to operate on individual MLO and CC views independently. More recently, researchers are looking into quantitatively combining information from multiple views to obtain better detection performance (Wei et al. 2011, Samulski and Karssemeijer 2011). This is again a step in the right direction as radiologists often make decisions on the presence of an abnormality based on their assessment of multiple views.

21.1.1.3 Improving CADe Specificities

Low mammographic CADe specificity is mainly attributed to false positives that are generated due to the inability of the CADe system to correctly disambiguate between pixels belonging to normal tissue and cancerous tissue. As illustrated in Figure 21.1, the second stage of a CADe system is responsible for the specificity of the system. In the second stage, features are extracted at each candidate pixel detected in the first stage. These features usually include texture features and quantitative shape descriptors that best describe the lesion (Sampat et al. 2005, Giger et al. 2008). Once the features are extracted, a classifier is used to learn a model for discriminating between true lesion pixels and false lesion pixels. The classification model is learned from a separate set of training images. A seemingly obvious strategy to reduce the false positives and improve on the specificity is to learn good classification models. However, the models that are learned from the data are essentially as good as the features that are extracted. If the distribution of features extracted at normal and cancerous

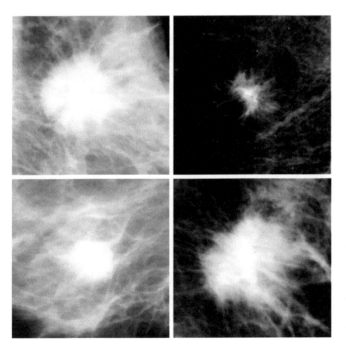

FIGURE 21.2 Examples of four different spiculated masses as seen on mammography to illustrate their varied appearance.

pixels overlap, then it becomes very hard to arrive at a correct classification. A natural solution to this problem is to carefully learn probabilistic models of the lesion attributes from the data. Such models could be useful for not only developing detection algorithms as described in the previous subsection, but also for classifying candidate pixel locations detected in the first stage as lesion or nonlesion.

The specificity of double reading using CADe could also be improved via sophisticated prompting strategy. Present-day CADe systems employ a simple prompting strategy such as a crosshair (an asterisk-like mark) for drawing the radiologist's attention to the suspicious location in the image. It is important to note here that in clinical practice, the radiologist first reads the mammogram, performs an interpretation, and then invokes the CADe system to make an interpretation. The radiologists have an option to either change their interpretation based on what the CADe points out, or to go with their original interpretation. If the CADe system is able to provide a richer visual prompt than the simple crosshair, then it could prove to be a far more compelling evidence for the radiologists to either go with their original interpretation or change their interpretation based on the CADe prompt. An example of a richer visual prompt is the annotation of converging radial linear structures (e.g., spicules) along with a crosshair pointing to the centroid of the mass. There has been some recent work in this area (for example, see Muralidhar et al. 2010), but lot more needs to be done. For example, with the emergence of stereo display and high-resolution color monitors, it would be interesting to explore 3-D and color prompting strategies. Visual prompting and how radiologists react to the prompts also crosses into the area of human–computer interaction. CADe systems could benefit by leveraging ideas from that discipline.

21.1.1.4 Improved Evaluation of CADe Systems

One of the major problems with the evaluation of new CADe systems in laboratory settings is the lack of benchmark mammographic data sets, small sample size, and a limited number of readers serving as the goal standard. Many laboratory studies have utilized the Digital Database for Screening Mammography (DDSM) (Heath et al. 1998, 2001). The DDSM is the largest publicly available database comprising of digitized mammograms. However, despite databases such as DDSM being used in many laboratory studies, it is hard to assess the relative performance of the different CADe systems, as there are differences in the cases that have been used for training, validation, and testing of the systems.

The clinical studies are usually conducted with commercial CADe systems manufactured by vendors such as Hologic, Inc. (Bedford, MA) and iCAD, Inc. (Nashua, NH). While the U.S. FDA (2012) has a regulatory process in place for the premarket approval of commercial CADe systems, it is not clear if the same benchmark data sets have to be used in evaluating the systems. Further, most of the published studies in literature do not give details regarding the version of the CADe software used, and how these versions differ from earlier versions of the software.

Another problem with present-day mammographic databases such as DDSM is that these are comprised only of digitized mammograms. Digital mammography is becoming ubiquitous these days, and there are no benchmark data sets of digital mammograms that are available for the development and evaluation of CADe systems. A well-defined benchmark data set comprising of digital mammograms with a clear specification of the training, validation, and testing partitions will go a long way in ensuring consistency in the evaluation of CADe systems in the laboratory setting. Such benchmark data sets, when periodically updated with newer cases, could also play a role in the regulatory process associated with commercial CADe systems.

The issue of small sample size in the evaluation of mammographic CADe systems is problematic as it invariably leads to an overlap of content in the training, validation, and testing data sets, thereby violating the basic requirement of independent and identically distributed samples in the design of machine-learning-based systems. As a result, evaluation of CADe systems with small samples invariably introduces problems associated with bias and overfitting. While the design of large benchmark data sets will ameliorate this problem, there is a need to statistically understand the implications of CADe evaluation with small samples. The problem of bias also rears up when a limited number of radiologists are involved as readers in a reference panel to ascertain the gold standard. Researchers have begin to identify these problems, for example, see the works of Choudhury et al. (2010) and Way et al. (2010), and much more needs to be done.

21.1.1.5 Online Adaptation of CADe Systems

As described in the previous subsection, a key component of the CADe system is the machine-learning model (or classifier) that is learned from a separate set of training images. The classification model in a typical CADe system is a supervised model such as a support vector machine, an artificial neural network, or a logistic regression model. Such supervised classification models require instances of both positive (e.g., mammograms with lesions) and negative (normal mammograms) classes in the training data set, and for these models to generalize well on previously unseen images, the training set must comprise of a reasonable sample of images that are representative of both the classes. However, invariably, the supervised models are learned from a finite set of positive and negative class mammograms that do not span the set of all possible mammographic lesion appearances and all possible normal mammograms. When a CADe system is deployed in clinical practice, the classification model is held fixed over time, and there is no online learning to periodically adapt the system to new data. In an online learning setup, a supervised model periodically receives new data and makes a prediction using the model learned from previous training data. The model is also provided with the correct label of the new data, and the model parameters are updated based on the examples in the new data for which a wrong prediction was made. Hence, the model periodically receives feedback on its performance and can improve over time. This process is illustrated in Figure 21.3

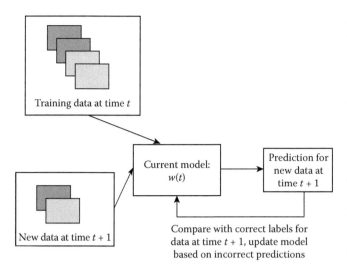

FIGURE 21.3 Schematic illustrating the online learning process. The dark and light shades of gray denote the positive and negative instances.

and has been successfully used by the machine learning community in applications such as content recommendation in Internet search.

The lack of online adaptation of CADe systems to new mammographic images was a point that was also raised by Berry (2011) in his editorial on the study of Fenton et al. (2011). Online adaptation seems a natural thing to do for CADe systems; after all, human beings learn over time from the different experiences they encounter in their lives, and computers have to be no different. However, as pointed out by Berry (2011), online adaptation will raise interesting questions on the CADe regulatory process. How can we be sure that the adapted model helps radiologists achieve similar sensitivities and specificities as the previous model that was approved by the FDA for clinical use? There are no answers yet for such questions; however that should not prevent the CADe community from exploring ways to make the systems more intelligent.

21.2 The Role of CDSS for Diagnosis in Mammography

CDSS is being investigated for mammography to help radiologists make better decisions on patient management after the mammograms have been interpreted. For example, the CDSS could make a recommendation on the BI-RADS® assessment (D'Orsi et al. 2003) to the radiologist. The assessment could be that a certain patient might need an immediate biopsy due to a highly suspicious lesion, or might need additional imaging in the form of MRI or ultrasound or a short-term follow-up.

Presently, CDSS for mammography are under development and investigation in laboratory settings and have not been approved for clinical use by the FDA. Most CDSS employ a magnifying glass approach in which regions of interest (ROIs) are extracted from the image around suspicious pixels (Sampat et

al. 2005). These ROIs are analyzed in detail by extracting quantitative features and learning supervised classification models by training on known BI-RADS assessment ratings. The holy grail of CDSS would be a tumor signature associated with each lesion that would reliably reveal the follow-up action to be performed on the patient.

21.2.1 Beyond Detection and Diagnosis

CDSS in mammography also has an important role to play in areas beyond conventional lesion detection and diagnosis. For instance, CDSS could be developed for personalized mammographic decision support for automatic assessment of an individual's breast cancer risk. For instance, the density of the parenchymal pattern as evident on a mammogram has long been hypothesized to be an indicator of breast cancer risk (Wolfe 1976). Given that mammography today is largely digital, algorithms can be developed that could reliably estimate the breast cancer risk of an individual based on the parenchymal patterns. However, the relationship among the density of parenchymal patterns, the heterogeneity of breast tissue, and breast cancer risk is not fully understood. There has been some recent work on this topic (e.g., Brandt et al. 2011), and this continues to be an avenue of exciting research. Furthermore, CDSS systems could also take into account information from multiple sources such as an individual's personal and family history along with the automatic quantification of the parenchymal pattern as evident on a mammogram or other imaging modalities to better assess the risk of breast cancer.

Another area where CDSS has a role to play in mammography is the training of radiology residents. Training of residents in mammography has historically been conducted via interaction with experienced mammographers and personal supervision during image interpretation. However, the advent of CDSS brings with it the opportunity of developing systems that could learn from an individual's error pattern and score cases in increasing order of difficulty and present it to the resident undergoing training. Preliminary work on this concept was recently reported by Mazurowski et al. (2010) and this remains an extremely interesting avenue for future research.

21.3 The Future of CADe and CADx for X-Ray-Based Breast Imaging

Mammography has been the standard first choice modality for screening asymptomatic women for breast cancer. However, mammography is not perfect. The 3-D to 2-D projection introduces anatomical noise due to overlapping out-of-plane tissue structures, and this makes it hard for radiologists to accurately detect abnormalities. Further, anatomical noise could also create false visual relationships between different healthy tissue structures suggesting the presence of cancer, when in reality none exists. Hence, mammography is known to suffer from low specificity and positive predictive value. CADe systems have shown to be very promising in helping radiologists to detect early signs of breast cancer as demonstrated by a number of independent laboratory and clinical studies. However, as described in the previous subsections, these systems are not perfect and there are several avenues for improvement.

To overcome the problems posed by anatomical noise during mammographic image interpretation, the breast imaging community has been actively developing new x-ray-based 3-D breast imaging modalities such as breast tomosynthesis (Niklason et al. 1997) (Chapter 4), stereo mammography (Getty et al. 2008) (Chapter 3), and breast computed tomography (CT) (Boone et al. 2006) (Chapter 5). A breast tomosynthesis system was recently, in 2011, approved by the FDA for clinical use in the United States. Stereo mammography and breast CT systems are currently the subject of clinical trials in the United States, although stereo mammography has been approved for clinical use in Europe. All the three modalities are very promising in that they provide a 3-D view of the anatomical structures in the breast. Karellas and Vedantham (2008) have recently written a review that talks about these new developments in 3-D x-ray-based imaging.

The role of CDSS, will be all that more important for interpreting the 3-D imaging data. Indeed, the initial clinical observer studies conducted on breast tomosynthesis strongly suggest the utility of CADe for helping radiologists interpret the large volume of data that can be potentially generated when tomosynthesis is routinely used in clinical practice (Good et al. 2008, Gur et al. 2009). The same holds true for stereo mammography and breast CT. CDSS research for the 3-D x-ray-based breast imaging modalities is at a very nascent stage and the review by Muralidhar et al. (2011) provides a summary of this initial research. For breast tomosynthesis, current CDSS research has focused on the development of algorithms for the detection of masses and microcalcifications and also to gain an understanding on the utility of both projection images and reconstructed slices for the detection and characterization of breast lesions (Sahiner et al. 2012, Chan et al. 2010, Singh et al. 2008).

An interesting question that is yet to be answered is whether CDSS systems could leverage from the interplay between mammography and the various 3-D breast imaging modalities. For example, can a classification model that is learned to detect breast lesions on mammographic image data be transferred to breast tomosynthesis or stereo mammography? This problem bears strong similarities to problems in "transfer learning," an area of machine learning that deals with transfer of knowledge from one domain to another. Hence, it is even more important that the future CDSS systems developed for x-ray-based breast imaging keep this interplay between the different modalities in mind.

21.4 CDSS for MRI and Ultrasound

Breast MRI and ultrasound have been used as adjuvant imaging technologies to mammography for achieving higher breast cancer detection sensitivity and for reducing the number of unnecessary biopsies. Breast MRI has received a lot of attention because of its ability to detect cancers not visible due to dense breast tissue on mammography (Lehman and Smith 2009). Ultrasound has been mainly used for differentiating cysts from solid lesions, and for characterization of masses, staging, and guiding biopsies (Kopans 2006).

Most computer-based systems for breast MRI and ultrasound are used as diagnostic aids to help radiologists interpret the images. Breast MRI is routinely performed with a contrast agent and is hence often referred to as dynamic contrast-enhanced MRI. As a result, breast MRI yields both anatomical and functional information. CDSS systems for MRI and ultrasound are used mainly for analyzing the images and for quantifying the textural and morphologic properties of suspicious regions of the image. In the case of MRI, the systems are also used for quantifying the kinetic properties of the suspicious regions from the contrast enhancement curves. The reader is referred to the article by Muralidhar et al. (2011) for a more detailed review on the current state of CDSS in breast MRI.

Current CDSS research for breast MRI has also focused on the development of algorithms for automatically quantifying breast density (Nie et al. 2010). Studies have been conducted comparing 3-D ultrasound and breast MRI for breast density estimation (Moon et al. 2011). Accurate estimation of breast density and its relationship with breast cancer risk is an interesting open problem for most modalities. CDSS systems for characterizing breast masses on ultrasound are also gaining in importance in the research community. Strategies based on content-based retrieval and nonlinear dimensionality reduction for characterization of masses have been recently developed for breast ultrasound (Cho et al. 2011, Jamieson et al. 2010). Preliminary work has also been conducted to analyze the repeatability of CDSS for ultrasound (Drukker et al. 2010).

CDSS for mammography have received a lot more attention than the computer-based systems for breast MRI and ultrasound. The main reason for this is that mammography is the first choice modality for screening asymptomatic women. Hence, it is absolutely critical to make the right decision during mammographic screening. Failure to detect an abnormality could prove to be fatal. On the other hand, incorrectly detecting an abnormality when none exists, and recommending a follow-up diagnostic workup could prove to be emotionally very stressful for the patient and her family members. When a woman is referred for breast MRI or ultrasound imaging, she comes in with a suspicion of an abnormality being present and one that could not be resolved on mammography. The follow-up imaging helps ascertain the presence or absence of any abnormalities. The chances that an abnormality is missed at this stage are fairly remote. Yet, the role of CDSS is equally important in helping radiologists interpret MRI and ultrasound images to arrive at the right diagnosis.

Similar issues exist with CDSS systems for breast MRI and ultrasound as with CADe systems for mammography. There is a need to come up with benchmark data sets for developing and evaluating CDSS systems for breast MRI and ultrasound.

Further, for breast MRI, it is important that the CDSS systems be developed such that they are independent of the imaging protocol that is employed (e.g., fat-saturated vs. nonfat-saturated breast MRI acquisition). Ideally, the CDSS system developed should be a multimodality system. One should not have to duplicate the effort involved in developing a system for one modality to make it work on another modality. This should be achievable since the ultimate goal of the CDSS system is to make a decision on the patient follow-up action irrespective of the modality being used. Essentially, the task remains the same. The core component of a CDSS system is the classification model that is learned from the data. Learning models from one domain (in this case, the domain refers to a modality) and transferring to another domain is certainly an avenue for future research. Like with mammography CADe, online adaptation of CDSS systems is also an interesting direction to explore. Learning various probabilistic models from image statistics, lesion properties, and patient follow-up actions might hold the key to a true, multimodality CDSS system.

21.5 Conclusions

Present-day CDSS systems for interpreting breast imaging data are far from perfect. However, one could also argue that humans have their limitations and are far from perfect when it comes to interpreting breast imaging data. This is especially true for mammography. The question should be, "Does the performance of a computer-based system fall within the range of performance one would observe when a large number of radiologists interpret the same image as the computer-based system?" Humans learn as they as progress, and so can computer-based systems. We have provided our perspective on some of these issues and future directions to explore to make CDSS systems more reliable and robust. There is no reason not to believe that these systems cannot be made better. After all, the supercomputer Watson outperformed knowledgeable humans in the Jeopardy challenge, and this was seen as a remarkable achievement in the artificial intelligence community. With sophisticated data analysis techniques, there is hope that a CDSS system would eventually perform as well as experienced radiologists and help reduce breast cancer mortality rates.

References

Baker JA, Rosen EL, Lo JY, Gimenez EI, Walsh R, Soo MS. 2003. Computer-aided detection (CAD) in screening mammography: Sensitivity of commercial CAD systems for detecting architectural distortion. AJR. Am J Roentgenol 181: 1083–1088.

Banik S, Rangayyan RM, Desautels JEL. 2011. Detection of architectural distortion in prior mammograms. IEEE Trans Med Imag 30: 279–294.

Bassett LW, Bent C, Sayre JW, Marzan R, Verma A, Porter C. 2011. Breast imaging training and attitudes: Update survey of senior radiology residents. Am J Roentgenol 197: 263–269.

Berry DA. 2011. Computer-assisted detection and screening mammography: Where's the beef? J Natl Cancer Inst 103: 1139–1141.

Birdwell RL. 2009. The preponderance of evidence supports computer-aided detection for screening mammography. Radiology 253: 9–16.

Boone JM, Kwan ALC, Yang K, Burkett GW, Lindfors KK, Nelson TR. 2006. Computed tomography for imaging the breast. Journal of Mammary Gland Biology and Neoplasia 11: 103–111.

Brandt SS, Karemore G, Karssemeijer N, Nielsen M. 2011. An anatomically oriented breast coordinate system for mammogram analysis. IEEE Trans Med Imag 30: 1841–1851.

Burgess AE, Jacobson FL, Judy PF. 2001. Human observer detection experiments with mammograms and power-law noise. Med Phys 28: 419–437.

Burhenne LW et al. 2000. Potential contribution of computer-aided detection to the sensitivity of screening mammography. Radiology 215: 554–562.

Chan HP et al. 2010. Characterization of masses in digital breast tomosynthesis: Comparison of machine learning in projection views and reconstructed slices. Med Phys 37: 3576–3586.

Cho HC et al. 2011. Similarity evaluation in a content-based image retrieval (CBIR) CADx system for characterization of breast masses on ultrasound images. Med Phys 38: 1820–1831.

Choudhury KR, Paik DS, Yi CA, Napel S, Roos J, Rubin GD. 2010. Assessing operating characteristics of CAD algorithms in the absence of a gold standard. Med Phys 37: 1788–1795.

D'Orsi CJ, Bassett LW, Berg WA et al. 2003. BI-RADS: Mammography. Reston, VA: American College of Radiology.

Drukker K, Pesce L, Giger M. 2010. Repeatability in computer-aided diagnosis: Application to breast cancer diagnosis on sonography. Med Phys 37: 2659–2669.

FDA U.S. 2012. Guidance for Industry and Food and Drug Administration Staff: Computer-Assisted Detection Devices Applied to Radiology Images and Radiology Device Data—Premarket Notification [510(k)] Submissions.

Fenton JJ et al. 2011. Effectiveness of computer-aided detection in community mammography practice. J Natl Cancer Inst 103: 1152–1161.

Freer TW, Ulissey MJ. 2001. Screening mammography with computer-aided detection: Prospective study of 12,860 patients in a community breast center. Radiology 220: 781–786.

Getty DJ, D'Orsi CJ, Pickett RM. 2008. Stereoscopic digital mammography: Improved accuracy of lesion detection in breast cancer screening. Lect Notes Comput Sci 5116: 74–79.

Giger ML, Chan HP, Boone J. 2008. Anniversary paper: History and status of CAD and quantitative image analysis: The role of Medical Physics and AAPM. Med Phys 35: 5799–5820.

Good WF et al. 2008. Digital breast tomosynthesis: A pilot observer study. Am J Roentgenol 190: 865–869.

Gromet M. 2008. Comparison of computer-aided detection to double reading of screening mammograms: Review of 231,221 mammograms. Am J Roentgenol 190: 854–859.

Gur D et al. 2009. Digital Breast Tomosynthesis: Observer performance study. Am J Roentgenol 193: 586–591.

Heath M, Bowyer KW, Kopans D, Moore R, Kegelmeyer P Jr. 2001. The digital database for screening mammography. In Yaffe MJ, ed. 5th International Workshop on Digital Mammography. Toronto, Canada: Med Phys Publishing. pp. 212–218.

Heath M, Bowyer KW, Kopans D. 1998. Current status of the digital database for screening mammography. In Karssemeijer N, Thijssen M, Hendriks JH, eds. Digital Mammography. Dordrecht: Kluwer Academic Publishers. pp. 457–460.

Jamieson AR, Giger ML, Drukker K, Li H, Yuan Y, Bhooshan N. 2010. Exploring nonlinear feature space dimension reduction and data representation in breast CADx with Laplacian eigenmaps and t-SNE. Med Phys 37: 339–351.

Karellas A, Vedantham S. 2008. Breast cancer imaging: A perspective for the next decade. Med Phys 35: 4878–4897.

Karssemeijer N, te Brake GM. 1996. Detection of stellate distortions in mammograms. IEEE Trans Med Imag 15: 611–619.

Kim SJ, Moon WK, Kim SY, Chang JM, Kim SM, Cho N. 2010. Comparison of two software versions of a commercially available computer-aided detection (CAD) system for detecting breast cancer. Acta Radiologica. Epub ahead of print.

Kopans DB. 2006. Breast ultrasound. In Breast Imaging. Philadelphia: Lippincott, Williams & Wilkins. pp. 555–606.

Lehman CD, Smith RA. 2009. The role of MRI in breast cancer screening. J Natl Comprehensive Cancer Network 7: 1109–1115.

Li L, Mao F, Qian W, Clarke LP. 1997. Wavelet transform for directional feature extraction in medical imaging. In Image Processing, 1997: Proceedings, International Conference on, Vol. 503, pp. 500–503.

Liberman L, Abramson AF, Squires FB, Glassman JR, Morris EA, Dershaw DD. 1998. The breast imaging reporting and data system: Positive predictive value of mammographic features and final assessment categories. AJR. Am J Roentgenol 171: 35–40.

Liu S, Babbs CF, Delp EJ. 2001. Multiresolution detection of spiculated lesions in digital mammograms. IEEE Trans Image Process 10: 874–884.

Mazurowski MA, Baker JA, Barnhart HX, Tourassi GD. 2010. Individualized computer-aided education in mammography based on user modeling: Concept and preliminary experiments. Med Phys 37: 1152–1160.

Moon WK et al. 2011. Comparative study of density analysis using automated whole breast ultrasound and MRI. Med Phys 38: 382–389.

Muralidhar GS, Bovik AC, Markey MK. 2011. Computer-aided detection and diagnosis for 3D x-ray based breast imaging. In Suzuki K, ed. Machine Learning in Computer-Aided Diagnosis: Medical Imaging Intelligence and Analysis. IGI-Global.

Muralidhar GS, Whitman GJ, Haygood TM, Stephens TW, Bovik AC, Markey MK. 2010. Evaluation of stylus for radiographic image annotation. J Digit Imag 23: 701–705.

Muralidhar GS et al. 2010. Snakules: A model-based active contour algorithm for the annotation of spicules on mammography. IEEE Trans Med Imag 29: 1768–1780.

Muralidhar GS et al. 2011. Recent advances in computer-aided diagnosis in breast MRI Mount Sinai. J Med. submitted.

Nie K, Chang D, Chen J-H, Hsu C-C, Nalcioglu O, Su M-Y. 2010. Quantitative analysis of breast parenchymal patterns using 3D fibroglandular tissues segmented based on MRI. Med Phys 37: 217–226.

Niklason LT et al. 1997. Digital tomosynthesis in breast imaging. Radiology 205: 399–406.

Orel SG, Kay N, Reynolds C, Sullivan DC. 1999. BI-RADS categorization as a predictor of malignancy. Radiology 211: 845–850.

Qian W, Li L, Clarke L, Clark RA, Thomas J. 1999. Digital mammography: Comparison of adaptive and nonadaptive CAD methods for mass detection. Acad Radiol 6: 471–480.

Qian W et al. 1995. Tree structured wavelet transform segmentation of microcalcifications in digital mammography. Med Phys 22: 1247–1254.

Rangayyan RM, Ayres FJ. 2006. Gabor filters and phase portraits for the detection of architectural distortion in mammograms. Med Bio Eng Comput 44: 883–894.

Reiser I, Lee S, Nishikawa RM. 2011. On the orientation of mammographic structure. Med Phys 38: 2011.

Sadaf A, Crystal P, Scaranelo A, Helbich T. 2009. Performance of computer-aided detection applied to full-field digital mammography in detection of breast cancers. Eur J Radiol. In press.

Sahiner B et al. 2012. Computer-aided detection of clustered microcalcifications in digital breast tomosynthesis: A 3D approach. Med Phys 39: 28–39.

Sampat MP, Bovik AC, Whitman GJ, Markey MK. 2008. A model-based framework for the detection of spiculated masses on mammography. Med Phys 35: 2110–2123.

Sampat MP, Markey MK, Bovik AC. 2005. Computer-aided detection and diagnosis in mammography. In Bovik AC, ed. Handbook of Image and Video Processing. Academic Press. pp. 1195–1217.

Sampat MP et al. 2006. The reliability of measuring physical characteristics of spiculated masses on mammography. Brit J Radiol 79: S134–S140.

Samulski M, Karssemeijer N. 2011. Optimizing case-based detection performance in a multiview CAD system for mammography. IEEE Trans Med Imag 30: 1001–1009.

Singh S, Tourassi GD, Baker JA, Samei E, Lo JY. 2008. Automated breast mass detection in 3D reconstructed tomosynthesis volumes: A featureless approach. Med Phys 35: 3626–3635.

Skaane P, Kshirsagar A, Stapleton S, Young K, Castellino RA. 2007. Effect of computer-aided detection on independent double reading of paired screen-film and full-field digital screening mammograms. Am J Roentgenol 188: 377–384.

Way TW, Sahiner B, Hadjiiski LM, Chan H-P. 2010. Effect of finite sample size on feature selection and classification: A simulation study. Med Phys 37: 907–920.

Wei J et al. 2011. Computer-aided detection of breast masses: Four-view strategy for screening mammography. Med Phys 38: 1867–1876.

Wing P, Langelier MH. 2009. Workforce shortages in breast imaging: Impact on mammography utilization. Am J Roentgenol 192: 370–378.

Wolfe JN. 1976. Breast patterns as an index of risk for developing breast cancer. Am J Roentgenol 126: 1130–1137.

Yang SK et al. 2007. Screening mammography-detected cancers: Sensitivity of a computer-aided detection system applied to full-field digital mammograms. Radiology 244: 104–111.

Index

Page numbers followed by f and t indicate figures and tables, respectively.

T - #0249 - 111024 - C0 - 280/210/15 - PB - 9780367576646 - Gloss Lamination